Lecture Notes in Computer Science 3860

Commenced Publication in 1973
Founding and Former Series Editors:
Gerhard Goos, Juris Hartmanis, and Jan van Leeuwen

T0226189

David Pointcheval (Ed.)

Topics in Cryptology – CT-RSA 2006

The Cryptographers' Track at the RSA Conference 2006
San Jose, CA, USA, February 13-17, 2006
Proceedings

 Springer

Volume Editor

David Pointcheval
CNRS
ENS/DI
45, rue d'Ulm, 75005 Paris, France
E-mail: David.Pointcheval@ens.fr

Library of Congress Control Number: 2005937532

CR Subject Classification (1998): E.3, G.2.1, D.4.6, K.6.5, K.4.4, F.2.1-2, C.2, J.1

LNCS Sublibrary: SL 4 – Security and Cryptology

ISSN   0302-9743
ISBN-10   3-540-31033-9 Springer Berlin Heidelberg New York
ISBN-13   978-3-540-31033-4 Springer Berlin Heidelberg New York

Springer is a part of Springer Science+Business Media

springer.com

© Springer-Verlag Berlin Heidelberg 2006
Printed in Germany

Typesetting: Camera-ready by author, data conversion by Scientific Publishing Services, Chennai, India
Printed on acid-free paper   SPIN: 11605805   06/3142   5 4 3 2 1 0

Preface

The RSA® Conference, with over 15,000 attendees, as well as over 225 sponsors and exhibitors, is the largest computer security event of the year. The Cryptographers' Track is one of the many parallel tracks. These proceedings contain the papers presented during the sixth edition. The tradition indeed started in 2001, and is by now well established: the Cryptographers' Track at the RSA Conference is among the major events in cryptography.

There were 72 submitted contributions, of which 22 were selected for presentation. They cover all aspects of cryptography (symmetric and asymmetric cryptography, constructions and attacks, new trends). In addition, the program includes two invited talks, by Xiaoyun Wang on "Cryptanalysis of Hash functions and Potential Dangers," and Philip MacKenzie on "Passwords Will Not Die: How Cryptography Can Help Deal with Them."

All the submissions were reviewed by at least three members of the Program Committee. I am very grateful to the 24 members for their hard and conscientious work. Many thanks to the 89 external reviewers:

Masayuki Abe	Eiichiro Fujisaki	Miodrag Mihaljevic
Kazumaro Aoki	Jun Furukawa	Kazuhiko Minematsu
Giuseppe Ateniese	David Galindo	Fabian Monrose
Roberto Avanzi	Shai Halevi	Paul Montague
Zuzana Beerliová	Helena Handschuh	Steve Myers
Olivier Billet	Chris Heneghan	David Naccache
Alex Biryukov	Thomas Holenstein	Antonio Nicolosi
Ian Blake	Fumitaka Hoshino	Satoshi Obana
Colin Boyd	Yong Ho Hwang	Satomi Okazaki
Eric Brier	Toshiyuki Isshiki	Katsuyuki Okeya
Aniello Castiglione	Ellen Jochemsz	Francis Olivier
Juyoung Cha	Antoine Joux	Roger Oyono
Aldar Chan	Ari Juels	Dan Page
Liqun Chen	Charanjit Jutla	Jung Hyung Park
Kookrae Cho	Aggelos Kiayias	Kun Peng
Scott Contini	Hiroaki Kikuchi	Krzysztof Pietrzak
Paolo D'Arco	Tetsutarou Kobayashi	Dominik Raub
Jintai Ding	Tadayoshi Kohno	Yasuyuki Sakai
Christophe Doche	Hugo Krawczyk	Kouichi Sakurai
Orr Dunkelman	Sandeep Kumar	Werner Schindler
Matthias Fitzi	Tanja Lange	Jae Woo Seo
Pierre-Alain Fouque	Jung Wook Lee	Jong Hoon Shin
Jacques J.A. Fournier	Barbara Masucci	Igor Shparlinski
Kouichi Fujisaki	Alexander May	Ron Steinfeld

Mike Szydlo	Karine Villegas	Christopher Wolf
Yael Tauman Kalai	Shabsi Walfish	Alex Yampolskiy
Isamu Teranishi	Huaxiong Wang	Yeon Hyeong Yang
Toshio Tokita	Xiaofeng Wang	Yiqun Lisa Yin
Michael Tunstall	Bogdan Warinschi	Jeong Il Yoon
Frederik Vercauteren	Benne de Weger	

Note that these proceedings contain the revised versions of the selected papers. Since the revisions were not checked again before publication, the authors (and not the committee) bear full responsibility of the contents of their papers.

I also would like to thank Jacques Beigbeder for maintaining the submission and webreview servers, and Duong Hieu Phan for the fast set up of the review phase. The submission software was written by Chanathip Namprempre, and the webreview system by Wim Moreau and Joris Claessens. Many thanks to Burt Kaliski for interfacing with the RSA conference organizers, and to Alfred Hofmann at Springer for the production of this volume.

Finally, I wish to thank all the authors who submitted papers, and the authors of accepted papers for sending their final versions on time.

November 2005

David Pointcheval
Program Chair
CT-RSA 2006

Organization

RSA Conference 2006 was organized by RSA Security Inc. and its partner organizations around the world. The Cryptogaphers' Track at RSA Conference 2006 was organized by RSA Laboratories (http://www.rsasecurity.com).

Program Chair

David Pointcheval   CNRS/ENS, France

Program Committee

Eli Biham	Technion, Israel
Xavier Boyen	Voltage, USA
Benoît Chevallier-Mames	Gemplus, France
Anand Desai	NTT MCL, USA
Yvo Desmedt	University College London, UK
Yevgeniy Dodis	New York Univ., USA
Steven Galbraith	Royal Holloway University of London, UK
Rosario Gennaro	IBM T.J. Watson Research Center, USA
Henri Gilbert	France Telecom R&D, France
Martin Hirt	ETH Zurich, Switzerland
Nick Howgrave-Graham	NTRU Cryptosystems, USA
Markus Jakobsson	Indiana Univ., USA
Jonathan Katz	Univ. of Maryland, USA
Kwangjo Kim	ICU, Korea
Pil Joong Lee	POSTECH, Korea
Arjen Lenstra	Lucent Technologies, USA & TU Eindhoven, The Netherlands
Javier Lopez	Univ. of Malaga, Spain
Tatsuaki Okamoto	NTT, Japan
Josef Pieprzyk	Macquarie Univ., Australia
Guillaume Poupard	DCSSI Crypto Lab, France
Bart Preneel	K.U. Leuven, Belgium
Kazue Sako	NEC, Japan
Ivan Visconti	Univ. di Salerno, Italy
Moti Yung	RSA Labs & Columbia Univ., USA

Table of Contents

Signatures

Side-Channel Attacks

CCA Encryption

Message Authentication

Block Ciphers

Multi-party Computation

Cache Attacks and Countermeasures: The Case of AES

Dag Arne Osvik[1], Adi Shamir[2], and Eran Tromer[2]

[1] dag.arne@osvik.no
[2] Department of Computer Science and Applied Mathematics,
Weizmann Institute of Science, Rehovot 76100, Israel
{adi.shamir, eran.tromer}@weizmann.ac.il

Abstract. We describe several software side-channel attacks based on inter-process leakage through the state of the CPU's memory cache. This leakage reveals memory access patterns, which can be used for cryptanalysis of cryptographic primitives that employ data-dependent table lookups. The attacks allow an unprivileged process to attack other processes running in parallel on the same processor, despite partitioning methods such as memory protection, sandboxing and virtualization. Some of our methods require only the ability to trigger services that perform encryption or MAC using the unknown key, such as encrypted disk partitions or secure network links. Moreover, we demonstrate an extremely strong type of attack, which requires knowledge of neither the specific plaintexts nor ciphertexts, and works by merely monitoring the effect of the cryptographic process on the cache. We discuss in detail several such attacks on AES, and experimentally demonstrate their applicability to real systems, such as OpenSSL and Linux's `dm-crypt` encrypted partitions (in the latter case, the full key can be recovered after just 800 writes to the partition, taking 65 milliseconds). Finally, we describe several countermeasures for mitigating such attacks.

Keywords: side-channel attack, cache, memory access, cryptanalysis, AES.

1 Introduction

1.1 Overview

Many computer systems concurrently execute programs with different privileges, employing various partitioning methods to facilitate the desired access control semantics. These methods include kernel vs. userspace separation, process memory protection, filesystem permissions and `chroot`, and various approaches to virtual machines and sandboxes. All of these rely on a model of the underlying machine to obtain the desired access control semantics. However, this model is often idealized and does not reflect many intricacies of actual implementation.

In this paper we show how a low-level implementation detail of modern CPUs, namely the structure of memory caches, causes subtle indirect interaction between processes running on the same processor. This leads to cross-process information leakage. In essence, the cache forms a shared resource which all processes

D. Pointcheval (Ed.): CT-RSA 2006, LNCS 3860, pp. 1–20, 2006.

compete for, and it thus affects and is affected by every process. While the *data* stored in the cache is protected by virtual memory mechanisms, the *metadata* about the contents of the cache, and hence the memory access patterns of processes using that cache, is not fully protected.

We describe several methods an attacker can use to learn about the memory access patterns of another process. These are classified into methods that affect the state of the cache and then measure the effect on the running time of the encryption, and methods that investigate the state of the cache after or during encryption. The latter are found to be particularly effective and noise-resistant.

We demonstrate the cryptanalytic applicability of these methods to the Advanced Encryption Standard (AES, [11]) by showing a known-plaintext (or known-ciphertext) attack that performs efficient full key extraction. For example, an implementation of one variant of the attack performs full AES key extraction from the `dm-crypt` system of Linux using only 800 accesses to an encrypted file, 65ms of measurements and 3 seconds of analysis; attacking simpler systems, such as "black-box" OpenSSL library calls, is even faster at 13ms and 300 encryptions.

One variant of our attack has the unusual property of performing key extraction *without knowledge of either the plaintext or the ciphertext*. This is an unusually strong form of attack in which an unprivileged process can, just by accessing its own memory space, obtain bits from a secret AES key used by another process, without any (explicit) communication between the two. This too is demonstrated experimentally.

Implementing AES in a way that is impervious to this attack, let alone developing an efficient generic countermeasure, appears non-trivial; in Section 5, various countermeasures are described and analyzed.

Many details and variants have been omitted due to space constraints; see `http://www.wisdom.weizmann.ac.il/~tromer/cache` for an extended version.

1.2   Related Works

The possibility of cross-process leakage via cache state has been mentioned in several previous works. It was considered in 1992 by Hu [7] in the context of intentional transmission via covert channels. In 1998, Kelsey et al. [8] mentioned the prospect of "attacks based on cache hit ratio in large S-box ciphers". In 2002, Page [9] described theoretical attacks using cache misses, but assumed the ability to identify cache misses with very high temporal resolution; its applicability in realistic scenarios is unclear. In 2003, Tsunoo et al. [15] described attacks using timing effects due to collisions in the memory lookups *inside* the cipher, as opposed to the cipher-attacker collisions we investigate.

Concurrently with but independently of our work, Bernstein [2] describes attacks on AES that exploit timing variability due to cache effects; his attack can be seen as a variant of our Evict+Time measurement method (see Section 3.4). The main difference is that [2] does not use an explicit model of the cache and active manipulation, but rather relies only on the existence of some consistent statistical timing pattern due to various uncontrolled memory access effects. The resulting attack is simpler and more portable, but have several shortcomings.

First, it requires reference measurements of encryption under *known* key in an identical configuration, and these are often not readily available (e.g., a user may be able to write data to an encrypted filesystem, but creating a reference filesystem with a known key is a privileged operation). Second, the attack of [2] relies on timing the encryption and thus, similarly to our Evict+Time method, seems impractical on many real systems due to excessively low signal-to-noise ratio; our alternative methods (Sections 3.5 and 4) address this. Third, even when the attack of [2] works, it requires a much higher number of analyzed encryptions.[1]

Also concurrently with but independently of our work, Percival [14] describes a cache-based attack on RSA for processors with simultaneous multithreading. The measurement method is similar to one variant of our asynchronous attack (Section 4), but the cryptanalytic aspect is very different since the algorithms and time scales involved in RSA encryption are very different from those of AES. Both [2] and [14] contain discussions of countermeasures against the respective attacks, and some of these are also relevant to our attacks (see Section 5).

Koeune and Quisquater [6] described a timing attack on a "bad implementation" of AES which uses its algebraic description in a "careless way" (namely, using a conditional branch in the MixColumn operation). That attack is not applicable to common software implementations, but should be taken into account in regard to certain countermeasures against our attack (see Section 5.2).

Leakage of memory access information has also been considered in other contexts, yielding theoretical [5] and practical [16][17] mitigation methods; these are discussed in Section 5.3.

2   Preliminaries

2.1   Memory and Cache Structure

Modern processors use one or more levels of *set-associative memory cache*. Such a cache consists of storage cells called *cache lines*, each consisting of B bytes. The cache is organized into S *cache sets*, each containing W cache lines[2], so overall the cache contains $S \cdot W \cdot B$ bytes. The mapping of memory addresses into the cache is limited as follows. First, the cache holds copies of aligned blocks of B bytes in main memory, which we will term *memory blocks*; when a cache miss occurs, a full memory block is copied into one of the cache lines. Second, each memory block may be cached only in a specific cache set; specifically, the memory block starting at address a can be cached only in the W cache lines belonging to cache set $\lfloor a/B \rfloor$ mod S. See Figure 1(a). Thus, the memory blocks are partitioned into S classes, where the blocks in each class contend for the cache lines in a single cache set.

---

[1] In our experiments the attack code of [2] failed to get a signal from `dm-crypt` even after a 10 hours run, whereas in an identical setup our Prime+Probe performed full key recovery using 65ms of measurements.

[2] In common terminology, W is called the *associativity* and the cache is called W-*way associative*.

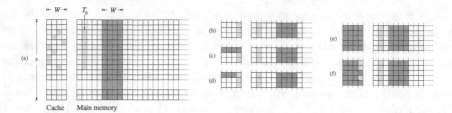

Fig. 1. (a) Schematic of a set-associative cache. The light gray blocks represent a cached AES lookup table. The dark gray blocks represent the attacker's memory. States (b)-(d) depict Evict+Time and (e)-(f) depict Prime+Probe (see Section 3).

2.2   Memory Access in AES Implementations

This paper focuses on AES (see Section 6.1 for a discussion of other ciphers). Performance-oriented implementations on 32-bit (or higher) processors typically use the following formulation, as prescribed in the Rijndael AES submission [4].[3]

Several lookup tables are precomputed once by the programmer or during system initialization. There are 8 such tables, $T_0, \ldots, T_3$ and $T_0^{(10)}, \ldots, T_3^{(10)}$, each containing 256 4-byte words. The contents of the tables, defined in [4], are inconsequential for most of our attacks.

During key setup, a given 16-byte secret key $\boldsymbol{k} = (k_0, \ldots, k_{15})$ is expanded into 10 round keys[4], $\boldsymbol{K}^{(r)}$ for $r = 1, \ldots, 10$. Each round key is divided into 4 words of 4 bytes each: $\boldsymbol{K}^{(r)} = (K_0^{(r)}, K_1^{(r)}, K_2^{(r)}, K_3^{(r)})$. The 0-th round key is just the raw key: $K_j^{(0)} = (k_{4j}, k_{4j+1}, k_{4j+2}, k_{4j+3})$ for $j = 0, 1, 2, 3$. The details of the rest of the expansion are mostly inconsequential.

Given a 16-byte plaintext $\boldsymbol{p} = (p_0, \ldots, p_{15})$, encryption proceeds by computing a 16-byte intermediate state $\boldsymbol{x}^{(r)} = (x_0^{(r)}, \ldots, x_{15}^{(r)})$ at each round r. The initial state $\boldsymbol{x}^{(0)}$ is computed by $x_i^{(0)} = p_i \oplus k_i$ ($i = 0, \ldots, 15$). Then, the first 9 rounds are computed by updating the intermediate state as follows, for $r = 0, \ldots, 8$:

$$
\begin{aligned}
(x_0^{(r+1)}, x_1^{(r+1)}, x_2^{(r+1)}, x_3^{(r+1)}) &\leftarrow T_0[x_0^{(r)}] \oplus T_1[x_5^{(r)}] \oplus T_2[x_{10}^{(r)}] \oplus T_3[x_{15}^{(r)}] \oplus K_0^{(r+1)} \\
(x_4^{(r+1)}, x_5^{(r+1)}, x_6^{(r+1)}, x_7^{(r+1)}) &\leftarrow T_0[x_4^{(r)}] \oplus T_1[x_9^{(r)}] \oplus T_2[x_{14}^{(r)}] \oplus T_3[x_3^{(r)}] \oplus K_1^{(r+1)} \\
(x_8^{(r+1)}, x_9^{(r+1)}, x_{10}^{(r+1)}, x_{11}^{(r+1)}) &\leftarrow T_0[x_8^{(r)}] \oplus T_1[x_{13}^{(r)}] \oplus T_2[x_2^{(r)}] \oplus T_3[x_7^{(r)}] \oplus K_2^{(r+1)} \\
(x_{12}^{(r+1)}, x_{13}^{(r+1)}, x_{14}^{(r+1)}, x_{15}^{(r+1)}) &\leftarrow T_0[x_{12}^{(r)}] \oplus T_1[x_1^{(r)}] \oplus T_2[x_6^{(r)}] \oplus T_3[x_{11}^{(r)}] \oplus K_3^{(r+1)}
\end{aligned}
\tag{1}
$$

Finally, to compute the last round (1) is repeated with $r = 9$, except that $T_0, \ldots, T_3$ is replaced by $T_0^{(10)}, \ldots, T_3^{(10)}$. The resulting $\boldsymbol{x}^{(10)}$ is the ciphertext. Compared to the algebraic formulation of AES, here the lookup tables account for the combination of SHIFTROWS, MIXCOLUMNS and SUBBYTES operations; the change of lookup tables for the last is due to the absence of MIXCOLUMNS.

---

[3] Some implementations use variant with a different table layouts; see Section 5.2.
[4] We consider AES with 128-bit keys. The attacks can be adapted to longer keys.

2.3 Notation

We treat bytes interchangeably as integers in $\{0, \ldots, 255\}$ and as elements of $\{0, 1\}^8$ that can be XORed. Let δ denote the cache line size B divided by the size of each table entry (usually 4 bytes); on most platforms of interest we have $\delta = 16$. For a byte y and table T_ℓ, we will denote $\langle y \rangle = \lfloor y/\delta \rfloor$ and call this *the memory block of y* in T_ℓ. The significance of this notation is as follows: two bytes y, z fulfill $\langle y \rangle = \langle z \rangle$ iff, when used as lookup indices into the same table T_ℓ, they would cause access to the same memory block[5]; they would therefore be impossible to distinguish based only on a single memory access. For a byte y and table T_ℓ, we say that an AES encryption with given inputs *accesses the memory block of y in T_ℓ* if, according to the above description of AES, at some point in the encryption there will be some table lookup to $T_\ell[z]$ where $\langle z \rangle = \langle y \rangle$.

In Section 3 we will show methods for discovering (and taking advantage of the discovery) whether the encryption code, invoked as a black box, accesses a given memory block. To this end we define the following predicate: $Q_k(p, \ell, y) = 1$ iff the AES encryption of the plaintext p under the encryption key k accesses the memory block of index y in T_ℓ at least once throughout the 10 rounds.

Also in Section 3, our measurement procedures will sample *measurement score* from a distribution $M_k(p, \ell, y)$ over $\mathbb{R}$. The exact definition of $M_k(p, \ell, y)$ will vary, but it will approximate $Q_k(p, \ell, y)$ in the following rough sense: for a large fraction of the keys k, all tables ℓ and a large fraction of the indices x, for random plaintexts and measurement noise, the expectation of $M_k(p, \ell, y)$ is larger when $Q_k(p, \ell, y) = 1$ than when $Q_k(p, \ell, y) = 0$.

3 Synchronous Known-Data Attacks

3.1 Overview

The first family of attacks, termed *synchronous attacks*, is applicable in scenarios where the plaintext or ciphertext is known and the attacker can operate synchronously with the encryption on the same processor, by using (or eavesdropping upon) some interface that triggers encryption under an unknown key. For example, a Virtual Private Network may allow an unprivileged user to send data packets through a secure channel. This lets the user trigger encryption of plaintexts that are mostly known (up to some uncertainties in the packet headers), and our attack would thus, under some circumstances, enable any such user to discover the key used by the VPN to protect all users' packets. As another example, consider the Linux `dm-crypt` and `cryptoloop` services. These allow the administrator to create a virtual device which provides encrypted storage into an

---

[5] We assume that the tables are aligned on memory block boundaries, which is usually the case. Non-aligned tables would *benefit* our attacks by leaking an extra bit per key byte in the first round. We also assume for simplicity that all tables are mapped into distinct cache sets; this holds with high probability on many systems (and our practical attacks also handle some exceptions).

underlying physical device, and typically a normal filesystem is mounted on top of the virtual device. If a user has write permissions to *any* file on that filesystem, he can use it to trigger encryptions of known chosen plaintext, and using our attack he is subsequently able to discover the encryption key used for the underlying device. We have experimentally demonstrated the latter attack, and showed it to reliably extract the full AES key using about 65ms of measurements (involving just 800 write operations) followed by 3 seconds of analysis.

The full attack obtains a set of random samples, and then performs off-line cryptanalysis. The latter proceeds by hypothesis testing: we guess small parts of the key, use the guess to predict memory accesses, and check whether the predictions are consistent with the collected data. In the following we first describe the cryptanalysis in an idealized form using the predicate Q, and adapt it to the noisy measurements of M. We then show two different methods for obtaining these measurements, detail some experimental results and outline possible variants and extensions.

3.2   One-Round Attack

The simplest known-plaintext synchronous attack exploits the fact that in the first round, the accessed table indices are simply $x_i^{(0)} = p_i \oplus k_i$ for all $i = 0, \ldots, 15$. Thus, given knowledge of the plaintext byte p_i, any information on the accessed index $x_i^{(0)}$ directly translates to information on key byte k_i. The basic attack, in idealized form, is as follows.

Suppose that we obtain samples of the ideal predicate $Q_{\boldsymbol{k}}(\boldsymbol{p}, \ell, y)$ for some table ℓ, arbitrary table indices y and known but random plaintexts $\boldsymbol{p}$. Let k_i be a key byte such that the first encryption round performs the access "$T_\ell[x_i^{(0)}]$", i.e., such that $i \equiv \ell \pmod 4$. Then we can discover the partial information $\langle k_i \rangle$ about k_i, by testing candidate values $\tilde{k}_i$ and checking them as follows. Consider the samples that fulfill $\langle y \rangle = \langle p_i \oplus \tilde{k}_i \rangle$. These samples will be said to be *useful for* $\tilde{k}_i$, and we can reason about them as follows. If indeed $\langle k_i \rangle = \langle \tilde{k}_i \rangle$ then we will always have $Q_{\boldsymbol{k}}(\boldsymbol{p}, \ell, y) = 1$ for useful samples, since the table lookup "$T_\ell[x_i^{(0)}]$" will indeed access the memory block of y in T_ℓ. Conversely, if $\langle k_i \rangle \neq \langle \tilde{k}_i \rangle$ then we are assured that "$T_\ell[x_i^{(0)}]$" will *not* access the memory block of y; however, during the full encryption process there will be $4 \times 9 - 1 = 35$ more accesses to T_ℓ. Those 35 accesses are affected by other plaintext bytes, so (for sufficiently random plaintexts) the probability that the encryption will not access that memory block in any round is $(1 - \delta/256)^{35}$. By definition, that is also the probability of $Q_{\boldsymbol{k}}(\boldsymbol{p}, \ell, y) = 0$, and in the common case $\delta = 16$ it is approximately 0.104. Thus, after receiving a few dozen useful samples we can identify a correct $\langle \tilde{k}_i \rangle$ — namely, the one for which $Q_{\boldsymbol{k}}(\boldsymbol{p}, \ell, y) = 1$ whenever $\langle y \rangle = \langle p_i \oplus \tilde{k}_i \rangle$. Applying this test to each key byte k_i separately, we can thus determine the top $\log_2(256/\delta) = 4$ bits of every key byte k_i (when $\delta = 16$), i.e., half of the key. Note that this is the maximal amount of information that can be extracted from the memory lookups of the first round, since they are independent and each can be distinguished only up to the size of a memory block.

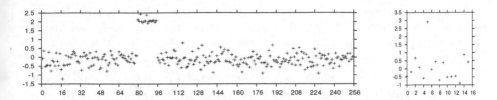

Fig. 2. Candidate scores for a synchronous attack using Prime+Probe measurements, analyzing a `dm-crypt` encrypted filesystem on Linux 2.6.11 running on an Athlon 64, after analysis of 30,000 (left) or 800 (right) triggered encryptions. The horizontal axis is $\tilde{k}_5 = p_5 \oplus y$ (left) or $\langle \tilde{k}_5 \rangle$ (right) and the vertical axis is the average measurement score over the samples fulfilling $y = p_5 \oplus \tilde{k}_5$ (in units of clock cycles). The high nibble of $k_5 = 0\mathrm{x}50$ is easily gleaned.

In reality, we do not have the luxury of the ideal predicate, and have to deal with measurement score distributions $M_{\boldsymbol{k}}(\boldsymbol{p}, \ell, y)$ that are correlated with the ideal predicate but contain a lot of (possibly structured) noise. For example, we will see that $M_{\boldsymbol{k}}(\boldsymbol{p}, \ell, y)$ is often correlated with the ideal $Q_{\boldsymbol{k}}(\boldsymbol{p}, \ell, y)$ for some ℓ but is uncorrelated for others (see Figure 4). We thus proceed by averaging over many samples. As above, we concentrate on a specific key x_i and a corresponding table ℓ. Our measurement will yield samples of the form (p, y, m) consisting of arbitrary table indices y, random plaintexts $\boldsymbol{p}$, and measurement scores m drawn from $M_{\boldsymbol{k}}(\boldsymbol{p}, \ell, y)$. For a candidate key value $\tilde{k}_i$ we define the *candidate score of* $\tilde{k}_i$ as the expected value of m over the samples useful to $\tilde{k}_i$ (i.e., conditioned on $y = p_i \oplus \tilde{k}_i$). We estimate the candidate score by taking the average of m over the samples useful for $\tilde{k}_i$. Since $M_{\boldsymbol{k}}(\boldsymbol{p}, \ell, y)$ approximates $Q_{\boldsymbol{k}}(\boldsymbol{p}, \ell, y)$, the candidate score should be noticeably higher when $\langle \tilde{k}_i \rangle = \langle k_i \rangle$ than otherwise, allowing us to identify the value of k_i up to a memory block.

Indeed, on a variety of systems we have seen this attack reliably obtaining the top nibble of every key byte. Figure 2 shows the candidate scores in one of these experiments (see Sections 3.5 and 3.6 for details); the $\delta = 16$ key byte candidates $\tilde{k}_i$ fulfilling $\langle \tilde{k}_i \rangle = \langle k_i \rangle$ are easily distinguished.

3.3   Two-Rounds Attack

The above attack narrows each key byte down to one of δ possibilities, but the table lookups in the first AES round can not reveal further information. For the common case $\delta = 16$, the key still has 64 unknown bits. We thus proceed to analyze the 2nd AES round, exploiting the non-linear mixing in the cipher to reveal additional information. Specifically, we employ four specific[6] equations, derived from the Rijndael specification [4], which express the indices $x_2^{(1)}, x_5^{(1)}, x_8^{(1)}$ and $x_{15}^{(1)}$ used in four of the table lookups in the 2nd round. For example, we have

$$x_2^{(1)} = s(p_0 \oplus k_0) \oplus s(p_5 \oplus k_5) \oplus 2 \bullet s(p_{10} \oplus k_{10}) \oplus 3 \bullet s(p_{15} \oplus k_{15}) \oplus s(k_{15}) \oplus k_2 \quad (2)$$

where $s(\cdot)$ denotes the S-box function and $\bullet$ denotes multiplication over GF(256).

---

[6] These are special in that they involve just 4 unknown quantities (see below).

Consider equation (2) above, and suppose that we obtain samples of the ideal predicate $Q_{\boldsymbol{k}}(\boldsymbol{p}, \ell, y)$ for table $\ell = 2$, arbitrary table indices y and known but random plaintexts $\boldsymbol{p}$. We already know $\langle k_0 \rangle, \langle k_5 \rangle, \langle k_{10} \rangle, \langle k_{15} \rangle$ and $\langle k_2 \rangle$ from attacking the first round, and we also know the plaintext. The unknown low bits of k_2 (i.e., $k_2 \bmod \delta$), affect only the low bits of $x_2^{(1)}$, (i.e., $x_2^{(1)} \bmod \delta$), and these do not affect which memory block is accessed by "$T_2[x_2^{(1)}]$". Thus, the only unknown bits affecting the memory block accessed by "$T_2[x_2^{(1)}]$" are the lower $\log_2 \delta$ bits of k_0, k_5, k_{10} and k_{15}. This gives a total of δ^4 (i.e., 2^{16} for $\delta = 2^4$) possibilities for candidate values $\tilde{k}_0, \tilde{k}_5, \tilde{k}_{10}, \tilde{k}_{15}$, which are easily enumerated. We can identify the correct candidate as follows, thereby completing the recovery of these four key bytes.

Identification of a correct guess is done by a generalization of the hypothesis-testing method used for the one-round attack. For each candidate guess, and each sample, $Q_{\boldsymbol{k}}(\boldsymbol{p}, \ell, y)$ we evaluate (2) using the candidates $\tilde{k}_0, \tilde{k}_5, \tilde{k}_{10}, \tilde{k}_{15}$ while fixing the unknown low bits of k_2 to an arbitrary value. We obtain a predicted index $\tilde{x}_2^{(1)}$. If $\langle y \rangle = \langle \tilde{x}_2^{(1)} \rangle$ then we say that this sample is *useful* for this candidate, and reason as follows. If the guess was correct then $\langle y \rangle = \langle \tilde{x}_2^{(1)} \rangle = \langle x_2^{(1)} \rangle$ and thus "$T_2[x_2^{(1)}]$" causes an access to the memory block of y in T_2, whence $Q_{\boldsymbol{k}}(\boldsymbol{p}, \ell, y) = 1$ by definition. Otherwise we have $k_i \neq \tilde{k}_i$ for some $i \in \{0, 5, 10, 15\}$ and thus

$$x_2^{(1)} \oplus \tilde{x}_2^{(1)} = c \bullet s(p_i \oplus k_i) \oplus c \bullet s(p_i \oplus \tilde{k}_i) \oplus \cdots$$

for some $c \in \{1, 2, 3\}$. Since $\boldsymbol{p}$ is random the remaining terms are independent of the first two. By a differential property of the AES S-box, it follows that the probability that "$T_2[x_2^{(1)}]$" does not cause an access to the memory block of y in T_2 is at least $(1 - \delta/256)^3$. Each of the other 35 accesses to T_2 performed during the encryption will access the memory block of y in T_2 with probability $\delta/256$. Hence, $Q_{\boldsymbol{k}}(\boldsymbol{p}, \ell, y) = 0$ with probability greater than $(1 - \delta/256)^{3+35}$, so to eliminate all the wrong candidates out of the δ^4 we need about $\log \delta^{-4} / \log(1 - \delta/256 \cdot (1 - \delta/256)^{38})$ samples. This amounts to about 2056 samples[7] when $\delta = 16$.

Similarly, each of the other three equations above lets us guess the low bits of four distinct key bytes, so taken together they reveal the full key. While we cannot reuse samples between equations since they refer to different tables ℓ, we can reuse samples between the analysis of the first and second round. Thus, if we had access to the ideal predicate Q we would need a total of about 8220 samples and a run-time complexity of $4 \cdot 2^{16} \cdot 2056 \approx 2^{29}$ simple tests to extract the full AES key.

In reality we get only measurement scores from the distributions $M_{\boldsymbol{k}}(\boldsymbol{p}, \ell, y)$ that approximate the ideal predicate $Q_{\boldsymbol{k}}(\boldsymbol{p}, \ell, y)$. Similarly to the one-round attack, we proceed by computing, for each candidate $\tilde{k}_i$, a candidate score obtained by averaging the measurement scores of all samples useful to $\tilde{k}_i$. We then pick the $\tilde{k}_i$ having the largest measurement score. The number of samples required to

---

[7] With some of our measurement methods the attack requires only a few hundred *encryptions*, since each encryption provides samples for multiple y.

reliably obtain all key bytes by this method is, in some experimentally verified settings, only about 7 times larger than the ideal (see Section 3.6).

3.4   Measurement Via Evict+Time

One method for extracting measurement scores is to manipulate the state of the cache before each encryption, and observe the execution time of the subsequent encryption. Recall that we assume the ability to trigger an encryption and know when it has begun and ended. We also assume knowledge of the memory address of each table T_ℓ, and hence of the cache sets to which it is mapped.[8] We denote these (virtual) memory addresses by $V(T_\ell)$. In a chosen-plaintext setting, the measurement routine proceeds as follows given a table ℓ, index y into ℓ and plaintext p:

(a) Trigger an encryption of p.
(b) *(evict)* Access some W memory addresses, at least B bytes apart, that are congruent to $V(T_\ell) + y \cdot B/\delta$ modulo $S \cdot B$.
(c) *(time)* Trigger a second encryption of p and time it. This is the measurement score.

The rationale for this procedure is as follows. Step (a) ensures that all table memory blocks accessed during the encryption of p are cached[9]; this is illustrated in Figure 1(b). Step (b) then accesses memory blocks, in the attacker's own memory space, that happen to be mapped to the same cache set as the memory block of y in T_ℓ. Since it is accessing W such blocks in a cache with associativity W, we expect these blocks to completely replace the prior contents of the cache. Specifically, the memory block of index y in the encryption table T_ℓ is now not in cache; see Figure 1(c). When we time the duration of the encryption in (c), there are two possibilities. If $Q_k(p, \ell, y) = 1$, that is if the encryption of the plaintext p under the unknown encryption key k accesses the memory block of index y in T_ℓ, then this memory block will have to be re-fetched from memory into the cache, leading to Figure 1(d). This fetching will slow down the encryption. Conversely, if $Q_k(p, \ell, y) = 0$ then this memory fetch will not occur. Thus, all other things being equal, the expected encryption time is larger when $Q_k(p, \ell, y) = 1$. The gap is on the order of the timing difference between a cache hit and a cache miss.

Figure 3 demonstrates experimental results. The bright diagonal corresponds to samples where $\langle y \rangle \oplus \langle p_0 \rangle = \langle k_0 \rangle = 0$, for which the encryption in step (c) always suffers a cache miss.

This measurement method is easily extended to a case where the attacker can trigger encryption with plaintexts that are known but not chosen (e.g., by sending network packets to which an uncontrolled but guessable header is added). This is done by replacing step (a) above with one that simply triggers encryptions of arbitrary plaintexts in order to cause *all* table elements to be loaded into cache.

---

[8] Also, as before, the cache sets of all tables are assumed to be distinct.
[9] Unless the triggered encryption code has excessive internal cache contention.

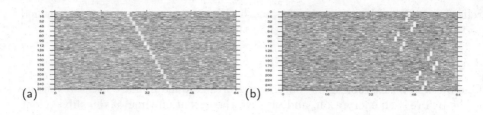

Fig. 3. Timings (lighter is slower) in Evict+Time measurements on a 2GHz Athlon 64, after 10,000 samples, attacking a procedure that executes an encryption using OpenSSL 0.9.8. The horizontal axis is the evicted cache set (i.e., $\langle y \rangle$ plus an offset due to the table's location) and the vertical axis is p_0 (left) or p_5 (right). The patterns of bright areas reveal high nibble values of 0 and 5 for the corresponding key byte values.

The weakness of this measurement method is that, since it relies on timing the triggered encryption operation, it is very sensitive to variations in the operation. In particular, triggering the encryption (e.g., through a kernel system call) typically executes additional code, and thus the timing may include considerable noise due to sources such as instruction scheduling, conditional branches and cache contention. Indeed, using this measurement method we were able to extract full AES keys from an artificial service doing AES encryptions using OpenSSL library calls, but not from more typical "heavyweight" services. For the latter, we invoked the alternative measurement method described in the next section.

3.5   Measurement Via Prime+Probe

This measurement method tries to discover the set of memory blocks read by the encryption *a posteriori*, by examining the state of the cache after encryption. This method proceeds as follows. The attacker allocates a contiguous byte array $A[0, \ldots, S \cdot W \cdot B - 1]$, with start address congruent mod $S \cdot B$ to the start address of T_0.[10] Then, given a plaintext p, it obtains measurement scores for all tables ℓ and all indices y and does so using a *single* encryption:

(a) *(prime)* Read a value from every memory block in A.
(b) Trigger an encryption of p.
(c) *(probe)* For every table $l = 0, \ldots 3$ and index $y = 0, \delta, 2\delta, \ldots, 256 - \delta$:
 - Read the W memory addresses $A[1024\ell + 4y + tSB]$ for $t = 0, \ldots, W-1$. The total time it takes to perform these reads is the measurement score, i.e., our sample of $M_k(p, \ell, y)$.

Step (a) completely fills the cache with the attacker's data; see Figure 1(e). The encryption in step (b) causes partial eviction; see Figure 1(f). Step (c) checks, for each cache set, whether the attacker's data is still present after the encryption:

---

[10] For simplicity, here we assume this address is known, and that T_0, T_1, T_2, T_3 are contiguous.

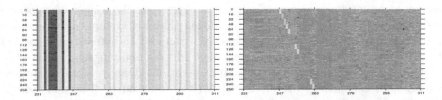

Fig. 4. Prime+Probe attack using 30,000 encryption calls on a 2GHz Athlon 64, attacking Linux 2.6.11 `dm-crypt`. The horizontal axis is the evicted cache set (i.e., $\langle y \rangle$ plus an offset due to the table's location) and the vertical axis is p_0. Left: raw timings (lighter is slower). Right: after subtraction of the average timing of the cache set. The bright diagonal reveals the high nibble of $p_0 = 0x00$.

cache sets that were accessed by the encryption in step (b) will incur cache misses in step (c), but cache sets that were untouched by the encryption will not, and thus induces a timing difference.

Crucially, the attacker is timing a simple operation performed by *itself*, as opposed to a complex encryption service with various overheads executed by someone else (as in the Evict+Time approach); this is considerably less sensitive to timing variance, and oblivious to time randomization or canonization (which are frequently proposed countermeasures against timing attacks; see Section 5). Another benefit lies in inspecting all cache sets simultaneously after each encryption, so that each encryption effectively yields $4 \cdot 256/\delta$ samples of measurement score, rather than a single sample.

An example of the measurement scores obtained by this method, for a real cryptographic system, are shown in Figure 4. Note that to obtain a visible signal it is necessary to normalize the measurement scores by subtracting, from each sample, the average timing of its cache set; this is because different cache sets are affected differently by auxiliary memory accesses (e.g., stack and I/O buffers) during the system call.

3.6   Experimental Results

We have tested the synchronous attacks against AES in various settings. To have an initial "clean" testing environment for our attack code, we started out using OpenSSL library calls as black-box functions, pretending we have no access to the key. In this setting, and with full knowledge of the relevant virtual and physical address mappings, using Prime+Probe measurements we recover the full 128-bit AES key after only 300 encryptions on Athlon 64, and after 16,000 encryptions on Pentium 4E. In the same setting, but without any knowledge about address mappings (and without any attempt to discover it systematically) we still recover the full key on Athlon 64 after 8,000 encryptions.

We then set out to test the attacks on a real-life encrypted filesystem. We set up a Linux `dm-crypt` device, which is a virtual device which uses underlying storage (here, a loopback device connected to a regular file) and encrypts all data at the sector level (here, using 128-bit AES encryptions in ECB mode). On top of this we create and mount an ordinary ext2 filesystem. We trigger encryptions by performing writes to an ordinary file inside that file system, after opening it in O_DIRECT mode; each write consisted of a random 16-byte string repeated 32 times. Running this with knowledge about address mappings, we succeed in extracting the full key after just 800 write operations done in 65ms (including the analysis of the cache state after each write), followed by 3 seconds of off-line analysis. Data from two analysis stages for this kind of attack are shown in Figure 4 (for visual clarity, the figures depict a larger number of samples).

The Evict+Time measurements (Figure 3) let us recover the secret key using about 500,000 samples when attacking OpenSSL on Athlon 64. Gathering the data takes about half a minute of continuous measurement, more than three orders of magnitude slower than the attacks based on Prime+Probe.

These results required handling several practical complications, whose details are omitted for brevity. For example, the memory addresses of the encryption tables are in general not known to the attacker and need to be identified. Most processors employ a multi-level cache hierarchy involving several parameter sets and timing gaps, which can be exploited. The distinction between virtual and physical memory addresses affects the mapping of memory blocks to cache sets and the way the latter are accessed. Various machine-specific tricks are needed to obtain high-resolution, low-latency time measurements.

3.7   Variants and Extensions

There are many possible extensions to the basic techniques described above. For example, variants of the above techniques can also be applied in known-ciphertext (as opposed to known-plaintext) setting, by analyzing the last rounds instead of the first ones. The two-rounds attack can be made more efficient and noise-resilient by analyzing further equations. On some processors, timing variability leaks information on memory accesses with resolution better than δ (e.g., due to cache bank collisions), hence analysis of the first round via Evict+Time can yield additional key bits.

We believe this attack can be converted into a remote attack on a network-triggerable cryptographic network process (e.g., IP/Sec or OpenVPN). The cache manipulation can be done remotely, for example by triggering accesses to the network stack's TCP connection table, but its efficiency remains to be evaluated.

4   Asynchronous Attacks

4.1   Overview

While the synchronous attack presented in the previous section leads to very efficient key recovery, it is limited to scenarios where the attacker has some interaction with the encryption code which allows him to obtain known plaintexts and execute code synchronously before and after encryption. We now proceed to describe a class of attacks that eliminate these prerequisites. The attacker will execute his own program on the same processor as the encryption program, but without any explicit interaction such as inter-process communication or I/O, and the only knowledge assumed is about a non-uniform distribution of the plaintexts or ciphertexts (rather than their specific values). Essentially, the attacker will ascertain patterns of memory access performed by other processes just by performing and measuring accesses to its own memory. This attack is more constrained in the hardware and software platforms to which it applies, but it is very effective on certain platforms, such as processors with simultaneous multithreading.

4.2 One-Round Attack

The basic form of this attack works by obtaining a statistical profile of the frequency of cache set accesses. The means of obtaining this will be discussed in the next section, but for now we assume that for each table T_ℓ and each memory block $n = 0, \ldots, 256/\delta - 1$ we have a *frequency score* value $F_\ell(n) \in \mathbb{R}$, that is strongly correlated with the relative frequencies. For a simple but common case, suppose the attacker process is performing AES encryption of English text, in which most bytes have their high nibble set to 6 (i.e., lowercase letters a through p). Since the actual table lookups performed in round 1 of AES are of the form "$T_\ell[x_i^{(0)}]$" where $x_i^{(0)} = p_i \oplus k_i$, the corresponding frequency scores $F_\ell(n)$ will have particularly large values when $n = 6 \oplus \langle k_i \rangle$ (assuming $\delta = 16$). Thus, just by finding the n for which $F_\ell(n)$ is large and XORing them with the constant 6, we get the high nibbles $\langle k_i \rangle$.

Note, however, that we cannot distinguish the order of different memory accesses to the same table, and thus cannot distinguish between key bytes k_i involved in the first-round lookup to the same table ℓ. There are four such key bytes per table (for example, k_0, k_5, k_{10}, k_{15} affect T_0; see Section 2.2). Thus, when the four high key nibbles $\langle k_i \rangle$ affecting each table are distinct (which happens with probability $((16!/12!)/16^4)^4 \approx 0.2$), the above reveals the top nibbles of all key bytes but only up to four disjoint permutations of 4 elements. Overall this gives $64/\log_2(4!^4) \approx 45.66$ bits of key information, somewhat less than the one-round synchronous attack. When the high key nibbles are not necessarily disjoint we get more information, but the analysis of the signal is somewhat more complex.

More generally, suppose the attacker knows the first-order statistics of the plaintext; these can usually be determined just from the type of data being encrypted (e.g., English text, numerical data in decimal notation, machine code or database records). Specifically, suppose that for $n = 0, \ldots, 256/\delta - 1$ the attacker knows $R(n) = \Pr[\langle p_i \rangle = n]$, i.e., the histogram of the plaintext bytes truncated into blocks of size δ (the probability is over all plaintext blocks and all bytes i inside each block). Then the partial key values $\langle k_i \rangle$ can be identified by finding those that yield maximal correlation between $F_\ell(n)$ and $R(n \oplus \langle k_i \rangle)$.

4.3 Measurements

One measurement method exploits the simultaneous multithreading feature available in some high-performance processors (e.g., Pentium and Xeon processors with HyperThreading). This feature allows concurrent execution of multiple processes on the same physical processor, with instruction-level interleaving and parallelism. When the attacker process runs concurrently with its victim, it can analyze the latter's memory accesses in real time; in particular, it can gather statistics such as the frequency scores $F_\ell(n) \in \mathbb{R}$. This can be done via a variant of the Prime+Probe measurements of Section 3.5, as follows.

For each cache set, the attacker thread runs a loop which closely monitors the time it takes to repeatedly load a set of memory blocks that exactly fills that cache set, i.e., W memory blocks mapped to that cache set (similarly to step (c) of the Prime+Probe measurements).[11] As long as the attacker is alone in using the cache set, all accesses hit the cache and are very fast. However, when the victim thread accesses a memory location which maps to the set being monitored, that causes one of the attacker's cache lines to be evicted from cache and replaced by a cache line from the victim's memory. This leads to one or (most likely) more cache misses for the attacker in subsequent loads, and slows him down until his memory once more occupies all the entries in the set. The attacker thus measures the time over an appropriate number of accesses and computes their average, giving us the frequency score $F_\ell(n)$.

4.4   Experimental Results

Attacking a series of processes encrypting English text with the same key using OpenSSL, we effectively retrieve 45.7 bits of information[12] about the key after gathering timing data for about 1 minute. Timing data from one of the runs is shown in Figure 5.

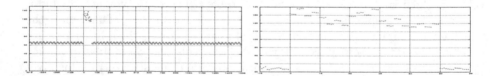

Fig. 5. Frequency scores for OpenSSL AES encryption of English text. Horizontal axis: cache set. Timings performed on 3GHz Pentium 4E with HyperThreading. To the right we zoom in on the AES lookup tables; the pattern corresponds to the top nibbles of the secret key 0x004080C0105090D02060A0E03070B0F0.

4.5   Variants and Extensions

This attack vector is quite powerful, and has numerous possible extensions, such as the following.

The second round can be analyzed using higher-order statistics on the plaintext, yielding enough key bits for exhaustive search.

---

[11] Due to the time-sensitivity and effects such as prefetching and instruction reordering, getting a significant signal requires a carefully crafted architecture-specific implementation of the measurement code.

[12] For keys with distinct high nibbles in each group of 4; see Section 4.1.

If measurements can be made to detect order of accesses (which we believe is possible with appropriately crafted code), the attacker can analyze more rounds as well as extract the unknown permutations from the first round. Moreover, if the temporal resolution suffices to observe adjacent rounds, then it becomes possible to recover the key without even known plaintext *distribution*.

We have demonstrated the attack on a Pentium 4E with HyperThreading, but it can also be performed on other platforms without relying on simultaneous multithreading. The key is for the attacker to execute its own code midway through an encryption, and this can be achieved by exploiting the interrupt mechanism. For example, the attacker can predict RTC or timer interrupts and yield the CPU to the encrypting process a few cycles before such an interrupt; the OS scheduler is invoked during the interrupt, and if dynamic priorities are set up appropriately in advance then the attacker process will regain the CPU and can analyze the state of the cache to see what the encrypting process accessed during those few cycles. On multi-core processors, shared caches can lead to inter-core attacks; in SMP systems, cache coherency mechanisms may be exploitable.

As in the synchronous case, one can envision remote attack variants that take advantage of data structures to which accesses can be triggered and timed through a network (e.g., the TCP state table).

5 Countermeasures

In the following we list several potential methods for mitigating the information leakage, focusing on those that can be implemented in software. As these methods have different trade-offs and are architecture- and application-dependent, we cannot recommend a single recipe for all implementors. Many of these methods are also applicable to primitives other than AES. See the extended version of this paper for further discussion.

5.1 Avoiding Memory Accesses

Our attacks exploit the effect of memory access on the cache, and would thus be completely mitigated by an implementation that does not perform any table lookups. For AES, the lookup tables have concise algebraic descriptions, but performance is degraded by over an order of magnitude[13]. Another approach is that of bitslice implementations [3], which employ a description of the cipher in terms of bitwise logical operations, and vectorize these operations across wide registers. For AES, we expect (but have not yet verified) that amortized performance would be comparable to that of a lookup-based implementation.

5.2 Alternative Lookup Tables

There are alternative formulations of AES, using a smaller set of tables. We have considered the most common implementation, employing four 1024-byte tables

---

[13] This kind of implementation has also been attacked through the timing variability in some implementations [6].

for the main rounds. Variants have been suggested with one or two 256-byte table, one 1024-byte table, or one 2048-byte table. Generally, the smaller the tables the slower the encryption.

In regard to the synchronous attacks considered in Section 3, smaller tables necessitate more measurements by the attacker, but the synchronous attacks remain feasible for all but the (slow) 256-byte table variant. In regard to the asynchronous of Section 4, if the attacker can sample at intervals on the order of single table lookups (which is architecture-specific) then these alternative representations provide no appreciable security benefit.

5.3   Data-Oblivious Memory Access Pattern

Instead of avoiding table lookup, one could employ them but ensure that the pattern of accesses to the memory is completely oblivious to the data passing through the algorithm. Most naively, to implement a memory access one can read *all* entries of the relevant table, in fixed order, and use just the one needed. This induces significant slowdown, even after some possible relaxations.

Goldreich and Ostrovsky [5] gave a generic program transformation for hiding memory accesses, which is quite satisfactory from an (asymptotic) theoretical perspective. However, its concrete overheads in time and memory size appear too high for most applications. Xhuang, Zhang, Lee and Pande [16][17] addressed this from a practical perspective and proposed several techniques which are more efficient, but require non-trivial hardware support in the processor or memory system and do not provide perfect security in the general case.

5.4   Application-Specific Algorithmic Masking

There is extensive literature about side-channel attack countermeasures for hardware ASIC and FPGA implementations. Some of them are algorithmic masking techniques which may be adapted to software (for AES, see e.g. [13] and the references within). However, these are designed to protect only against first-order analysis, i.e., against attacks that measure some aspect of the state only at one point in the computation, and our asynchronous attacks do not fall into this category. Moreover, the security proofs consider leakage only of specific intermediate values, which do not correspond to the ones leaking through accessed memory addresses. Lastly, every AES masking method we are aware of has either been shown to be insecure even for its original setting (let alone ours), or is significantly slower in software than a bitsliced implementation.

5.5   Cache State Normalization and Process Blocking

Against the synchronous attacks of Section 3, it suffices to simply normalize the state of the cache just before encryption (to prevent the initial cache state from affecting the encryption, as in Evict+Time) and just after the encryption (to prevent the encryption from affecting the final cache state, as in Prime+Probe). Normalization can be achieved, for example, by loading all lookup tables into the cache (the attack of [2] may remain applicable). However, this method provides little protection against the asynchronous attacks of Section 4. To fully protect against those, during the encryption one would have to disable interrupts and stop simultaneous threads (and possibly, other SMP processors). This would degrade performance and reliability.

5.6   Disabling Cache Sharing

To protect against software-based attacks, it would suffice to prevent cache state effects from spanning process boundaries. Alas, practically this is very expensive to achieve. On a single-threaded processor, it would require flushing all caches during every context switch. On a processor with simultaneous multithreading, it would also require the logical processors to use separate logical caches, statically allocated within the physical cache; some modern processors do not support such a mode. One would also need to consider the effect of cache coherency mechanisms in SMP configurations. A relaxed version would activate the above means only for specific processes, or specific code sections, marked as sensitive.

5.7   Static or Disabled Cache

One brutal countermeasure against the cache-based attacks is to completely disable the CPU's caching mechanism; the effect on performance would be devastating. An alternative is to activate a "no-fill" mode where the cache is used but not updated (i.e., eviction is disabled). We are not aware of any processor that provides the necessary facilities with reasonable overhead. In some cases it may be possible to delegate the encryption to a co-processor with the necessary properties. For example, the SPE cores in IBM's Cell processor can be used as a cryptographic co-processor[14].

5.8   Dynamic Table Storage

The cache-based attacks observe memory access patterns to learn about the table lookups. Instead of eliminating these, we may try to decorrelate them. For example, one can use many copies of each table, placed at various offsets in memory, and have each table lookup (or small group of lookups) use a pseudo-randomly chosen table. Somewhat more compactly, one can use a single table, but pseudorandomly move it around memory several times during each encryption. Another variant is to mix the order of the table elements several times during each encryption.

5.9   Hiding the Timing

The attacks rely on timing information, and thus could be foiled by its absence. One may try to add noise to the observed timings by adding random delays

---

[14] The Cell's parallelism and abundance of wide registers (which can be utilized for bitslicing) appears attractive for cryptographic and cryptanalytic applications.

to measured operations, by normalizing all operations to a fixed time, or by limiting the system clock resolution or accuracy. Effective elimination of the timing information has a high cost in performance or in system capabilities.

5.10 Selective Round Protection

The attacks we described detect and analyse memory accesses in the first two rounds (for known input) or last two rounds (for known output). To protect against these specific attacks it suffices to protect those four rounds by the means given above while using the faster, unprotected implementation for the internal rounds.[15] Other cryptanalytic attacks (e.g., using differential cryptanalysis) can still be applied to the internal rounds, but their complexity is higher.

5.11 Operating System Support

Several of the above suggestions require privileged system operation. In some scenarios and platforms, these countermeasures may be superior (in efficiency or safety) to any method that can be achieved by user processes. Operating systems may thus provide cryptographic primitives to user programs, as part of their functionality. A more flexible approach is to provide a "sensitive section" service, which executes user code under a specific promise (e.g., no context switching or simultaneous multithreading) and, in case the promise must be violated, provides graceful recovery (e.g., by flushing the caches) and reports the failure to the user.

6 Conclusions and Implications

6.1 Vulnerable Cryptographic Primitives

We have demonstrated efficient side-channel attacks on the AES cipher, in software. Some variants of our attack do not even require known plaintext or ciphertext, and have no direct interaction with the analyzed process other than running on the same CPU.

Beyond AES, such attacks are potentially applicable to any implementation of a cryptographic primitive that performs data-dependent memory accesses. The efficiency of the attack depends heavily on the structure of the cipher and chosen implementation, but heuristically, large lookup tables increase the effectiveness of all attacks, as do large lookup entries; having few accesses to each table helps the synchronous attacks, whereas the related property of having temporally infrequent accesses to each table helps the asynchronous attack.

For example, DES is vulnerable when implemented using large lookup tables. Cryptosystems based on large-integer modular arithmetic, such as RSA, is

---

[15] This was suggested to us by Intel Corp.

vulnerable in some implementations (see [14]). The same potentially applies to ECC-based cryptosystems.

Primitives that are normally implemented without lookup tables, such as bitsliced Serpent [1] and the SHA family [12], are impervious to the attacks described here. However, to protect against timing attacks one should scrutinize implementations for use of instructions whose timing is data-dependent (e.g., bit shifts and multiplications on some platforms) and for data-dependent execution branches (which may be analyzed through data cache access, instruction/trace cache access or timing). Note that timing variability could be measured by an unrelated process running on the same machine, by a variant of the asynchronous attack, through the effect on the scheduling of memory accesses.

6.2   Vulnerable Systems

At the system level, cache state analysis is of concern in essentially any case where process separation is employed in the presence of malicious code. Beyond the demonstrated case of encrypted filesystems, this includes many multi-user systems, as well as web browsing and DRM applications. Disturbingly, virtual machines and sandboxes offer little protection, since for the asynchronous attack the attacker needs only the ability to access his own memory and measure time. Thus, the attack may cross the boundaries supposedly enforced by FreeBSD jail(), VMware[16], Xen, NGSCB, the Java Virtual Machine and plausibly even scripting language interpreters. Remote attacks are in principle possible, and if proven efficient could pose serious threats to secure network connections such as IP/Sec and OpenVPN. Finally, while we have focused our attention on cryptographic systems (in which even small amount of leakage can be devastating), the leakage also occurs in non-cryptographic systems and may thus leak sensitive information directly.

6.3   Mitigation

We have described a variety of countermeasures against cache state analysis attacks. However, none of these unconditionally mitigates the attacks while offering performance close to current implementations. Thus, finding an efficient solution that is application- and architecture-independent remains an open problem. In evaluating countermeasures, one should pay particular attention to the asynchronous attacks, which on some platforms allow the attacker to obtain (a fair approximation of) the full transcript of memory accesses done by the cryptographic code.

Acknowledgements.   We are indebted to Ernie Brickell, Jean-Pierre Seifert and Michael Neve of Intel Corp. for insightful discussions and proposal of several countermeasures, to Daniel J. Bernstein for suggesting the investigation of remote attacks, and to Eli Biham for directing us to reference [7].

---

[16] This compromises the system described in a recent NSA patent [10].

References

1. R. J. Anderson, E. Biham, L. R. Knudsen, *Serpent: A proposal for the Advanced Encryption Standard*, AES submission, 1998, http://www.cl.cam.ac.uk/~rja14/serpent.html
2. D. Bernstein, *Cache-timing attacks on AES*, preprint, 2005, http://cr.yp.to/papers.html#cachetiming
3. E. Biham, *A fast new DES implementation in software*, proc. FSE 1997, LNCS 1267, 260–272, Springer, 1997
4. J. Daemen, V. Rijmen, *AES Proposal: Rijndael*, version 2, AES submission, 1999, http://csrc.nist.gov/CryptoToolkit/aes/rijndael/Rijndael-ammended.pdf
5. O. Goldreich, R. Ostrovsky, *Software protection and simulation on oblivious RAMs*, Journal of the ACM, vol. 43 no. 3, 431–473, 1996
6. F. Koeune, J. Quisquater, *A timing attack against Rijndael*, technical report CG-1999/1, Université catholique de Louvain, http://www.dice.ucl.ac.be/crypto/tech_reports/CG1999_1.ps.gz
7. Wei-Ming Hu, *Lattice scheduling and covert channels*, IEEE Symposium on Security and Privacy, 52–61, IEEE, 1992
8. J. Kelsey, B. Schneier, D. Wagner, C. Hall, *Side channel cryptanalysis of product ciphers*, proc. 5th European Symposium on Research in Computer Security, LNCS 1485, 97–110, Springer-Verlag, 1998
9. D. Page, *Theoretical use of cache memory as a cryptanalytic side-channel*, technical report CSTR-02-003, Department of Computer Science, University of Bristol, 2002, http://www.cs.bris.ac.uk/Publications/pub_info.jsp?id=1000625
10. Robert V. Meushaw, Mark S. Schneider, Donald N. Simard, Grant M. Wagner, *Device for and method of secure computing using virtual machines*, US patent 6,922,774, 2005
11. National Institute of Standards and Technology, Advanced Encryption Standard (AES) (FIPS PUB 197), 2001
12. National Institute of Standards and Technology, Secure Hash Standard (SHS) (FIPS PUB 180-2), 2002
13. E. Oswald, S. Mangard, N. Pramstaller, V. Rijmen, *A side-channel analysis resistant description of the AES S-box*, proc. FSE 2005, Springer-Verlag, to appear
14. C. Percival, *Cache missing for fun and profit*, BSDCan 2005, Ottawa, 2005; see http://www.daemonology.net/hyperthreading-considered-harmful/
15. Y. Tsunoo, T. Saito, T. Suzaki, M. Shigeri, H. Miyauchi, *Cryptanalysis of DES implemented on computers with cache*, proc. CHES 2003, LNCS 2779, 62-76, 2003
16. X. Zhuang, T. Zhang, H. S. Lee, S. Pande, *Hardware assisted control flow obfuscation for embedded processors*, proc. Intl. Conference on Compilers, Architectures and Synthesis for Embedded Systems, 292-302, ACM, 2004
17. X. Zhuang, T. Zhang, S. Pande, *HIDE: An Infrastructure for Efficiently protecting information leakage on the address bus*, proc. Architectural Support for Programming Languages and Operating Systems, 82–84, ACM, 2004

Related-Key Impossible Differential Attacks on 8-Round AES-192

Eli Biham[1], Orr Dunkelman[1,*], and Nathan Keller[2]

[1] Computer Science Department, Technion, Haifa 32000, Israel
{biham, orrd}@cs.technion.ac.il
[2] Einstein Institute of Mathematics, Hebrew University,
Jerusalem 91904, Israel
nkeller@math.huji.ac.il

Abstract. In this paper we examine the strength of AES against the related-key impossible differential attack, following the work of Jakimoski and Desmedt [12]. We use several additional observations to substantially improve the data and time complexities of their attacks. Amongst our results, we present a related-key attack on 7-round AES-192 with data complexity of 2^{56} chosen plaintexts (instead of 2^{111}). Our attack on 8-round AES-192 has data complexity of $2^{68.5}$ chosen plaintexts (instead of 2^{88}). The time complexities of our attacks is also substantially lower than the time complexities of previous attacks.

Keywords: AES, related-key differentials, impossible differentials.

1 Introduction

The *Advanced Encryption Standard* [9] is a 128-bit block cipher with variable key length (128, 192, and 256-bit keys are allowed). Since its selection, AES gradually became one of the most worldwide used block ciphers. Therefore, a constant evaluation of its security with respect to various cryptanalytic techniques is required. AES was already analyzed in many papers, each using different attacks [5, 6, 8, 10, 11, 12].

Related-key attacks [1] consider the information that can be extracted from two encryptions using related (but unknown) keys. In the attack, the attacker uses weaknesses of the encryption function and of the key schedule algorithm to derive information on the unknown keys. Related-key differential attacks [13] study the development of differences in two encryptions under two related keys and use them to derive the actual values of the keys. Usually the attacker exploits differential relations that hold with a relatively high probability, like in ordinary differential attacks [4]. However, differential relations holding with a very low (or zero) probability can also be used [2, 3, 12]. In this case, the attack is called *related-key impossible differential attack*.

---

* The research presented in this paper was supported by the Clore scholarship programme.

D. Pointcheval (Ed.): CT-RSA 2006, LNCS 3860, pp. 21–33, 2006.

In this paper we examine the security of AES against related-key impossible differential attacks. We concentrate on the 192-bit key version of AES (AES-192) since in this variant the diffusion of the key schedule is slower than in the other versions and thus the potential vulnerability to related-key attacks is bigger.

The relatively weak key schedule of AES-192 has inspired much research: In [12] Jakimoski and Desmedt presented a related-key differential attack applicable up to a 6-round AES-192 (out of the 12 rounds). An improved version of the attack (also presented in [12]) uses truncated differentials and is applicable up to a 7-round version. In addition, Jakimoski and Desmedt [12] devised several related-key impossible differential attacks that are applicable up to an 8-round AES-192. In [11] Hong et al. presented a related-key rectangle attack applicable up to an 8-round AES-192. The best known related-key attack on AES-192 was devised by Biham et al. [5] and it is applicable to a 9-round variant of the cipher.

For comparison, the best attack on AES-192 not under the related-key model is a SQUARE attack presented in [10]. It can attack up to 8 rounds of AES-192, using almost the entire code book. The time complexity of this attack is 2^{188} encryptions.

In this paper we present several new related-key impossible differential attacks. The attacks use the 5.5-round impossible differential suggested by Jakimoski and Desmedt [12]. However, by making additional observations on the behavior of the key schedule, we can reduce the data complexity of our attacks by a factor of 2^{55} for the 7-round attack, and by a factor of $2^{19.5}$ for the 8-round attack. The time complexity is also reduced significantly. We summarize our results along with previously known results in Table 1.

This paper is organized as follows: In Section 2 we give a brief description of AES. In Section 3 we describe the new related-key attack on 7-round AES-192. In Section 4 we extend the 7-round attack to attacks on 8-round AES-192. Finally, Section 5 summarizes this paper.

Table 1. Summary of the Previous Attacks and of Our New Attacks

Cipher	Number of Rounds	Complexity		Number of Keys	Attack Type & Source
		Data	Time		
AES-192	7	2^{92} CP	2^{186}	1	Imp.Diff. [8]
(12 rounds)	7	$19 \cdot 2^{32}$ CP	2^{155}	1	SQUARE [10]
	8	$2^{128} - 2^{119}$ CP	2^{188}	1	SQUARE [10]
	7	2^{111} RK-CP	2^{116}	2	RK Imp.Diff. [12]
	8	2^{88} RK-CP	2^{183}	2	RK Imp.Diff. [12]
	8	$2^{86.5}$ RK-CP	$2^{86.5}$	4	RK Rectangle [11]
	9	2^{86} RK-CP	2^{125}	256	RK Rectangle [5]
	7	2^{56} RK-CP	2^{94}	32	RK Imp.Diff.;Sect. 3
	8	2^{116} RK-CP	2^{134}	32	RK Imp.Diff.;Sect. 4
	8	2^{92} RK-CP	2^{159}	32	RK Imp.Diff.;Sect. 4
	8	$2^{68.5}$ RK-CP	2^{184}	32	RK Imp.Diff.;Sect. 4

RK – Related-key, CP – Chosen plaintext,
Time complexity is measured in encryption units

2   Description of AES

The advanced encryption standard [9] is an SP-network that supports key sizes of 128, 192, and 256 bits. The 128-bit plaintexts are treated as byte matrices of size 4x4, where each byte represents a value in $GF(2^8)$. An AES round applies four operations to the state matrix:

- SubBytes (SB) – applying the same 8x8 S-box 16 times in parallel on each byte of the state,
- ShiftRows (SR) – cyclic shift of each row (the i'th row is shifted by i bytes to the left),
- MixColumns (MC) – multiplication of each column by a constant 4x4 matrix over the field $GF(2^8)$, and
- AddRoundKey (ARK) – XORing the state and a 128-bit subkey.

The MixColumns operation is omitted in the last round, and an additional AddRoundKey operation is performed before the first round (using a whitening key). As all other works on AES, we shall assume that reduced-round variants also have the MixColumns operation omitted from the last round.

The number of rounds depends on the key length: 10 rounds for 128-bit keys, 12 rounds for 192-bit keys, and 14 rounds for 256-bit keys. The rounds are numbered $0, \ldots, Nr - 1$, where Nr is the number of rounds ($Nr \in \{10, 12, 14\}$). For sake of simplicity we shall denote AES with n-bit keys by AES-n, i.e., AES with 192-bit keys (and thus with 12 rounds) is denoted by AES-192.

The key schedule of AES-192 takes a 192-bit key and transforms it into 13 subkeys of 128 bits each. The subkey array is denoted by $W[0, \ldots, 51]$, where each word of $W[\cdot]$ consists of 32 bits. The first six words of $W[\cdot]$ are loaded with the user supplied key. The remaining words of $W[\cdot]$ are updated according to the following rule:

- For $i = 6, \ldots, 51$ do
  - If $i \equiv 0 \bmod 6$ then $W[i] = W[i - 6] \oplus SB(W[i - 1] \lll 8) \oplus RCON[i/6]$,
  - else $W[i] = W[i - 1] \oplus W[i - 6]$.

where $RCON[\cdot]$ is an array of predetermined constants, and $\lll$ denotes rotation of the word by 8 bits to the left.

The best known attack on AES-192 is a SQUARE attack on 8 rounds [10]. The attack requires almost the entire code book ($2^{128} - 2^{119}$ chosen plaintexts) and has a time complexity equivalent to 2^{188} encryptions. The SQUARE attack applied to 7-round AES-192 requires $19 \cdot 2^{32}$ chosen plaintexts and has a time complexity of 2^{155} encryptions.

The best impossible differential attack on AES-192 is on 7-round AES-192 [8]. Its data complexity is 2^{92} chosen plaintexts and its time complexity is 2^{186} encryptions.

There are several related-key attacks on AES-192. A related-key impossible differential attack on an 8-round variant is presented in [12]. This attack requires 2^{88} related-key chosen plaintexts and has a running time of 2^{183} encryptions. The attack uses two related keys.

A related-key rectangle attack on 8-round AES-192 using four related keys is presented in [11]. It requires $2^{86.5}$ chosen plaintexts (encrypted under four keys) and has a time complexity equivalent to $2^{86.5}$ encryptions.

Another related-key rectangle attack on AES-192 is presented in [5]. This attack can be applied up to nine rounds using 2^{86} related-key chosen plaintexts encrypted under 256 keys. Its time complexity is 2^{125} encryptions.

The related-key attacks exploit a weakness in the key schedule algorithm of AES-192. Unlike AES-128 and AES-256, the key schedule algorithm of AES-192 applies a nonlinear component (SubBytes) once every six key words (or once every round and a half), instead of once every four key words (once every round). This leads to the introduction of better and longer related-key differentials.

2.1   Notations Used in the Paper

In our attacks we use the following notations: x_i^I denotes the input of round i, while x_i^S, x_i^{Sh}, x_i^M, and x_i^O denote the intermediate values after the application of SubBytes, ShiftRows, MixColumns, and AddRoundKey operations of round i, respectively. Of course, the relation $x_{i-1}^O = x_i^I$ holds.

We denote the subkey of round i by subscript k_i, and the first (whitening) key is k_{-1}, i.e., the subkey of the first round is k_0. In some cases, we are interested in interchanging the order of the MixColumns operation and the subkey addition. As these operations are linear they can be interchanged, by first XORing the data with an equivalent key and only then applying the MixColumns operation. We denote the equivalent subkey for the changed version by w_i, i.e., $w_i = MC^{-1}(k_i)$.

We denote the z'th column of x_i by $x_{i,Col(z)}$, i.e., $w_{0,Col(0)} = MC^{-1}(k_{0,Col(0)})$. We also denote the byte in the y'th row and the z'th column of the state matrix x (of round i) by byte $x_{i,y,z}$ where $y, z \in \{0, 1, 2, 3\}$. For example, $x_{2,0,3}^M$ denotes the fourth byte in the first row of the intermediate value after the application of the MixColumns transformation in round 2. Another notation for bytes of some intermediate state x_i is an enumeration $\{0, 1, 2, \ldots, 15\}$ where the byte $x_{i,y,z}$ corresponds to byte $4z + y$ of x_i.

In the paper we also use the notation $x_i = ((x_{i,Col(0)}), (x_{i,Col(1)}), (x_{i,Col(2)}), (x_{i,Col(3)}))$. The column j of x_i is represented as $(x_{i,0,j}, x_{i,1,j}, x_{i,2,j}, x_{i,3,j})$.

3   Related-Key Impossible Differential Attacks on 7-Round AES-192

3.1   A 5.5-Round Related-Key Impossible Differential of AES-192

First we recall the related-key impossible differential presented in [12] that we use in our attacks. The impossible differential starts at the middle of round 2 and ends just after round 7. Note that in [12] the differential was used in rounds 0–4 (including the whitening key).

Consider rounds 2–7 of AES-192. Throughout the attack we assume that the subkey differences in these six rounds and the surrounding rounds are as

Table 2. Subkey Differences Required for the 5.5-Round Impossible Differential

Round (i)	$\Delta k_{i,Col(0)}$	$\Delta k_{i,Col(1)}$	$\Delta k_{i,Col(2)}$	$\Delta k_{i,Col(3)}$
-1	$(0,0,0,f)$	$(0,0,0,0)$	$(a,0,0,0)$	$(a,0,0,0)$
0	$(a,0,0,0)$	$(a,0,0,0)$	$(0,0,0,0)$	$(0,0,0,0)$
1	$(a,0,0,0)$	$(0,0,0,0)$	$(a,0,0,0)$	$(0,0,0,0)$
2	$(0,0,0,0)$	$(0,0,0,0)$	$(a,0,0,0)$	$(a,0,0,0)$
3	$(0,0,0,0)$	$(0,0,0,0)$	$(0,0,0,0)$	$(0,0,0,0)$
4	$(a,0,0,0)$	$(0,0,0,0)$	$(0,0,0,0)$	$(0,0,0,0)$
5	$(0,0,0,0)$	$(0,0,0,0)$	$(a,0,0,0)$	$(a,0,0,0)$
6	$(a,0,0,0)$	$(a,0,0,0)$	$(0,0,0,b)$	$(0,0,0,b)$
7	$(a,0,0,b)$	$(0,0,0,b)$	$(a,0,0,b)$	$(0,0,0,b)$
8	$(0,0,c,b)$	$(0,0,c,0)$	$(a,0,c,b)$	$(a,0,c,0)$
9	$(0,0,c,b)$	$(0,0,c,0)$	$(0,d,c,b)$	$(0,d,0,b)$

a,b,c,d, and f are non-zero byte differences.

presented in Table 2. We shall address the conditions on the difference between the keys to achieve these subkey differences later.

The related-key impossible differential is of 5.5 rounds, and is built in a miss-in-the-middle manner [2]. A 4.5-round related-key differential with probability 1 is "concatenated" to a 1-round related-key differential with probability 1, in the inverse direction, where the intermediate differences contradict one another. The 5.5-round related-key impossible differential is

$$\Delta x_2^M = ((0,0,0,0),(0,0,0,0),(a,0,0,0),(a,0,0,0)) \not\rightarrow$$
$$\Delta x_7^O = ((?,?,?,?),(0,0,0,b),(?,?,?,?),(?,?,?,?)),$$

where ? denotes any value.

The first 4.5-round differential is obtained as follows: The input difference $\Delta x_2^M = ((0,0,0,0),(0,0,0,0),(a,0,0,0),(a,0,0,0))$ is canceled by the subkey difference at the end of round 2. The zero difference $\Delta x_3^I = 0$, is preserved through all the operations until the AddRoundKey operation of round 4, and hence $\Delta x_4^M = 0$. The subkey difference in k_4 becomes the data difference, i.e., $\Delta x_5^I = ((a,0,0,0),(0,0,0,0),(0,0,0,0),(0,0,0,0))$. This difference is in a single byte, and thus, the difference after the first three operations of round 5 is in all the four bytes of a column, i.e., $\Delta x_5^M = ((y,z,w,v),(0,0,0,0),(0,0,0,0),(0,0,0,0))$ where y,z,w,v are unknown non-zero byte values. After the subkey addition this difference becomes $\Delta x_5^O = ((y,z,w,v),(0,0,0,0),(a,0,0,0),(a,0,0,0))$.

This difference evolves after the SubBytes and ShiftRows of round 5 into $\Delta x_6^{Sh} = ((y',0,0,0),(0,0,0,v'),(a',0,w',0),(a'',z',0,0))$, where y',z',w',a', and a'' are unknown non-zero values. Hence, $\Delta x_6^M =((N,N,N,N),(N,N,N,N),(?,?,?,?),(?,?,?,?))$ where N denotes non-zero differences (possibly distinct). Finally, after the key addition this difference evolves to $\Delta x_6^O = ((?,N,N,N),(?,N,N,N),(?,?,?,?),(?,?,?,?))$.

Hence, the input difference $\Delta x_2^M = ((0,0,0,0),(0,0,0,0),(a,0,0,0),(a,0,0,0))$ evolves with probability one into a non-zero difference in bytes 1,2,3,5,6, and 7 of x_6^O. The propagation of the differences is shown in Figure 1.

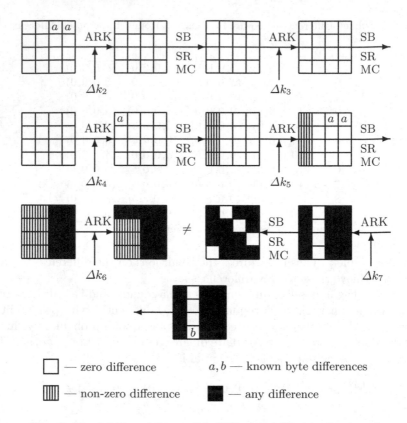

☐ — zero difference a, b — known byte differences

▦ — non-zero difference   ■ — any difference

Fig. 1. The 5.5-Round Impossible Differential Used in the Attack

The second differential ends after round 7 with output difference $\Delta x_7^O = ((?, ?, ?, ?), (0, 0, 0, b), (?, ?, ?, ?), (?, ?, ?, ?))$. When rolling back this difference through the AddRoundKey operation, we get the difference $\Delta x_7^M = ((?, ?, ?, ?), (0, 0, 0, 0), (?, ?, ?, ?), (?, ?, ?, ?))$, which leads to a zero output difference of the MixColumns operation in the second column. Hence, the input difference to round 7 is $\Delta x_7^I = ((?, ?, ?, 0), (0, ?, ?, ?), (?, 0, ?, ?), (?, ?, 0, ?))$. This difference contradicts the first differential as with probability one $x_{6,3,0}^O = x_{7,3,0}^I$ has a non-zero difference while the second differential predicts that this byte has a zero difference with probability 1. This contradiction is emphasized in Figure 1.

3.2   A 7-Round Related-Key Impossible Differential Attack

Using the above impossible differential we can attack a 7-round variant of AES-192. We attack rounds 2–8 of the cipher, using a pair of related keys that has the subkey differences described earlier. Our attack is based on the following two observations:

1. If the input difference of the differential holds, then the plaintext difference in eight of the 16 bytes is known, while in the other eight bytes almost any

difference can be used. Thus, our attack can use structures in order to bypass round 2.

2. It is sufficient to guess only one subkey byte of the last round ($k_{8,3,2}$) in order to check out whether the output difference of the impossible differential holds.

We note that due to the special structure of the key schedule, the best round to start the attack with is round 2 of the original AES.

For sake of simplicity, we currently assume that the values of a, b, c and f are known, i.e., we have two related keys K_1 and K_2 with the required subkey differences. This does not hold, but we shall deal with this issue later.

In order to make the attack faster we first perform a precomputation: For all the 2^{64} possible pairs of values of the two last columns of x_2^M, i.e., $x_{2,Col(2)}^M$ and $x_{2,Col(3)}^M$ with difference $((a,0,0,0),(a,0,0,0))$, compute the values of the eight bytes $1, 2, 6, 7, 8, 11, 12$, and 13 of x_2^I. Store the pairs of 8-byte values in a hash table H_p indexed by the XOR difference in these bytes.

The algorithm of the attack is as follows:

1. Generate two pools S_1 and S_2 of m plaintexts each, such that for each plaintext pair $P_1 \in S_1$ and $P_2 \in S_2$, $P_1 \oplus P_2 = (?,0,0,?),(?,?,0,0),(a,?,?,0)$, $(0,0,?,?)$, where "?" denotes any byte value.

2. Ask for the encryption of the pool S_1 under K_1, and of the pool S_2 under K_2. Denote the ciphertexts of the pool S_1 by T_1, and the encrypted ciphertexts of the pool S_2 by T_2.

3. For all ciphertexts $C_2 \in T_2$ compute $C_2^* = C_2 \oplus ((0,0,0,0),(0,0,0,0),$ $(0,0,0,b)$, $(0,0,c,0))$.

4. Insert all the ciphertexts $C_1 \in T_1$ and the values $\{C_2^*|C_2 \in T_2\}$ into a hash table indexed by bytes 1,4, and 14.

5. Guess the value of the subkey byte $k_{8,3,2}$ and perform the followings:

   (a) Initialize a list A of the 2^{64} possible values of the bytes $1, 2, 6, 7, 8, 11, 12$, and 13 of k_1.

   (b) Decrypt the byte $x_{8,3,2}$ in all the ciphertexts to get the intermediate values before the subkey addition at the end of round 7.

   (c) For every pair C_1, C_2^* in the same bin of the hash table, check whether the corresponding intermediate values are equal. If no, discard the pair.

   (d) For every remaining pair C_1, C_2^* consider the corresponding plaintext pair and compute $P_1 \oplus P_2$ in the eight bytes $1, 2, 6, 7, 8, 11, 12$, and 13. Denote the resulting value by P'.

   (e) Access the bin P' in H_p, and for each pair (x, y) in that bin remove from the list A the values $P_1 \oplus x$ and $P_1 \oplus y$, where P_1 is restricted to eight bytes (plaintext bytes $1, 2, 6, 7, 8, 11, 12$, and 13).[1]

   (f) If A is not empty, output the values in A along with the guess of $k_{8,3,2}$

---

[1] Not all entries contain values. It is expected that only 36% of these entries suggest values to discard. However, once an entry of H_p is non-empty, it suggests at least two values to discard.

The total amount of possible pairs C_1, C_2^* is m^2. The filtering in Step 4 is done using a 24-bit condition, thus, we expect about $2^{-24}m^2$ pairs in every bin of the hash table. In Step 5 we have an additional 8-bit filtering (for every possible value of $k_{8,3,2}$ separately) and therefore about $2^{-32}m^2$ pairs remain for a given subkey guess of $k_{8,3,2}$. Each pair deletes 1 subkey candidate on average out of the 2^{64} candidates. Hence, after $m' = 2^{-32}m^2$ pairs the expected number of remaining subkeys is $2^{64}(1 - 1/2^{64})^{m'}$. For $m' = 2^{70}$ the expected number is about e^{-20} and we can expect that only the right subkey remains. Moreover, for wrong guesses of $k_{8,3,2}$ no subkey is expected to remain. Hence, we get the value of 72 subkey bits. In order to get $m' = 2^{70}$ we need $m = 2^{51}$ chosen plaintexts in each of the two pools.

The time complexity of the attack is dominated by Step 5(e). In this step m' pairs are analyzed, leading to one memory access on average to H_p, and one memory access to A. This step is repeated 2^8 times (once for any guess of $k_{8,3,2}$). Therefore, the time complexity is 2^{79} memory accesses, which are equivalent to about 2^{73} encryptions. The precomputation requires about 2^{62} encryptions and the required memory is about 2^{69} bytes. The data complexity of the attack is 2^{52} chosen plaintexts.

Note that in the attack we assumed that the values of a, b, c, and f are known. We deal with these values and add the required corrections in the attack in the next subsection.

3.3   Overcoming the Nonlinearity of the Key Schedule Algorithm

Our attack uses a pair of related keys such that the subkey differences between them are presented in Table 2. However, due to the nonlinearity of the key schedule there is no key difference that can assure these subkey differences. In particular, while the value a can be chosen by the attacker, the values b and c that are results of application of SubBytes transformation are unknown given the initial key difference. This problem was already dealt in [12]. The solution we present in Section 4 is similar to the one presented in [12].

In our attack we have an additional problem: Since we have a round before the differential (instead of adding the round after the differential, as was done in [12]), we cannot choose $\Delta k_2 = ((0,0,0,0),(0,0,0,0),(a,0,0,0),(a,0,0,0))$ to be the first four columns of the key difference.

The key difference is $\Delta k = ((0,0,0,f),(0,0,0,0),(a,0,0,0),(a,0,0,0),$ $(a,0,0,0), (a,0,0,0))$, and as noted before, the value f is unknown. In comparison, in [12] there are no unknown bytes in the key difference since the attack starts in round 0 of AES. Due to the differential properties of the SubBytes transformation, f can assume 127 values with approximately the same probability.

The values of b and c are unknown but since the both of them are results of application of the SubBytes transformation, we know that given a, there are only 127 possible values of b, and given b there are only 127 possible values of c. Hence, we can repeat the attack for all the values of b and for every value of b, for all the values of c. The expected number of remaining wrong suggestions in the original attack (about e^{-20}) assures that for wrong guesses of b and c, with high probability no subkey will be suggested.

Therefore, the total time complexity of the attack is multiplied by 2^{21} since we repeat the attack for all the possible values of f, b, c. The data and memory requirements remain unchanged.

However, if we just try all the possible values of f, we need to encrypt the plaintexts under 128 related keys since changing the value of f changes the key difference. We can partially solve this problem by using *structures of keys*. We take two structures of 16 keys each such that the difference between two keys in different structures is $\Delta k = ((0, 0, 0, ?), (0, 0, 0, 0), (a, 0, 0, 0), (a, 0, 0, 0),$ $(a, 0, 0, 0), (a, 0, 0, 0))$, for a random value ?. Such structures are achieved by fixing all the bytes except for one in all the keys and choosing the value of this byte randomly. The structures induce 256 pairs of keys, and for each pair of keys, we perform the attack described above. Since the differences in the byte marked by ? are random, the probability that after the SubBytes transformation the difference will be a for at least one pair is approximately $1 - (1 - 1/256)^{256} = 1 - 1/e = 0.63$. Hence, with probability 0.63 we will get the required subkey differences for at least one pair of keys and for this pair the attack succeeds.

We can improve the time complexity by a factor of 2 by performing the attack only for those pairs of keys for which the difference in the marked byte can be transformed to the difference a by the SubBytes transformation. There are 127 such differences and thus the attack is expected to be performed only 127 times.

The total complexity of the attack is therefore the following: The data complexity is 2^{52} plaintexts encrypted under 16 keys each, or a total of 2^{56} chosen plaintexts, the time complexity is 2^{94} encryptions and the required memory is 2^{69} bytes.

4   Three 8-Round Impossible Differential Attacks

In this section we present three attacks on 8-round AES-192. All the three attacks are based on the 7-round attack and the main difference between them is a time-data trade-off.

Consider an 8-round version of AES-192 starting with round 2. In all the 8-round attacks we guess part of the last round subkey (k_9), peel off the last round and apply the 7-round attack. For the description of the attacks it is more convenient to change the order of the MixColumns and the AddRoundKey operations at the end of round 8. As mentioned earlier this is done by replacing the subkey k_8 with an equivalent subkey w_8. Note that since the subkey difference Δk_8 is known, the difference between the corresponding equivalent subkeys Δw_8 is also known.

In the 7-round attack we have to check whether the difference in three bytes in the beginning of round 8 is zero and whether the difference in one specific byte is b. A zero difference at the beginning of round 8 remains such a difference until the end of the round (up to the MixColumns operation), and thus we have to check whether the difference in the corresponding three bytes in the beginning of the last round is zero. For the fourth byte, we compute its difference of the pair at the beginning of round 8.

4.1 The 8-Round Attacks

The attack can be performed in one out of three possible ways:

1. Guess 12 bytes of the last round subkey (k_9) and partially decrypt these bytes in the last round. The difference in the remaining four bytes is unknown. To know this difference without guessing more subkey material, we treat only ciphertext pairs that have zero difference in these bytes. This condition allows us to use only 2^{-32} of the possible ciphertext pairs, but this price is well worth it. As the difference Δx_8^O is known, we check whether the difference in bytes 1,4, and 14 is zero. Then, we guess one subkey byte ($w_{8,3,2}$) and continue partial decryption to find out whether the difference b holds. If all the required differences hold then this ciphertext pair can be used to discard wrong subkey guesses like in the 7-round attack.

   In this variant of the attack, we guess a total of 168 subkey bits. This leads to a very high time complexity, but to a relatively low data complexity.

2. Guess eight bytes of k_9 and use only the pairs for which the difference in the eight ciphertext bytes which are XORed with an unguessed subkey is zero. Again, after partially decrypting the ciphertexts, we guess the byte $w_{8,3,2}$ and then we are able to check the differences in the four required bytes.

   In this variant, we guess 136 subkey bits, but only a portion of 2^{-64} of the pairs can be used in the attack and thus the data complexity is higher.

3. Guess only four bytes of k_9 and use only the pairs for which the difference in the 12 ciphertext bytes that are XORed with an unguessed subkey is zero. After the partial decryption, we guess the key byte $w_{8,3,2}$ in order to check whether the impossible differential can be "satisfied".

   In this variant of the attack, we guess only 104 subkey bits, leading to a substantially lower time and memory requirements. On the other hand, we use a portion of only 2^{-96} of the possible pairs, which increases the data complexity.

Since the attacks are similar, we present in detail only the first attack. The complexities of all the three attacks are summarized in Table 1.

Just like before, we assume that the values a, b, c, d and f are known. We shall address this issue after the attack.

In the first version of the attack we guess the values of bytes 0,2,3,5,6,7,8,9,10, 12,13, and 15 of k_9 and byte $w_{8,3,2}$. The values of these subkey bytes allow us to partially decrypt the last round in Columns 0, 2, and 3 where byte $x_{8,3,2}^O$ is also partially decrypted through round 8.[2] Then, we can perform the 7-round attack for every guess. Note that we can also choose other three columns to guess as long as Column 2 is included. Our choice is optimal when the values of b, c, d are not known.

---

[2] Since we analyze only pairs for which the difference in bytes 1,4,11, and 14 of the ciphertexts is zero, we know also the difference in $x_{8,Col(1)}^O$.

The attack algorithm is as follows:

1. Generate two pools S_1 and S_2 of m plaintexts each, such that for each plaintext pair $P_1 \in S_1$ and $P_2 \in S_2$, $P_1 \oplus P_2 = (?, 0, 0, ?), (?, ?, 0, 0), (a, ?, ?, 0),$ $(0, 0, ?, ?)$.
2. Ask for the encryption of the pool S_1 under K_1, and of the pool S_2 under K_2. Denote the ciphertexts of the pool S_1 by T_1, and similarly the ciphertexts of the pool S_2 by T_2.
3. For all ciphertexts $C_2 \in T_2$ compute $C_2^* = C_2 \oplus ((0, 0, 0, 0), (0, 0, 0, 0),$ $(0, 0, 0, b), (0, 0, 0, 0))$.
4. Insert all the ciphertexts $C_1 \in T_1$ and the values $\{C_2^* | C_2 \in T_2\}$ into a hash table indexed by bytes 1,4, and 14.
5. For every guess of the 12 bytes 0,2,3,5,6,7,8,9,10,12,13, and 15 of k_9 and $w_{8,3,2}$ do:
   (a) Initialize a list A of the 2^{64} possible values of the bytes $1, 2, 6, 7, 8, 11, 12,$ and 13 of the subkey k_1.
   (b) Partially decrypt the last two rounds in all of the ciphertexts to obtain $x_{9,1,0}^I, x_{9,2,3}^I$ and $x_{8,3,1}^I$. For all the ciphertexts of T_2, XOR the value of the byte $x_{8,3,1}^I$ with b.
   (c) For all the pairs $C_1 \in T_1, C_2 \in T_2$, such that C_1 and C_2^* collide in the hash table check whether the difference in the three computed bytes equals zero. Otherwise, discard the pair.
   (d) For every remaining pair, consider the corresponding pair of plaintexts and compute $P_1 \oplus P_2$ restricted to the eight bytes $1, 2, 6, 7, 8, 11, 12,$ and 13.
   (e) Access H_p in the entry $P_1 \oplus P_2$ (restricted to the eight bytes) and for every pair (x, y) in the same bin compute the values $P_1 \oplus x$ and $P_1 \oplus y$. Delete these values from the list A.
   (f) If A is not empty, output the guess for the 13 bytes and the list A.

4.2   Analysis of the Attack

The analysis of the attack is similar to the analysis of the 7-round attack. We start with $m = 2^{63.5}$ plaintexts in each pool. The plaintexts compose 2^{127} possible pairs. After the initial filtering 2^{95} pairs remain. For every guess of the 104 bits in the last rounds, about 2^{71} pairs remain after the second filtering. Each pair discards one possible value for the subkey of round 1 on average. Therefore, the probability that some wrong subkey guess remains is at most $2^{64} e^{-128} = 2^{-120}$. Therefore, the expected number of subkey suggestions (for the 168 subkey bits) is approximately $2^{-120} 2^{104} = 2^{-16}$. Hence, with a high probability only the right value remains. The remaining subkey bits can be found using auxiliary techniques.

The time complexity of the attack is dominated by the time complexity of Steps 5(d) and 5(e). For every guess of the 104 bits, we try the 2^{71} possible pairs and for each of these pairs we perform two memory accesses on average. Thus, the time complexity of this stage is about 2^{176} memory accesses, which are equivalent to about 2^{170} encryptions.

Hence, the data complexity of the attack (if b, c, d, and f are known) is $2^{64.5}$ chosen plaintexts, the time complexity is about 2^{170} encryptions and the required memory is about 2^{69} bytes.

However, the values of b, c, d and f are unknown and if we repeat the attack for all the possible guesses, the complexity will be more than the complexity of exhaustive key search.

Here we can use again the differential properties of the key schedule algorithm. We observe that the value of d is determined by the value $k_{9,2,1}$ and the value c is determined by $k_{7,3,3} = k_{9,3,0} \oplus k_{9,3,1}$. All of these subkey bytes are guessed in the beginning of the attack. Hence, for every guess of the 104 bits we have to repeat the attack only for all the possible values of b and f. As in the 7-round attack, the values of f are obtained by using structures of keys. Note that due to the low expected number of remaining subkey candidates for a single application of the attack (2^{-16}), we expect that when the attack is applied 2^{14} times, only a few subkey candidates remain.

Hence, the total complexity of the attack is as follows: The data complexity is $2^{63.5}$ chosen plaintexts encrypted under 32 keys each (or a total of $2^{68.5}$ chosen plaintexts), the time complexity is 2^{184} encryptions and the memory complexity is about 2^{69} bytes.

As mentioned before, we can perform the attack when discarding more pairs in exchange for guessing less subkey material in round 9. By considering only the ciphertext pairs with zero difference in two columns (instead of only one), we reduce the time complexity of the attack to 2^{159}. On the other hand the data complexity is increased to 2^{92} chosen plaintexts. Another possible trade-off is to consider only ciphertext pairs with zero difference in three columns. This leads to an attack that requires a total of 2^{116} chosen plaintexts and has a running time equivalent to 2^{134} encryptions. The complexity of the attacks can be found in Table 1.

5   Summary and Conclusions

In this paper we have presented several new related-key impossible differential attacks on 7-round and 8-round AES-192. The data and time complexities are summarized in Table 1. Our attacks significantly improve the attacks presented in [12], but use different properties of the key schedule of AES-192. Hence, if one could combine the attacks together, then an attack on 9-round AES-192 faster than exhaustive search may be found. However, we could not find such combination at this stage.

In our attack we perform the key recovery in the round before the differential, whereas in [12] only the rounds after the differential are attacked. As a result, our attack has to overcome the nonlinearity of the key schedule. This is achieved by using 32 keys from two *structures of keys* based on the differential properties of the key schedule algorithm.

We conclude that our paper joins a series of papers identifying problems in the key schedule algorithm of AES, and more precisely, in the key schedule algorithm of AES-192. This may be of a concern for the long term security of AES, even though at the moment none of the attacks succeeds in retrieving the key of the full AES-192 better than exhaustive key search.

References

1. Eli Biham, *New Types of Cryptanalytic Attacks Using Related Keys*, Journal of Cryptology, vol. 7, number 4, pp. 229–246, Springer-Verlag, 1994.
2. Eli Biham, Alex Biryukov, Adi Shamir, *Miss in the Middle Attacks on IDEA and Khufu*, proceedings of Fast Software Encryption 6, Lecture Notes in Computer Science 1636, pp. 124–138, Springer-Verlag, 1999.
3. Eli Biham, Alex Biryukov, Adi Shamir, *Cryptanalysis of Skipjack Reduced to 31 Rounds*, Advances in Cryptology, proceedings of EUROCRYPT '99, Lecture Notes in Computer Science 1592, pp. 12–23, Springer-Verlag, 1999.
4. Eli Biham, Adi Shamir, *Differential Cryptanalysis of the Data Encryption Standard*, Springer-Verlag, 1993.
5. Eli Biham, Orr Dunkelman, Nathan Keller, *Related-Key Boomerang and Rectangle Attacks*, Advances in Cryptology, proceedings of EUROCRYPT 2005, Lecture Notes in Computer Science 3557, pp. 507–525, Springer-Verlag, 2005.
6. Alex Biryukov, *The Boomerang Attack on 5 and 6-round AES*, proceedings of Advanced Encryption Standard 4, Lecture Notes in Computer Science 3373, pp. 11–16, Springer-Verlag, 2005.
7. Alex Biryukov, David Wagner, *Slide Attacks*, proceedings of Fast Software Encryption 6, Lecture Notes in Computer Science 1636, pp. 245–259, Springer-Verlag, 1999.
8. Raphael Chung-Wei Phan, *Impossible Differential Cryptanalysis of 7-round Advanced Encryption Standard (AES)*, Information Processing Letters, Vol. 91, Number 1, pp. 33-38, Elsevier, 2004.
9. Joan Daemen, Vincent Rijmen *The design of Rijndael: AES — the Advanced Encryption Standard*, Springer-Verlag, 2002.
10. Niels Ferguson, John Kelsey, Stefan Lucks, Bruce Schneier, Mike Stay, David Wagner, Doug Whiting, *Improved Cryptanalysis of Rijndael*, proceedings of Fast Software Encryption 8, Lecture Notes in Computer Science 1978, pp. 213–230, Springer-Verlag, 2001.
11. Seokhie Hong, Jongsung Kim, Guil Kim, Sangjin Lee, Bart Preneel, *Related-Key Rectangle Attacks on Reduced Versions of SHACAL-1 and AES-192*, proceedings of Fast Software Encryption 12, Lecture Notes in Computer Science 3557, pp. 368–383, Springer-Verlag 2005.
12. Goce Jakimoski, Yvo Desmedt, *Related-Key Differential Cryptanalysis of 192-bit Key AES Variants*, proceedings of Selected Areas in Cryptography 2003, Lecture Notes in Computer Science 3006, pp. 208–221, Springer-Verlag, 2004.
13. John Kelsey, Bruce Schneier, David Wagner, *Related-Key Cryptanalysis of 3-WAY, Biham-DES, CAST, DES-X, NewDES, RC2, and TEA*, proceedings of Information and Communication Security 1997, Lecture Notes in Computer Science 1334, pp. 233–246, Springer-Verlag, 1997.

Session Corruption Attack and Improvements on Encryption Based MT-Authenticators*

Xiaojian Tian and Duncan S. Wong

Department of Computer Science,
City University of Hong Kong, Hong Kong
{xjtian, duncan}@cityu.edu.hk

Abstract. Bellare, Canetti and Krawczyk proposed a security model (BCK-model) for authentication and key exchange protocols in 1998. The model not only reasonably captures the power of practical attackers but also provides a modular approach to the design of secure key exchange protocols. One important element in this approach is the MT-authenticator. An MT-authenticator transforms a message transmission protocol for an ideally authenticated network to an equivalent protocol for a real, unauthenticated network such that all attacks that can be launched in the unauthenticated network can also be launched in the authenticated network. In this paper, we show that the proof of the encryption-based MT-authenticator proposed in their paper is flawed, which leads to their encryption-based MT-authenticator insecure. An attack called session corruption attack can be launched successfully against the MT-authenticator in the unauthenticated network but not against the corresponding message transmission protocol in the authenticated network. To thwart this attack, we propose several improved techniques and two new encryption-based MT-authenticators.

Keywords: MT-authenticator, BCK-model, CK-model, Verifiable Encryption.

1 Introduction

Key Exchange (KE, for short) and Authentication protocols are among the most important cryptographic primitives for securing distributed computing. Despite the fact that a vast number of protocols have been proposed (see [8] for a detailed survey), many of them have subsequently been found to be vulnerable to various kinds of attacks. To provide greater assurance for the security of such protocols, one approach, which seems to be quite promising and is receiving great attention from protocol designers and analysts, is to rigorously define a security model and then show that a secure protocol meets all the claims specified in the model.

---

* The work was supported by a grant from the Research Grants Council of the Hong Kong Special Administrative Region, China (Project No. 9040904 (RGC Ref. No. CityU 1161/04E)).

D. Pointcheval (Ed.): CT-RSA 2006, LNCS 3860, pp. 34–51, 2006.

A good security model can appropriately capture a large collection of practical attacks that could be launched in the real world by defining the exact capabilities of adversaries in a modeled communication network.

Bellare and Rogaway [4] proposed a security model (BR-model) for symmetric-key based KE protocols in 1993. Their work was later extended to the three-party case and the asymmetric key setting [5, 7, 6]. When compared with another approach described below, BR-model is more difficult to use since even a small change in the protocol may need a new proof and there is no systematic way to identify or reuse building blocks for simplifying the construction and security analysis of new protocols.

In 1998, Bellare, Canetti and Krawczyk [3] (BCK-model) proposed a different approach which treats authentication and key exchange separately. Later, Canetti and Krawczyk [12] (CK-model) extended their work and changed the definition of secure KE from simulation-based to indistinguishability-based. One of the major advantages of using these models is that some building blocks in these models can be reused when constructing new protocols. In [3, 12], a building block called MT-authenticator is proposed. An MT-authenticator transmits a single message from one party to another party. By applying an MT-authenticator to each message flow of a secure protocol for authenticated network, an equivalent protocol for unauthenticated network can be built. The MT-authenticator can also be reused for constructing some other protocols. Since MT-authenticators are the essential modules used to transmit messages for a protocol in the unauthenticated network, their security decides the final security of the protocol.

1.1 Contributions

In the BCK-model, an adversary can corrupt parties as well as individual *sessions* of a party. If a session is corrupted, the adversary will learn the internal state associated with that particular session only while cannot obtain the internal state of other sessions of the party. This attack is called session corruption attack. The significance of such attack depends on what an adversary can obtain from such an attack, so it is important to define explicitly what parameters are included in the session state of a party. Unfortunately, the encryption-based MT-authenticator proposed in [3] is ambiguous in this aspect. We find that this MT-authenticator could only be made secure and its security proof could only be valid after making a number of assumptions about the session states of a party. But unfortunately these assumptions are not mentioned in the original paper, so an incautious implementation of the protocol may very likely neglect these assumptions and cause the protocol insecure in practice. Based on the plain description of the encryption-based MT-authenticator in [3], we show that this MT-authenticator suffers from the session corruption attack. The corresponding proof of this MT-authenticator given under the BCK-model also fails to consider this attack.

As defined in the BCK-model, a secure MT-authenticator should emulate a single message transmission (MT) protocol in unauthenticated networks and any attacks that can be launched in an unauthenticated network can already be done in an authenticated network. However, we find that the session corruption attack can be launched successfully against the encryption-based MT-authenticator in the unauthenticated network but not in the authenticated network. From the corresponding security analysis given in [3], we find the problem and explain how the session corruption attack has been overlooked in their security proof.

The session corruption attack in the BCK-model is further categorized into two attacks in the CK-model [12]: session-state reveal and session-output query. In the context of CK-model, we can see that the encryption-based MT-authenticators mentioned above are vulnerable to the attack of session-state reveal.

To solve the problem, we propose several improved techniques and design two new encryption-based MT-authenticators. The first one uses an application-specific decryptor to thwart the session corruption attack. The second one uses a more standard decryptor but it requires an additional proof system. This authenticator can also be viewed as the first verifiable encryption based MT-authenticator.

As informed by one of the anonymous reviewers, Canetti and Krawczyk [12–Sec. 5.3] discussed a similar problem in the context of the encryption-based key exchange protocol and made explicit specification on the session state to ensure the security of that protocol. They also remarked that the encryption-based key exchange protocol is an adaption of the encryption-based MT-authenticator proposed in [3], so their discussion also applies to the latter. However, they did not point out the flaws in the security proof of the encryption-based MT-authenticator in [3]. Their assumption on the session state may also be too restrictive in practice. Comparing with their remark, in this paper, we give more details on the session corruption attack and point out the problem in the corresponding security proof for the encryption-based MT-authenticator in [3]. We also propose two new MT-authenticators that rely on more natural assumptions that we will explain later in this paper.

Organization. We review the BCK-model in Sec. 2 and describe the session corruption attack against the encryption-based MT-authenticator proposed in [3] in Sec. 3. In Sec. 4, we propose several improved techniques and two new encryption-based MT-authenticators.

2   BCK-Model

In the BCK-model [3], a message-driven protocol is an interactive process residing in a party. The protocol (or the code of the protocol) consists of one or more subroutines (or sub-protocols). Several subroutines are first invoked by the party with some initial state that includes the protocol input, random coins and the party's identity. Some of these invocations may be classified as sessions. Each invoked subroutine can then wait for activations. An activation can be caused by two types of events: external requests that model information coming from other processes run by the party, or messages coming from the network. Upon activation, the subroutine processes the incoming data with its current state, generates new internal state, and prepares outgoing messages to the network and external requests to other processes run by the party. In addition, an output value is generated. Once the activation is complete, the subroutine waits for the next activation. Notice that the output value is appended to the one generated during the previous activation.

Each party has its local state and each subroutine of a party also has its distinct local state which is independent of local states of other subroutines. Thus each session is independent of other sessions and sessions can be identified by distinct session IDs.

In a system of n parties, $P_1, \cdots, P_n$, each of the parties is assumed to be running a copy of a message-driven protocol π which consists of one or more subroutines (which can be invoked for one or more sessions). The protocol π has an initialization function I that models the initial phase of authenticated information exchange among parties.

The BCK-model defines two games/models with two different types of adversaries running in them. The two games are authenticated-links model (AM) and unauthenticated-links model (UM). Activations of subroutines of π in each party are controlled and scheduled by an adversary. In AM, an adversary $\mathcal{A}$ is restricted to deliver messages faithfully, and each message can only be delivered once. $\mathcal{A}$ cannot inject or modify messages, but can choose not to deliver messages. $\mathcal{A}$ can also issue external requests. Furthermore, $\mathcal{A}$ can issue a party corruption query to any party in the system and learn the entire current state of that party. The entire current state also includes the local states of all the subroutines. To allow the adversary to compromise the security of only some of the sessions within some party, $\mathcal{A}$ is allowed to issue a session corruption query which lets $\mathcal{A}$ learn only the local/internal state associated with a particular session. Once $\mathcal{A}$ corrupts a session, $\mathcal{A}$ is free to forge any messages with the session ID of the corrupted session specified as the origin of the messages. Note that the long-term private keys of a party are not part of its session state. As remarked in [12], this is a fundamental requirement in a model that differentiates session corruptions from total corruptions. For each of these two corruption queries, a special value is added to the party's output specifying that the party or the particular session identified by its session ID has been corrupted. In this paper, we focus on discussing issues related the session corruption query.

In UM, an adversary $\mathcal{U}$ has all the capabilities of the AM-adversary $\mathcal{A}$. In addition to that, $\mathcal{U}$ is not limited to delivering messages faithfully as $\mathcal{A}$ does in AM. Instead, $\mathcal{U}$ can also activate subroutines of π in any parties with arbitrary incoming messages, even with fake messages that have never been sent. $\mathcal{U}$ can make the same set of queries in UM as $\mathcal{A}$ does in AM. Throughout this paper, we assume that all parties including the adversaries are probabilistic polynomial time Turing machines.

The global output of running a protocol is the concatenation of the cumulative outputs of all the parties and the adversary in each of the models. Let $AUTH_{\pi,\mathcal{A}}$ be the global output of running a message-driven protocol π in AM and $UNAUTH_{\pi,\mathcal{U}}$ be that in UM.

Definition 1 ([3], Def. 1). *Let π and π' be n-party message-driven protocols in AM and UM, respectively. We say that π' emulates π in unauthenticated* *networks if for any UM-adversary $\mathcal{U}$ there exists an AM-adversary $\mathcal{A}$ such that $AUTH_{\pi,\mathcal{A}}$ and $UNAUTH_{\pi',\mathcal{U}}$ are computationally indistinguishable.*

Since the authentication in AM is explicitly ensured, if π' emulates π in unauthenticated networks, the authentication in UM is also ensured.

Definition 2 (Authenticator, [3], Def. 2). *A compiler $\mathcal{C}$ is an algorithm that takes for input descriptions of protocols and outputs descriptions of protocols. An authenticator $\mathcal{C}$ is a compiler that for any protocol π, the protocol $\mathcal{C}(\pi)$ emulates π in unauthenticated networks.*

In [3], a general technique for designing an authenticator $\mathcal{C}$ is proposed: First, a message transmission protocol (MT) designed to work in AM is given. Upon activation within a party P_i on external request (P_j, s, m), P_i sends the message (P_i, P_j, s, m) to party P_j, and outputs "P_i sent m to P_j in session s". Upon receipt of a message (P_i, P_j, s, m), P_j outputs "P_j received m from P_i in session s". Second, an MT-authenticator λ is designed that emulates protocol MT in unauthenticated networks. Finally, given a protocol π, for each message that π sends, λ is invoked and activated with external request for sending that message to the specified recipient. When λ outputs, for example, "P_j received m from P_i in session s", π is activated with incoming message m from P_i.

We assume that each message transmitted in the network contains the identities of the sender and the receiver, as well as the session IDs of the sender's session and the receiver's session. Throughout this paper, we also assume that the sender and the receiver share the same session ID and each message contains only one copy of it. When the identities of the sender and the receiver are implicitly specified in the context, we omit them in our description. In the original paper of Bellare et al. [3], the authors use m to denote a message which comprises both the actual message content and a session ID. This causes certain inconvenience in presentation, as an attacker may choose to modify only the session ID or only the actual message content. In this paper, we adopt a more explicit approach. We use m (m', m^*, etc) to denote only the actual content of a message. It does not include sender or receiver identity, or the session ID.

3   MT-Authenticator and Session Corruption Attack

In addition to the BCK-model, Bellare et al. proposed several MT-authenticators in [3]. One of them is based on public key encryption operation. In short, we call it the BCK encryption-based MT-authenticator and denote it as λ_{ENC}.

$$A \rightarrow B : s, m$$
$$A \leftarrow B : s, m, ENC_{e_A}(N_B)$$
$$A \rightarrow B : s, m, MAC_{N_B}(s, m, B)$$

ENC_{e_A} denotes the encryption algorithm of the public key encryption scheme under the public key of A and MAC_{N_B} denotes the message authentication under an authentication key N_B. It is assumed that MAC is secure against chosen message attack.

When party A is activated by an external request to send a message m to party B in session s, a two-party protocol $\hat{\lambda}_{ENC}$ is invoked that proceeds as follows. First, A sends "message: s, m" to B and outputs "A sent m to B in session s". B then chooses a random challenge $N_B \in_R \{0,1\}^k$, and sends "challenge: $s, m, ENC_{e_A}(N_B)$" to A. Upon receipt of "challenge: s, m, c" from B, A decrypts c to get N_B and sends "mac: $s, m, MAC_{N_B}(s, m, B)$" to B. Upon receipt of "mac: s, m, v" from A, B first checks whether $v = MAC_{N_B}(s, m, B)$. If yes, it accepts m and outputs "B received m from A in session s"; otherwise, it rejects this message and terminates this invocation of $\hat{\lambda}_{ENC}$ (i.e. terminates the session s within B).

Remark: In any part of $\hat{\lambda}_{ENC}$, if anyone of A and B is activated with an invalid or unexpected external request or incoming message that is not specified above, the activation will be ignored. For example, in the second message flow, upon receipt of "challenge: s, m, c" from B, A actually checks whether it has sent m to B in session s before sending out the mac response. If not, A should ignore the challenge. In [3], this is not explicitly mentioned although we believe that the authors may have assumed to handle this case in some similar method. This handling method is important and should be mentioned. For instance, suppose A does not check whether it has sent m to B before decrypting c and sending out the mac response, then A will answer all types of challenges and an adversary can make use of A as a MAC oracle for any encrypted authentication key. In the rest of this paper, we therefore assume that this handling method is always included in each of the protocols to be discussed.

3.1   Session Corruption Attack Against λ_{ENC}

In [3], λ_{ENC} is claimed to be secure in the sense that λ_{ENC} emulates protocol MT in unauthenticated networks. However, we find that this authenticator is subject to the session corruption attack. Using this attack, we can construct a UM-adversary $\mathcal{U}$ and show that there is no AM-adversary $\mathcal{A}$ such that $AUTH_{MT,\mathcal{A}}$ and $UNAUTH_{\lambda_{ENC},\mathcal{U}}$ are computationally indistinguishable. We illustrate this attack in Fig. 1.

In this attack, $\mathcal{U}$ activates A to establish a session s with B, $\mathcal{U}$ also impersonates as A to establish another session s^* with B. We use '$s.i$' ('$s^*.i$') to denote the i-th message in the session s (s^*). The message flow marked with a '$\diamond$' denotes the message forged by the adversary $\mathcal{U}$. Detailed description of the attack is as follows:

1. $\mathcal{U}$ activates party A with an external request to send a message m to party B in session s.
2. A sends m to B, and outputs "A sent m to B in session s".
3. B chooses $N_B \in_R \{0,1\}^k$, and sends "challenge: $s, m, ENC_{e_A}(N_B)$" to A. However, this message is intercepted and blocked by $\mathcal{U}$.

---

s.1) $A \rightarrow B : s, m$

s.2) $A \leftarrow B : s, m, ENC_{e_A}(N_B)$
     $\mathcal{U}$ intercepts and blocks this message.

◇ $s^*.1)$ $A \rightarrow B : s^*, m^*$

$s^*.2)$ $A \leftarrow B : s^*, m^*, ENC_{e_A}(N_B^*)$
     $\mathcal{U}$ intercepts and blocks this message.

◇ s.2) $A \leftarrow B : s, m, ENC_{e_A}(N_B^*)$
     $\mathcal{U}$ issues **session corruption** query for session s in A and gets N_B^*.

◇ $s^*.3)$ $A \rightarrow B : s^*, m^*, MAC_{N_B^*}(m^*, B)$

Fig. 1. Session corruption attack on BCK Encryption-based MT-authenticator

4. $\mathcal{U}$ activates party B with "message: s^*, m^{*}" in another session s^*, alleging that it comes from A.
5. B chooses $N_B^* \in_R \{0, 1\}^k$, and sends "challenge: $s^*, m^*, ENC_{e_A}(N_B^*)$" to A.
6. $\mathcal{U}$ intercepts this message and replaces s^* with s and m^* with m. $\mathcal{U}$ then sends "challenge: $s, m, ENC_{e_A}(N_B^*)$" to A in session s, alleging that it comes from B.
7. A decrypts $ENC_{e_A}(N_B^*)$ to get N_B^*. Now N_B^* is in the local internal state of session s in party A.
8. $\mathcal{U}$ issues a **session corruption** query for session s in party A and gets N_B^*.
9. $\mathcal{U}$ calculates $v^* = MAC_{N_B^*}(s^*, m^*, B)$ and sends "mac: s^*, m^*, v^{*}" to B, alleging that it comes from A.
10. B verifies that $v^* = MAC_{N_B^*}(s^*, m^*, B)$. He then accepts m^* and outputs "B received m^* from A in session s^{*}".

The attack (or the UM-adversary $\mathcal{U}$) makes the output of the authenticator, $UNAUTH_{\lambda_{ENC},\mathcal{U}}$, contain "$B$ received m^* from A in session s^{*}", but without the corresponding "A sent m^* to B in session s^{*}" in it.

To see that there is no AM-adversary $\mathcal{A}$ which can make a computationally indistinguishable output $AUTH_{MT,\mathcal{A}}$ for protocol MT, first we note that $\mathcal{U}$ does not corrupt party A in UM. Hence $\mathcal{A}$ cannot corrupt A in AM either, otherwise these events will be recorded in A's output and make $AUTH_{MT,\mathcal{A}}$ and $UNAUTH_{\lambda_{ENC},\mathcal{U}}$ distinguishable. Second, if $\mathcal{A}$ does not corrupt A while B outputs "B received m^* from A in session s^{*}" in AM, there are only two possible cases:

1. A has sent m^* to B in session s^*, or
2. A does not send m^* to B in session s^*. By definition of AM, this can only happen if $\mathcal{A}$ has issued a **session corruption** query for session s^* in A and added a 'fake' message m^* into the authenticated network (this requires that the origin of m^* must be the corrupted session s^*).

However, no matter in which case, $AUTH_{MT,\mathcal{A}}$ and $UNAUTH_{\lambda_{ENC},\mathcal{U}}$ are always distinguishable. Therefore, λ_{ENC} cannot emulate protocol MT in unauthenticated networks.

Difference Between Simulated Interaction and Real Interaction. In [3–Proposition 5], a proof is given for the security of λ_{ENC}. One essential step in the proof is to use an UM-adversary $\mathcal{U}$ to construct an encryption-aided MAC forger $\mathcal{F}$:

> Let e^*, d^* be the encryption and decryption keys of the underlying public key encryption scheme, respectively. The inputs to $\mathcal{F}$ are e^* and $f = ENC_{e^*}(N^*)$ where $N^* \in_R \{0,1\}^k$. $\mathcal{F}$ has access to two oracles: $\mathcal{DO}$ and $\mathcal{MO}$. $\mathcal{DO}$ is a decryption oracle that decrypts messages (that are different from f) under the decryption key d^*; and $\mathcal{MO}$ is a MAC oracle that computes a MAC tag $MAC_{N^*}(m)$ on message m under N^*. The goal of $\mathcal{F}$ is to output a pair $(\hat{m}, MAC_{N^*}(\hat{m}))$ where $\mathcal{MO}$ was not queried on $\hat{m}$.

In the construction, $\mathcal{F}$ runs $\mathcal{U}$ on a simulated interaction with a set of parties running λ_{ENC}. First, $\mathcal{F}$ chooses and distributes keys for the imitated parties according to the initialization function I, with the exception that the public encryption key associated with some party P^*, chosen at random among all the imitated parties, is replaced with e^*. Next, $\mathcal{F}$ randomly chooses a message m^* out of all messages such that some party $\tilde{P}$ was activated with "message: s^*, m^*" from P^*. $\mathcal{F}$ then has $\tilde{P}$ respond with "challenge: s^*, m^*, f" (that is, $\tilde{P}$'s encrypted challenge is N^*). Finally, if $\mathcal{U}$ activates $\tilde{P}$ with incoming message "mac: s^*, m^*, c" from P^*, then $\mathcal{F}$ outputs $((s^*, m^*, \tilde{P}), c)$ and halts, hoping that $c = MAC_{N^*}(s^*, m^*, \tilde{P})$. During the simulation, the following cases will be handled specially:

1. If party P^* is corrupted, then the simulation is aborted and $\mathcal{F}$ is failed.
2. If P^* receives a challenge c different from f, $\mathcal{F}$ asks $\mathcal{DO}$ for the decryption.
3. If P^* is activated with incoming challenge "challenge: s, m, f" from some party Q such that
   (a) $Q \neq \tilde{P}$ or $s \neq s^*$ or $m \neq m^*$, then $\mathcal{F}$ asks $\mathcal{MO}$ for $MAC_{N^*}(s, m, Q)$;
   (b) otherwise, that is, if $Q = \tilde{P}$, $s = s^*$ and $m = m^*$, the simulation is aborted and $\mathcal{F}$ is failed.

For other queries made by $\mathcal{U}$, $\mathcal{F}$ answers them accordingly.

The proof claims that $\mathcal{U}$'s view of the interaction with $\mathcal{F}$ (conditioned on the event that $\mathcal{F}$ doesn't abort the simulation) is distributed identically to $\mathcal{U}$'s view of a real interaction with an unauthenticated network. However, this is not true.

Consider the following situation. Suppose that firstly P^* is activated to send a message m to $\tilde{P}$ in session s, then $\mathcal{U}$ impersonates P^* and sends m^* to $\tilde{P}$ in another session s^*. According to the construction of $\mathcal{F}$ described above, $\mathcal{F}$ has $\tilde{P}$ respond with "challenge: s^*, m^*, f". Suppose $\mathcal{U}$ intercepts this message, changes s^* to s, m^* to m, and sends "challenge: s, m, f" to P^*. This is the case 3a mentioned above. If $\mathcal{U}$ now issues a session corruption query for session s of P^*, $\mathcal{U}$ would not find N^* in the internal state of session s as $\mathcal{F}$ cannot ask $\mathcal{DO}$ for the decryption of f.

In a real interaction with an unauthenticated network, $\mathcal{U}$ would find N^* in the internal state of session s. Hence $\mathcal{F}$ cannot provide $\mathcal{U}$ with an ideal simulation. This case is overlooked in the proof of [3–Proposition 5].

4   Improvements

Before proposing our approaches for improving the BCK encryption-based MT-authenticator, we first describe an unsuccessful attempt to improve it. Its main idea is to use double encryption. In this attempt, the second flow of λ_{ENC} is modified below.

$$A \leftarrow B : s, m, \ ENC_{e_A}(ENC_{e_A}(N_B), s, m, B).$$

Upon receipt of this message from B, A first calls the decryption oracle to get $\omega = ENC_{e_A}(N_B), s, m, B$, then checks the triple (s, m, B), and proceeds only when they all fit. A then calls the decryption oracle again on ω and gets N_B. This approach seems to work, but it is actually still vulnerable to the session corruption attack. An adversary may compromise this approach in two steps. First, it uses the method as illustrated in Sec. 3.1 to get $\omega = ENC_{e_A}(N_B)$. Second, it uses the very same method again to get N_B. In this attack, it issues two session corruption queries to two different sessions. This example shows again the powerfulness of the session corruption attack. Furthermore, it is easy to see that, this approach does not work no matter how many times N_B is encrypted.[1]

Another approach to improve the BCK encryption-based MT-authenticator is to explicitly require that N_B is not part of the state information of the involved session of party A. Instead, N_B should be handled in some secure part of A which cannot be revealed by any types of allowable queries. Under this assumption, the proof of [3–Proposition 5] will become valid. Actually a similar approach was also used by Canetti and Krawczyk in [12–Sec. 5.3] and by Yang et al. in [17] to ensure the security of an encryption-based key exchange protocol. While this approach solves the problem, it uses a somewhat unnatural assumption. In the following, we suggest two other methods that also thwart the session corruption attack and use more natural assumptions.

4.1 λ_{ENC1}

λ_{ENC1} is illustrated in the following diagram and described as follows.

$$A \rightarrow B : s, m$$
$$A \leftarrow B : s, m, \ ENC_{e_A}(N_B, s, h(m), B)$$
$$A \rightarrow B : s, m, N_B$$

Let k be a security parameter. Besides, there is an additional hash function denoted by $h : \{0, 1\}^* \rightarrow \{0, 1\}^k$ but no MAC scheme is needed. For simplicity,

---

[1] The BCK encryption-based authenticator could be proven secure in a variant of the BCK-model in which events are 'atomic', which is in the sense that session corruption attack is not allowed to occur once after a session received an activation and before the session completes its process triggered by the activation and sends out a message or has its output value ready. Also the value of N_B is also assumed to be destroyed before the end of the event. However, we believe that it is hard to convince that events must be atomic as we cannot be sure that attackers will not or would not break into a session while a party or a session is in the middle of something.

we assume that the length of session ID and that of a party's identity are all k bits long. Since the encryption algorithm ENC will be used in λ_{ENC1} to encrypt a message which is composed of a k-bit random challenge, a session ID, a hash value and a party's identity, we require that the message space of the encryption algorithm ENC defined by the public key of any party in the system should contain $\{0,1\}^{4k}$. In practice, the requirement could be different due to the variations of the domains of session IDs and party identities.

1. When party A is activated by an external request to send a message m to party B in session s, A sends "message: s, m" to B and outputs "A sent m to B in session s".
2. Upon receipt of "message: s, m" from A, B randomly chooses $N_B \in_R \{0,1\}^k$, computes $c = ENC_{e_A}(N_B, s, h(m), B)$, and sends "challenge: s, m, c" to A.
3. Upon receipt of "challenge: s, m, c" from B, A checks whether it has sent m to B in session s. If not, the challenge is ignored; otherwise, it activates a process $\mathcal{D}_A$ to decrypt c. $\mathcal{D}_A$ is called an application-specific decryption box:

   > $\mathcal{D}_A$ is a special routine/process in party A. When given a ciphertext $\hat{c}$, $\mathcal{D}_A$ decrypts it to get a k-bit value $\hat{N}_B$, a session ID $\hat{s}$, a hash value $\hat{h}$ and a party identity $\hat{B}$. If the decrypted values are invalid (that is, anyone of them is not in the corresponding domain), $\mathcal{D}_A$ simply informs the calling session of A that $\hat{c}$ is invalid. Otherwise, these values are stored in the internal state of $\mathcal{D}_A$ which is independent to any of the sessions in A. In other words, a session corruption query cannot reveal the internal state of $\mathcal{D}_A$. However, it will be revealed if A is corrupted. When these values are stored in the internal state of $\mathcal{D}_A$, the calling session of A that activates $\mathcal{D}_A$ for decrypting $\hat{c}$ can then ask $\mathcal{D}_A$ for each of these four values separately.

4. If $\mathcal{D}_A$ informs that c is invalid, A will stop handling the challenge anymore. Otherwise, assume that the four values that $\mathcal{D}_A$ gets by decrypting c are N_B', s', h', and B'. A first asks $\mathcal{D}_A$ for s', h' and B', and checks whether $s' = s$, $h' = h(m)$ and $B' = B$. If any of the checks does not pass, A rejects this challenge by stop handling it anymore. Note that in this stage, A's session state includes s', h' and B' but not N_B'.
5. If all checks are passed, A then asks $\mathcal{D}_A$ for the value of N_B', and sends "response: s, m, N_B'" to B. Note that A's session state contains N_B' in this stage.
6. Upon receipt of "response: s, m, N_B'" from A, B checks whether $N_B' = N_B$. If yes, B accepts this message and outputs "B received m from A in session s", otherwise it rejects this message.

Theorem 1. *Assume the encryption scheme in use is IND-CCA2 secure. Then protocol λ_{ENC1} emulates protocol MT in unauthenticated networks.*

Proof is given in Appendix A.

λ_{ENC1} requires the initiator of a message transmission to equip a special implementation of the decryption algorithm, called application-specific decryption box. In practice, the application-specific decryption box can be imagined as a a hardware 'blackbox' which carries out the decryption requests made by the initiator. Examples of such a blackbox include a smartcard, a personal cryptographic token or even a special hardware cryptographic accelerator. In the following, we propose another encryption-based MT-authenticator which does not need the application-specific decryption box. Instead, the standard/conventional implementation of the decryption algorithm on the initiator will be fine. We denote this MT-authenticator by λ_{ENC2}. The protocol needs a verifiable encryption scheme such as [10] which is constructed from a public key encryption scheme and a zero-knowledge proof system. In other words, we can consider λ_{ENC2} as a Verifiable Encryption (VE) Based MT-authenticator.

4.2 λ_{ENC2}

Before describing our VE-based MT-authenticator λ_{ENC2}, we first briefly review some concepts and notations of a special honest-verifier zero-knowledge proof system and a VE scheme.

A Special Honest-Verifier Zero-Knowledge (Special HVZK) proof system is an interactive process between two parties, a prover $\mathcal{P}$ and a verifier $\mathcal{V}$. $\mathcal{P}$ proves to $\mathcal{V}$ that he knows some knowledge while $\mathcal{V}$ obtains nothing except the fact that $\mathcal{P}$ really knows that knowledge. The protocol is restricted to three moves. The common input of $\mathcal{P}$ and $\mathcal{V}$ is y and an additional secret input of $\mathcal{P}$ is x. In the first move, $\mathcal{P}$ sends a 'commitment' α to $\mathcal{V}$. In the second move, $\mathcal{V}$ sends a 'challenge' β back to $\mathcal{P}$. In the third move, $\mathcal{P}$ sends a 'response' ξ to $\mathcal{V}$. There must exist a simulator Sim that on input y and any 'challenge' $\tilde{\beta}$, outputs a 'commitment' and a 'response' $\tilde{\alpha}$ and $\tilde{\xi}$ such that the distribution of the triple $(\tilde{\alpha}, \tilde{\beta}, \tilde{\xi})$ is indistinguishable from the triple (α, β, ξ) obtained from a real interaction of $\mathcal{P}$ and $\mathcal{V}$ for which $\beta = \tilde{\beta}$. By using the notations introduced by Camenisch and Stadler [11], a proof system can be conveniently represented. For example,

$$PK[x, z : A = g^x h^z]$$

denotes the proof of knowledge of integers x and z such that $A = g^x h^z$. The convention is that, the elements listed before the colon are $\mathcal{P}$'s secrets to be proven, while all other parameters are known to $\mathcal{V}$. The Special HVZK proof system can be converted into a signature of knowledge using the Fiat-shamir heuristic [13].

A verifiable encryption (VE) scheme [16, 1, 9, 2, 10] for a relation $\mathcal{R}$ is a protocol that allows a prover to convince a verifier that a ciphertext is an encryption of a value w under a given public key such that $(w, \delta) \in \mathcal{R}$ for a given δ, while no more information about w is leaked. We use a VE scheme which is composed modularly of an encryption scheme and a proof system. The encryption scheme should be IND-CCA2 secure [15] and the proof system should be sound and Special HVZK [1, 10]. The VE scheme proposed by Camenisch and Shoup [10] is in this type.

In our construction, we need a one-way function family (or a collection of one-way functions) [14] $F = \{f_\gamma : W \to \Delta\}_{\gamma \in \Gamma}$ where Γ is the set of keys/indices of f_γ. Assume that $f_i \neq f_j$ for any $i \neq j$. F also induces a family of relations $\{\mathcal{R}_\gamma \in W \times \Delta\}_{\gamma \in \Gamma}$ indexed by $\gamma \in \Gamma$. A relation $\mathcal{R}_\gamma$ is defined by

$$(w, \delta) \in \mathcal{R}_\gamma \ \text{ if } \ \delta = f_\gamma(w)$$

where w is called a *witness* of δ. We assume that there exists a VE scheme for such a family of relations $\{\mathcal{R}_\gamma\}_{\gamma \in \Gamma}$.

We are now ready to describe our VE-based MT-authenticator λ_{ENC2}. In the following, We describe a general construction of λ_{ENC_2}. Due to the page limitation, we skip describing a concrete example. But readers can readily build one using Camenisch and Shoup's VE scheme [10].

The Generic Construction of λ_{ENC2}. Suppose a VE scheme for a relation family $\mathcal{R}$ with respect to a one-way function family F described above is specified. Similar to λ_{ENC} and λ_{ENC1}, the encryption algorithm of the VE scheme is denoted by ENC. Suppose there are n parties in the unauthenticated network. The initialization function I of λ_{ENC_2} first invokes, once for each party, the key generation function of the VE scheme with security parameter k. Let e_i and d_i be the public encryption key and the private decryption key of party P_i, respectively. The public information I_0 is all the encryption keys. The private information I_i of P_i is d_i. Assume that the message space defined by any public key contains $\{0,1\}^k$.

As in λ_{ENC1}, we assume that all messages sent by a sender to a receiver in a particular session are different. λ_{ENC2} proceeds as follows.

1. When some party A is activated by an external request to send a message m to party B in session s, A sends "message: s, m" to B and outputs "A sent m to B in session s".
2. Upon receipt of "message: s, m" from A, B then randomly chooses a challenge $N_B \in_R \{0,1\}^k$, computes $c = ENC_{e_A}(N_B)$ and $\delta = f_{h(A,B,s,m)}(N_B)$ where $h : \{0,1\}^* \to \Gamma$ is a collision-resistant hash function, and sends "challenge: s, m, c, δ" to A. As noted before, we assume that the message space defined by any public key corresponding to the VE scheme contains $\{0,1\}^k$. Also note that this is equivalent to saying that $(N_B, \delta) \in \mathcal{R}_{h(A,B,s,m)}$.
3. Upon receipt of this message from B, A first checks whether it has sent m to B in session s. If not, it stops handling this challenge; otherwise it requests B to initiate the proof system of the VE scheme for proving that the message encrypted in c under the public key e_A is a witness of δ with respect to the relation $\mathcal{R}_{h(A,B,s,m)}$. Note that A does not use its private key in this subprotocol nor decrypt c to get N_B. The proof can be started when B sends the challenge to A.
4. If A accepts B's proof, it then decrypts c to get N_B and sends "response: s, m, N_B" to B.
5. Upon receipt of "response: s, m, N_B" from A, B first checks whether N_B is the challenge it has sent in response to the message m from A in session s. If not, B rejects this message; otherwise B accepts this message and outputs "B received m from A in session s".

λ_{ENC2} is illustrated as follows.

$$A \rightarrow B : s, m$$
$$A \leftarrow B : s, m, c = ENC_{e_A}(N_B), \quad \delta = f_{h(A,B,s,m)}(N_B),$$
$$PK[N_B : c = ENC_{e_A}(N_B) \wedge (N_B, \delta) \in \mathcal{R}_{h(A,B,s,m)}]$$
$$A \rightarrow B : s, m, N_B$$

Theorem 2. *Assume that the verifiable encryption scheme in use is secure such that the underlying encryption scheme is IND-CCA2 and the proof system is sound and Special HVZK in the sense of [10], and that F in use is a one-way function family described as above. Then protocol λ_{ENC2} emulates protocol MT in unauthenticated networks.*

The proof is given in Appendix B.

Remark 1: The four elements in the hash function are crucial to the security of λ_{ENC2}. If some of them are missing, attacks could be launched by a UM-adversary $\mathcal{U}$. For example, if 'B' and 's' are missing in the hash function, $\mathcal{U}$ can launch the following attack. $\mathcal{U}$ first impersonates A and sends a message m to B in session s. After intercepting B's challenge, $\mathcal{U}$ then has A send the same message m to another party C in a session s', and forwards B's challenge to A, alleging that it is the challenge of C. A will accept the challenge, decrypt the ciphertext in the challenge, and send the plaintext to C. $\mathcal{U}$ intercepts the plaintext and forwards to B.

Acknowledgements. We would like to thank the anonymous reviewers and David Pointcheval for their helpful comments and suggestions.

References

1. N. Asokan, V. Shoup, and M. Waidner. Optimistic fair exchange of digital signatures. In *Proc. EUROCRYPT 98*, pages 591–606. Springer-Verlag, 1998. LNCS Vol. 1403.
2. F. Bao. An efficient verifiable encryption scheme for encryption of discrete logarithms. In *Proc. Smart Card Research and Applications (CARDIS) 1998*, pages 213–220. Springer-Verlag, 2000. LNCS Vol. 1820.
3. M. Bellare, R. Canetti, and H. Krawczyk. A modular approach to the design and analysis of authentication and key exchange protocols. In *Proc. 30th ACM Symp. on Theory of Computing*, pages 419–428. ACM, May 1998.
4. M. Bellare and P. Rogaway. Entity authentication and key distribution. In *Proc. CRYPTO 93*, pages 232–249. Springer-Verlag, 1994. LNCS Vol. 773.
5. M. Bellare and P. Rogaway. Provably secure session key distribution – the three party case. In *Proc. 27th ACM Symp. on Theory of Computing*, pages 57–66, Las Vegas, 1995. ACM.
6. S. Blake-Wilson, D. Johnson, and A. Menezes. Key agreement protocols and their security analysis. In *Sixth IMA International Conference on Cryptography and Coding*, pages 30–45. Springer-Verlag, 1997. LNCS Vol. 1355.

7. S. Blake-Wilson and A. Menezes. Entity authentication and authenticated key transport protocols employing asymmetric techniques. In *Security Protocols Workshop*, pages 137–158. Springer-Verlag, 1997. LNCS Vol. 1361.
8. C. Boyd and A. Mathuria. *Protocols for Authentication and Key Establishment.* Springer-Verlag, 2003.
9. J. Camenisch and I. Damgård. Verifiable encryption, group encryption, and their applications to separable group signatures and signature sharing schemes. In *Proc. ASIACRYPT 2000*, pages 331–345. Springer-Verlag, 2000. LNCS Vol. 1976.
10. J. Camenisch and V. Shoup. Practical verifiable encryption and decryption of discrete logarithms. In *Proc. CRYPTO 2003*, pages 126–144. Springer-Verlag, 2003. LNCS Vol. 2729.
11. J. Camenisch and M. Stadler. Efficient group signature schemes for large groups. In *Proc. CRYPTO 97*, pages 410–424. Springer-Verlag, 1997. LNCS Vol. 1294.
12. R. Canetti and H. Krawczyk. Analysis of key-exchange protocols and their use for building secure channels. In *Proc. EUROCRYPT 2001*, pages 453–474. Springer-Verlag, 2001. LNCS Vol. 2045. http://eprint.iacr.org/2001/040/.
13. A. Fiat and A. Shamir. How to prove yourself: Practical solutions to identification and signature problems. In *Proc. CRYPTO 86*, pages 186–199. Springer-Verlag, 1987. LNCS Vol. 263.
14. O. Goldreich. *Foundations of Cryptography Basic Tools.* Cambridge University Press, 2001.
15. C. Rackoff and D. R. Simon. Non-interactive zero-knowledge proof of knowledge and chosen ciphertext attack. In *Proc. CRYPTO 91*, pages 433–444. Springer, 1992. LNCS Vol. 576.
16. M. Stadler. Publicly verifiable secret sharing. In *Proc. EUROCRYPT 96*, pages 191–199. Springer-Verlag, 1996. LNCS Vol. 1070.
17. G. Yang, D. Wong, and X. Deng. Efficient anonymous roaming and its security analysis. In *Third International Conference on Applied Cryptography and Network Security (ACNS 2005)*, pages 334–349. Springer-Verlag, 2005. LNCS Vol. 3531.

A   Proof of Theorem 1

Proof. For any UM-adversary $\mathcal{U}$, we show that there exists an AM-adversary $\mathcal{A}$ such that $AUTH_{MT,\mathcal{A}}$ and $UNAUTH_{\lambda_{ENC1},\mathcal{U}}$ are computationally indistinguishable.

$\mathcal{A}$ proceeds as follows. $\mathcal{A}$ runs $\mathcal{U}$ on a simulated interaction with a set of parties running λ_{ENC1}. First $\mathcal{A}$ chooses and distributes encryption and decryption keys of the public key encryption scheme. Then $\mathcal{A}$ proceeds the simulation as follows:

1. When $\mathcal{U}$ activates an imitated party A' with an external request to send a message m to another party B' in session s', $\mathcal{A}$ activates party A with an external request to send m to party B in a corresponding session s in an authenticated network. As a consequence, the message (A, B, s, m) is added to a set M of undelivered messages [3]. As assumed, no message will appear twice in M.

2. When an imitated party B' outputs "B' received m from A' in session s'", $\mathcal{A}$ activates session s of B on incoming message m provided (A, B, s, m) is in the set M. Furthermore, (A, B, s, m) is now deleted from M.

3. When $\mathcal{U}$ corrupts an imitated party A', $\mathcal{A}$ also corrupts the corresponding party A in the authenticated network. $\mathcal{A}$ then hands all the internal states of A' to $\mathcal{U}$. Specifically, $\mathcal{A}$ hands the decryption key of A' to $\mathcal{U}$.

4. When $\mathcal{U}$ corrupts a session s' within an imitated party A', $\mathcal{A}$ also corrupts the corresponding session s within A in the authenticated network. $\mathcal{A}$ hands the current session state of s' within A' to $\mathcal{U}$. If there is no corresponding session within A in the authenticated network, then $\mathcal{A}$ first issues an external request to A to establish a session s, then corrupts that session.

5. $\mathcal{A}$ outputs whatever $\mathcal{U}$ outputs.

Let $\mathbf{E}$ denote the event that an imitated party B' outputs "B' received m from A' in session s'" while A' and its session s' were not corrupted, but there is no (A, B, s, m) currently in the set M of undelivered messages in the authenticated network. When $\mathbf{E}$ happens, either A was not activated for sending m to B in the corresponding session s or B has already had the same output before. In the first case, it implies that A' was not activated (by an external request) for sending m to B' in session s' (according to step 1 of the simulation above). In the second case, if the simulation does not fail, it implies that B' has already had the same output before (according to step 2 of the simulation above). Note that A' does not send m to B' in session s' for more than once. In this event, we say that $\mathcal{U}$ broke party A' in session s'.

We note that, only when $\mathbf{E}$ happens, $\mathcal{A}$ cannot proceed with the simulation shown above (step 2). Our purpose is to show that, $\mathbf{E}$ only happens with negligible probability. We show it by contradiction. If $\mathbf{E}$ happens with a non-negligible probability, then we can construct a distinguisher $\mathcal{D}$ that breaks the indistinguishability (IND-CCA2) of the public key encryption scheme.

Let e, d be the public encryption and private decryption keys of the public key encryption scheme generated with security parameter k. $\mathcal{D}$ plays the following game with a simulator $\mathcal{S}$. $\mathcal{D}$ is given the encryption key e and a decryption oracle $\mathcal{DO}$ under decryption key d. $\mathcal{D}$ picks two messages $\mathsf{m}_0, \mathsf{m}_1$ from the message space defined by ENC under e and gives these messages to $\mathcal{S}$. $\mathcal{S}$ then chooses $b \in_R \{0, 1\}$ and returns $c^* = ENC_e(\mathsf{m}_b)$ to $\mathcal{D}$. After receiving c^*, $\mathcal{D}$ can adaptively query $\mathcal{DO}$ with any ciphertext except c^*. Finally $\mathcal{D}$ outputs a bit b' as his guess of b. Below is the construction of $\mathcal{D}$.

Construction of $\mathcal{D}$. $\mathcal{D}$ runs $\mathcal{U}$ in a simulated interaction with n parties $P_1, \cdots, P_n$ running λ_{ENC1} and answers all $\mathcal{U}$'s queries according to the protocol specification. First, $\mathcal{D}$ randomly chooses a party denoted by P^*, and chooses and distributes the encryption and decryption keys to all parties according to function I, with the exception that the encryption key of P^* is set to be e. Next, $\mathcal{D}$ randomly chooses a message m^* out of all messages such that some party $\tilde{P}$ was activated by an incoming message "message: s^*, m^*" from P^*. If there is no such m^*, $\mathcal{D}$ fails the simulation and outputs a random bit b' as his guess of b. Otherwise, when $\mathcal{U}$ delivers m^* to $\tilde{P}$ in session s^*, $\mathcal{D}$ chooses two values $N_0, N_1 \in_R \{0, 1\}^k$, constructs two messages:

$$\mathsf{m}_0 = N_0||s^*||h(m^*)||\tilde{P}$$
$$\mathsf{m}_1 = N_1||s^*||h(m^*)||\tilde{P}$$

and gives them to the simulator $\mathcal{S}$. After $\mathcal{S}$ returns the challenging ciphertext c^*, $\mathcal{D}$ has $\tilde{P}$ respond with "challenge: s^*, m^*, c^*". Finally, if $\mathcal{U}$ activates $\tilde{P}$ with incoming message "response: s^*, m^*, N", $\mathcal{D}$ checks whether $N \in \{N_0, N_1\}$. If yes, $\mathcal{D}$ outputs b' such that $N = N_{b'}$, otherwise, $\mathcal{D}$ simply picks a random value $b' \in_R \{0,1\}$, outputs it and halts. During the simulation, the following cases will be handled specially:

1. When party P^* receives a challenge c other than c^*, $\mathcal{D}$ simulates the behavior of the application-specific decryption box according to the protocol specification. The decryption of c is done by querying $\mathcal{DO}$.
2. If party P^* is activated with incoming challenge "challenge: s, m, c^*" from some party Q such that
   (a) $Q \neq \tilde{P}$ or $s \neq s^*$ or $m \neq m^*$, then $\mathcal{D}$ has P^* stop handling the challenge and ignore it. In this state, if session s of P^* is corrupted, $\mathcal{D}$ has to return $(s^*, h(m^*), \tilde{P})$ and all other internal state information of session s of P^* to $\mathcal{U}$, but not N_b. N_b is not in the internal state of session s of P^* as in step 4 of λ_{ENC1} (page 43), this incoming challenge will not pass all the checks. P^* will therefore not proceed to step 5 of λ_{ENC1}.
   (b) otherwise, that is, if $Q = \tilde{P}$, $s = s^*$ and $m = m^*$, $\mathcal{D}$ fails the simulation and randomly picks a bit b' as its output.
3. If P^* or session s^* of P^* is corrupted by $\mathcal{U}$, $\mathcal{D}$ fails and outputs a random bit b'.

For other queries made by $\mathcal{U}$, $\mathcal{D}$ answers them accordingly.

First note that all capabilities of $\mathcal{U}$, namely party corruption, session corruption and sub-routine (session) activations, are simulated correctly by $\mathcal{D}$. More importantly, no matter at which point, conditioned on the event that $\mathcal{D}$ does not fail in the simulation, the internal states of all sessions of all imitated parties are identical to that of a real interaction of $\mathcal{U}$ with an unauthenticated network. In addition, since P^* is randomly chosen, $\mathcal{U}$'s view of the interaction with $\mathcal{D}$, conditioned on the event that $\mathcal{D}$ does not fail in the simulation, is identically distributed to $\mathcal{U}$'s view of a real interaction with an unauthenticated network.

Let $\mathbf{E}^*$ be the event that $\mathbf{E}$ occurs and that $\mathcal{U}$ broke P^* in session s^* (with m^*). Since P^* is randomly chosen by $\mathcal{D}$ and the event $\mathbf{E}^*$ and the failure of $\mathcal{D}$ (that is, when P^* is corrupted, session s^* of P^* is corrupted, there is no activation of incoming message from P^*, or P^* is activated with incoming challenge "challenge: s^*, m^*, c^*".) never occur at the same simulation, we have the event $\mathbf{E}^*$ occurs with probability at least ϵ/ℓ where ϵ is the probability of event $\mathbf{E}$ and ℓ is the number of activations with incoming messages in the simulation.

When $\mathbf{E}^*$ happens, $\mathcal{U}$ must have activated $\tilde{P}$ with incoming message "response: s^*, m^*, N_b", so $\mathcal{D}$ gets the right N_b with probability at least ϵ/ℓ. In this case, $\mathcal{D}$ outputs the right b'. Furthermore, since N_0 and N_1 are randomly picked from $\{0,1\}^k$, $\mathcal{U}$ activates $\tilde{P}$ with incoming message "response: s^*, m^*, N_{1-b}" with probability at most 2^{-k} due to random guessing. In this case, $\mathcal{D}$ outputs the wrong b'. In other cases, i.e, if $N \notin \{N_0, N_1\}$, $\mathcal{D}$ outputs the right b' with probability $1/2$. So the overall probability that $\mathcal{D}$ outputs the right b' is at least $\epsilon/\ell + 1/2(1 - \epsilon/\ell - 2^{-k}) = 1/2 + 1/2(\epsilon/\ell - 2^{-k})$. In addition, $\mathcal{D}$ has never asked $\mathcal{DO}$ for decrypting c^*. Consequently, $\mathcal{D}$ has successfully broken the encryption scheme with probability at least $1/2 + 1/2(\epsilon/\ell - 2^{-k})$ which means the advantage of $\mathcal{D}$ over random guessing is at least $1/2(\epsilon/\ell - 2^{-k})$. □

B   Proof of Theorem 2

Proof. Since the verifiable encryption (VE) scheme in use is assumed to be secure such that the underlying encryption scheme is IND-CCA2 and the underlying proof system is sound as well as Special HVZK, the VE scheme must also be secure against adaptive ciphertext-only attack. That is, given a ciphertext associated with the transcript of zero-knowledge proof, it should be infeasible for an adversary to obtain the message encrypted in the ciphertext even a decryption oracle $\mathcal{DO}$ is given that can be queried in any adaptive way provided the targeting ciphertext is not queried.

Up to the definition of event **E**, the proof is identical to that of Theorem 1. Suppose **E** happens with non-negligible probability, then we can construct a verifier $\mathcal{V}$ that breaks the ciphertext-only security of the VE scheme.

Construction of $\mathcal{V}$. $\mathcal{V}$ runs $\mathcal{U}$ in a simulated interaction with n parties $P_1, \cdots, P_n$ running λ_{ENC2} and answers all $\mathcal{U}$'s queries according to the protocol specification. First, $\mathcal{V}$ randomly chooses a party P^*, and chooses and distributes the encryption and decryption keys to all parties according to function I, with the exception that the encryption key of P^* is set to e, which is the public key of the targeting VE scheme. The decryption key of P^* is unknown to $\mathcal{V}$. Next, $\mathcal{V}$ randomly chooses a message m^* out of all messages such that some party $\tilde{P}$ was activated by an incoming message "message: s^*, m^*" from P^*. If there is no such m^*, $\mathcal{V}$ fails the simulation and halts. Otherwise, when $\mathcal{U}$ delivers m^* to $\tilde{P}$ in session s^*, $\mathcal{V}$ launches a run of the VE scheme with a prover $\mathcal{P}$ as follows:

> $\mathcal{V}$ computes $t = h(P^*, \tilde{P}, s^*, m^*)$ and interacts with $\mathcal{P}$ for one run of the VE scheme for relation $\mathcal{R}_t$. $\mathcal{P}$ randomly chooses a value $N^* \in_R \{0,1\}^k$, calculates $c^* = ENC_e(N^*)$ and $\delta^* = f_t(N^*)$, and sends (c^*, δ^*) to $\mathcal{V}$. $\mathcal{V}$ then has $\tilde{P}$ respond with "challenge: s^*, m^*, c^*, δ^*" to P^*. $\mathcal{P}$ then launches the proof system of the VE scheme with $\mathcal{V}$ for
>
> $$PK[N^* : c^* = ENC_e(N^*) \wedge (N^*, \delta^*) \in \mathcal{R}_t].$$

$\mathcal{V}$ also has $\tilde{P}$ launch the proof system with P^*. In the proofing interaction, $\mathcal{V}$ has $\tilde{P}$ send the 'commitment' that $\mathcal{P}$ sends to $\mathcal{V}$ to P^*, and forwards the 'challenge' that $\tilde{P}$ receives from P^* to $\mathcal{P}$, and has $\tilde{P}$ respond with the same 'response' that $\mathcal{V}$ receives from $\mathcal{P}$ to P^*.

Finally, if $\mathcal{U}$ activates $\tilde{P}$ with incoming message "response: s^*, m^*, N" from P^*, $\mathcal{V}$ checks if $(N, \delta^*) \in \mathcal{R}_t$. If yes, $\mathcal{V}$ outputs N and halts. Otherwise, $\mathcal{V}$ fails the simulation and halts. During the simulation, the following cases will be handled specially:

1. If P^* receives a challenge c other than c^* and the corresponding session of P^* is expecting the receipt of a challenge, $\mathcal{V}$ simulates the behavior of P^* according to the protocol specification by checking if the associated message has been sent before in the session and then launching the proof system of the VE scheme (step 3 on page 45). If the proof is passed, c is decrypted by querying $\mathcal{DO}$ and the simulation proceeds.
2. If P^* is activated with incoming message "challenge: s, m, c^*, δ^*" from some party Q such that
   (a) $Q \neq \tilde{P}$ or $s \neq s^*$ or $m \neq m^*$, then $\mathcal{V}$ has P^* stop handling the challenge and ignore it. This is because it is negligible to have the incoming message being a valid ciphertext of the VE scheme due to the assumption that F is a one-way function family and that h is collision-resistant;
   (b) otherwise, $\mathcal{V}$ proceeds to simulate session s^* of P^* in the run of the VE proof system (step 3 on page 45). If the proof is not passed, $\mathcal{V}$ has P^* stop handling the challenge. Otherwise, $\mathcal{V}$ fails the simulation and halts.
3. If P^* or session s^* of P^* is corrupted by $\mathcal{U}$, $\mathcal{V}$ fails the simulation and halts.

For other queries made by $\mathcal{U}$, $\mathcal{V}$ answers them accordingly.

Note that all capabilities of $\mathcal{U}$, namely party corruption, session corruption and sub-routine (session) activations, are simulated correctly by $\mathcal{V}$. Also, no matter at which point, conditioned on the event that $\mathcal{V}$ does not fail in simulation, the internal states of all sessions of all imitated parties are identical to that of a real interaction of $\mathcal{U}$ with an unauthenticated network. In addition, since P^* is randomly chosen, $\mathcal{U}$'s view of the interaction with $\mathcal{V}$, conditioned on the event that $\mathcal{V}$ does not fail in the simulation, is identically distributed to $\mathcal{U}$'s view of a real interaction with an unauthenticated network.

Let $\mathbf{E}^*$ be the event that $\mathbf{E}$ occurs and that $\mathcal{U}$ broke P^* in session s^*. Since P^* is randomly chosen by $\mathcal{V}$ and the event $\mathbf{E}^*$ and the failure of $\mathcal{V}$ (that is, when P^* is corrupted, session s^* of P^* is corrupted, there is no activation of incoming message from P^*, or P^* is activated with incoming challenge "challenge: s^*, m^*, c^*, δ^*" with the proof initiated by $\tilde{P}$ is passed.) never occur at the same simulation, we have the event $\mathbf{E}^*$ occurs with probability at least ϵ/ℓ where ϵ is the probability of event $\mathbf{E}$ and ℓ is the number of activations with incoming messages in the simulation.

When $\mathbf{E}^*$ happens, $\mathcal{U}$ must have activated $\tilde{P}$ with incoming message "response: s^*, m^*, N^*". So $\mathcal{V}$ gets the decryption of c^* with probability at least ϵ/ℓ. In addition, $\mathcal{V}$ has never asked $\mathcal{DO}$ for decrypting c^*. Consequently, $\mathcal{V}$ has successfully broken the VE scheme with probability at least ϵ/ℓ.

$\square$

Fair Identification

Omkant Pandey[1], Julien Cathalo[2,*], and Jean-Jacques Quisquater[2]

[1] Department of Computer Science, UCLA
omkant@cs.ucla.edu
[2] UCL Crypto Group, Belgium
{cathalo, quisquater}@dice.ucl.ac.be

Abstract. This paper studies a new problem called fair identification: given two parties, how should they identify each other in a fair manner. More precisely, if both parties are honest then they learn each other's identity, and if anyone is cheating then either both of them learn each other's identity or no one learns no information about the identity of the other. We propose a security model and a provably secure optimistic fair identification protocol.

1 Introduction

Suppose that Alice and Bob are interested in knowing each other but each of them is hesitating in revealing his/her identity first. How should they identify each other so that fairness is guaranteed for both parties i.e. if both of them are honest, they learn each other's identity; if any one of them cheats then either both of them learn each other's identity or no one learns no information about the identity of the other. This problem is termed as *fair identification*.

We will be interested in a protocol which ensures the following:

- If A and B are honest, both of them learn each other's identity.
- If anyone is cheating, either no one learns anything about the identity of the other or both of them learn each other's identity.
- Identities of A and B remain secret to an outsider against active attacks.
- A third party is needed only in case of disputes i.e. when cheating occurs (in other words, the protocol should be optimistic).

Harder variants of this problem are also possible: for example, one could consider concurrent attacks instead of active attacks, but in this paper we will focus on active attacks only.

Several problems similar to fair identification have been studied in the literature. It is thus a natural approach to examine a few kinds of cryptographic primitives used to solve such problems to see if they can trivially achieve fair identification.

---

* Supported by *Walloon Region / WIST-MAIS project*.

D. Pointcheval (Ed.): CT-RSA 2006, LNCS 3860, pp. 52–63, 2006.

- In a mutual authentication scheme [10, 6, 23], two parties authenticate each other, but fairness is not ensured.
- In a fair exchange of signatures protocol [1, 20, 16, 2], each party obtains the other's signature in a fair manner. There is a fundamental difference between exchanging a signature and exchanging one's identity. Fair exchange of signatures is based upon the concept of verifiability of signatures without actually completely revealing them [13, 3, 20]. But a successful verification always confirms the identity of the other party. Thus, fair exchange of digital signatures does not seem to provide any trivial solution to fair identification.
- Identity escrow schemes [25] allow an entity A to send some information to B that commits to A's identity, meaning that this information would allow an authorized third party to recover A's identity. A and B could run a fair identification protocol as follows:
    1. A runs the identity escrow protocol with B.
    2. B confirms his true identity to A.
    3. A confirms her true identity to B in similar way.

  If A is cheating, B can go to the escrow agent with transcripts of identity escrow protocol in step 1 to obtain A's identity. In this protocol, eavesdroppers can learn the identities of A and B. Simply encrypting the communication does not thwart active attacks. Thus this protocol does not satisfy the requirements for fair identification.

Since there seems to be no trivial way to achieve fair identification, we propose a new scheme. Rather than build a scheme from scratch, we use existing cryptographic primitives and combine them to design a fair identification scheme.

This paper is organized as follows. Section 2 lists the cryptographic primitives that we use as building blocks for our protocol. Section 3 defines a security model for fair identification. Section 4 describes our fair identification protocol. Section 5 discusses a few variants of the initial problem.

2   Building Blocks

In this section, we introduce the cryptographic primitives that we use as building blocks for our fair identification protocol. These building blocks are a signature scheme with a special feature, a public key encryption scheme, and a group signature scheme. This section is not necessary for understanding the security model of section 3 and hence can be skipped. However, it is necessary for the protocol (section 4).

2.1   Signatures with Key-Independent Coupons

Some signature schemes have the interesting feature that a part of the signature can be computed prior to the knowledge of the message to sign. They are sometimes called on-line/off-line signatures [28] or coupon-based signatures, and the pre-computed part is called the coupon.

In this paper, we additionally require that the coupon can be computed without knowing the signing key. We call such a scheme a signature scheme with key-independent coupons.

Many known signature schemes satisfy this requirement; in fact, any signature scheme obtained by applying the Fiat-Shamir heuristic [21] does (the coupon is the commitment of the corresponding identification scheme), like the Schnorr signature scheme [27] for example.

More formally, a signature scheme with key-independent coupons is a tuple of algorithms $SS = (\mathcal{K}_S, \mathcal{S}, \mathcal{V})$ satisfying the usual properties of a signature scheme where: $\mathcal{K}_S$, $\mathcal{S}$, and $\mathcal{V}$ are key-generation, signing and verification algorithms respectively. Additionally, $\mathcal{S}$ internally works as follows ($s =$ signing key and $\hat{st} =$ some state information, $m =$ message to be signed): Algorithm $\mathcal{S}(m, s)$: $\{(x, \hat{st}) \leftarrow \mathcal{S}_x(); y \leftarrow \mathcal{S}_y(x, \hat{st}, m, s); \text{Return}(x, y); \}$.

Here, $\mathcal{S}_x$ is the algorithm that generates the coupon and $\mathcal{S}_y$ is the algorithm that generates the remaining part of the signature.

NOTATION: Let *cert* denote the certificate for the public key of a signature scheme with key-independent coupons. We assume two algorithms Extract and Valid such that: Extract(*cert*) extracts the public key from *cert* and Valid(*cert*) verifies if *cert* is valid (output 1) or not (output 0). Furthermore, we assume that all certificates are of the same size and that all signatures are also equal in size (if not, add leading 0s to make them equal to the maximum possible size).

2.2   Public Key Encryption in Multi-user Setting

By $\mathcal{PE}$ we denote an IND-CCA secure public-key encryption scheme secure in a multi-user setting (Bellare et al [5]).

Recall that a public-key encryption scheme $\mathcal{PE} = (\mathcal{K}, \mathcal{E}, \mathcal{D})$ consists of three algorithms. The *key generation* algorithm $\mathcal{K}$ is a randomized algorithm that takes nothing as input and returns a pair (pk, sk) of matching public and secret keys; we write $(pk, sk) \xleftarrow{R} \mathcal{K}()$. The *encryption* algorithm $\mathcal{E}$ is a randomized algorithm that takes the public key pk and a *plaintext* M (from the message space $\text{MsgSp}(pk)$) to return a *ciphertext* C'; we write $C' \xleftarrow{R} \mathcal{E}_{pk}(M)$. The *decryption* algorithm $\mathcal{D}$ is a deterministic algorithm that takes the secret key sk and a ciphertext C' to return the corresponding plaintext M (or a special symbol $\perp$ if C' is invalid); we write $M \leftarrow \mathcal{D}_{sk}(C')$.

Two ideal examples of $\mathcal{PE}$ are: Cramer-Shoup [19, 5] and RSA-OAEP [9, 7, 29, 22].

2.3   Group Signatures as Verifiable Commitments to Identity

Group signatures allow a group member to sign anonymously on behalf of the group. If needed, the signature can be opened by a trusted third party, called group manager, to reveal the identity of the signer. For an in-depth discussion on group signatures see [14, 12, 4, 8, 11].

Informally, a group signature scheme $\mathcal{G} = (\mathsf{Setup}, \mathsf{Join}, \mathsf{Sign}, \mathsf{Verify}, \mathsf{Open})$ is a 5-tuple of algorithms, where:

$-\mathsf{Setup}$ is the algorithm which takes no input. It initializes the system and outputs the group public key GPK, secret data for the group manager, and any other parameters needed.

$-\mathsf{Join}$ is an interactive protocol executed between the group manager and a user (say A). As a result, A learns his secret data g_A for generating group signatures and the group manager might also learn some data to aid him later in opening the signatures if required.

$-\mathsf{Sign}$ is the signing algorithm. It takes as input the message m to be signed and the secret data g_A of any user A to produce a group signature σ_A.

$-\mathsf{Verify}$ is the algorithm to verify the correctness of a group signature on a given message. It takes as input the message m, the signature σ and the group public key GPK; it outputs 1 if σ is a valid group signature on m, and 0 otherwise.

$-\mathsf{Open}$ is the algorithm that only the group manager can use to identify the signer of a particular group signature. It takes as input the signature σ, manager's secret data for opening group signatures and perhaps some other information; its output is a proof identifying the signer of the signature.

For notation, $\sigma_A(m)$ will denote the group signature of A on message m. When only σ_A is written, it means that it is a group signature of A on whatever message and that opening it would identify A as its signer.

The following properties are desirable for group signatures: *Correctness* - group signatures produced using Sign are always accepted by Verify. *Unforgeability* - only group members can sign efficiently; *Anonymity* - given a group signature, it is computationally hard to identify its signer for everyone but the group manager; *Unlinkability* - deciding whether two valid group signatures were computed by the same group member, is computationally hard; *Exculpability* - Neither a group member nor the group manager can produce a group signature on behalf of any other member. *Openability* - The group manager is always able to open and identify the actual signer of a valid group signature; and *Coalition-Resistance* - A colluding subset of group members (even if comprised of the entire group) cannot generate a group signature that the group manager cannot open.

Let us explain why and how we use group signatures in our scheme. Our approach to design a fair identification scheme is inspired by the way fair exchange of signature schemes are built. In order to fairly exchange signatures, users needs a way to commit to their signatures such that a third party can reveal the signature if needed (this was formalized by Dodis and Reyzin [20] and called a verifiably committed signature scheme). Similarly, in order to fairly exchange identities, users need a way to commit to their identities. Group signatures solve this problem; they can be seen as "verifiable commitments to identity", or VCI for short.

3   A Security Model for Fair Identification

In this section we introduce the security notions that we require for a fair identification scheme.

3.1   Setting

Parties A and B are willing to fairly identify each other. The trusted third party is T. In order to simplify the description of the protocol, T will have two functions: certificate generation and dispute resolution. We restrict ourselves to the case of active attacks meaning that each player communicates with only one player at a time.

We need to consider coalition attacks, where the adversary is allowed to form coalitions with any number of users. We shall treat each coalition of adversaries as a *single* adversary who will be considered as identified if any one in the coalition is identified with overwhelming probability. This approach of identifying at least one adversary has been widely used in traitor tracing [15, 17, 18, 24] and also in group signatures (see the coalition resistance property).

Because now each coalition can be replaced by a single adversary, we can assume that *users form no coalitions* at all.

3.2   Canonical Protocol

We now present a canonical protocol for identification. For the sake of simplicity, we omit exchanges that happen before B commits to his identity. The sketch of this protocol is the following: in the first step, B commits to his identity. In the second step, A reveals and proves her identity. In the third step, B reveals and proves his identity.

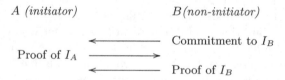

Fig. 1. Canonical fair identification protocol

3.3   Definitions

In a real world scenario, the adversary might interact with any of the users at will, eavesdrop, modify/stop the data, etc. To simulate all these actions, we use oracles.

An *identity oracle* $\mathcal{O}_i$ for identity I_i essentially simulates the behavior of a honest user whose identity is I_i and hence knows the required secret s_i to prove its identity and any other secret data necessary for generating its own VCI. Each oracle is capable of executing a fair identification protocol with any other oracle or user.

Definition 1 (Identity Oracles). *An identity oracle $\mathcal{O}_i$, is the simulation of a honest user with identity I_i equipped with all necessary secrets required to execute the fair identification protocol. Besides the messages of the fair identification protocol, the oracle understands the following instructions – here, $\mathbb{O}$ is either an identity oracle $\mathcal{O}_j \neq \mathcal{O}_i$ or the player who issues these instructions:*

- *START(O): when this instruction is issued to $\mathcal{O}_i$, it starts executing a fair identification protocol with* $\mathbb{O}$.
- *TRANSCRIPT(O): when this instruction is issued to $\mathcal{O}_i$, it provides the issuer with the transcripts of a fair identification protocol run between $\mathcal{O}_i$ and* $\mathbb{O}$.

In both cases, $\mathcal{O}_i$ will be the initiator. During any live session, the oracle either sends an appropriate message or waits for an appropriate message. It stops whenever an invalid message arrives or if the protocol completes successfully.

The START instruction allows simulating live sessions whereas TRANSCRIPT instruction simulates access to the old transcripts. The time for START operation will be one unit. When an adversary eavesdrops and gathers transcripts, gathering one transcript in real world lasts as long as one fair identification run between two parties. Once the adversary has all the transcripts it needs, it can access them in constant time. But for this, he must still gather and store transcripts, and the time taken for that should actually be counted. Thus, the time for obtaining an answer for a TRANSCRIPT instruction is essentially the run-time for one protocol run.

By q_i, denote the total number of all those instructions (START, TRANSCRIPT), in which the identity oracle $\mathcal{O}_i$ appears (either initiator or non-initiator). Let q_s denote the maximum value of q_i over all i, i.e. $q_i \leq q_s, \forall i, 0 \leq i \leq u - 1$. We will expect the protocol to be secure for large values of q_s.

Now we are ready to present a game for the adversary and formally define the notion of fairness. For the rest of the paper, C denotes the adversary and his identity is I_C. By ε we denote an empty string. We assume that there are $u + 1$ users in the system including the adversary. Excluding C, there are u users with identities denoted by $I_0, I_1, \ldots I_{u-1}$. To simulate them, we assume u identity oracles: $\mathcal{O}_0, \mathcal{O}_1, \ldots \mathcal{O}_{u-1}$. By S we represent the set $\{\mathcal{O}_0, \mathcal{O}_1, \ldots \mathcal{O}_{u-1}\}$ and access to S means access to each of its element oracles.

Definition 2 (Fairness-game). *Let $\mathcal{O}_0, \mathcal{O}_1, \ldots \mathcal{O}_{u-1}$ be the identity oracles corresponding to the identities $I_0, I_1, \ldots I_{u-1}$ and $S = \{\mathcal{O}_0, \mathcal{O}_1, \ldots \mathcal{O}_{u-1}\}$. Adversary C is asked to choose any two identities, say I_0 and I_1 of his choice from the set $\{I_0, I_1, \ldots I_{u-1}\}$. Now one of these two identities is selected at random, and represented by I_b where $b \overset{R}{\leftarrow} \{0, 1\}$. An identity oracle Ch with identity I_b is generated as a challenge oracle for the adversary. The adversary is given access to Ch and S. Adversary knows the identities of all the oracles in S but not of Ch. The adversary can instruct these oracles and ask for transcripts of any communication and/or for opening new sessions with anyone. It is mandatory that each oracle executes only one session at a time, and that it can be involved in no more than q_s instructions during the entire game.*

Oracle Ch differs from other identity oracles in one manner: it keeps a state bit ID, initialized to 0 at the start of the game, which changes automatically along the game progress, as follows:

- *Initiator case: when Ch is initiator in a session, a fresh session-specific state $St = (\tilde{a}, \tilde{b})$ is created at the start of the session, where $\tilde{a}$ is a bit-string with* *initial value ε and $\tilde{b}$ is a single bit with initial value 0. If V is the* VCI *received by Ch in the first step and V is valid, then $\tilde{a}$ is set to V. Value of $\tilde{b}$ becomes 1 if an invalid message arrives in the third step or if this step never occurs. If the third step succeeds, then let I' be the identity whose proof was accepted by Ch in this step. Then,*

$$\text{ID} = 1 \ \textit{if} \ \begin{cases} I' = I_C \\ or \\ St = (\text{VCI}_C, 1) \ \textit{i.e.} \ \tilde{a} = \text{VCI}_C \ \textit{and} \ \tilde{b} = 1 \end{cases}$$

- *Non-initiator case: if Ch is the non-initiator, consider the second step of canonical* FIP. *This step succeeds if the initiator provides appropriate data. If the initiator does not provide appropriate data, then at a later point in time, oracle might receive this data from T during the dispute resolution when initiator approaches T. Be it through any of these cases, let I' represent the identity whose proof is accepted by the oracle. Then,*

$$\text{ID} = 1 \ \textit{if} \ I' = I_C$$

Adversary wins the game if at the end of the game it outputs a bit b' such that: $(b' = b) \wedge (\text{ID} = 0)$.

Definition 3 (Fairness). *Let $\mathcal{W}$ denote the event that adversary C wins the fairness-game. An* FIP *is said to ensure fairness if any polynomial time adversary has only negligible advantage in fairness-game, where the advantage of the adversary is defined as,*

$$\mathbf{Adv}_C^{Ch,S} = 2 \cdot \Pr[\mathcal{W}] - 1 = 2 \cdot \Pr[(b' = b) \wedge (\text{ID} = 0)] - 1$$

Informally, the idea behind the fairness-game is that adversary is allowed to pick up any two oracles he would like to attack. One of these oracles is picked at random as challenge oracle for the adversary and the adversary is asked to guess the identity of the challenge oracle with probability acceptably larger than $\frac{1}{2}$.

In this process, the adversary is not allowed to give away his own identity. This is formalized by ID. In the initiator case, $\tilde{b} = 1$ means that the oracle detects cheating and hence will approach the third party. Adversary will be identified only if V is his own VCI. Hence, $St = (\text{VCI}_C, 1)$ means that C is identified through the third party. In both cases, $I' = I_C$ simply means that adversary himself gave out his identity. Thus ID is just like a flag which when set, represents that C's identity is disclosed to the challenge oracle.

4   A Fair Identification Protocol

In this section, we describe a fair identification protocol. We start by giving a sketch of our protocol. First, A generates and sends the coupon of his signature. Now, B, in the second step, generates a group signature on A's coupon and sends it to A. A now has the VCI of B and hence it now sends his identity and the remaining part of the signature. B verifies the signature and then sends his identity and signature to A. To remain anonymous to outsiders, they encrypt the communication using temporary keys (using $\mathcal{PE}$). To avoid active attacks, these keys are signed under the group signature of B.

Whenever not mentioned, the security parameter and other system parameters should be assumed implicitly available. INITIAL-SETUP is needed for each user to learn his corresponding secrets and to designate the trusted third party. More users can join at any time. EXCHANGE is the main protocol for exchanging identities. RESOLVE is the last component of our FIP, required only in case of dispute. Let T denote the trusted third party which will also take the role of group manager in the group signature scheme.

INITIAL-SETUP:

System specific parameters:

1. T declares: (a) A secure signature scheme with key-independent coupons $SS = (\mathcal{K}_S, \mathcal{S}, \mathcal{V})$, and (b) An IND-CCA secure public-key encryption scheme $\mathcal{PE} = (\mathcal{K}, \mathcal{E}, \mathcal{D})$ to be used by each player, whenever needed.
2. Decide a secure group signature scheme to be used, $\mathcal{G} = $ (Setup, Join, Sign, Verify, Open) with T being the group manager. T creates an instance of $\mathcal{G}$ by running the procedure Setup. T learns the secrets corresponding to the group manager and let GPK be the group public key.

User specific parameters:

1. Each user U first decides his public and private keys, I_U and s_U respectively, for the signature scheme SS by running $\mathcal{K}_S$. U proves to T that I_U is his public key.
2. Now U runs the Join protocol of $\mathcal{G}$, with T, to learn his group specific secret g_U needed to produce the group signatures.
3. T generates a certificate $cert_U$ mentioning that person with public key I_U is registered with T.

EXCHANGE:

Parties A (initiator) and B (non-initiator) execute the following steps to identify each other:

1. A: $(x_A, \hat{st}) \leftarrow \mathcal{S}_x()$, $(pk_A, sk_A) \leftarrow \mathcal{K}()$. A sends x_A, pk_A to B.
2. B: $(pk_B, sk_B) \leftarrow \mathcal{K}()$, $\sigma_B \leftarrow \mathsf{Sign}(x_A\|pk_A\|pk_B, g_B)$. Send pk_B, σ_B to A.
3. A: If $\mathsf{Verify}(x_A\|pk_A\|pk_B, \sigma_B, GPK) = 1$ then
    $$y_A \leftarrow \mathcal{S}_y(x_A, \hat{st}, \sigma_B, s_A), \ \psi_A \leftarrow \mathcal{E}_{pk_B}(y_A, cert_A), \text{ Send } \psi_A \text{ to } B.$$
    else STOP.
4. B: $(y_A, cert_A) \leftarrow \mathcal{D}_{sk_B}(\psi_A)$, $XY_A \leftarrow (x_A, y_A)$, Verify $cert_A$ and if valid, extract I_A.
    If $\mathcal{V}(\sigma_B, XY_A, I_A) = 1$ then
    $$XY_B \leftarrow \mathcal{S}(\sigma_B, s_B), \ \psi_B \leftarrow \mathcal{E}_{pk_A}(XY_B, cert_B), \text{ Send } \psi_B \text{ to } A.$$
    else STOP.

Final test, A: $(\mathrm{XY}_B, cert_B) \leftarrow \mathcal{D}_{sk_A}(\psi_B)$, Verify $cert_B$ and if valid, extract I_B and check that $\mathcal{V}(\sigma_B, \mathrm{XY}_B, I_B) = 1$. If tests do not succeed, A goes to T to execute RESOLVE.

RESOLVE:

> Party A claiming to be cheated, presents σ_B to T and the corresponding message components x_A, pk_A, pk_B, full signature XY_A on σ_B and identifies herself as A to T. T sends a proof identifying the signer (B) of σ_B to A and sends the full signature XY_A to B.

For a security proof of this protocol, please see the full version of this paper [26].

5   Extensions and Future Work

In this section, we discuss a few variants to the problem of fair identification. While not all of these variations might find a practical application, they constitute an interesting challenge.

5.1   Transferable Proofs of Identity

One could imagine a scenario where users want to fairly exchange *transferable* proofs of identity. In this case, the precise statement of the problem will be: how should two parties identify each other so that either each party learns a transferable proof of the other's identity or none of them learns nothing about the identity of the other.

In our fair identification protocol, the proofs of identity that users get if the protocol ends normally are indeed transferable (because those proofs are signatures), but it does not necessarily mean that our protocol is a fair exchange of identity proofs. This property can be ensured if group signatures can be opened in a transferable manner, i.e. when the third party opens the group signature, it outputs a transferable proof that identifies its signer. Though it might not be possible for every group signature scheme, it is indeed the case in modern schemes. For example, in ACJT scheme [4], this proof is actually an interactive zero-knowledge proof of equality of two discrete logarithms, between the group manager and the party interested in identifying the signer. This can be made transferable by applying the Fiat-Shamir heuristic. Thus, if cheating occurs in our protocol, the cheated party can also get the transferable proof which will be nothing but the group signature together with a transferable proof identifying its signer (obtained from T).

Now let us consider a stronger requirement: the proofs obtained with the help of the third party should be at least computationally indistinguishable from the proofs obtained directly from the concerned player. This requirement has an interesting link with the fair exchange of signatures where signatures opened by the third party should be indistinguishable from signatures computed by the signer.

Such indistinguishability cannot be achieved using our protocol. It would be interesting to see whether existing protocols for fair exchange of signatures can be modified to provide fair identification too.

5.2   Nontransferable Proofs of Identity

If the fair identification protocol between A and B ended successfully, A revealed her identity to B, so B can claim to another user that he performed an exchange with A. But A might want to be sure that B cannot *prove* that they indeed met. A protocol that would ensure this for A and B would be a fair exchange of nontransferable proofs of identities.

More precisely, if A and B execute the fair identification protocol and identify each other, then after the protocol completes, transcripts of the run prove nothing to anyone – transcripts could have very well been simulated by A (or B) himself.

Our protocol cannot be used to guarantee nontransferable proofs of identity. One might think of using interactive proof protocols instead of signatures to achieve this goal. Although this could work, neither this construction nor its security proof are trivial.

5.3   Conditional Fair Identification

Assume a situation where the parties know in advance whom they want to identify fairly. More precisely, A is willing to achieve fair identification only with B, and B is willing to achieve fair identification only with A. With very little tweaking, our protocol should work for this kind of problem. However a more general situation is the following one: A is willing to achieve fair identification with B if and only if B satisfies *some condition*, say C_B. Similarly, B is willing to do fair identification with A if and only if A satisfies some condition, say C_A. We term it as conditional fair identification. Our protocol may work for this, depending upon what these conditions are.

5.4   Perfect Fairness

In this paper, the coalition of adversaries was considered to be one single adversary. Following an approach used in traitor tracing schemes, we aimed at identifying at least one user in the coalition. However, a stronger notion of fairness is possible where one could aim at identifying *each* player in the coalition. Precisely, perfect fairness for A would mean that if A is identified by any player B in a particular run of the protocol, A will also identify *exactly* B. With this perfect fairness notion, it should not matter whether B is in a coalition or not. In order to achieve perfect fairness, it is necessary to use non-transferable proofs; otherwise, just one person could interact with A and obtain a transferable proof of identifying A and then show it to everyone in the coalition. Perfect fairness seems to be the most challenging property to ensure.

Acknowledgements

The authors wish to thank Benoît Libert and Judyta Stachniak for their suggestions and ideas and Sylvie Baudine for her permanent English support.

References

1. N. Asokan, V. Shoup, and M. Waidner. Optimistic Fair Exchange of Digital Signatures. In K. Nyberg, editor, *Advances in Cryptology - Eurocrypt 98*, volume 1403 of *Lecture Notes in Computer Science*, pages 591–606. Springer-Verlag, 1998.
2. N. Asokan, V. Shoup, and M. Waidner. Optimistic Fair Exchange of Digital Signatures. *IEEE Journal on Selected Areas in Communication*, 18(4):593–610, 2000.
3. G. Ateniese. Efficient Verifiable Encryption (and Fair Exchange) of Digital Signatures. In G. Tsudik, editor, *Sixth ACM Conference on Computer and Communication Security and Privacy*, pages 138–146. ACM, November 1999.
4. G. Ateniese, J. Camenisch, M. Joye, and G. Tsudik. A Practical and Provably Secure Coalition-Resistant Group Signature Scheme. In M. Bellare, editor, *Advances in Cryptology - CRYPTO 2000*, volume 1880 of *Lecture Notes in Computer Science*, pages 255–270. Springer, 2000.
5. M. Bellare, A. Boldyreva, and S. Micali. Public-key Encryption in a Multi-User Setting: Security Proofs and Improvements. In B. Preneel, editor, *Advances in Cryptology - Eurocrypt'00*, volume 1807 of *Lecture Notes in Computer Science*. Springer, 2000.
6. M. Bellare, R. Canetti, and H. Krawczyk. A modular approach to the design and analysis of authentication and key-exchange protocols. In *Proceedings of the 30th annual Symposium on the Theory of Computing – STOC*, pages 419–428. ACM Press, 1998.
7. M. Bellare, A. Desai, D. Pointcheval, and P. Rogaway. Relations among notions of security for public-key encryption schemes. In *Advances in Cryptology - Crypto'98*, volume 1462 of *Lecture Notes in Computer Science*. Springer, 1998.
8. M. Bellare, D. Micciancio, and B. Warinschi. Foundations of group signatures: Formal definitions, simplified requirements, and a construction based on general assumptions. In E. Biham, editor, *Advances in Cryptology - EUROCRYPT 2003*, volume 2656 of *Lecture Notes in Computer Science*, pages 614–629. Springer, 2003.
9. M. Bellare and P. Rogaway. Optimal Asymmetric Encryption – How to Encrypt with RSA. In *Advances in Cryptology - Eurocrypt'94*, Lecture Notes in Computer Science. Springer, 1995.
10. M. Bellare and P. Rogaway. Provably secure session key distribution: the three party case. In *Proceedings of the 27th annual Symposium on the Theory of Computing – STOC*, pages 57–66. ACM Press, 1995.
11. D. Boneh, X. Boyen, and H. Shacham. Short Group Signatures. In *Advances in Cryptology - CRYPTO 2004*, Lecture Notes in Computer Science. Springer, 2004.
12. J. Camenisch and M. Michels. A Group Signature Scheme with Improved Efficiency. In K. Ohta and D. Pei, editors, *Advances in Cryptology - ASIACRYPT '98*, volume 1514 of *Lecture Notes in Computer Science*, pages 160–174. Springer, 1998.
13. J. Camenisch and V. Shoup. Practical Verifiable Encryption and Decryption of Discrete Logarithms. In *Advances in Cryptology – CRYPTO*, LNCS. Springer, 2003.

14. J. Camenisch and M. Stadler. Efficient Group Signature Schemes for Large Groups. In B.S. Kaliski Jr., editor, *Advances in Cryptology - CRYPTO '97*, volume 1294 of *Lecture Notes in Computer Science*, pages 410–424. Springer, 1997.

15. H. Chabbane, D.H. Phan, and D. Pointcheval. Public Traceability in Traitor Tracing Schemes. In R. Cramer, editor, *Advances in Cryptology – Eurocrypt'05*, volume 3494 of *LNCS*. Springer, 2005.

16. L. Chen, C. Kudla, and K.G. Paterson. Concurrent Signatures. In C. Cachin and J. Camenisch, editors, *Advances in Cryptology - EUROCRYPT 2004*, volume 3027 of *Lecture Notes in Computer Science*, pages 287–305. Springer, 2004.

17. B. Chor, A. Fiat, and M. Naor. Tracing traitor. In Y. Desmedt, editor, *Advances in Cryptology – Crypto'94*, volume 839 of *LNCS*, pages 257–270. Springer, 1994.

18. B. Chor, A. Fiat, M. Naor, and B. Pinkas. Tracing traitor. *IEEE Transaction on Information Theory*, 46(3):893–910, 2000.

19. R. Cramer and V. Shoup. A practical public key cryptosystem provably secure against adaptive chosen ciphertext attack. In *Advances in Cryptology - Crypto'98*, Lecture Notes in Computer Science. Springer, 1998.

20. Y. Dodis and L. Reyzin. Breaking and Repairing Optimistic Fair Exchange from PODC 2003. In M. Yung, editor, *ACM Workshop on Digital Rights Management (DRM)*, pages 47–74, 2003.

21. A. Fiat and A. Shamir. How to prove yourself : practical solutions of identification and signature problems. In G. Brassard, editor, *Advances in Cryptology - Proceedings of CRYPTO '86*, volume 263 of *Lecture Notes in Computer Science*, pages 186–194. Springer-Verlag, 1987.

22. E. Fujisaki, T. Okamoto, D. Pointcheval, and J. Stern. RSA-OAEP is Secure under the RSA Assumption. Available on eprint archive – http://eprint.iacr.org/2000/061.

23. M. Jakobsson and D. Pointcheval. Mutual authentication and key-exchange protocols for low power devices. In *Financial Cryptography*, pages 178–195. Springer, 2001.

24. A. Kiayias and M. Yung. Traitor tracing with constant transmission rate. In *Advances in Cryptology – Eurocrypt'02*, volume 2332 of *LNCS*, pages 450–465. Springer, 2002.

25. J. Kilian and E. Petrank. Identity Escrow. In *Advances in Cryptology - CRYPTO '98*, volume 1642 of *Lecture Notes in Computer Science*, pages 169–185. Springer, 1998.

26. O. Pandey, J. Cathalo, and J.-J. Quisquater. Fair Identification, 2005. Full version available at http://www.cs.ucla.edu/~omkant.

27. C. P. Schnorr. Efficient Signature Generation by Smart Cards. *Journal of Cryptology*, 4(3):161–174, 1991.

28. A. Shamir and Y. Tauman. Improved Online/Offline Signature Schemes. In *Advances in Cryptology - CRYPTO 2001*, volume 2139 of *Lecture Notes in Computer Science*, pages 355–367. Springer, 2001.

29. V. Shoup. OAEP Reconsidered. Available on eprint archive – http://eprint.iacr.org/2000/060.

Efficient Doubling on Genus 3 Curves over Binary Fields[*]

Xinxin Fan[1], Thomas Wollinger[2], and Yumin Wang[1]

[1] State Key Lab of Integrated Service Networks,
Xidian University, Xi'an, P.R. China
`xxfan@mail.xidian.edu.cn`
`ymwang@xidian.edu.cn`
[2] Communication Security Group (COSY),
Ruhr-Universitäet Bochum, Germany
`wollinger@crypto.rub.de`

Dedicated to my supervisor Prof. Yumin Wang on occasion of his 70th birthday.

Abstract. The most important and expensive operation in a hyperelliptic curve cryptosystem (HECC) is the scalar multiplication by an integer k, i.e., computing an integer k times a divisor D on the Jacobian. Using some recoding algorithms for the scalar, we can reduce the number of divisor class additions during the process of computing the scalar multiplication. On the other side, the divisor doublings will stay the same for all kinds of scalar multiplication algorithms. In this paper we accelerate the divisor doublings for genus 3 HECC over binary fields by using special types of curves. Depending on the degree of h, our explicit formulae only require $1I + 11M + 11S$, $1I + 13M + 13S$, $1I + 20M + 12S$ and $1I + 26M + 11S$ for divisor doublings in the best case, respectively. Especially, for the case of deg $h = 1$, our explicit formula improve the recent result in [GKP04] significantly by saving $31M$ at the cost of extra $7S$. In addition, we discuss some cases which are not included in [GKP04].

By constructing birational transformation of variables, we derive explicit doubling formulae for special types of equations of the curve. For each type of curve, we analyze how many field operations are needed. So far no attack on any of the all curves suggested in this paper is known, even though some cases are very special. Our results allow to choose curves from a large variety which have extremely fast doubling needing only one third the time of an addition in the best case. Furthermore, an actual implementation of the new formulae on a Pentium-M processor shows their practical relevance.

Keywords: Genus 3 Hyperelliptic Curve, Explicit Doubling Formulae, Fast Arithmetic, Binary Fields.

1 Introduction

In 1988, Neal Koblitz suggested for the first time the generalization of elliptic curves to curves of higher genus for cryptographic use, namely hyperelliptic

---

[*] Supported by the National NKBRSF '973' Program of China (Grant No.G19990 35803).

curves [Kob88, Kob89]. Because of their short operand size compared to other public-key schemes, HECC seem well suitable for small processor architectures, where memory and speed are constrained. During the last decade, elliptic curve cryptosystems (ECC) [Kob87, Mil86] have been extensively studied from both a pure and applied perspective. However, HECC did not obtain a lot of attention until recent years. There has been a major effort in improving the group operations and in implementing HECC on different processors. Using explicit formulae instead of Cantor algorithm has reduced sharply the complexity of arithmetic in the ideal class group of hyperelliptic curves and obtained fast implementation in software [PWGP03, WPW+03, Lan03, GMA+04, Ava04, Wol04, FWW05] and hardware platform [BCLW02, Cla02, EMY04, KWC+04].

For all kinds of cryptographic protocols based on ECC or HECC, the computation of scalar multiplication is the main operation. Scalar multiplication algorithms include usually divisor class additions, doublings and perhaps some precomputations. By some recoding methods for the scalar, we can reduce the number of divisor class additions. However, we cannot decrease the number of divisor class doublings in the general case (for some special curves, it is possible). Therefore, divisor class doublings will become the crucial step for the performance of the entire cryptosystem. Improving the arithmetic of doubling has a direct impact on the efficiency of the whole system.

In [LS04, BD04], the authors discussed different isogeny classes for genus 2 curves over $GF(2^n)$, n is the degree of the extension field. They gave a complete study of all cases of defining equation of the curve and made a trade-off between speed-up and special parameters. Although we can use Koblitz curves to accelerate the computation of scalar multiplication [GLS00, Lan04], there are only 6 and 24 different isogenie classes for genus 2 and 3 binary curves, respectively. Hence the choice of curves is rather limited. In order to enlarge the range of selecting curves, we will address some special kinds of genus 3 curves defined over the extension field in this paper.

For genus 3 curves over $GF(2^n)$, Pelzl et al. [PWGP03] discussed a very special type of curves with $h(x) = 1$ and gave efficient doubling explicit formulae. In [GKP04], the authors proposed efficient algorithms to compute the resultant of two polynomials and of the inverse of one polynomial modulo another, and improved the overall complexity of the addition and doubling algorithms for both even and odd characteristics. Their explicit formulae are applicable to almost all hyperelliptic curves of genus 3. By using a birational transformation of the form $(x, y) \mapsto (\lambda x + \mu, \nu y)$, they discuss five possible types of curves for even characteristic case.

In this article, we generalize the ideas proposed in [LS04] to the genus 3 case and improve the results in [GKP04] further. We construct the isomorphic transformations first to achieve as many zero coefficients as possible, and then make strong use of the defining equation of the curve to derive explicit doubling formulae. We discuss some special curves which can lead to fast computation of doubling a divisor class. Finally, we combine the new doubling explicit formulae with the NAF method to compute the scalar multiplication, and give detailed experiment results.

The remainder of the paper is organized as follows: Section 2 states a brief mathematical background related to genus 3 hyperelliptic curves over binary fields. Section 3 describes the Harley's algorithm for doubling a divisor class. In section 4, 5, 6, and 7 we derive the new explicit doubling formulae for the genus 3 curves according to the different degree of $h(x)$. Section 8 summarizes our contributions. Finally, we present our experimental results in Section 9 and conclude with a discussion of our results in Section 10.

2   Genus 3 Hyperellitpic Curves

In this section we present the representation of the divisor class group elements for genus 3 hyperelliptic curves over characteristic two finite fields. More mathematical background and more details about hyperelliptic curves can be found in the literature [Can87, Kob89, MWZ96, CFA$^+$05].

Let $GF(2^n)$ be a binary finite field with 2^n elements. A hyperelliptic curve C of genus 3 over $GF(2^n)$ with one point at infinity is defined by an equation of the form

$$(*) \qquad\qquad C : Y^2 + h(X)Y = f(X),$$

where $h(X) = h_3X^3 + h_2X^2 + h_1X + h_0 \in GF(2^n)[X]$ is a polynomial of degree ≤ 3, $f(X) = X^7 + f_6X^6 + f_5X^5 + f_4X^4 + f_3X^3 + f_2X^2 + f_1X + f_0 \in GF(2^n)[X]$ is a monic polynomial of degree 7, and the curve C has no sigular point.

The equation $(*)$ defining a hyperelliptic curves C of genus 3 is unique up to a change of coordinates of the form

$$(x, y) \longrightarrow (\alpha^2 x + \beta, \alpha^7 y + t(x)),$$

Where $\alpha, \beta \in GF(2^n)$ with $\alpha \neq 0$ and $t(x) \in GF(2^n)[x]$ with $\deg t \leq 3$ [Loc94]. Let $P_i = (x_i, y_i) \in \overline{GF(2^n)} \times \overline{GF(2^n)}$ be a point on the curve C and P_∞ be a point at infinity, where $\overline{GF(2^n)}$ is the algebraic closure of $GF(2^n)$. The inverse of $P_i = (x_i, y_i)$ is the point $-P_i = (x_i, y_i + h(x_i))$. P is called a ramification point if $P = -P$ holds. A divisor is a formal sum of points: $D = \Sigma m_iP_i, m_i \in Z$. A semi-reduced divisor is given by $D = \Sigma m_iP_i - (\Sigma m_i)P_\infty$, where $m_i \geq 0$ and $P_i \neq -P_j$ for $i \neq j$, and semi-reduced divisor D is called reduced if $\Sigma m_i \leq 3$ holds for genus 3 hyperelliptic curves. The divisor class group $J_C(GF(2^n))$ of C forms a finite abelian group and therefore we can construct cryptosystems based on discrete logarithm problems on the Jacobian of C. Any equivalent class D in $J_C(GF(2^n))$ can be represented by a pair of polynomials (u, v), which satisfies the following conditions [Mum84]:

$$u(x) = \prod (x + x_i)^{m_i}, v(x_i) = y_i, \deg v < \deg u, u \mid v^2 + hv + f.$$

3 Harley's Algorithm for Divisor Class Doublings

In [GH00], the authors noticed that one can reduce the number of operations by distinguishing between possible cases according to the properties of the input divisors. They proposed an efficient algorithm (Using many computational algebra tricks such as Karatsuba Multiplication, Chinese Remainder Theorem and Newton Iteration) to compute in the Jacobian of hyperelliptic curves. For a complete description about explicit formulae for group operations we refer to [Wol04].

In this paper we concentrate on the doublings for the genus 3 curves in the most significant case where the input divisor $[u(x), v(x)]$ has full degree and u and h do not have a common factor. Therefore, we assume from now on

$$D = [u(x), v(x)], \quad \deg u(x) = 3, \quad \text{resultant } [u(x), h(x)] \neq 0.$$

Let $u(x) = x^3 + u_2 x^2 + u_1 x + u_0, v(x) = v_2 x^2 + v_1 x + v_0$. Using the following Harley's algorithm, we can double a divisor class on a Jacobian:

- Step 1. Compute the resultant r of u and h;
- Step 2. Compute the almost inverse $inv = r/h \bmod u = inv_2 x^2 + inv_1 x + inv_0$;
- Step 3. Compute $z = ((f - hv - v^2)/u) \bmod u = z_2 x^2 + z_1 x + z_0$;
- Step 4. Compute $s' = z \cdot inv \bmod u = s_2' x^2 + s_1' x + s_0'$;
- Step 5. Compute $s = (s'/r)$ and make s monic: $s = x^2 + s_1 x + s_0$;
- Step 6. Compute $G = s \cdot u = x^5 + g_4 x^4 + g_3 x^3 + g_2 x^2 + g_1 x + g_0$;
- Step 7. Compute $u' = u^{-2}\{[G + (r/s_2')v]^2 + (r/s_2')hG + (r/s_2')^2(hv - f)\} = x^4 + u_3' x^3 + u_2' x^2 + u_1' x + u_0'$;
- Step 8. Compute $v' = -[G(s_2'/r) + h + v] \bmod u' = v_3' x^3 + v_2' x^2 + v_1' x + v_0'$;
- Step 9. Reduce $u' : u'' = (f - v'h - v'^2)/u' = x^3 + u_2'' x^2 + u_1'' x + u_0''$;
- Step 10. Compute $v'' = -(v' + h) \bmod u'' = v_2'' x^2 + v_1'' x + v_0''$.

We now study the different cases of the equations depending on the degree of h because the actual execution of the Harley's algorithm depends on the coefficients of the curve. We will present the explicit formulae for four different cases: $\deg h = 0$, $\deg h = 1$, $\deg h = 2$ and $\deg h = 3$. In the two latter cases, we try to find special curves which can lead to a significant speedup. The major speedup is obtained by simplifying and canceling r in the expressions. For hyperelliptic curves of genus 3 and defined over fields of characteristic two there exist no supersingular cases [Gal01, RS02], i.e. for genus 3 HECC we can take special curves with h constant. Using these special curves, we can obtain the explicit formulae with low complexity and better performance regarding the number of required field operations for the execution of the group operations.

4 Case deg $h = 0$

In this section we assume $\deg h = 0$. One can obtain an isomorphic curve where $f_6 = f_5 = f_4 = f_2 = 0$ and h_0 is divided by any α^7. To improve the efficiency of

HECC, we hope that the coefficient h_0 is 'small' in an isomorphic curve, which allows the multiplication with it to be performed via additions. Hence we will choose α^7 such that $\frac{h_0}{\alpha^7}$ is 'small' in the practical use. If we choose finite fields $GF(2^n)$ with $n \equiv 1(mod\ 3)$ or $n \equiv 2(mod\ 3)$ there are no elements $\alpha \in GF(2^n)$ such that $\alpha^7 = 1$ (the unit element of $GF(2^n)$). Therefore, there is always an α such that $\alpha^7 = h_0$. For $n \equiv 0(mod\ 3)$ this happens with probability $1/7$. We obtain the isomorphic curve by using the following birational transformation of variables and dividing the equation by α^{14}:

$$Y \leftarrow \alpha^7 \widetilde{Y} + a\widetilde{X}^3 + b\widetilde{X}^2 + c\widetilde{X}, X \leftarrow \alpha^2 \widetilde{X} + d$$

where $d = \sqrt{f_5},\ a = \alpha^6\sqrt{f_6 + d},\ b = \alpha^4\sqrt{f_4 + f_5 \cdot d + f_6 \cdot d^2 + d^3}$ and $c = \alpha^2\sqrt{f_2 + f_3 \cdot d + f_6 \cdot d^4 + d^5 + h_0 \cdot b}$. Therefore we obtain a curve of the form $Y^2 + h_0Y = X^7 + f_3X^3 + f_1X + f_0$, usually with $h_0 = 1$. Adding a constant term to the substitution of $\widetilde{Y}$ one can achieve $f_0 = 0$ with probability $1/2$. Hence, there are only two parameters f_3, f_1 as opposed to five in the general case showing that the type is indeed special.

With the new curve coefficients the expression r and s will simplify to:

$$r = h_0^3, s_2' = h_0^2z_2, s_1' = h_0^2z_1, s_0' = h_0^2z_0.$$

We note that
$$u_3' = 0, u_2' = s_1^2 = (s_1'/s_2')^2 = (z_1/z_2)^2,$$
$$u_1' = (r/s_2')^2 = h_0^2(z_2^{-1})^2, u_0' = s_0^2 = (s_0'/s_2')^2 = (z_0/z_2)^2,$$

and
$$v_3' = (u_2' + g_3)(s_2'/r) = h_0^{-1}(u_2'z_2 + z_0 + u_2z_1 + u_1z_2),$$
$$v_2' = (g_4u_2'+u_1'+g_2)(s_2'/r)+v_2 = h_0^{-1}[(u_2z_2+z_1)u_2'+u_1'z_2+u_2z_0+u_1z_1+u_0z_2]+v_2,$$
$$v_1' = (g_4u_1' + u_0' + g_1)(s_2'/r) + v_1 = h_0^{-1}[(u_2z_2 + z_1)u_1' + u_0'z_2 + u_1z_0 + u_0z_1] + v_1,$$
$$v_0' = (g_4u_0' + g_0)(s_2'/r) + h_0 + v_0 = h_0^{-1}[(u_2z_2 + z_1)u_0' + u_0z_0] + h_0 + v_0.$$

Since $f + hv + v^2 = uz + u^2x$ we also have that

$$f_0 + h_0v_0 + v_0^2 = u_0z_0,$$
$$f_1 + h_0v_1 = u_1z_0 + u_0z_1 + u_0^2,$$
$$h_0v_2 + v_1^2 = u_2z_0 + u_1z_1 + u_0z_2,$$
$$f_3 = z_0 + u_2z_1 + u_1z_2 + u_1^2,$$
$$v_2^2 = z_1 + u_2z_2,$$
$$0 = u_2^2 + z_2.$$

Using the equations above, we can calculate cheaply u_2', u_0' and v_3', v_2', v_1', v_0' as follows:
$$u_2' = (z_1/z_2)^2 = [(v_2^2 + u_2z_2)/z_2]^2 = (v_2^2z_2^{-1})^2 + u_2^2,$$

$$u_0' = (z_0/z_2)^2 = [(f_3 + u_1^2 + u_2 z_1 + u_1 z_2)/z_2]^2 = [(f_3 + u_1^2)z_2^{-1}]^2 + u_1^2 + u_2^2 u_2',$$

$$v_3' = h_0^{-1}(u_2' z_2 + f_3 + u_1^2),$$

$$v_2' = h_0^{-1}(v_2^2 u_2' + v_1^2) + h_0 z_2^{-1},$$

$$v_1' = h_0^{-1}[(v_2^2 + z_2)(u_1' + u_0') + v_2^2 u_0' + f_1 + u_0^2] + h_0 z_2^{-1},$$

$$v_0' = h_0^{-1}(v_2^2 u_0' + f_0 + v_0^2) + h_0.$$

We give the doubling formulae for this case in Table 1. The operations are counted for the case $h_0 = 1, h_0^{-1}$ is 'small' (multiplication with h_0^{-1} are not counted), and arbitrary h_0. Both h_0^2 and h_0^{-1} are precomputed.

Table 1. Doubling deg $h = 0$, deg $u = 3$

Input	$[u,v], u = x^3 + u_2 x^2 + u_1 x + u_0, v = v_2 x^2 + v_1 x + v_0; h_0^2, h_0^{-1}$			
Output	$[u'',v''] = 2[u,v]$			
Step	Expression	$h_0 = 1$	h_0^{-1} small	h_0 arbitrary
1	Compute $\tilde{u} = u^2$ and $\tilde{v} = v^2$:	$6S$	$6S$	$6S$
	If $u_2 = 0$ then call the Cantor algorithm			
	$\tilde{u}_2 = u_2^2, \tilde{u}_1 = u_1^2, \tilde{u}_0 = u_0^2, \tilde{v}_2 = v_2^2, \tilde{v}_1 = v_1^2, \tilde{v}_0 = v_0^2;$			
2	Compute $u' = x^4 + u_3' x^3 + u_2' x^2 + u_1' x + u_0'$:	$1I, 3M, 3S$	$1I, 4M, 3S$	$1I, 4M, 3S$
	$t_1 = f_3 + \tilde{u}_1, t_2 = f_1 + \tilde{u}_0, t_3 = f_0 + \tilde{v}_0;$			
	$invz_2 = \tilde{u}_2^{-1}, u_3' = 0, u_2' = (\tilde{v}_2 invz_2)^2 + \tilde{u}_2;$			
	$u_1' = h_0^2(invz_2)^2, u_0' = (t_1 invz_2)^2 + \tilde{u}_1 + \tilde{u}_2 u_2';$			
3	Compute $v' = v_3' x^3 + v_2' x^2 + v_1' x + v_0'$:	$4M$	$5M$	$9M$
	$v_3' = h_0^{-1}(u_2' \tilde{u}_2 + t_1), v_2' = h_0^{-1}(\tilde{v}_2 u_2' + \tilde{v}_1) + h_0 invz_2;$			
	$v_1' = h_0^{-1}[(\tilde{v}_2 + \tilde{u}_2)(u_1' + u_0') + \tilde{v}_2 u_0' + t_2] + h_0 invz_2;$			
	$v_0' = h_0^{-1}(\tilde{v}_2 u_0' + t_3);$			
4	Reduce u', i.e. $u'' = x^3 + u_2'' x^2 + u_1'' x + u_0''$:	$1M, 2S$	$1M, 2S$	$1M, 2S$
	$u_2'' = v_3'^2, u_1'' = u_2', u_0'' = u_2'' u_2' + v_2'^2 + u_1';$			
5	Compute $v'' = v_2'' x^2 + v_1'' x + v_0''$:	$3M$	$3M$	$3M$
	$v_2'' = v_2' + v_3' u_2'', v_1'' = v_1' + v_3' u_1'', v_0'' = v_0' + v_3' u_0'';$			
Sum		$1I, 11M, 11S$	$1I, 13M, 11S$	$1I, 17M, 11S$

Remark 1. For the case of $h_0 = 1$, we obtain the same explicit formula as in [GKP04] (only save several field additions).

5   Case deg $h = 1$

In this section we discuss the case of deg $h = 1$. One can obtain an isomorphic curve where $f_6 = f_4 = h_0 = 0$ and h_1 is divided by any α^5. We will choose α^5 such that $\frac{h_1}{\alpha^5}$ is 'small' in the practical use. If we choose finite fields $GF(2^n)$ with n not being divided by 4 there are no elements $\alpha \in GF(2^n)$ such that $\alpha^5 = 1$. Therefore, there is always an α such that $\alpha^5 = h_0$. For $n \equiv 0 (mod\ 4)$

this happens with probability $1/5$. We obtain the isomorphic curve by using the following birational transformation of variables and dividing the equation by α^{14}:

$$Y \leftarrow \alpha^7 \widetilde{Y} + s\widetilde{X}^3 + t\widetilde{X}^2, X \leftarrow \alpha^2 \widetilde{X} + \beta$$

where $\beta = \frac{h_0}{h_1}, s = \alpha^6 \sqrt{f_6 + \beta}$ and $t = \alpha^4 \sqrt{f_4 + f_5\beta + f_6\beta^2 + \beta^3 + h_1\sqrt{f_6 + \beta}}$. Hence, we obtain a curve of the form $Y^2 + h_1XY = X^7 + f_5X^5 + f_3X^3 + f_2X^2 + f_1X + f_0$, usually with $h_1 = 1$. Adding a linear factor to the substitution of $\widetilde{Y}$ one can achieve $f_2 = 0$ with probability $1/2$. A constant term leads to $f_1 = 0$. Therefore, there are only three free parameters f_5, f_3, f_0.

With the new curve coefficients the expression r and s will simplify to:

$$r = u_0h_1^3, s_2' = z_0h_1^2, s_1' = (u_2z_0 + u_0z_2)h_1^2, s_0' = (u_1z_0 + u_0z_1)h_1^2,$$

$$rs_2' = u_0z_0h_1^5, s_2 = \frac{s_2'}{r} = \frac{z_0}{u_0h_1}.$$

In this case, we have that

$$u_3' = 0, u_2' = s_1^2 = (s_1'/s_2')^2 = (u_2 + u_0 \cdot \frac{z_2}{z_0})^2,$$

$$u_1' = (r/s_2')^2 = h_0^2u_0^2(z_0^{-1})^2, u_0' = s_0^2 = (s_0'/s_2')^2 = (u_1 + u_0 \cdot \frac{z_1}{z_0})^2,$$

and

$$v_3' = (u_2' + g_3)(s_2'/r) = h_1^{-1}[z_2 \cdot (u_0 \cdot \frac{z_2}{z_0}) + z_1 + u_2z_2],$$

$$v_2' = (g_4u_2' + u_1' + g_2)(s_2'/r) + v_2 = h_1^{-1}[z_2u_2' + \frac{h_1}{s_2} + u_2z_1 + u_1z_2 + z_0] + v_2,$$

$$v_1' = (g_4u_1' + u_0' + g_1)(s_2'/r) + h_1 + v_1 = h_1^{-1}[\frac{1}{h_1s_2}(\frac{z_2h_1}{s_2} + z_1^2) + u_2z_0 + u_1z_1 + u_0z_2] + v_1,$$

$$v_0' = (g_4u_0' + g_0)(s_2'/r) + v_0 = h_1^{-1}(z_2u_0' + u_1z_0 + u_0z_1) + v_0.$$

Since $f + hv + v^2 = uz + u^2x$ we also obtain that

$$f_0 + v_0^2 = u_0z_0 \quad (= rs_2'/h_1^5),$$

$$f_1 + h_1v_0 = u_1z_0 + u_0z_1 + u_0^2,$$

$$f_2 + h_1v_1 + v_1^2 = u_2z_0 + u_1z_1 + u_0z_2,$$

$$f_3 + h_1v_2 = z_0 + u_2z_1 + u_1z_2 + u_1^2,$$

$$v_2^2 = z_1 + u_2z_2,$$

$$f_5 = u_2^2 + z_2.$$

Using the equations above, we can calculate cheaply u_2', u_0' and v_3', v_2', v_1', v_0' as follows:

$$u_2' = (u_2 + u_0 \cdot \frac{z_2}{z_0})^2 = (u_2 + u_0 \cdot \frac{z_2}{u_0h_1s_2})^2 = (u_2 + \frac{z_2}{h_1s_2})^2,$$

$$u_0' = (u_1 + u_0 \cdot \frac{z_1}{z_0})^2 = (u_1 + u_0 \cdot \frac{z_1}{u_0 h_1 s_2})^2 = (u_1 + \frac{z_1}{h_1 s_2})^2,$$

$$v_3' = h_1^{-1}(\frac{z_2^2}{h_1 s_2} + v_2^2),$$

$$v_2' = h_1^{-1}(z_2 u_2' + \frac{h_1}{s_2} + f_3 + u_1^2),$$

$$v_1' = h_1^{-1}[\frac{1}{h_1 s_2}(z_2 \cdot \frac{h_1}{s_2} + z_1^2) + f_2 + v_1^2],$$

$$v_0' = h_1^{-1}(z_2 u_0' + f_1 + u_0^2).$$

We note that $f_0 + v_0^2 = u_0 \cdot z_0 = r \cdot s_2'/h_1^5$, so it is very cheap to calculate $r \cdot s_2'$ as the exact coefficients of z are not necessary. In Table 2, we present the doubling formula for this case. The operations are counted for the case $h_1 = 1, h_1^{-1}$ is 'small' (multiplication with h_1^{-1} are not counted), and arbitrary h_1. Both h_1^2 and h_1^{-1} are precomputed. In Step 2 the inversion and multiplication with k_0 can also be replaced by a division as the inverse is not used later on.

Table 2. Doubling deg $h = 1$, deg $u = 3$

Input	$[u,v], u = x^3 + u_2 x^2 + u_1 x + u_0, v = v_2 x^2 + v_1 x + v_0; h_1^2, h_1^{-1}$			
Output	$[u'', v''] = 2[u,v]$			
Step	Expression	$h_1 = 1$	h_1^{-1} small	h_1 arbitrary
1	Compute rs_2': $k_0 = u_0^2, z_2 = f_5 + u_2^2, t_1 = v_2^2;$ $z_1 = t_1 + u_2 z_2, w_0 = f_0 + v_0^2 \ (= rs_2'/h_1^5);$ If $w_0 = 0$ then call the Cantor algorithm	$1M, 4S$	$1M, 4S$	$1M, 4S$
2	Compute $1/h_1 s_2$ and s_1, s_0: $w_1 = (1/w_0) \cdot k_0 \ (= 1/h_1 s_2), k_1 = z_2 w_1;$ $k_2 = z_1 w_1, s_1 = u_2 + k_1, s_0 = u_1 + k_2;$	$1I, 3M$	$1I, 3M$	$1I, 3M$
3	Compute $u' = x^4 + u_3' x^3 + u_2' x^2 + u_1' x + u_0'$: $w_2 = h_1^2 w_1 \ (= h_1/s_2), u_3' = 0;$ $u_2' = s_1^2, u_1' = w_2 w_1, u_0' = w_2 + s_0^2;$	$3S$	$2M, 2S$	$2M, 2S$
4	Compute $v' = v_3' x^3 + v_2' x^2 + v_1' x + v_0'$: $v_3' = h_1^{-1}(z_2 k_1 + t_1), v_2' = h_1^{-1}(z_2 u_2' + w_2 + f_3 + u_1^2);$ $v_1' = h_1^{-1}[w_1(z_2 w_2 + z_1^2) + f_2 + v_1^2];$ $v_0' = h_1^{-1}(z_2 u_0' + f_1 + u_0^2);$	$5M, 4S$	$5M, 4S$	$9M, 4S$
5	Reduce u', i.e. $u'' = x^3 + u_2'' x^2 + u_1'' x + u_0''$: $u_2'' = v_3'^2, u_1'' = f_5 + u_2';$ $u_0'' = f_4 + u_2'' u_2' + v_2'^2 + u_1' + h_1 v_3';$	$1M, 2S$	$2M, 2S$	$2M, 2S$
6	Compute $v'' = v_2'' x^2 + v_1'' x + v_0''$: $v_2'' = v_2' + v_3' u_2'', v_1'' = v_1' + v_3' u_1'' + h_1, v_0'' = v_0' + v_3' u_0'';$	$3M$	$3M$	$3M$
Sum		$1I, 13M, 13S$	$1I, 16M, 12S$	$1I, 20M, 12S$

Remark 2. The algorithm in [GKP04] needs $1I, 44M, 6S$ to compute the divisor class doubling. However, our explicit formula requires only $1I, 13M, 13S$ in this case. Compared with the explicit formula in [GKP04], our formula saves $31M$ at the cost of extra $7S$ (a squaring is usually more efficient than a multiplication in binary fields).

6   Case deg $h = 2$

If h is of degree two then we cannot make any of its coefficients zero in general. In this section we will discuss special curves with $h_1 = 0$, that is, the curves having the form $Y^2 + (h_2 X^2 + h_0)Y = X^7 + f_6 X^6 + f_5 X^5 + f_4 X^4 + f_3 X^3 + f_2 X^2 + f_1 X + f_0$, which allows for a significant speedup. By making a change of coordinates we can obtain $f_5 = f_3 = f_2 = h_0 = 0$ and h_2 is divided by any α^3. We will choose α^3 such that $\frac{h_2}{\alpha^3}$ is 'small' in the practical use. If, as usual, one choose finite $GF(2^n)$ with n odd there are no non-trivial cubic roots of the unity. Hence, there is always an α such that $\alpha^3 = h_2$. For even n this happens with probability $1/3$. The isomorphic curve is obtained by using the following birational transformation of variables and dividing the equation by α^{14}:

$$Y \leftarrow \alpha^7 \widetilde{Y} + m\widetilde{X}^3 + s\widetilde{X} + t, X \leftarrow \alpha^2 \widetilde{X} + \beta$$

where $\beta = \sqrt{\frac{h_0}{h_2}}, m = \alpha^6 \cdot \frac{f_5 + \beta^2}{h_2}, s = \alpha^2 \cdot \frac{f_3 + \beta^4}{h_2}$ and $t = \frac{h_2^2 (f_2 + f_3 \beta + f_6 \beta^4 + \beta^5) + f_3^2 + \beta^8}{h_2^3}$.
Hence, we obtain a curve of the form $Y^2 + h_2 X^2 Y = X^7 + f_6 X^6 + f_4 X^4 + f_1 X + f_0$, usually with $h_2 = 1$. Adding a quadratic factor to the substitution of $\widetilde{Y}$ one can achieve $f_4 = 0$ with probability $1/2$. Accordingly, there are only three free parameters f_6, f_1, f_0.

Then the expressions for r and s will simplify to:

$$r = u_0^2 h_2^3, s_2' = (u_1 z_0 + u_0 z_1) h_2^2, s_1' = [u_2(u_1 z_0 + u_0 z_1) + u_0 z_0] h_2^2,$$

$$s_0' = [u_1(u_1 z_0 + u_0 z_1) + u_0(u_2 z_0 + u_0 z_2)] h_2^2,$$

$$s_1 = \frac{s_1'}{s_2'} = u_2 + k_1, s_0 = \frac{s_0'}{s_2'} = u_1 + k_2,$$

where $k_1 = \frac{u_0 z_0}{u_1 z_0 + u_0 z_1}$ and $k_2 = \frac{u_0(u_2 z_0 + u_0 z_2)}{u_1 z_0 + u_0 z_1}$. In this case, we have that

$$u_3' = 0, u_2' = s_1^2, u_1' = \frac{r}{s_2'}(h_2 + \frac{r}{s_2'}) = h_2^2 w_1(1 + w_1),$$

$$u_0' = \frac{r}{s_2'}[h_2(u_2 + s_1) + \frac{rf_6}{s_2'}] + s_0^2 = h_2^2 w_1(k_1 + f_6 w_1) + s_0^2,$$

where $w_1 = \frac{u_0^2}{u_1 z_0 + u_0 z_1}$ and

$$v_3' = (u_2' + g_3)(s_2'/r) = h_2^{-1}[z_2 + \frac{(u_0 z_0)^2}{u_0^2(u_1 z_0 + u_0 z_1)}],$$

$$v_2' = (g_4 u_2' + u_1' + g_2)(s_2'/r) + h_2 + v_2$$

$$= h_2^{-1}[z_1 + u_2 z_2 + \frac{(u_0 z_0)k_1^2}{u_0^2}] + h_2 w_1 + v_2,$$

$$v_1' = (g_4 u_1' + u_0' + g_1)(s_2'/r) + v_1$$

$$= h_2^{-1}[z_0 + u_1 z_2 + u_2 z_1 + \frac{(u_2 z_0 + u_0 z_2)^2}{u_1 z_0 + u_0 z_1}] + (h_2 w_1)(f_6 + k_1) + v_1,$$

$$v_0' = (g_4 u_0' + g_0)(s_2'/r) + v_0$$

$$= h_2^{-1}[u_2 z_0 + u_1 z_1 + u_0 z_2 + \frac{(u_0 z_0)k_2^2}{u_0^2}] + (h_2 k_1)(k_1 + f_6 w_1) + v_0.$$

And since $f + hv + v^2 = uz + u^2(x + f_6)$ we also have that

$$f_0 + v_0^2 = u_0 z_0 + f_6 u_0^2,$$

$$f_1 = u_1 z_0 + u_0 z_1 + u_0^2,$$

$$h_2 v_0 = u_2 z_0 + u_1 z_1 + u_0 z_2 + f_6 u_1^2,$$

$$h_2 v_1 = z_0 + u_2 z_1 + u_1 z_2 + u_1^2,$$

$$f_4 + h_2 v_2 + v_2^2 = z_1 + u_2 z_2 + f_6 u_2^2,$$

$$0 = u_2^2 + z_2.$$

We use the equations above to calculate k_1, k_2, w_1 and v_3', v_2', v_1', v_0' cheaper:

$$k_1 = \frac{f_0 + v_0^2 + f_6 u_0^2}{f_1 + u_0^2}, k_2 = \frac{u_0(h_2 v_0 + u_1 z_1 + f_6 u_1^2)}{f_1 + u_0^2}, w_1 = \frac{u_0^2}{f_1 + u_0^2},$$

$$v_3' = h_2^{-1}[z_2 + \frac{(f_0 + v_0^2 + f_6 u_0^2)^2}{u_0^2(f_1 + u_0^2)}],$$

$$v_2' = h_2^{-1}[f_4 + v_2^2 + f_6 u_2^2 + \frac{(f_0 + v_0^2 + f_6 u_0^2)k_1^2}{u_0^2}] + \frac{h_2 u_0^2}{f_1 + u_0^2},$$

$$v_1' = h_2^{-1}[u_1^2 + \frac{(u_1 z_1 + f_6 u_1^2 + h_2 v_0)^2}{f_1 + u_0^2}] + \frac{h_2 u_0^2(f_6 + k_1)}{f_1 + u_0^2},$$

$$v_0' = h_2^{-1}[f_6 u_1^2 + \frac{(f_0 + v_0^2 + f_6 u_0^2)k_2^2}{u_0^2}] + (h_2 k_1)(k_1 + \frac{u_0^2 f_6}{f_1 + u_0^2}).$$

We note that $r \cdot s_2' = u_0^2 \cdot (u_1 z_0 + u_0 z_1) \cdot h_2^5 = u_0^2 \cdot (f_1 + u_0^2) \cdot h_2^5$, so it is very cheap to calculate $r \cdot s_2'$ since we do not need to know the exact coefficients of z. We describe the doubling formula for this case in Table 3. The operations are counted for the case $h_2 = 1, h_2^{-1}$ is 'small' (multiplication with h_2^{-1} are not counted), and arbitrary h_2. Both h_2^2 and h_2^{-1} are precomputed.

Table 3. Doubling deg $h = 2$, $h_1 = 0$, deg $u = 3$

Input	$[u, v], u = x^3 + u_2 x^2 + u_1 x + u_0, v = v_2 x^2 + v_1 x + v_0; h_2^2, h_2^{-1}$			
Output	$[u'', v''] = 2[u, v]$			

Step	Expression	$h_2 = 1$	h_2^{-1} small	h_2 arbitrary
1	Precomputation: $\tilde{u}_2 = u_2^2, \tilde{u}_1 = u_1^2, \tilde{u}_0 = u_0^2, z_2 = f_4 + v_2^2 + f_6 \tilde{u}_2;$ $z_1 = z_2 + h_2 v_2 + \tilde{u}_2 u_2, t_1 = f_0 + v_0^2 + f_6 \tilde{u}_0;$ $t_2 = f_6 \tilde{u}_1, t_3 = t_2 + h_2 v_0 + u_1 z_1, t_4 = f_1 + \tilde{u}_0;$ If $t_4 = 0$ then call the Cantor algorithm	$5M, 5S$	$7M, 5S$	$7M, 5S$
2	Compute s_1, s_0: $t_5 = (t_4 \tilde{u}_0)^{-1}, t_6 = t_4 t_5, t_7 = \tilde{u}_0 t_5;$ $t_8 = t_1 t_6, k_1 = t_1 t_7, \tilde{k}_2 = t_3 t_7;$ $k_2 = u_0 \tilde{k}_2, s_1 = u_2 + k_1, s_0 = u_1 + k_2;$	$1I, 7M$	$1I, 7M$	$1I, 7M$
3	Compute $u' = x^4 + u_3' x^3 + u_2' x^2 + u_1' x + u_0'$: $w_1 = \tilde{u}_0 t_7, w_2 = h_2^2 w_1, u_2' = s_1^2, u_1' = w_2(1 + w_1);$	$1M, 2S$	$3M, 1S$	$3M, 1S$
4	Compute $v' = v_3' x^3 + v_2' x^2 + v_1' x + v_0'$: $v_3' = h_2^{-1}(\tilde{u}_2 + t_1^2 t_5), v_2' = h_2^{-1}(z_2 + t_8 k_1^2) + h_2 w_1;$ $v_1' = h_2^{-1}(\tilde{u}_1 + \tilde{k}_2 t_3) + (h_2 w_1)(f_6 + k_1);$ $v_0' = h_2^{-1}(t_2 + t_8 k_2^2) + (h_2 k_1)(k_1 + f_6 w_1);$	$7M, 3S$	$9M, 3S$	$13M, 3S$
5	Reduce u', i.e. $u'' = x^3 + u_2'' x^2 + u_1'' x + u_0''$: $u_2'' = f_6 + v_3'^2, u_1'' = u_2' + h_2 v_3';$ $u_0'' = f_4 + u_2' u_2 + u_1' + v_2'^2 + h_2 v_2';$	$1M, 2S$	$3M, 1S$	$3M, 1S$
6	Compute $v'' = v_2'' x^2 + v_1'' x + v_0''$: $v_2'' = v_2' + v_3' u_2'' + h_2, v_1'' = v_1' + v_3' u_1'';$ $v_0'' = v_0' + v_3' u_0'';$	$3M$	$3M$	$3M$
Sum		$1I, 24M, 12S$	$1I, 32M, 10S$	$1I, 36M, 10S$

Remark 3. For the general case with $h(x) = h_2 x^2 + h_1 x + h_0$, the authors in [GKP04] use a birational transformation to make the curve's coefficient f_6 zero. Their algorithm needs $1I, 52M, 8S$ to compute the divisor class doubling. Using special curves with $h(x) = h_2 x^2 + h_0$, our explicit formula requires only $1I, 24M, 12S$ for $h_2 = 1$. In the formulae presented in Table 3 there are four counted multiplications with f_6 which are cheaper when f_6 is 'small'.

7   Case deg $h = 3$

When h is of degree three, we cannot also make any of its coefficients zero in general. We will show that special curves with $h_2 = h_1 = h_0 = 0$ can obtain excellent performance in this section. We can construct a change of coordinates to make $f_5 = f_4 = f_3 = 0$ and $h_3 = 1$. The isomorphic curve is obtained by using the following birational transformation of variables and dividing the equation by h_3^{14}:

$$Y \leftarrow h_3^7 \widetilde{Y} + h_3^3 f_5 \widetilde{X}^2 + \frac{f_4 h_3^2 + f_5^2}{h_3} \widetilde{X} + \frac{f_3}{h_3}, X \leftarrow h_3^2 \widetilde{X}$$

Hence we obtain a curve of the form $Y^2 + X^3 Y = X^7 + f_6 X^6 + f_2 X^2 + f_1 X + f_0$. Adding a cubic factor to the substitution of $\tilde{Y}$ one can achieve $f_6 = 0$ with probability $1/2$. Thereby, there are only three free parameters f_2, f_1, f_0.

Then the expressions for r and s will simplify to:

$$r = u_0^3, s_2' = u_0(u_2 z_0 + u_1 z_1 + u_0 z_2) + u_1^2 z_0,$$

$$s_1' = u_2[u_0(u_2 z_0 + u_1 z_1 + u_0 z_2)] + u_0(u_1 z_0 + u_0 z_1),$$

$$s_0' = u_1[u_0(u_2 z_0 + u_1 z_1 + u_0 z_2)] + u_0[u_2(u_1 z_0 + u_0 z_1) + u_0 z_0],$$

$$s_1 = \frac{s_1'}{s_2'} = u_2 + k_1, s_0 = \frac{s_0'}{s_2'} = u_1 + k_2,$$

where $k_1 = \frac{u_0(u_1 z_0 + u_0 z_1)}{u_0(u_2 z_0 + u_1 z_1 + u_0 z_2) + u_1^2 z_0}$ and $k_2 = \frac{u_0[u_2(u_1 z_0 + u_0 z_1) + u_0 z_0]}{u_0(u_2 z_0 + u_1 z_1 + u_0 z_2) + u_1^2 z_0}$. In this case, we have that

$$u_3' = 0, u_2' = s_1^2 + \frac{r}{s_2'}, u_1' = \frac{r}{s_2'}(k_1 + \frac{r}{s_2'}), u_0' = \frac{r}{s_2'}(k_2 + u_2 k_1 + \frac{r f_6}{s_2'}) + s_0^2,$$

$$v_3' = (u_2' + g_3)(s_2'/r) + 1 = \frac{u_0 z_0}{u_0^2} + \frac{(u_1 z_0 + u_0 z_1)^2}{u_0^2(u_2 z_0 + u_1 z_1 + u_0 z_2) + u_1^2(u_0 z_0)},$$

$$v_2' = (g_4 u_2' + u_1' + g_2)(s_2'/r) + v_2 = \frac{(u_1 z_0 + u_0 z_1)(u_2' + u_2^2)}{u_0^2} + z_2 + k_1 + \frac{r}{s_2'} + v_2,$$

$$v_1' = (g_4 u_1' + u_0' + g_1)(s_2'/r) + v_1 = \frac{k_2[u_2(u_1 z_0 + u_0 z_1) + u_0 z_0] + u_2^2(u_0 z_0)}{u_0^2} +$$

$$k_1(k_1 + \frac{r}{s_2'}) + (k_2 + u_2 k_1 + \frac{r f_6}{s_2'}) + (z_1 + u_2 z_2) + v_1,$$

$$v_0' = (g_4 u_0' + g_0)(s_2'/r) + v_0 = \frac{(u_1 z_0 + u_0 z_1)(u_0' + u_1^2)}{u_0^2} + (z_0 + u_2 z_1 + u_1 z_2) + v_0.$$

And since $f + hv + v^2 = uz + u^2(x + f_6)$ we also have that

$$f_0 + v_0^2 = u_0 z_0 + f_6 u_0^2,$$

$$f_1 = u_1 z_0 + u_0 z_1 + u_0^2,$$

$$f_2 + v_1^2 = u_2 z_0 + u_1 z_1 + u_0 z_2 + f_6 u_1^2,$$

$$v_0 = z_0 + u_2 z_1 + u_1 z_2 + u_1^2,$$

$$v_1 + v_2^2 = z_1 + u_2 z_2 + f_6 u_2^2,$$

$$v_2 = u_2^2 + z_2.$$

Using the equations above, we can calculate $k_1, k_2, \frac{r}{s_2'}$ and v_3', v_2', v_1', v_0' cheaper:

$$k_1 = \frac{u_0^2(f_1 + u_0^2)}{u_0^2(f_2 + v_1^2 + f_6 u_1^2) + u_1^2(f_0 + v_0^2 + f_6 u_0^2)},$$

$$k_2 = \frac{u_0^2[u_2(f_1 + u_0^2) + (f_0 + v_0^2 + f_6 u_0^2)]}{u_0^2(f_2 + v_1^2 + f_6 u_1^2) + u_1^2(f_0 + v_0^2 + f_6 u_0^2)},$$

$$\frac{r}{s_2'} = \frac{u_0^4}{u_0^2(f_2 + v_1^2 + f_6 u_1^2) + u_1^2(f_0 + v_0^2 + f_6 u_0^2)},$$

$$v_3' = \frac{f_0 + v_0^2 + f_6 u_0^2}{u_0^2} + \frac{(f_1 + u_0^2)^2}{u_0^2(f_2 + v_1^2 + f_6 u_1^2) + u_1^2(f_0 + v_0^2 + f_6 u_0^2)},$$

$$v_2' = \frac{(f_1 + u_0^2)(u_2' + u_2^2)}{u_0^2} + k_1 + \frac{r}{s_2'} + u_2^2,$$

$$v_1' = \frac{k_2[u_2(f_1 + u_0^2) + (f_0 + v_0^2 + f_6 u_0^2)] + u_2^2(f_0 + v_0^2 + f_6 u_0^2)}{u_0^2} +$$

$$k_1(k_1 + \tfrac{r}{s_2'}) + (k_2 + u_2 k_1 + \tfrac{r f_6}{s_2'}) + (v_2^2 + f_6 u_2^2),$$

$$v_0' = \frac{(f_1 + u_0^2)(u_0' + u_1^2)}{u_0^2} + u_1^2.$$

We note that $r \cdot s_2' = u_0^2 \cdot [u_0^2 \cdot (u_2 z_0 + u_1 z_1 + u_0 z_2) + u_1^2 \cdot (u_0 z_0)] = u_0^2 \cdot [u_0^2 \cdot (f_2 + v_1^2 + f_6 u_1^2) + u_1^2 \cdot (f_0 + v_0^2 + f_6 u_0^2)]$. Therefore, we can calculate $r \cdot s_2'$ cheaply without knowing the exact coefficients of z. We present the explicit formula for this case in Table 4.

Table 4. Doubling deg $h = 3$, $h_2 = h_1 = h_0 = 0$, deg $u = 3$

Input	$[u, v], u = x^3 + u_2 x^2 + u_1 x + u_0, v = v_2 x^2 + v_1 x + v_0;$	
Output	$[u'', v''] = 2[u, v]$	
Step	Expression	Cost
1	Precomputation:	$5M, 6S$
	$\tilde{u}_2 = u_2^2, \tilde{u}_1 = u_1^2, \tilde{u}_0 = u_0^2, \tilde{v}_2 = v_2^2, \tilde{v}_1 = v_1^2, \tilde{v}_0 = v_0^2, t_1 = f_0 + \tilde{v}_0 + f_6 \tilde{u}_0;$	
	$t_2 = f_2 + \tilde{v}_1 + f_6 \tilde{u}_1, t_3 = \tilde{u}_0 t_2 + \tilde{u}_1 t_1, t_4 = f_1 + \tilde{u}_0, t_5 = u_2 t_4 + t_1;$	
	If $t_3 = 0$ then call the Cantor algorithm	
2	Compute s_1, s_0:	$1I, 6M$
	$t_6 = (t_3 \tilde{u}_0)^{-1}, t_7 = t_3 t_6, t_8 = \tilde{u}_0 t_6, t_9 = \tilde{u}_0 t_8;$	
	$k_1 = t_4 t_9, k_2 = t_5 t_9, s_1 = u_2 + k_1, s_0 = u_1 + k_2;$	
3	Compute $u' = x^4 + u_3' x^3 + u_2' x^2 + u_1' x + u_0'$:	$5M, 2S$
	$w_4 = \tilde{u}_0 t_9, u_2' = s_1^2 + w_4, t_{10} = k_1 + w_4, u_1' = w_4 t_{10};$	
	$t_{11} = k_2 + u_2 k_1 + f_6 w_4, u_0' = s_0^2 + w_4 t_{11};$	
4	Compute $v' = v_3' x^3 + v_2' x^2 + v_1' x + v_0'$:	$10M, 1S$
	$v_3' = t_1 t_7 + t_4^2 t_8, t_{12} = t_4 t_7, v_2' = t_{12}(u_2' + \tilde{u}_2) + k_1 + w_4 + \tilde{u}_2;$	
	$v_1' = k_1 t_{10} + t_{11} + (k_2 t_5 + \tilde{u}_2 t_1) t_7 + \tilde{v}_2 + f_6 \tilde{u}_2, v_0' = t_{12}(u_0' + \tilde{u}_1) + \tilde{u}_1;$	
5	Reduce u', i.e. $u'' = x^3 + u_2'' x^2 + u_1'' x + u_0''$:	$1M, 2S$
	$u_2'' = f_6 + v_3' + v_3'^2, u_1'' = u_2' + v_2', u_0'' = u_2'' u_2' + u_1' + v_2'^2 + v_1';$	
6	Compute $v'' = v_2'' x^2 + v_1'' x + v_0''$:	$3M$
	$v_2'' = v_2' + (v_3' + 1) u_2'', v_1'' = v_1' + (v_3' + 1) u_1'', v_0'' = v_0' + (v_3' + 1) u_0'';$	
Sum		$1I, 30M, 11S$

Remark 4. In [GKP04], the authors discuss two types of curves with $h_2 = 0$ and $f_6 = 0$, respectively. Their doubling formulae cost $1I, 63M, 9S$ and $1I, 64M, 5S$ for these two different cases. We note that using special curves with $h(x) = h_3 x^3$ can lead to fast computation of a divisor class doubling. We derive the new explicit doubling formula which needs only $1I, 30M, 11S$. In addition, there are four counted multiplications with f_6 which can be computed cheaply when f_6 is 'small' in the formulae.

8   Summary

Depending on the degree of h, we derived the corresponding explicit formulae which can compute the doublings fast in the previous sections. For h of degree 0 and 1 the case f_6 not small does not apply since we make it zero by isomorphic transformations. We also find the fast doubling formulae for the special curves when the degree of h is 2 and 3. All results are summarized in Table 5.

Table 5. Overview

$h(x)$	$h(x) = h_0$		
h_i	$h_0 = 1$	h_0^{-1} small	h_0 arb.
cost	$1I, 11M, 11S$	$1I, 13M, 11S$	$1I, 17M, 11S$
$h(x)$	$h(x) = h_1 x$		
h_i	$h_1 = 1$	h_1^{-1} small	h_1 arb.
cost	$1I, 13M, 13S$	$1I, 16M, 12S$	$1I, 20M, 12S$
$h(x)$	$h(x) = h_2 x^2$		
h_i	$h_2 = 1$	h_2^{-1} small	h_2 arb.
f_6 small	$1I, 20M, 12S$	$1I, 28M, 10S$	$1I, 32M, 10S$
f_6 arb.	$1I, 24M, 12S$	$1I, 32M, 10S$	$1I, 36M, 10S$
$h(x)$	$h(x) = x^3$		
f_6 small	$1I, 26M, 11S$		
f_6 arb.	$1I, 30M, 11S$		

9   Experimental Results

In order to test the performance of our new doubling formulae, we implemented genus 3 HECC over three binary fields. Due to the attack proposed by Thériault [Thé03], we should select at least 56-bit finite fields in order to obtain the same security as a 160-bit elliptic curve cryptosystem. We used the binary fields $GF(2^{59})$, $GF(2^{61})$ and $GF(2^{63})$. For $GF(2^{59})$ and $GF(2^{61})$, we used the minimal weight irreducible pentanomial $x^{59} + x^7 + x^4 + x^2 + 1$ and $x^{61} + x^5 + x^2 + x + 1$ to construct finite fields, respectively. However, for $GF(2^{63})$, we used the minimal weight irreducible trinomial $x^{63} + x + 1$ as field extension. Efficient algorithms summarized in [Pel02] were used to perform the field arithmetic. In addition, We used the NAF method to perform the scalar multiplication. All tests are implemented on a Pentium-M @1.5 GHz processor and with C programming language. The experimental results were depicted in the following three bar graphs.

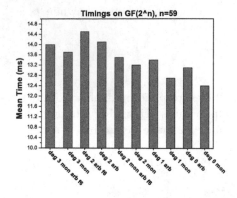

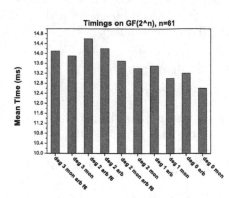

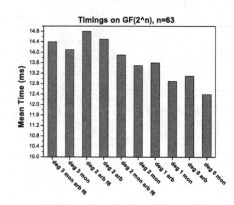

In the graphs above we include the following ten cases respectively:

- deg 3 mon arb f_6: The case where deg $h = 3, h_2 = h_1 = h_0 = 0, f_6 \neq 0$ and $h_3 = 1$;
- deg 3 mon: The case where deg $h = 3, h_2 = h_1 = h_0 = 0, f_6 = 0$ and $h_3 = 1$;
- deg 2 arb f_6:   The case where deg $h = 2, h_1 = h_0 = 0, f_6 \neq 0$;
- deg 2 arb: The case where deg $h = 2, h_1 = h_0 = 0, f_6 = 0$;
- deg 2 mon arb f_6: The case where deg $h = 2, h_1 = h_0 = 0, f_6 \neq 0$ and $h_2 = 1$;
- deg 2 mon: The case where deg $h = 2, h_1 = h_0 = 0, f_6 = 0$ and $h_2 = 1$;
- deg 1 arb: The case where deg $h = 1, h_0 = 0$;
- deg 1 mon: The case where deg $h = 1, h_0 = 0$ and $h_1 = 1$;
- deg 0 arb: The case where deg $h = 0$;
- deg 0 mon: The case where deg $h = 0$ and $h_0 = 1$;

10 Conclusion and Outlook

We have discussed how to accelerate the computation of divisor class doublings for genus 3 hyperelliptic curves defined over binary fields and given explicit formulae for four special types of curves. Compared with the results in [PWGP03, GKP04], our explicit formulae have reduced some field operations further and shown excellent performance on a Pentium-M processor. The divisor addition formulae depend far less on the coefficients of h. In [GKP04], the authors have improved the corresponding addition explicit formulae according to the degree of h.

In this paper we restricted our attentions to affine coordinate system. However, the authors of this paper have obtained inversion-free explicit formulae for genus 3 hyperelliptic curves [FWW05]. For genus 3 HECC defined over binary fields, the authors gave only inversion-free explicit formulae for special curves with $h(x) = 1$. Hence, how to extended the idea of this paper to projective coordinate system and improve the formulae in [FWW05] will be the next logical step to accelerate the implementation for genus 3 HECC.

Acknowledgements. First of all I express my deepest gratitude to my supervisor Prof. Yumin Wang for everything he did. Furthermore, thanks go to Dr. Kiumars Kaveh for offering me their research paper [GKP04]. I would also like to thank anonymous referees for their very useful suggestions and comments that improved the presentation of this paper.

References

[Ava04]   R. M. Avanzi. Aspects of Hyperelliptic Curves over Large Prime Fields in Software Implementations. In M. Joye and J.-J. Quisquater, editors, *Workshop on Cryptographic Hardware and Embedded Systems - CHES 2004*, volume LNCS 3156, pp. 148-162, Springer-Verlag, 2004.

[BCLW02]   N. Boston, T. Clancy, Y. Liow, and J. Webster. Genus Two Hyperelliptic Curve Coprocessor. In B. S. Kaliski, Ç. K. Koç, and C. Paar, editors, *Cryptographic Hardware and Embedded Systems - CHES 2002*, volume LNCS 2523, pp. 529 - 539. Springer-Verlag, 2002. Updated version available at http://www.cs. umd.edu/ clancy/docs/hec-ches2002.pdf.

[BD04]   B. Byramjee and S. Duqesne. Classification of genus 2 curves over F_{2^n} and optimazation of their arithmetic. *Cryptology ePrint Archieve*, Report 2004/107, 2004. http://eprint.iacr.org/.

[Can87]   D.G.Cantor. Computing In The Jacobian Of A Hyperelliptic Curve. *Math. Comp.* 48:95-101, 1987.

[CFA+05]   H.Cohen, G.Frey, R.Avanzi, C.Doche, T.Lange, K. Nguyen and F.vercauteren. *Handbook of Elliptic and Hyperelliptic Curve Cryptography*, Chapman Hall/CRC, 2005.

[Cla02]   T. Clancy. *Analysis of FPGA-based Hyperelliptic Curve Cryptosystems*. Master's thesis, University of Illinois Urbana-Champaign, December 2002.

[EMY04]   G. Elias, A. Miri and T.H. Yeap. High-Performance, FPGA-Based Hyperelliptic Curve Cryptosystems. In *The Proceeding of the 22nd Biennial Symposium on Communications*, May 2004, Queen's University, Kingston, Ontario, Canada.

[FWW05]   X. Fan, T. Wollinger, and Y. Wang. Inversion-Free Arithmetic on Genus 3 Hyperelliptic Curves and Its Implementations. *International Conference on Information Technology: Coding and Computing - ITCC*, pp. 642-647, IEEE Computer Society, 2005.

[Gal01]   S. D. Galbraith. Supersingular Curves in Cryptography. In *Advance in Cryptograpy - Asiacrypt 2001*, LNCS 2248, Springer-Verlag, pp. 495-513, 2001.

[GH00]   P. Gaudry and R. Harley. Counting Points on Hyperelliptic Curves over Finite Fields. In *ANTS-IV*, ser. LNCS 1838, W.Bosma, Ed. Berlin: Springer-Verlag, pp. 297-312, 2000.

[GKP04]   C. Guyot, K. Kaveh, V.M. Patankar. Explicit Algorithm for The Arithmetic on The Hyperelliptic Jacobians of Genus 3. *Journal of Ramanujan Mathematical Society*, 19 (2004), No.2, 119-159.

[GLS00]   C. Günther, T. Lange, and A. Stein. Speeding up the Arithmetic on Koblitz Curves of Genus Two. In *Selected Areas in Cryptography - SAC 2000*, Volume 2012, Lecture Notes in Computer Science, pp. 106-117, Springer-Verlag, 2000.

[GMA+04]   M. Gonda, K. Matsuo, K. Aoki, J. Chao and S. Tsujii. Improvements Of Addition Algorithm On Genus 3 Hyperelliptic Curves And Their Implementations. In Proc. of SCIS 2004, Japan, 2004.

[Kob87]   N. Koblitz. Elliptic curve cryptosystems. *Mathematics of Computation*, 48:203-209, 1987.

[Kob88]   N. Koblitz. A Family of Jacobians Suitable for Discrete Log Cryptosystems. In Shafi Goldwasser, editor, *Advances in Cryptology - Crypto '88*, LNCS 403, pp. 94-99, Berlin, 1988. Springer-Verlag.

[Kob89]   N. Koblitz. Hyperelliptic Cryptosystems. In Ernest F.Brickell, editor, *Journal of Cryptology*, pp. 139-150, 1989.

[KWC+04]   H. Kim, T. Wollinger, Y. Choi, K. Chung and C. Paar. Hyperelliptic Curve Coprocessors on a FPGA, In *Workshop on Information Security Applications - WISA*, volume LNCS 3325, pp. 360-374, Springer-Verlag, 2004.

[Lan03]   T. Lange. Formulae for Arithmetic on Genus 2 Hyperelliptic Curves. *Jounal of AAECC*, Septemper 2003.

[Lan04]   T. Lange. Koblitz Curve Cryptosystems. *Finite Fields and Their Applications*, 2004. to appear.

[Loc94]   P. Lockhart. On the discriminant of a hyperelliptic curve. In *Tran. Amer. Math. Soc.* 342, 2, pp. 729-752, 1994.

[LS04]   T. Lange and M. Stevens. Efficient Doubling on Genus Two Curves over Binary Fields. In H.Handschuh and A.Hasan, editors, *Eleventh Annual Workshop on Selected Areas in Cryptography - SAC 2004*, volume LNCS 3357, pp. 170-181, Springer-Verlag, 2005.

[Mil86]   V. Miller. Uses of Elliptic Curves in Cryptography. In H. C. Williams, editor, *Advances in Cryptology - CRYPTO '85*, LNCS 218, pp. 417-426, Berlin, Germany, Springer-Verlag, 1986.

[MWZ96]   A. Menezes, Y. Wu and R. Zuccherato. An Elementary Introduction to Hyperelliptic Curve. *Technical Report CORR 96-19*, University of Waterloo, 1996, Canada. Available at http://www.cacr.math.uwaterloo.ca

[Mum84]   D. Mumford. *Tata Lectures on Theta II*, Progress in Mathematics 43,
             Birkhäuser, 1984.

[Pel02]   J. Pelzl. *Hyperelliptic Cryptosystems on Embedded Microprocessor*. Mas-
             ter's thesis, Department of Electronical Engineering and Information Sci-
             ences, Ruhr-Universitaet Bochum, Bochum, Germany, September 2002.

[PWGP03]   J. Pelzl, T. Wollinger, J. Guajardo and C. Paar. Hyperellip-
             tic Curve Cryptosystems: Closing The Performance Gap To ellip-
             tic Curve (Update), *Cryptology ePrint Archieve*, Report 2003/026,
             http://eprint.iacr.org/, 2003

[PWP03]   J. Pelzl, T. Wollinger, and C. Paar. Low Cost Security: Explicit Formulae
             for Genus-4 Hyperelliptic Curves. In M. Matsui and R. Zuccherato, edi-
             tors, *Tenth Annual Workshop on Selected Areas in Cryptography - SAC
             2003*, volume LNCS 3006, pp. 1-16, Springer-Verlag, 2003.

[RS02]   K. Rubin and A. Silverberg. Supersingular abelian varieties in cryptology.
             In *Advance in cryptology - Crypto'2002*, volume 2442 of Lecture Notes
             in Computer Science, pp. 336-353, Springer-Verlag, 2002.

[Thé03]   N.Thériault. Index calculus attack for hyperelliptic curves of small genus.
             Advances in Cryptology - ASIACRYPT'03, G.Goos, J.Hartmanis, and
             J.van Leeuwen, Eds. Berlin: Springer Verlag, 2003, pp. 79 - 92, LNCS
             2894.

[Wol04]   T. Wollinger. *Software and Hardware Implementation of Hyperelliptic
             Curve Cryptosystems*. Europäischer Universitätsverlag, 3-86515-025-X,
             2004.

[WPW+03]   T. Wollinger, J. Pelzl, V. Wittelsberger, C. Paar, G. Saldamli, and Ç.
             K. Koç. Elliptic & hyperelliptic curves on embedded μp. *ACM Transac-
             tions in Embedded Computing Systems (TECS)*, 2003. Special Issue on
             Embedded Systems and Security.

Another Look at Small RSA Exponents

M. Jason Hinek

School of Computer Science, University of Waterloo,
Waterloo, Ontario, N2L-3G1, Canada
mjhinek@alumni.uwaterloo.ca

Abstract. In this work we consider a variant of RSA whose public and private exponents can be chosen significantly smaller than in typical RSA. In particular, we show that it is possible to have private exponents smaller than $N^{1/4}$ which are resistant to all known small private exponent attacks. This allows for instances of RSA with short CRT-exponents and short public exponents. In addition, the number of bits required to store the private key information can be significantly reduced in this variant.

1 Introduction

The RSA cryptosystem [20] is currently the most widely known and widely used public key cryptosystem. One of the main drawbacks of using RSA, however, is the large computational costs for encryption and decryption. Since these costs are linear in the bitsize of the encrypting and decrypting exponents, respectively, the costs can be reduced by using smaller exponents. Of course, there are some security risks if the exponents are chosen poorly so care must be taken.

There has been a significant amount of research on the security of RSA when the encrypting or decrypting exponent is small. In 1990, Wiener [26] showed that private exponents smaller than $N^{1/4}$ are insecure as they can efficiently be recovered. In 1998, Boneh, Durfee & Frankel [5, 6] showed that instances of RSA with extremely small public exponents are vulnerable to a partial key-exposure attack in which the entire private exponent can be computed with knowledge of only 1/4 of its least significant bits. In addition, there are many other small private exponent [3, 1] and partial key-exposure attacks [2, 11] on RSA as well.

In this work we consider a variant of RSA whose public and private exponents can be chosen significantly smaller than what is normally allowed with typical RSA. This variant, which we call *common prime RSA*, consists of choosing the RSA primes p and q so that $g = \gcd(p - 1, q - 1)$ has a large prime factor and defining the RSA public and private exponents as inverses modulo $\mathrm{lcm}(p - 1, q - 1)$ instead of $\phi(N) = (p - 1)(q - 1)$. In this scenario, it is possible to have private exponents much smaller than $N^{1/4}$ which are resistant to Wiener's small private exponent attack and the other known lattice based attacks. And, using a modification of Sun & Yang's method for constructing balanced RSA exponents [24], it is also possible to have public and private exponents which are simultaneously both less than $N^{0.4}$.

D. Pointcheval (Ed.): CT-RSA 2006, LNCS 3860, pp. 82–98, 2006.

The idea of using RSA primes with $\mathrm{lcm}(p-1, q-1)$ having a large prime factor is not new. In 1990 it was suggested by Wiener [26] as a defence to his continued fraction attack on small private exponent RSA and also by Girault [13] in an identity-based identification scheme (in which the common prime factor is made public). It was also suggested by Lim & Lee [16] in 1995 to improve server-aided RSA computations but this was later shown to be insecure by McKee & Pinch [19].

The rest of this work proceeds as follows. In Section 2, we give motivation for our work and define our variant of RSA. Section 3 gives an overview of the methods and tools that will be used in Section 4 to attack our variant. We briefly conclude the paper in Section 5.

2   Common Prime RSA

2.1   Motivation

Consider an instance of RSA with public key (e, N) and private key (d, N) where the RSA modulus $N = pq$ is the product of balanced primes and the encryption/decryption exponents are chosen to satisfy $ed \equiv 1 \pmod{\mathrm{lcm}(p-1, q-1)}$. This equivalence corresponds to the equation $ed = 1 + k(p-1)(q-1)/g$, where k is some integer, $g = \gcd(p-1, q-1)$ and $\gcd(k, g) = 1^1$. In 1990, it was shown by Wiener [26–equation 29] that if

$$kdg < \frac{pq}{\frac{3}{2}(p+q)}, \tag{1}$$

then $k/(dg)$ is one of the convergents in the continued fraction expansion of e/N. With the assumption that g is small, Wiener proceeded to show that N can be factored using this information. When the RSA primes are chosen randomly the value of g will most likely be 2 or some other very small number and this is the scenario in which his work is mainly focused. When g is larger, Wiener acknowledges that his attack is less effective and concludes the discussion of this case with "*However, choosing* $\gcd(p-1, q-1)$ *to be large may cause other problems.*" It is this quote that is the original motivation of this work. We are interested in identifying what these other problems might be and if there are instances of RSA with private exponent $d < N^{1/4}$ that are safe to use. To this end, we introduce a variant of RSA called common prime RSA.

2.2   Common Prime RSA

In the original presentation of RSA [20], the public and private exponents are chosen to be inverses of each other modulo $\phi(N) = (p-1)(q-1)$. It is now common to define these exponents modulo $\mathrm{lcm}(p-1, q-1)$ where it is assumed that $g = \gcd(p-1, q-1)$ is quite small (usually 2). We will call instances of RSA

---

[1] We assume k and g are relatively prime so that they correspond to the same values of k and g in [26]. They do not necessarily possess this property in general.

in which the exponents are defined in this way *typical RSA*. We also assume that the primes are balanced (i.e., p and q have the same bit-length).

In this work we consider a variant of RSA in which the primes have a special structure. In particular, for some large prime g let $p = 2ga + 1$ and $q = 2gb + 1$ be balanced primes with the restrictions that $\gcd(a, b) = 1$ and $h = 2gab + a + b$ is also prime. The first restriction ensures that $\gcd(p - 1, q - 1) = 2g$ while the second ensures that $(pq - 1)/2 = gh$ is a semiprime roughly the same size as pg. We will call primes p and q satisfying the above properties *common primes*, since $p - 1$ and $q - 1$ share a large common prime factor. We define *common prime RSA* to be any instance of RSA that uses balanced common primes and defines the public/private exponents modulo $\operatorname{lcm}(p - 1, q - 1) = 2gab$ with the added condition that the integer k in the equation $ed = 1 + k2gab$ is relatively prime to $2g$.

For notational convenience, we define $\gamma \in \mathbb{R}$ so that $g = N^\gamma$. Since we only consider instances of RSA with balanced primes we have that $0 < \gamma < 1/2$ (i.e., $g < N^{1/2}$). In Appendix A, we present an algorithm that, given n and γ, can be used to generate balanced common primes p and q such that $\gcd(p-1, q-1)/2$ is a $\lceil \gamma n \rceil$-bit prime. One drawback of using common prime RSA, however, is that the cost of generating the primes is significantly larger than with typical RSA, as can be seen in the following table:

γ	0.3	0.325	0.35	0.375	0.4	0.425	0.45	0.475
average time (sec)	155	317	213	251	268	301	127	236
standard deviation	204	320	140	138	178	328	98	196

Here the average time needed to generate common primes for various values of γ are shown for a 1024-bit modulus. The computations were done on a Sun Fire V440 server with four UltraSPARC IIIi processors with 8 GB of memory each running at 1.062 GHz. We should be able to significantly reduce the time for common prime generation using sieving techniques when finding a and b. This will be addressed in the full version of the paper.

2.3 Small Exponents and Other Fast Variants of RSA

The cost of encryption and decryption are linear in the bit-length of the public and private exponents, respectively. When a random public or private exponent is chosen in typical RSA, both exponents will be with very high probability roughly the same size of $\phi(N) \approx N$. This follows since the exponents are defined as inverses of each other modulo $\phi(N)$ and the vast majority of possible values for the exponents will be close in size to $\phi(N)$. Since common prime RSA exponents are defined modulo $\operatorname{lcm}(p - 1, q - 1) = 2gab$, the size of e and d will each be no larger than $\operatorname{lcm}(p - 1, q - 1) \approx N^{1-\gamma}$ and so when a random public or private exponent is chosen it is expected that each will be roughly $N^{1-\gamma}$. Thus, for random exponents we expect that computations for common prime RSA will be decreased by a factor of $1/(1 - \gamma)$ when compared to typical RSA.

To reduce encryption costs for typical RSA, the public exponent can be chosen to be very small (typically $2^{16} + 1$). This results in a random looking private exponent roughly the size of $\phi(N)$. Using common prime RSA with such a short public exponent will result in a private exponent roughly the size of $N^{1-\gamma}$, thus reducing the cost for decryption by a factor of $1/(1-\gamma)$. To reduce decryption costs for typical RSA, the private exponent can be chosen to be much smaller than $\phi(N)$. This results in a random looking public exponent roughly the size of $\phi(N)$. For typical RSA, the private exponent must be larger than $N^{0.292}$, however, to avoid the small private exponent attack of Boneh & Durfee [3, 4]. As will be shown later, this restriction on the small private exponent is weakened in common prime RSA. Thus, using a small private exponent with common prime RSA allows for a reduction in encryption costs by factor of $1/(1-\gamma)$ and also allows for a reduction in decryption costs by allowing smaller (safe) private exponents.

Using a modified version of the key generation method outlined by Sun & Yang [24], it is also possible to construct public/private exponent pairs such that $ed \approx N^{1-\gamma+2\omega}$ for some security parameter ω (e.g., $N^\omega = 2^{56}$). In this scenario it possible to generate public and private exponents which are simultaneously both smaller than $N^{0.4}$. Due to page length restrictions, we leave further discussion of this topic to the full version of the paper.

Each of the above scenarios use the original decryption method for RSA (i.e., exponentiation modulo N). In practice, however, the Chinese Remainder Theorem (CRT) is often used to speed up decryption. For a given ciphertext c, the plaintext is recovered by first computing $m_p = c^{d_p} \bmod p$ and $m_q = c^{d_q} \bmod q$, where $d_p = d \bmod p - 1$ and $d_q = d \bmod q - 1$ are called the CRT-exponents, and then combining m_p and m_q together using CRT. Without parallel computations, this will decrease the time for decryption by a factor of 4 (since the size of both the exponents and the modulus are reduced by a factor of 2). Using common prime RSA with CRT for decryption for random exponents thus allows for a decrease in decryption time by a factor of 4 and also a decrease in encryption time by a factor of $1/(1-\gamma)$.

In order to further decrease the costs of typical RSA (using CRT), it is common to use a very small public exponent such as $e = 2^{16} + 1$. In this case the private exponent is essentially random and will be roughly the same size as $\phi(N)$. This significantly reduces the encryption costs while keeping the decryption costs the same as for random exponents. We will call this *small public exponent RSA*. In some situations, however, it is crucial to minimize the decryption costs. To achieve this, Wiener [26] suggested using small CRT-exponents with a large public exponent. Here e and d appear to be random exponents but the CRT-exponents are chosen small (typically 160 bits). This variant is called *rebalanced RSA-CRT* (see Boneh & Shacham [7] for details). Combining common prime RSA with rebalanced RSA then allows for a decrease in encryption costs by a factor of $1/(1-\gamma)$, while keeping the decryption costs the same.

The transfer of costs from decryption to encryption in rebalanced RSA-CRT (from small public exponent RSA) is fairly drastic. In 2005, key generation algorithms that allow for more intermediate results were independently proposed by Galbraith, Heneghan & McKee [12] and Sun & Wu [23]. Here, public exponents significantly smaller than $\phi(N)$ can be used with small CRT-exponents. There is a trade-off between encryption and decryption costs here though. Decreasing the size of one (public exponent or CRT-exponents) comes at the expense of increasing the other. Following Sun, Hinek & Wu [22], we will refer to all of these as *generalized rebalanced RSA-CRT*. In common prime RSA, when the private exponent is smaller than $N^{1/2}$ both of the CRT-exponents are equal to the private exponent (i.e., $d_p = d_q = d$). Thus, we can view common prime RSA with small private exponent ($d < N^{1/2}$) as another instance of generalized rebalanced RSA when CRT is used for decryption.

There are other fast variants of RSA which we do not consider in this work. In particular, we do not consider mutli-factor RSA (which includes multi-prime RSA and the Takagi family schemes) or variants with with unbalanced primes. For descriptions of some of these see Boneh & Shacham [7] (for multi-factor RSA) or May [17] (for unbalanced RSA with small CRT-exponents).

2.4   Wiener's Attack

Since the motivation for common prime RSA originates from Wiener's continued fraction attack we revisit his attack in the scenario of common prime RSA. The main result of his attack can be written in the following form.

Theorem 1. *Let N be a RSA modulus with balanced common primes. Given any valid common prime RSA public key (N, e) with private exponent $d < \frac{1}{2\sqrt{6}} N^{1/4-\gamma/2}$, we can factor N in polynomial time.*

This result is essentially given by Wiener [26], except for the details of factoring N when g is large. Wiener's analysis relies on the inequality $k > g$ to recover $\phi(N) = (p - 1)(q - 1)$ which can then be used to factor N. In common prime RSA, however, this inequality is not necessarily satisfied. We outline a method for arbitrary g here.

We begin with the key equation $ed = 1 + k2gab$. Since $\operatorname{lcm}(p - 1, q - 1) = \phi(N)/\gcd(p - 1, q - 1)$ we can write this equation as $ed = 1 + k\phi(N)/(2g)$. And, introducing Λ so that $\phi(N) = N - \Lambda$ we can finally write the key equation as $ed = 1 + (k/2g)(N - \Lambda)$. Since the primes are balanced, we have $\Lambda < 3N^{1/2}$. Notice that

$$\left| \frac{e}{N} - \frac{k}{2gd} \right| \leq \left| \frac{1}{dN} - \frac{k\Lambda}{2gdN} \right| \leq \frac{k\Lambda}{gdN} \leq \frac{3}{gN^{1/2}},$$

since $k\Lambda/(2g) > 1$ and $k < d$. Using a well known result from continued fractions, if $|e/N - k/(2gd)| < 1/(2(2gd)^2)$ then $k/(2gd)$ is one of the convergents in the continued fraction expansion of e/N. Notice that $d < \frac{1}{2\sqrt{6}} N^{1/4-\gamma/2}$ is a sufficient condition so that

$$\left| \frac{e}{N} - \frac{k}{2gd} \right| \leq \frac{3}{gN^{1/2}} < \frac{1}{2(2gd)^2}.$$

So, when $d < \frac{1}{2\sqrt{6}} N^{1/4-\gamma/2}$ we know that $k/(2gd)$ is one of the convergents in the continued fraction expansion of e/N. To find this convergent, we simply compute and test each convergent k'/D' of e/N. Since k, g and d are all relatively prime to each other we know that one of the convergents will satisfy $k' = k$ and $D' = gd$. Each convergent is tested by computing $\gcd(D', (N-1)/2 = gh)$. The correct convergent will yield g since $D' = gd$, $d < h$ and h is prime. We now know e, d, k and g. Substituting these values into $N - 1 = 2g(2gab + a + b)$ and the key equation $ed = 1 + k2gab$ yields two equations in two unknowns (a and b) which we can easily solve. Once a and b are known, we have factored N since $p = ga + 1$ and $q = gb + 1$. Since there are $O(\log N)$ convergents of e/N and each operation discussed above can be done in time polynomial in $\log N$, we have proven Wiener's result for common prime RSA.

If somehow the value of g is known, Wiener's attack can be slightly modified to factor N for any private exponent less than $cN^{1/4}$ for some small constant c. Thus, the advantage of having a large g is effectively removed. The modification, described in [26], simply consists of computing the convergents of $(ge)/N$ instead of e/N to find k/d.

It was also mentioned by Wiener [26], that the attack might work for private exponents that do not satisfy (1). That is, the attack might succeed when $d < N^{1/4+\rho}$ for some $\rho > 0$. In 2005 however, it was shown by Steinfeld et al. [21] that the bound is essentially tight. Thus, the probability of the attack succeeding when the private exponent does not satisfy (1) is negligible. This also holds for the common prime RSA case. Therefore, Wiener's continued fraction attack can be thwarted in common prime RSA provided that it is infeasible to obtain g and that the private exponent is chosen greater than $\frac{1}{2\sqrt{6}} N^{1/4-\gamma/2}$.

As with typical RSA, Wiener's continued fraction is only the starting point when considering the security with respect to small private exponent attacks. In particular, lattice-based attacks must be considered. In addition, other security issues must also be considered. These include factoring attacks that exploit the special structure of the primes and all security issues associated with generalized rebalanced RSA-CRT. In Section 4, we consider various attacks on common prime RSA that either factor the modulus or simply recover the private exponent.

3   Toolbox

In this section we collect some methods (tools) that we will use to attack common primes RSA. All of the methods will be used as a black-box.

3.1   Factoring

The best generic factoring method is the general number field sieve (NFS). Following Lenstra [15], we will use

$$L[n] = e^{1.923(\log n)^{1/3}(\log\log n)^{2/3}},$$

as the heuristic expected runtime of the NFS to factor a composite number n. The largest integer factored using the NFS, as of May 2005 [27], is RSA200, a 200-digit number (665 bits).

The elliptic curve method (ECM) for factoring can be substantially faster than the NFS if one of the prime factors of n is significantly smaller than $\sqrt{n}$. Again, following Lenstra [15], we use

$$E[n, p] = (\log_2 n)^2 e^{\sqrt{2}(\log p)^{1/2}(\log \log p)^{1/2}}, \tag{2}$$

as the heuristic expected runtime of the ECM to find a factor p of n. The largest factor found with the ECM, as of April 2005 [27], is 66 digits (220 bits).

3.2 Continued Fractions

When Wiener's continued fraction attacks fails there is no information gained except that the private exponent is too large for the method to work. In 1997, Verheul & van Tilborg showed that each of the convergents of e/N will yield some information about k and d regardless of the size of the private exponent. In particular, for balanced prime RSA it was shown [25–equation 16] that

$$\frac{k}{gd} = \frac{p_{j+1}U + (U\Delta + V)p_j}{q_{j+1}U + (U\Delta + V)q_j}, \tag{3}$$

where p_j/q_j is the j-th convergent of continued fraction expansion of e/N and U, V and Δ are unknowns. Their attack consists of computing the correct convergent (i.e., the correct j value) and then performing an exhaustive search on U, V and Δ. The (correct) convergent they choose is such that the sizes of U, V and Δ are minimized. As with Wiener's attack, it is assumed that g is small in their attack. Allowing for arbitrary g, their result given below becomes one of our tools.

Theorem 2. *Let N be a RSA modulus with balanced common primes. Given any valid common prime RSA public key (N, e) with private exponent d, one can factor N in time polynomial in $\log N$ and linear in $g d^2/\sqrt{N}$ $(= N^{2(\delta-(1/4-\gamma/2))})$.*

Since the proof of this theorem follows the work in [25] very closely we refer the reader to the original work for more details.

In 2004, Dujella [10] improved Verheul & van Tilborg's result by formulating new relations between the convergents which reduces the estimate on the search space for (U, V, Δ). The improvement does not change the asymptotic size of the search space though.

3.3 Lattice Basis Reduction

We use the following results that are based on Coppersmith's method for finding small solution to modular univariate polynomials and bivariate integer polynomials [8], as simplified by Howgrave-Graham [14] and Coron [9], respectively.

The first is a generalization of Coppersmith's result for univariate modular polynomials by May [18].

Theorem 3. *Let N be an integer of unknown factorization, which has a divisor $b \geq N^\beta$. Let $f_b(x)$ be a monic univariate polynomial of degree d with integer coefficients. In the limiting case of large N, all solutions x_0 of the modular equation $f_b(x) \equiv 0 \pmod{b}$ satisfying $|x_0| \leq N^{\beta^2/d}$, can be found using lattice basis reduction techniques.*

The next result is a generalization of the bivariate integer polynomial case to multivariate integer polynomials by Ernst et al. [11].

Theorem 4. *Consider the polynomial $F(x, y, z) = ax + by + cyz + d$ with integer coefficients. Given positive integers X, Y and Z let $W = \|F(xX, yY, zZ)\|_\infty$. In the limiting case of large X, Y, and Z, if*

$$X^{1+3\tau} Y^{2+3\tau} Z^{1+3\tau+3\tau^2} \leq W^{1+3\tau},$$

for some real $\tau \geq 0$, then three linearly independent polynomials with common root (x_0, y_0, z_0) satisfying $|x_0| \leq X$, $|y_0| \leq Y$, $|z_0| \leq Z$ and $F(x_0, y_0, z_0) = 0$ can be found using lattice basis reduction techniques (if such a root exists). If the three polynomials are also algebraically independent then x_0, y_0 and z_0 can be computed.

The final result that we consider is a straightforward generalization of Coppersmith's methods to linear polynomials (both modular and integer). The general result does not seem to be published, but is simple to prove for each value of m. We leave this proof for the full version of the paper.

Heuristic 1. *For some positive integer m consider a linear polynomial $f(x_1, \ldots, x_m) \in \mathbb{Z}[x_1, \ldots, x_m]$. Given m bounds $(X_1, \ldots, X_m)$ define $W = \|f(x_1 X_1, \ldots, x_m X_m)\|_\infty$ and let N be an integer of unknown factorization. In the limiting case of large X_i, N and W we have*

1. *All small roots $(y_1, \ldots, y_m)$ of $f(x_1, \ldots, x_m)$ modulo N can be found provided that $\prod_{i=1}^m X_i < N$ and $|y_i| < X_i$ for $i = 1, \ldots, m$.*
2. *All small roots $(y_1, \ldots, y_m)$ of $f(x_1, \ldots, x_m)$ over $\mathbb{Z}$ can be found provided that $\prod_{i=1}^m x_i < W$ and $|y_i| < X_i$ for $i = 1, \ldots, m$.*

The previous result is only a heuristic since there is no guarantee that the roots can be found. The lattice-based methods implicit in the result will generate m linearly independent polynomials all with the desired root over the integers (as in Theorem 4). Is it commonly assumed that these polynomials are also algebraically independent which would allow for the roots to be computed.

4 Attacking Common Prime RSA

In this section we present all the attacks on common prime RSA that we have considered. For each attack, if applicable, we will present parameters so that the expected complexity to mount the attack is roughly 2^ℓ operations for some $\ell > 0$. To obtain a desired complexity estimate, we might perform an exhaustive search to guess some bits of a certain unknown quantity and repeat an attack for each guess. Taking $\ell = 80$ approximates the expected complexity of factoring a 1024-bit RSA modulus. We take this complexity to be the base in which security is measured. For example, parameters for common prime RSA in which all attacks have (expected) complexity of at least 2^{80} will be considered safe. Generally, we omit the 80 and simply leave ℓ so that a higher level of security for future reference can easily be considered.

4.1   Factoring N with g

We consider the three cases in which g can be related to $a + b$ separately. That is, we consider $g > a + b$, $g = a + b$ and $g < a + b$ individually.

First we consider the case that $g > a + b$. Given N and g let $m = (N-1)/(2g)$ and $c = a + b$ so that $N = 2g(2gab + a + b) + 1$ can be written $m = 2gab + c$. Since $c = a + b < g$, reducing this equation modulo g gives $c = (m \bmod g)$. Substituting $b = c - a$ back into $N = 2g(2gab + a + b) + 1$ yields the following quadratic equation in the unknown variable $a : 2ga^2 - 2gca + (N-1)/(2g) - c = 0$. Thus, a is one of the solutions given by $\left(gc \pm \sqrt{2g^2c^2 + 2gc - N + 1}\right)/(2g)$. The correct a recovers $p = ga + 1$ which then gives $q = N/p$. There is no choice of $g > a + b$ that will avoid this attack if g is known.

In the unlikely event that g and $a + b$ are equal, we can compute a and b from $a + b = g$ and $ab = (N - 1)/(4g^2) - 1/2$ (obtained by rearranging the equation for N and replacing $a + b$ with g). The system of two equations in two unknowns is easily solved for a and b. The factorization of N follows since a, b and g are all known.

Lastly, we consider the case $g < a + b$. This scenario has already been considered by McKee & Pinch [19]. In particular, they present an attack which uses Shanks' baby-step giant-step method with an expected runtime $O(N^{1/4-\gamma})$. To obtain an expected complexity of at least 2^ℓ, it follows that g should be chosen so that $\gamma < 1/4 - \ell/n$.

4.2   Factoring N with g When d Is Small

We have already seen, in Section 2.4, that Wiener's continued fraction attack is easily modified to factor N when g is known and $d < N^{1/4}/(2\sqrt{6})$. We now show that N can be factored when g is known with significantly larger private exponents using lattice basis reduction techniques.

Consider g such that $1/4 - \ell/n \leq \gamma < 1/4$, for say $\ell = 80$, so that the factoring attacks of the previous subsection are no longer feasible. First notice that $M = \lceil (N - 1)/(4g^2) \rceil$ is a good (over) approximation of ab. To see this, we rearrange $N = 2g(2gab + a + b) + 1$ to obtain $(N - 1)/(4g^2) = ab + (a + b)(2g)$. Hence,

$$\left| \frac{N - 1}{4g^2} - ab \right| = \frac{a + b}{2g} \leq N^{1/2 - 2\gamma}.$$

Rounding $(N - 1)/(4g^2)$ up to the nearest integer will not change this. Next we define α to be the difference between ab and M. Thus, $\alpha = ab - M$ and

$|\alpha| \leq N^{1/2-2\gamma}$. Substituting $ab = M + \alpha$ into the key equation $ed = 1 + k(2gab)$ we obtain $ed = 1 + 2gk(M + \alpha)$, which leads to the polynomial

$$f(x, y, z) = ex - 2gMy - 2gyz - 1.$$

Notice that this polynomial has the root $(x_0, y_0, z_0) = (d, k, \alpha)$ over $\mathbb{Z}$. Defining the bounds $X = N^\delta$, $Y = N^\delta$ and $Z = N^{1/2-2\gamma}$, we see that $W = \|f(xX, yY, zZ)\|_\infty = N^{1+\delta-\gamma}$. So, applying the result of Theorem 4, we find that for large N the inequality $X^{1+3\tau}Y^{2+3\tau}Z^{1+3\tau+3\tau^2} \leq W^{1+3\tau}$ is satisfied whenever

$$\delta < \frac{(12\tau^2 + 6\tau + 2)\gamma + (-3\tau^2 + 3\tau + 1)}{2(2 + 3\tau)},$$

which becomes our sufficient condition for recovering (x_0, y_0, z_0). We are, of course, assuming that the polynomials found will be algebraically independent. Notice that bound on δ is linear in γ so that choosing $\gamma_{min} = 1/4 - \ell/n$ gives bound on δ that is sufficient for all values of $1/4 - \ell/n < \gamma < 1/4$. Some values of this bound (δ_{min}) and the corresponding value of τ used to obtain it (τ_{min}) are given in the following table using $\ell = 80$ and various common modulus sizes (n).

n	δ_{min}	τ_{min}
1024	0.452	0.7037
2048	0.524	1.1982
4096	0.581	1.9175
8192	0.618	2

4.3   Obtaining g (Factoring $N - 1$)

The last two subsections show that common prime RSA is insecure for many parameters if g is known. In this subsection, we show that obtaining g from N (and e) is difficult if g is chosen properly. The key to obtaining g is factoring the semiprime $(N - 1)/2 = gh$. Since $(N - 1)/2$ is essentially the same size as N, finding g using the NFS will be no easier than factoring N itself. Finding g with the ECM might be more fruitful, however, depending on how unbalanced g and h are. As a rough estimate, we will equate the heuristic runtime of the ECM, equation (2), with that of factoring a 1024-bit typical RSA modulus. That is, we require the expected runtime of the ECM to be at least 2^ℓ with $\ell = 80$. For a given modulus size, let γ_{ecm} be the bound for γ such that for any $\gamma < \gamma_{ecm}$, the expected runtime of the ECM is less than $2^\ell = 2^{80}$. Some values for γ_{ecm} for some common modulus sizes (n) are given in the following table.

n	γ_{ecm}	γ_{ecm}^+
1024	0.256	0.293
2048	0.128	0.146
4096	0.064	0.073
8192	0.032	0.037

All of the values of γ_{ecm} correspond a 263-bit g. Keeping in mind that the largest factor obtained by the ECM to date (as of April 2005) is a 220-bit number, this value of γ_{ecm} does not seem to offer much forward security. To try to account for this, we have also listed values of γ_{ecm}^{+} which correspond to values of g with 300 bits.

4.4   Factoring N with a and b

Notice that $N = 2g(2gab + a + b) + 1$ can also be written as $4abg^2 + 2(a + b)g - N + 1 = 0$. When a and b are known, we can simply solve this quadratic equation for g. In particular, g is given by the positive solution

$$g = \left(-(a + b) + \sqrt{a^2 - 2\,ab + b^2 + 4\,abN} \right) / (2ab).$$

It is unclear how a and b might be obtained from N (and e) other than by simply guessing. For an exhaustive search on a and b to have expected complexity at least 2^{ℓ}, it follows that g should be chosen so that $\gamma < 1/2 - \ell/(2n)$.

4.5   Factoring N

It has been shown by McKee & Pinch [19], that the special structure of the common prime RSA primes p and q lead to an efficient factoring method for N is g is large enough. Their method is a modification of Pollard's rho method. In particular, the usual random map $x \mapsto x^2 + 1 \bmod N$ is replaced with $x \mapsto x^{N-1} + 3 \bmod N$. Since $N - 1 = 2gh$ and $p - 1 = ga$ there can be at most a values of $x^{N-1} \bmod p$. Thus the expected number of steps is $O(\sqrt{a}) = O(N^{1/4 - \gamma/2})$. To obtain an expected complexity of at least 2^{ℓ}, it follows that g should be chosen so that $\gamma < 1/2 - 2\ell/n$.

4.6   Factoring N When d Is Small

In this section we consider methods that exploit the key equation when the private exponent is small. Both continued fraction and lattice-based attacks will be considered. As we shall see, the best method varies with the size of g: when g is small the continued fraction attacks work best, while lattice-based attacks work better for larger g.

Extended Continued Fraction Attack. Using the result of Theorem 2 (Verheul & van Tilborg's extension of Wiener's attack) we can estimate bounds on d so that the expected work required to factor N is 2^{ℓ} for $\ell \geq 80$. Since the complexity of the attack is linear in $N^{2(\delta - (1/4 - \gamma/2))}$, it follows that if $\delta > \frac{1}{4} + \frac{\ell}{n} - \frac{\gamma}{2}$, then the complexity is at least 2^{ℓ}.

Lattice-Based Attack. In this section we present some attacks based on Coppersmith's method for finding small solutions to univariate modular equations. In particular, we make use of Theorem 3 and Heuristic 1. As with most lattice-based attacks on RSA, we begin with the key equation. In this case it is $ed = 1 + k2gab$.

Following the ideas of May, we consider $\widehat{e} = e^{-1} \mod gh$. If the inverse does not exist we have found a factor of gh (i.e., we know g) and we can apply the results of Sections 4.1 and 4.2. So, without loss of generality, we assume that the inverse exists. Also, since $\widehat{e} = e^{-1} \mod gh$, there must exist an integer $\widehat{\alpha}$ such that $\widehat{e}e = 1 + \widehat{\alpha}gh$. Multiplying the key equation by $\widehat{e}$ and simplifying yields $d - \widehat{e} = (2\widehat{e}kab - \widehat{\alpha}dh)g$, which motivates the following univariate monic linear polynomial $f_g(x) = x - \widehat{e}$. Notice that $x_0 = d$ satisfies $f_g(x_0) \equiv 0 \pmod{g}$. Since g obviously divides $(N - 1)/2 = gh$ we can apply the results of Theorem 3, with modulus $(N - 1)/2$, $\beta = \gamma$ (as $g = N^\gamma > (N - 1)^\gamma$) and $d = 1$, to the polynomial $f_g(x)$. Thus, for large N, we can compute d provided $d < N^{\gamma^2}$.

Once d is recovered we can compute g since $\gcd(ed - 1 = k2gab, N - 1 = 2gh) = 2g$. This follows since none of a, b or k which are all smaller than h can divide h as it is prime. With d and g known, we then use lattice basis reduction techniques, again, to find k. As in Section 4.2, we write $ab = M + \alpha$ where $M = \lceil (N - 1)/(4g^2) \rfloor$ and $|\alpha| < N^{1/2 - 2\gamma}$ is unknown. Substituting this into the key equation and rearranging, we obtain $(ed - 1)/(2g) = kM + k\alpha$, where only k and α are unknown. This equation motivates the following linear bivariate polynomial $g(x, y) = Mx + y - (N - 1)/(2g)$, which has the root $(x_0, y_0) = (k, k\alpha)$. Defining the bounds $X = N^\delta$ and $Y = N^{\delta + 1/2 - 2\gamma}$, we obtain $W = \|g(xX, yY)\|_\infty = N^{1 + \delta - 2\gamma}$. Applying the results of Theorem 1, we find that a sufficient condition to recover $x_0 = k$ and $y_0 = k\alpha$ is given by $\delta < 1/2$ which is always true at this stage since we already knew that $\delta < \gamma^2 < 1/4$ when computing d. Substituting the known values of e, d, g and k into $N - 1 = 2g(2gab + a + b)$ and the key equation $ed = 1 + k2gab$ leads to two equations in two unknowns which we can easily solve. Once a and b are known, we have factored N since $p = ga + 1$ and $q = gb + 1$.

A simple partial key-exposure attack follows from the above result as well. If we write the private key as $d = d_2 2^{\ell_2} + d_1 2^{\ell_1} + d_0$ where everything but d_1 is known, then multiplying the equation

$$e(d_2 2^{\ell_2} + d_1 2^{\ell_1} + d_0) = 1 + k2gab,$$

with $(e2^{\ell_1})^{-1} \mod gh$ will result in a monic univariate linear polynomial that will yield d_1 provided $d_1 < N^{\gamma^2}$ (for large N). Of course, letting $d_2 = 0, \ell_1 = 0$ and $d_0 = 0$ recovers the small exponent case. In this partial key-exposure attack, it might not be possible to factor the modulus once the entire private exponent is recovered. When the private exponent d is smaller than $N^{1/2}$, the method described above will allow us to compute k and hence factor N. If $d > N^{1/2}$ the method is not guaranteed to work. Of course, once d has been recovered we have already broken the particular instance of RSA. We have now arrived at the following result.

Theorem 5. *Let N be a RSA modulus with common balanced primes. For any valid public key (N, e) with private exponent d, in the limit of large N, if all but $\gamma^2 \log_2 N$ contiguous bits of d are known then the rest of d can be found in time polynomial in $\log N$. If the private exponent also satisfies $d < N^{1/2}$ then N can also be factored in time polynomial in $\log N$.*

Let 2^{T_C} be the time it takes to mount the attack in Theorem 5 (essentially, the time to perform Coppersmith's method). In order to ensure that the expected runtime complexity of attack is 2^ℓ for $\ell \geq 80$ when performing an exhaustive search on the most significant bits of d (guessing d_2), it is sufficient that $\delta > \gamma^2 + (\ell - T_C)/n$. Here an exhaustive search on the most significant bits of d involves mounting the lattice-attack for each guess of d_2. As a rough estimate, we will use $T_C = 20$ which corresponds to about a million operations. This will underestimate the actual complexity of performing the lattice basis reduction for all lattice dimensions except very small ones, but this is acceptable since we are erring on the side of caution.

The second lattice-based attack we consider involves finding small solutions of the linear 4-variable polynomial $f(x, y, z, u) = e^2 x + ey - (N - 1)z + u$. This polynomial is obtained by considering the following two equations (both derived from the key equation)

$$ed = 1 + k(p - 1)\widehat{q} \quad \text{where} \quad \widehat{q} = (q - 1)/(2g), \text{ and}$$
$$ed = 1 + k(q - 1)\widehat{p} \quad \text{where} \quad \widehat{p} = (p - 1)/(2g).$$

There are many ways of multiplying these equations together, but the optimal form seems to be multiplying $(ed - 1 + k\widehat{q} = kp\widehat{q})$ with $(ed - 1 + k\widehat{p} = kq\widehat{p})$. After some rearrangement, we obtain

$$e^2 d^2 + e(k(\widehat{q} + \widehat{p}) - 2)d - k^2\widehat{p}\widehat{q}(N - 1) - k(\widehat{p} + \widehat{q}) - 1 = 0.$$

The polynomial f is obtained from this equation by letting each group of unknowns be a single variable. It is clear that the polynomial f has the root $(x_0, y_0, z_0, u_0) = (d^2, (k(\widehat{q} + \widehat{p}) - 2)d, k^2\widehat{p}\widehat{q}, -k(\widehat{p} + \widehat{q}) + 1)$. Defining the bounds $(X, Y, Z, U) = (N^{2\delta}, N^{2\delta+1/2-\gamma}, N^{2\delta+1-2\gamma}, N^{\delta+1/2-\gamma})$, we see that

$$W = \|f(xX, yY, zZ, uU)\|_\infty = N^{2+2\delta-2\gamma}.$$

So, applying the result of Heuristic 1, we see that a sufficient condition to recover the roots, assuming that the polynomials obtained are also algebraically independent, is given by $\delta < 2\gamma/5$. We can extend this to a partial key-exposure attack by guessing the ℓ high bits of $u_0 = -k(\widehat{p}+\widehat{q})+1$. Letting $\widehat{u}$ be the high bits of u_0, we define the new polynomial $f(x, y, z, u) = e^2 x + ey - (N-1)z - u + \widehat{u}$. The value of W remains unchanged, but the bound for u now becomes $U = N^{\delta+1/2-\gamma-\ell/n}$. Thus, the new sufficient condition for success becomes $\delta < 2\gamma/5 + \ell/(5n)$.

Many other polynomials were candidates for lattice-based attacks, but none resulted in a stronger attack. In the full version of the paper, we will list all the potential polynomials along with the sufficient conditions for each.

4.7   Attacks on Generalized Rebalanced RSA-CRT

When common prime RSA is used with a private exponent smaller then $N^{1/2}$, it can be viewed as an instance of generalized rebalanced RSA-CRT in which both CRT-exponents are equal to the private exponent. Therefore, we must also consider the security of generalized rebalanced RSA-CRT. Following the analysis of Sun, Hinek & Wu [22], we must ensure that each of the following inequalities is satisfied to ensure a secure instance of RSA:

$$5\delta + 2\alpha > 2 + \ell/n,$$
$$3\delta + 2\alpha > 3/2 + \ell/n \text{ or } \alpha < 1/4 - \ell/n,$$
$$6\delta + 3\alpha > 5/2 + \ell/n,$$
$$6\delta + 3\alpha > 7/3 - \ell/n - \sqrt{(2\ell/n)^2 - (4\ell/n) - (12\alpha\ell)/n + 6\alpha}/2 \text{ , and}$$
$$\delta > 2\ell/n.$$

Here α is the size of the public exponent ($e = N^\alpha$). When each of the inequalities is satisfied, any of the known attacks has expected complexity of at least 2^ℓ. For more details, see [22].

4.8   Summary

Here we summarize all of the conditions on δ and γ to ensure a secure instance of common prime RSA. If all of the inequalities given below are satisfied then the expected complexity for each of the attacks described in the previous subsections is at least 2^ℓ.

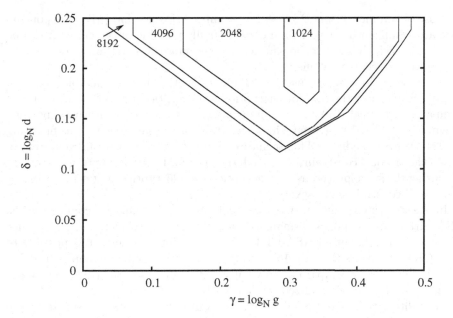

Fig. 1. Instances of common prime RSA which are safe to the attacks in this work. For each modulus size shown, instances (values of γ and δ) above the plot lines have expected attack complexities of 2^{80}. Any instances below the lines can be considered unsafe.

$$\delta > 1/4 + -\gamma/2 + \ell/n \qquad \text{(Extended Continued Fraction)}$$
$$\delta > \gamma^2 + (\ell - T_C)/n \qquad \text{(First Lattice Attack)}$$
$$\delta > 2\gamma/5 + \ell/(5n) \qquad \text{(Second Lattice Attack)}$$
$$\gamma < 1/2 - 2\ell/n \qquad \text{(Factoring } N \text{)}$$
$$\gamma > \gamma_{ecm}^+ \qquad \text{(Factoring } N - 1 \text{)}$$

When a small private exponent is used in common prime RSA along with a random looking public exponent, parameters of δ and γ satisfying each of the above inequalities will also satisfy all of the inequalities in the previous subsection for generalized rebalanced RSA-CRT. We leave the discussion of more general choices of exponents for the full version of the paper.

In Figure 1, we illustrate these inequalities using $\ell = 80$ and $T_C = 20$ for various modulus sizes. As can be seen in the figure, the region of safe parameter choices increases with increasing modulus size. This is expected of course, since the security parameter ($\ell = 80$) is fixed and corresponds to the work needed to factor a 1024-bit RSA modulus. Even though the region of safe instances for a 1024-bit modulus seems quite small, there are many actual instances, as the plot is using a logarithmic scale.

5   Conclusion

The use of common prime RSA seems promising. Public and private exponents that are significantly smaller than ones normally possible in typical RSA can be constructed which seem to be safe. Of course, more research must be done to scrutinize the security of this variant.

In the scenario in which using CRT for decryption is not available, common prime RSA reduces the cost for encryption or decryption by a factor of $1/(1-\gamma)$ compared to typical RSA for random exponents. In addition, private exponents as small as $N^{0.165}$ can be used which seem to offer the same security as factoring a 1024-bit RSA modulus. This scenario might occur in a constrained system in which the physical constraints outweigh the need for faster decryption. For example, using the same procedure for encryption and decryption might be necessary if memory constraints are severe.

In the scenario in which using CRT for decryption is available, common prime RSA with private exponents smaller than $N^{1/2}$ can be considered as an instance of generalized rebalanced RSA-CRT. For a 1024-bit modulus, it is possible to have CRT-exponents that are 169 bits ($N^{0.165}$) with a public exponent roughly $N^{0.674}$. This is very comparable to what is achievable with the other key generation algorithms for generalized rebalanced RSA-CRT (see [12, 22]).

In addition to the reduction in encryption and decryption costs, it is also possible to decrease the number of bits needed to store the private key. When the private exponent is smaller than $N^{1/2}$, the CRT-exponents are both equal to the private exponent and so only d needs to be stored. And, since p and q (and N) are easily computed from g, a and b, we can use (d, g, a, b) as a reduced private key. This reduced key contains all the necessary information for decryption, with or without using CRT, and requires $(\gamma + \delta)n$ fewer bits than storing (d_p, d_q, p, q).

Acknowledgements

The author would like thank Mark Giesbrecht, Ellen Jochemsz, Doug Stinson, Hung-Min Sun and the anonymous referees for their useful comments and suggestions, pointing out some relevant references and suggesting attacks.

References

1. J. Blömer and A. May. Low secret exponent RSA revisited. In *Cryptography and Lattices – Proceedings of CALC '01*, volume 2146 of *Lecture Notes in Computer Science*, pages 4–19. Springer-Verlag, 2001.
2. J. Blömer and A. May. New partial key exposure attacks on RSA. In *Advances in Cryptology – Proceedings of CRYPTO 2003*, volume 2729 of *Lecture Notes in Computer Science*, pages 27–43. Springer-Verlag, 2003.
3. D. Boneh and G. Durfee. Cryptanalysis of RSA with private key d less than $N^{0.292}$. In *Advances in Cryptology – Proceedings of EUROCRYPT '99*, volume 1592 of *Lecture Notes in Computer Science*, pages 1–11. Springer-Verlag, 1999.
4. D. Boneh and G. Durfee. Cryptanalysis of RSA with private key d less than $N^{0.292}$. *IEEE Transactions on Information Theory*, 46(4):1339–1349, July 2000.
5. D. Boneh, G. Durfee, and Y. Frankel. An attack on RSA given a small fraction of the private key bits. In *Advances in Cryptology – Proceedings of ASIACRYPT '98*, volume 1514 of *Lecture Notes in Computer Science*, pages 25–34. Springer-Verlag, 1998.
6. D. Boneh, G. Durfee, and Y. Frankel. Exposing an RSA private key given a small fraction of its bits. Revised and extended version of proceedings of ASIACRYPT '98 [5]. Available at http://crypto.stanford.edu/~dabo/abstracts, 2001.
7. D. Boneh and H. Shacham. Fast variants of RSA. *Cryptobytes*, 5(1):1–9, 2002.
8. D. Coppersmith. Small solutions to polynomial equations, and low exponent RSA vulnerabilities. *Journal of Cryptology*, 10(4):233–260, 1997.
9. J.-S. Coron. Finding small roots of bivariate integer polynomial equations revisited. In *Advances in Cryptology – Proceedings of EUROCRYPT 2004*, volume 3027 of *Lecture Notes in Computer Science*, pages 492–505. Springer-Verlag, 2004.
10. A. Dujella. Continued fractions and RSA with small secret exponent. *Tatra Mt. Math. Publ.*, 29:101–112, 2004.
11. M. Ernst, E. Jochemsz, A. May, and B. de Weger. Partial key exposure attacks on RSA up to full size exponents. In *Advances in Cryptology – Proceedings of EUROCRYPT 2005*, volume 3494 of *Lecture Notes in Computer Science*, pages 371–387. Springer-Verlag, 2005.
12. S. D. Galbraith, C. Heneghan, and J. F. McKee. Tunable balancing of RSA. In *Information Security and Privacy, 10th Australasian Conference, ACISP 2005*, volume 3574 of *Lecture Notes in Computer Science*, pages 280–292. Springer, 2005.
13. M. Girault. An identity-based identification scheme based on discrete logorithms modulo a composite number. In *Advances in Cryptology – Proceedings of EUROCRYPT '90*, volume 473 of *Lecture Notes in Computer Science*, pages 481–486. Springer-Verlag, 1991.
14. N. A. Howgrave-Graham. Finding small roots of univariate modular equations revisited. In *Cryptography and Coding, 6th IMA International Conference, Cirencester, UK, December 17-19, 1997, Proceedings*, volume 1355 of *Lecture Notes in Computer Science*, pages 131–142. Springer-Verlag, 1997.

15. A. K. Lenstra. Unbelievable security : Matching AES security using public key systems. In *Advances in Cryptology – Proceedings of ASIACRYPT 2001*, volume 2248 of *Lecture Notes in Computer Science*, pages 67–86. Springer-Verlag, 2001.
16. C. H. Lim and P. J. Lee. Security and performance of server-aided RSA computation protocols. In *Advances in Cryptology – Proceedings of CRYPTO '95*, volume 963 of *Lecture Notes in Computer Science*, pages 70–83. Springer-Verlag, 1995.
17. A. May. Cryptanalysis of unbalanced RSA with small CRT-exponent. In *Advances in Cryptology – Proceedings of CRYPTO 2002*, volume 2442 of *Lecture Notes in Computer Science*, pages 242–256. Springer-Verlag, 2002.
18. A. May. Secret exponent attacks on RSA-type schemes with moduli $N = p^r q$. In *Public Key Cryptograph - PKC 2004*, volume 2947 of *Lecture Notes in Computer Science*, pages 218–230. Springer-Verlag, 2004.
19. J. McKee and R. G. E. Pinch. Further attacks on server-aided RSA cryptosystems. Available at `http://citeseer.ist.psu.edu/388295.html`, 1998.
20. R. L. Rivest, A. Shamir, and L. Adleman. A method for obtaining digital signatures and public key cryptosystems. *Commun. of the ACM*, 21:120–126, 1978.
21. R. Steinfeld, S. Contini, H. Wang, and J. Pieprzyk. Converse results to the Wiener attack on RSA. In *Public Key Cryptography - PKC 2005*, volume 3386 of *Lecture Notes in Computer Science*, pages 184–198. Springer-Verlag, 2005.
22. H.-M. Sun, M. J. Hinek, and M.-E. Wu. On the design of rebalanced RSA-CRT. Technical Report CACR 2005-35, Centre for Applied Cryptographic Research, University of Waterloo, 2005. `http://www.cacr.math.uwaterloo.ca/`.
23. H.-M. Sun and M.-E. Wu. An approach towards rebalanced RSA-CRT with short public exponent. Cryptology ePrint Archive, Report 2005/053, 2005. `http://eprint.iacr.org/`.
24. H.-M. Sun and C.-T. Yang. RSA with balanced short exponents and its application to entity authentication. In *Public Key Cryptography - PKC 2005*, volume 3386 of *Lecture Notes in Computer Science*, pages 199–215. Springer-Verlag, 2005.
25. E. R. Verheul and H. C. A. van Tilborg. Cryptanalysis of 'less short' RSA secret exponents. *Appl. Algebra Eng. Commun. Comput.*, 8(5):425–435, 1997.
26. M. Wiener. Cryptanalysis of short RSA secret exponents. *IEEE Transactions on Information Theory*, 36(3):553–558, 1990.
27. P. Zimmerman. Integer factoring records. `http://www.loria.fr/~zimmerma/records/factor.html`, May 2005.

A   Algorithms

The following algorithm can be used to generate common primes.

Algorithm A.1: PRIMEGENERATION(n, γ)

$g \leftarrow 2 \times$ (a random $\lceil \gamma n \rceil$-bit prime)
repeat
    $a, b \leftarrow$ random $\lceil (\frac{1}{2} - \gamma)n - 1 \rceil$-bit positive integers
    $p \leftarrow ga + 1, \quad q \leftarrow gb + 1, \quad h \leftarrow gab + a + b$
until p, q, h are primes and $\gcd(a, b) = 1$
return p, q

Collision-Resistant Usage of MD5 and SHA-1 Via Message Preprocessing

Michael Szydlo[1] and Yiqun Lisa Yin[2]

[1] RSA Laboratories, Bedford, MA 01730
mszydlo@rsasecurity.com
[2] Independent Security Consultant
yiqun@alum.mit.edu

Abstract. A series of recent papers have demonstrated collision attacks on popularly used hash functions, including the widely deployed MD5 and SHA-1 algorithm. To assess this threat, the natural response has been to evaluate the extent to which various protocols actually depend on collision resistance for their security, and potentially schedule an upgrade to a stronger hash function. Other options involve altering the protocol in some way. This work suggests a different option. We present several simple message pre-processing techniques and show how the techniques can be combined with MD5 or SHA-1 so that applications are no longer vulnerable to the known collision attacks. For some applications, this may a viable alternative to upgrading the hash function.

Keywords: SHA-1, MD5, padding, hash collision, signature.

1 Introduction

The recent advances in cryptanalysis of hash functions have been spectacular, and the collision attacks on MD5 and SHA-1 are of particular practical importance since these algorithms are so widely deployed. To assess the threat, the first step is to re-examine which protocols actually depend on collision resistance for their security. The most common type of vulnerable application is the use of standard signatures to provide non-repudiation or certification services.

Applications which do not require collision resistance are unlikely to require changes in the near future as a result of these recent collision results. For those that do, changing the hash function is the simplest response, and the standardized SHA-2[21] family (which includes SHA-256) is the leading candidate for an upgrade. Although it has not received the same amount of analysis as earlier hash functions, SHA-256 is expected to be significantly stronger. There has been some progress analyzing SHA-256, for example [10] and [8]. These papers show that SHA-256 also has "local collisions" (defined in [5]) with probability between 2^{-9} and 2^{-39}. This implies that the security of SHA-256 is mainly hinged on its message pre-processing.

A second alternative is to re-design the protocols themselves, so they no longer rely on collision resistance of the hash function. This can be done on a case by

D. Pointcheval (Ed.): CT-RSA 2006, LNCS 3860, pp. 99–114, 2006.

case basis or in a more uniform fashion. For example, a recent Internet Draft [9] proposes to change signature scheme protocols by use of a primitive called "randomized hashing". There are some architectural advantages to changing the signature scheme in such a modular way, replacing each hash invocation with a random member of the hash family. Any protocol employing this solution will require a good source of randomness, and will also need to specify and manage the random hash family member. This kind of solution can certainly be considered viable when the additional resource of randomness is readily available.

This paper points out a third option: There are simple, alternate modes of using MD5 or SHA-1, in a manner which renders them no longer susceptible to the known collision attacks. These approaches essentially involve some light message pre-processing code to effectively *derive* a new hash function from an old one. Although the exact same standardized hash function is used, this technique can be viewed as an indirect but convenient way of effectively upgrading the hash function. Advantages include the fact that no additional resource of randomness is needed and no change to the output length or truncation is required. In the short term, some implementations might find this to be a compelling alternative which will serve to extend the useful life of MD5 or SHA-1.

Organization

In Sections 2 we present some background material on the hash functions under consideration, and in and Section 3 review the nature of the recent collision attacks. In Section 4 we further motivate and present the basic message pre-processing technique. In Section 5 we present the details of the construction. Focusing primarily on SHA-1, we analyze the security in light of the known collision attacks in Section 6. An alternate approach to message preprocessing is described in Section 8. In Section 9, we provide analysis specific to MD5. Final conclusions and recommendations are made in Section 10.

2   Background

2.1   The MDx Family of Hash Functions

The MDx family of hash functions includes MD4 (1990) and MD5 (1991), which were designed by Ron Rivest to be one-way and collision resistant. SHA-0 (1993), SHA-1, (1995) and SHA-2 (2001) were produced by the NSA and standardized by NIST and follow similar design principles as Rivest's algorithms. SHA-1 is currently the FIPS Secure hash standard [20], and is the most widely deployed hash function. An earlier version of this algorithm was SHA-0 [19], while the SHA-2 family is intended for higher security levels. Until recently, SHA-1 was considered to be as secure as its 160 bit output would allow, and MD5 also still enjoys significant deployment.

The general approach behind the design of these hash function involves the Merkle-Damgård *iterative structure*, (see [6, 15]), to allow arbitrary length messages. The algorithms divide the input into fixed length blocks and process the blocks sequentially by updating an initial state variable. Each block is combined with the previous state in a *compression function* to calculate an updated state, or *chaining variable*. When the entire message has been processed, the output is the final state. The state vectors for MD5 and SHA-1 are 128 bits and 160 bits, respectively.

Coron, et. al. [4] suggest a modification of the Merkle-Damgård chaining method. However, their work is orthogonal to ours since we focus on the compression function, except in the IV message dependent approach of Section 8.

The compression functions consist of two basic components, *message expansion* and *round operations*. The compression function of SHA-1 operates on 512-bit message blocks, and utilizes a 160 bit state variable, represented by five 32-bit words, denoted A, B, C, D, E. The block of 512 bits is expanded to 2560 bits, represented by 80 words of 32 bits. Each of these words is used to update the internal state in a *round update function*. MD5 follows a similar structure, but uses a 128 bit state variable, and has 64 rounds instead of 80.

2.2 Collision Attacks on MD5 and SHA-1

Successful cryptanalysis of these hash functions has generally focused on finding collisions, rather than on inverting the hash functions. Wang, et. al. announced real collisions for MD4, MD5, RIPEMD HAVAL-128 in 2004 and 2005 in [22, 23, 25], and also introduced *message modification techniques*. These results have been improved by Klima [13], and Naito et. al. [17], and as of writing, the complexity of locating a collision in MD4 and MD5 are approximately 2^2 and 2^{30}.

Regarding SHA-0 and SHA-1, early analysis in 1998 by Chabaud and Joux used differential methods (local collisions and disturbance vectors) to find a collision attack on SHA-0 of complexity 2^{61} [5]. Biham and Chen found near collisions on SHA-0 in complexity 2^{40} [1]. The work of Biham, Joux, and Chen included the first real collision of SHA-0 in [11, 3]. Additional work on reduced round versions of SHA-1 appeared in [2, 18, 16]. Recently Wang, Yin, and Yu described an improved attack on SHA-0 in [26], and finally, in [24] presented the first attack on the full SHA-1, where they show that finding collisions is at most of complexity 2^{69}. Improvements to these attacks were announced in [27] where the attack complexity has been reduced to 2^{63}.

2.3 From Random Collisions to Meaningful Collisions

An early critique had been the initial collisions found by researchers results have involved just a few message blocks or short binary strings, without enough structure to be considered "meaningful" collisions. However, meaningful collisions *can* be found for these hash functions, and regardless, general collision resistance is a real design goal of hash function construction. For example, Lenstra et. al. [14] have found collisions between two distinct X.509 certificates, and collisions between two properly formatted postscript documents has been exhibited in [7]. Each of these examples involved the MD5 hash function. Examining these two examples, we see that they exploit the relative freedom in the form of certificates and postscript documents, and one may still argue that collisions are likely difficult to produce among messages of a suitably restricted form.

3   Analyzing the Recent Collision Attacks

In this section, we analyze the nature of the recent collision attacks on the MDx family of the hash functions and motivate techniques that would be useful to thwart such attacks. Throughout this paper, we will use M to denote a message to be hashed. Both MD5 and SHA-1 break the message to be hashed into 512 bit blocks. When we need to refer to a single block we denote it m, and m is often partitioned into sixteen 32-bit message words denoted by $m_0, m_1, ..., m_{15}$. When referring to the expansion function internal to the compression function, we denote the expanded message block by w. In the case of SHA-1, w is partitioned into eighty 32-bit message words, denoted by $w_0, ... w_{79}$. In the case of MD5, w is expanded into 64 32-bit message words. We use C to denote the compression function, and H to denote the complete hash function.

3.1   Basic Ideas in the Collision Attacks

We first briefly review some of the basic ideas behind these attacks. Focusing on a single block, the general common strategy behind these collision attacks involves finding a message difference $\Delta(w) = w - w'$ between two expanded messages such that the probability that $C(m)$ equals $C(m')$ is higher than expected. This is possible when it can be arranged such that during the round computations of the blocks m and m' the state vectors never deviate significantly, and can be "corrected" with high enough probability.

The basic tool is the *local collision*, a series of a few rounds in which certain small differences in the expanded message words will be absorbed with reasonable probability. Due to the message expansion there will be many differing words of m and m', so these local collisions must be strung together. *Disturbance vectors* describe how the local collisions are joined. The entire sequence of differences in the state vectors is called a *differential path*. The overall success probability depends on the simultaneous satisfaction of a set of conditions for each local collision.

The structure of the various attacks consist of analysis of the local collisions, search for a low Hamming weight disturbance vector, a brute force search on input messages, and a variety of methods are used to boost the success probability, including specifying concrete conditions for the differential path, *message modification* so that some conditions always hold, and usage of two blocks to construct collisions from near collisions. We remark that the above summary most accurately describes the approaches for SHA-1, and the analysis of MD5 differs slightly.

3.2   Thwarting the Collision Attacks

From the summary of the attacks above we can see that there are several strategies which one might employ to attempt to prevent the success of these approaches. The most obvious approach is to attempt to prevent the existence of any "good" differential – a differential path that leads to (near) collisions and holds with probability greater than $2^{-n/2}$. An additional precaution would be to restrain the power of the message modification techniques, thereby significantly reducing the success probability of the attack. A third possibility is to consider situations in which the Merkle-Damgård iterative structure can not be exploited; for example if single message bits were to affect multiple blocks.

4   Message Pre-processing Techniques

In this section we describe the general message preprocessing framework, and discuss the *streaming* requirement that some applications may have.

4.1   Message Pre-processing Framework

The working assumption behind the general techniques we suggest for improving the collision resistance is that the underlying hash function itself will not be changed. Let M be a message string to be hashed, and let H be a standard hash function, such as MD5, or SHA-1. Our objective is to define a *derived* hash function H^* which calls H as a subroutine. Our proposal is simply to preprocess the message before it is hashed in a standard way. Formally, let $\phi : M \mapsto M^*$ be a *preprocessing* function mapping strings to strings. For each such function, a derived hash function H^* may be defined by

$$H^*(M) = H(\phi(M)).$$

Of course, we are interested in cases where ϕ is a relatively simple function, and the derived hash function H^* is collision resistant with respect to known attacks, even if H is not. The function ϕ must be chosen appropriately for a particular H to ensure that H^* is secure.

4.2   Streaming Data Requirement

Many applications are set up architecturally to incrementally digest a large message as it becomes available. For example, with SHA-1, applications can repeatedly make a *SHA-1Update* function call as portions of the message stream in. This requirement can be satisfied when the message pre-processing can also be performed in a streaming fashion, for example, by dividing the message into blocks and expanding each one. Formally, we call a ϕ a *local expansion* if ϕ can be defined by $\phi(m_0, m_1, \ldots m_k) = m_0^*, m_1^*, \ldots m_k^*$ where each m_i is of fixed length and $m_i^* = f(m_i)$ for some expansion function $f : \{0,1\}^l \to \{0,1\}^{l^*}$, where $l^* > l$. It is clear that when ϕ is a local expansion, the state of the preprocessing function can be stored in the message digest context, so that a derived update function could also call *SHA-1Update* as a subroutine.

5   Local Expansion Approaches

We now discuss two local expansion approaches to message preprocessing: message whitening and message interleaving.

5.1   Message Whitening

In this approach, the basic idea is to alter the message by inserting fixed characters at regular intervals. The motivation here is to decrease the flexibility in finding good message differentials. These fixed characters can be taken to be words filled with all zero bits, so we call the approach *whitening*. For a hash function with at 512-bit block size, sequential chunks of fewer than 512 bits can be expanded into a full 512 bits. For example each sequence of $(16 - t)$ 32-bit words $m = (m_0, m_1, \ldots m_{15-t})$ could be expanded to $m = (m_0, m_1, \ldots m_{15-t}, 0, \ldots, 0)$, where the last t words would be fixed as zeros. Each execution of the compression function effectively only process $(16 - t)$ message words rather than 16 message words, so it is easy to calculate the performance slowdown. This approach is also easy to implement, since such a preprocessing function ϕ is a local expansion, the streaming requirement would be met. From a security standpoint, the intuition is that processing fewer bits of message should allow the message to be better mixed within the calculation.

A variant of this approach may select specific words to whiten to further increase the difficulty of known attacks. Below, we discuss how whitening the middle two words of SHA-1 significantly reduces the effect of message modification techniques.

5.2   Message Self Interleaving

In this approach, the basic idea is to duplicate each message word so that each bit appears twice after the preprocessing. Assuming the entire message M is broken up into some number of 32-bit words: $M = (m_0, m_1, \ldots m_k)$, then the preprocessed message would be $\phi(m) = (m_0, m_0, m_1, m_1, \ldots m_k, m_k)$ where each word appears twice. As with the message whitening approach, message interleaving causes fewer message bits to be fed into each message block, causing better mixing. As ϕ is a local expansion, the streaming requirement is also met.

5.3   Generalized Local Expansion

The whitening and interleaving approaches discussed above have obvious minor variations, such as choice of which bits to whiten. The frequency of message interleaving could also be chosen word by word, rather than character by character. Both of these approaches, as well as the minor variants have the property that the local expansion is a linear function. Thus, one way to generalize is to consider an arbitrary linear function. Although we prefer simpler pre-processing functions, one could certainly consider non-linear functions as well, effectively using an arbitrary local expansion. Regardless of the specific function, these approaches all attempt to increase security by increasing the structure of each message block. This can make finding good differentials more difficult for the attacker, as well as disrupt message modification techniques.

6   Security Analysis of Local Expansion Approaches

In this section, we further discuss why the message pre-processing techniques described in the preceding section help prevent existing attacks. We focus our discussions on SHA-1.

6.1   Intuition

The message whitening and message interleaving both operate by increasing the structure within each block. For these approaches and their variants, we can simply view the derived hash function as a modification of the original hash function, except with a different message expansion rule.

Concretely, in the case of SHA-1, the message interleaving approach effectively takes as input 256 bits of data instead of 512, and expands them to the 80 words required by the SHA-1 round operations. The amount of data required by the whitening approach would depend on its calibration, i.e., how many bits or words were whitened. Intuitively, this means that fewer data bits are processed for each execution of the compression function, and hence the derived hash function can offer a better mixing of the data bits.

In the following, we provide more quantitative analysis of the two message pre-processing approaches by considering how they affect constructing good differentials and performing message modification, both of which are critical in existing collision attacks.

6.2   Insights from Coding Theory

One way to understand the effect of message whitening or message interleaving is to study the *code of expanded message words*. For hash functions which employ a linear message expansion rule the space of expanded messages is a linear code, so we have a tool to reason about the existence of low Hamming weight vectors. For example, for SHA-1 each block expands 16×32 bits into 80×32 bits. The expansion function $E\{0,1\}^{512} \rightarrow \{0,1\}^{2560}$ is defined word-wise by the recurrence relation

$$w_t = (w_{t-3} \oplus w_{t-8} \oplus w_{t-14} \oplus w_{t-16}) <<< 1. \tag{1}$$

For MD5, the original message is simply repeated 3 times, so the expanded message words of both MD5 and SHA1 can be viewed as linear codes of dimension 512. In either case, the code is generated by the 512 basis vectors $E(1,0,\ldots,0)$, $E(0,1,\ldots,0)$, ..., $E(0,\ldots,0,1)$.

Both the interleaving and whitening approaches work by restricting the form of the 512-bit input message block, thus restricting full code of expanded message words. The form of the whitened message is $m^* = (m_0, m_1, \ldots m_{15-t}, 0, \ldots 0)$ so the restricted code is $512 - 32t$ dimensional, generated by basis vectors corresponding to the non-whitened bits. The form of an interleaved message is $m^* = (m_0, m_0, m_1, m_1, \ldots, m_7, m_7)$, so this code is only 256 dimensional, generated by vectors of the form $(1, 0, \ldots; 1, 0, \ldots; 0; \ldots)$, where each generator consists of zeros except for two matching 1 bits. When we view the collision attacks as attempts to piece together local collisions in a manner consistent with this linear code, it becomes clear that reducing the dimension of the code will make these attacks less feasible.

Reducing Solutions to Linearized Hash Function: Another way to understand the whitening and message interleaving is in terms of the set of solutions to a linearized version of the hash function. This is the approach followed by Oswald and Rijmen in [18]. Rather than focus on local collisions, they analyze the difference between the linearized and actual SHA-1, so that each difference in the expanded message word yields one or more conditions which will be only probabilistically satisfied in the actual SHA-1. They search for low Hamming weight code words $\Delta(w)$ which are also solutions to the linearized SHA-1 equation. The solutions yielding an output of 160 zeros are defined by an additional 160 linear constraints (see [18] for details), so it is natural to consider the *code of linear solutions*, consisting of expanded message words which also satisfy these 160 constraints. This restricted code has dimension 512-160=352, and the collision attack first seeks a low Hamming weight code words, then a message pair such that the conditions will be satisfied.

In this framework, our message interleaving approach corresponds to the addition of 256 additional constraints, and the whitening approach corresponds to the addition of $32t$ additional constraints. Although there is no simple way to locate low Hamming weight codewords in an arbitrary code, the existence and number of lower weight words decreases as the *minimum relative distance* goes up. This ratio is simply the ratio of the code length to the code dimension, and equals $352/2560 = 7.27$ for the original code. This code, restricted with whitening parameter t, has dimension $352 - 32t$, so in case $t = 4$, the minimum relative distance is increase to 8.88. If, instead, the code is restricted by the interleaving approach, the dimension is reduced to $352 - 256 = 96$, so the minimum relative distance increases to 26.66. This heuristic does not preclude the existence of good differentials, but it does provide a useful metric for how restricting the form of messages will increase the difficulty of the known collision attacks.

6.3   Preventing Good Differentials

We now address more concretely the best known attacks on SHA-1. As discussed earlier, a major step for constructing a good differential path for SHA-1 is to find a disturbance vector with low Hamming weight. In this section, we consider how message pre-processing affects constructing good differentials.

First, we review some basic facts of the SHA-1 disturbance vectors. A disturbance vector dv is a set of 80 32-bit words dv_i ($i = 0, ..., 79$), and $dv_{i,j} = 1$ iff a local collision starts in step i bit j. Each local collision consists of 5 additional changes in the expanded message word, called *correction vectors* (See [5]). The correction vectors wc_1, wc_2, wc_3, wc_4, wc_5 are automatically linearly determined from dv, and the difference in the expanded messages is simply the sum $\Delta(w) = dv + \Sigma wc_i$ (mod 2). Although, only $\Delta(w)$ must be a code word (i.e. satisfy the recurrence relation), in practical attacks dv itself is taken to be a code word, so that the five $\{wc_i\}$ and $\Delta(w)$ are automatically code words. The

Hamming weight of dv, denoted by $HW(dv)$ is the central important factor in determining the success of the collision attacks, an estimate of the complexity of an attack on SHA-1 is about $2^{3HW(dv)}$.[1]

In the attacks on SHA-1 [24, 27], disturbance vectors of low Hamming weight were found by a heuristic search algorithm, and it is based on the following intuition: If we view dv as an 80-by-32 0-1 matrix, then the non-zero entries in a low Hamming dv are likely to be concentrated in one column. The search algorithm proceeds by first choosing a 16-bit column in the matrix and expanding backwards and forwards with message expansion. The best vector is then chosen among all possible choices for the column. Using this heuristic search, the lowest Hamming weight is reached when the 16-bit column takes the value $L = (100...000)$.

Now we are ready to analyze how the two pre-processing techniques affect finding low Hamming weight disturbance vectors. For the message whitening technique, each whitened message word m_i^* would yield an extra condition on the differential path, namely

$$\Delta m_i^* = 0. \tag{2}$$

For the interleaving techniques, the extra conditions on Δm^* are

$$\Delta m_{2i}^* = \Delta m_{2i+1}^*, \text{ for } i = 0, 1, ..., 7. \tag{3}$$

Experiments: We used the same heuristic search algorithm to find disturbance vectors for "SHA-1 with message pre-processing". Our assumption is that a good disturbance vector follows similar patterns as the ones for the original SHA-1. Starting with L, we computed 150 words of dv by expanding L forwards and backwards with the recurrence relation E and compute many words of Δw from dv. The words of this extended dv may be found in the rows of Table 5 in reference [24]. The next step is to pick 80 words from the computed Δw such as the above conditions due to whitening (or interleaving) are satisfied while keeping the Hamming weight as small as possible. Using the numbering of [24], and focusing on whitening with $t = 2$, we examined the values of Δw, and see that there are no two consecutive zero words before step 55, and there are no two consecutive words that are the same before step 53. This means that we have to shift down by 40 words when choosing a good disturbance vector, in order for Δw to satisfy the message pre-processing conditions. This would cause a significant increase in the Hamming weight of the vector, so these experimental results suggest that the Hamming weight of the disturbance vector (restricted to steps 21-80) would go from 25 (for SHA-1) to over 80. Even if advanced message modification such as that announced in [27] progresses to 32 steps, the hamming weight for the remaining 48 steps would be sufficient.

We remark that the conditions on Δm_i^* given in the above two equations are *necessary* conditions for the differential path to be constructed, since they are

---

[1] It was the introduction of message modification techniques in [24] that allowed the initial conditions in steps 1-20 to be automatically satisfied so that the limiting factor was actually the Hamming weight in the final 60 words of dv.

derived from the pair of input message words m_i and m_i'. This is in contrast to the three conditions on the disturbance vectors in the original attack on SHA-0 [5] as well several works on SHA-1 [3, 24]. Those conditions are for easier construction of a valid path from the disturbance vector, and so they are *not necessary* conditions. That's why these conditions can be removed as in the attack on the full SHA-1 [24]. However, the above conditions, due to message pre-processing, cannot be removed. Finally, the techniques of Jutla and Patthak [12] could be adapted to provide rigorous bounds on the hamming weights of the codes associated to whitened or interleaved message blocks.

6.4   Weakening Message Modification

In addition to preventing good differentials, the whitening and interleaving approaches also render the message modification techniques less effective, thereby increasing the complexity of existing collision attacks.

First, we briefly review the basics of the message modification techniques [25, 23]. For the MD4-family of hash functions, including MD5 and SHA-1, the round function has the following general form:

$$a_i = G(\text{input chaining variables}) + m_{i-1},$$

where a_i is the output chaining variable and m_{i-1} is the message word used in step i. Once the differential path has been constructed, it is easy to derive a set of sufficient conditions on a_i that ensure that all conditions on path hold. The conditions are of the form $a_{i,j} = v$, where v is 0 or 1. The main idea of the message modification techniques is simply to set $a_{i,j}$ to the correct bit v and then recompute $m_{i-1} = a_i - G()$. In other words, we can modify the message word in step i to make the condition on a_i to hold. This basic technique can be used for the first 16 steps since the message words are all independent of each other up this point. A simple variation of the basic technique is to modify the message words used in the two steps *before* step i (i.e., m_{i-2} or m_{i-3}) to achieve the same goal. This is particular useful when m_{i-1} cannot be modified due to other constraints. In addition, more advanced techniques, called multi-step message modification techniques, were introduced for dealing with computation beyond the first 16 steps. The improvements announced in [27] are achieved with such advanced message modification techniques.

Next, we analyze how message interleaving affects the effectiveness of message modification. Since $m_{2i}^* = m_{2i+1}^*$ (for $i = 0, 1, ..., 7$), the two consecutive message words have to be modified simultaneously, making it almost impossible to change any single bit. Now suppose a differential path P has already been chosen, and conditions on a_i have been determined. Since most of these conditions can no longer be made to hold through message modification, the complexity of the attack using path P would go up significantly.

In the case of whitening, the t whitened message words $m_i^*, m_{i+1}^*, ... m_{i+t-1}^*$ cannot be modified, since these message words are simply zero and independent of the input message. It is possible to modify a couple of message words immediately before the whitening step so that some of the conditions on a_i and a_{i+1} can still hold, but the effect can be weakened if we choose $t \geq 4$.

Targeted Whitening: For a given path P, it is good to choose the t consecutive message words that maximize the total number of conditions s in those steps. In the attack on SHA-1 [24], the conditions on a_i are given in Table 12. From the table, it is easy to see that if $t = 4$, and we whiten words 7 to 10, the total number of conditions is $s = 83$.

It is possible that the attacker could select a new differential path P' other than what was used in existing attacks on SHA-1 and MD5, and he could try to minimize the number of conditions associated with P' in the specified whitening steps. However, such an approach would likely not be very effective for the following reason: One special feature of the differential paths in existing attacks is that they are "front-loading" (with a lot of conditions in the first 20 steps) in order to minimize the number of conditions after step 20, which is directly related to the complexity of the attack. Hence, if the attacker selects P' that has fewer conditions in the first 20 steps, then it is very likely that P' would have more conditions later. This observation applies even more strongly when considering the improved attacks of [27] which extend message modification to additional steps.

7   Implementation Consideration

In this section, we consider practical implementation issues related to the message pre-processing proposal. For ease of discussion, we refer to the derived new hash function as SHApp, where "pp" stands for pre-processing[2]. We will consider issues related to programming implementation of SHApp as well as upper layer protocols that call SHApp as subroutines.

7.1   Programming Implementation

We propose two possible implementation options for SHApp. They vary only in terms of where pre-processing occurs in the code, and they are suitable for different applications.

Option 1: Pre-processing within SHA-1 Function. For most existing implementation of SHA-1, the hash computation on a given input is generally carried out by three functional calls as described below.[3]

```
SHAInit(context)
SHAUpdate(context, input, inputLen)
SHAFinal(digest, context)
```

---

[2] A more accurate name would be SHA1pp, but we omit the "1" so that it can be pronounced as "shap."

[3] The naming for the functions may vary slightly among implementations. For example, SHAUpdate may be called SHAadd etc. Despite this name variation, the functions accomplish essentially the same thing: the first one initializes the IV; the second one does proper padding and the main loop; the third one finalizes the computation and writes output.

We can implement the new hash function SHApp with the same sequence of functional calls as follows:

```
SHAppInit(context)
   // same as SHAInit
SHAppUpdate(context, input, inputLen)
{
    newInput = SHAppPreProcess (input)
    newInputLen = Length (newInput)
    SHAUpdate(context, newInput, newInputLen)
}
SHAppFinal(digest, context)
   // same as SHAFinal
```

Note that `SHAppUpdate` has exactly the *same* i/o interface as the original `SHAUpdate` in existing implementation. The pre-processing step is done as a private function that is invisible to upper layer protocols using SHApp. Due to the simplicity of whitening and interleaving, only a small amount of code is needed for implementing the `SHAppPreProcess` function.

Option 2: Pre-processing outside SHA-1 Function. For some applications, implementation of SHA-1 may be hard-coded, and hence it can be difficult to make internal changes to the code as described in option 1. In this case, pre-processing can be done entirely prior to calling the function SHA-1 as below.

```
SHApp(message)
{
    newMessage = SHAppPreProcess (message)
    SHA-1(newMessage)
}
```

Note that the original implementation of SHA-1 is used as a "black box" without changing anything inside. Again, there is no impact on the interface.

Interoperability. We remark that for both options, the result of the hash computation is the same for the same message. There is no interoperability issue between the two options. Hence implementers can simply choose the option that best suits their applications.

7.2 Protocols

From the discussions on programming implementation, we can see that SHApp have exactly the same input and output interface as the original SHA-1. Hence, replacing SHA-1 with SHApp in a protocol would not cause any upper layer changes other than replacing the *Algorithm Identifier*.

Newer digital signature schemes (e.g., RSA-PSS) have a "hierarchical" identifier, where the hash function is a parameter. For those schemes, the algorithm identifier for SHApp is sufficient.

For various older digital signature schemes, a new algorithm identifier is needed for both SHApp itself as well as the combination of SHApp with the specific signature scheme. The relevant standards organizations need to take care of the assignment for combinations of DSA, ECDSA, etc. For example, RSA Security can assign identifiers for SHApp and its combination with PKCS #1 v1.5. Depending on the standards, it may take little time or some amount of time for such assignments.

8   IV Message Dependence Approaches

In this section we describe a completely different approach does not involve a local expansion, but instead works by effectively ensuring that the initialization vector (IV) is message dependent.

8.1   Message Duplication

One way to cause the IV to be message dependent is to concatenate the message with itself before hashing. To simplify the explanation, we suggest first padding M so that it is a whole number of blocks. With this assumption, the pre-processing is simply $\phi(M) = M||M$, where $||$ denotes string concatenation. Let us examine the calculation halfway through, just after all the blocks of the first M have been processed. Notice that the full original message M is left to be processed, except that the intermediate IV chaining variable is a function of the message itself. This illustrates that an equivalent way to view this construction is as a regular hash of M where the starting IV chaining variable is a function of the message itself rather than constant.

8.2   Security Analysis

The IV message dependence approach increases security in a way completely different than the local expansion approach. Instead of affecting the blockwise compression function, they rely on the fact that the entire message must be processed twice within the framework of the Merkle-Damgård iterative chaining. Since each message bit is input to separate blocks, the previous attack strategies simply can not be applied. Instead, attacks on this variant would have to be of a completely different sort, and would not be able to focus on a single compression function, or on a few adjacent message blocks. This, or any other variant of the IV message dependent approach would also present an additional obstacle to automatically constructing collisions on long messages from single block collisions.

This approach is interesting because it is an extremely simple way of thwarting the known collision attacks for MD5 and SHA-1. However, a disadvantage with this approach is that the preprocessing function ϕ is not a local expansion, so it can not be effectively used with streaming data.

One might also consider alternate methods of achieving IV message dependence, for example by setting the initial starting IV value to be the first 160 bits of $H(M)$. However, this would not be not a "pure" preprocessing technique, and would require accessing the internals of the hash function itself, to set the IV value.

9   Analysis for MD5

SHA-1 was designed based on MD4 and MD5, and hence MD5 and SHA-1 are quite similar in terms of their structure and choices of mathematical operations. Consequently, the latest collision attacks on MD5 [25] and SHA-1 [24] also share some similarities. Therefore, most of the security analysis in preceding section also directly applies to MD5, including the general insight from coding theory, effects on message modification, and the IV-message dependency.

Here we point out some differences between the two hash functions and how they would affect the analysis. The main difference lies in the message expansion. For MD5, each message block expands 16×32 bits into 64×32 bits. The expansion function E operates by repeating and *re-ordering* the 16 message words 3 times. So the MD5 message expansion is much simpler than SHA-1, and hence offers less mixing.

The differential path P used in the latest attack on MD5 is also different from SHA-1, other than they are both "front-loading", and the MD5 analysis does not make use of an explicit disturbance vector. In the recent attack on MD5, the path was constructed by first finding a near collision that only involves the MSB in the second half and then deriving a more complicated collision path in the first half. For the chosen path, Δm_i is non-zero in steps 5, 12, and 15.

Message interleaving would result in 6 of the Δm_i to be non-zero, which would make the particular path P invalid. More importantly, interleaving would make message modification almost impossible. Note that there are over 200 conditions associated with P in the first 16 steps, and *all* these conditions need to be set true through message modification in order to reduce the complexity of the attack to about 2^{30+}. Therefore, message interleaving can significantly increase the complexity of existing attacks.

A similar argument can be carried out for message whitening, although a higher parameter of t would be required to rule out the availability of low Hamming weight Δm_i vectors. In this case, it seems more difficult to have a rigorous argument that the attacker cannot find a completely new path that would effectively target the particular whitening techniques.

10   Conclusions

In this paper we have considered several techniques to use SHA-1 and MD5 in a more collision resistant manner. The simplest approach which we have discussed in this paper is the message whitening approach. The word-wise message interleaving is also quite simple, and has very similar security properties. These approaches are both easy to implement, support streaming message digesting, and are amenable to analysis with respect to the known differential attacks.

The IV message dependent approaches are appealing due to their immunity to single-block collision attack approaches, but have the drawback that they are not convenient for message streaming.

For practical applications wishing to improve SHA-1 use, we suggest the use of message whitening pre-processing with parameter $t \geq 4$, so that 12 words of the message are expanded into 16. This results in a performance slowdown of 25 percent. An even more secure alternative would be the message interleaving, although it results in a slowdown of 50 percent. For MD5, our recommendation is to use the message interleaving approach, or in case the application does deal with small data items (such as certificates), the IV message dependence approach.

Our solutions can be viewed as a general purpose, safer, collision resistant way of using MD5, and SHA-1. Due to their simplicity, we contend that such an approach can be appealing for practitioners who wish to increase security in the short term, without changing the underlying hash function at all.

Relationship to Hash Function Design: The solution in this paper is not intended to be a complete replacement for an appropriate, timely hash function update nor for improved hash function design. On the other hand, our proposal has something in common with proposals for enhancing the security of SHA-like hash designs such as [12] in that we also focus on the code of expanded message words.

Future Improvements on Collision Attacks: With respect to the attacks of [24, 27], both whitenening with parameter $t \geq 4$ and message interleaving techniques still yield a derived hash function for which collisions can not be found with effort below 2^{80}. Although it is impossible to predict the improvements in collision attacks we make a few comments on the robustness of our techniques. In general, the message pre-processing we propose makes will apply to other attacks of the same genre, because: (1) message modification in general is much harder, (2) multi-step message modification techniques are almost impossible, and (3) the constraints on Δm are targeted at preventing effective "front-loading" of differential paths.

Acknowledgments

The authors would like to thank Scott Contini, Russ Housley, Burt Kaliski, Jim Randall, Ron Rivest, Moti Yung, and the anonymous reviewers for helpful comments. Special thank to Paul Hoffman for discussions on implementation issues.

References

1. E. Biham and R. Chen. *Near Collisions of SHA-0*. In Advances in Cryptology – Crypto'04 , Springer-Verlag, August 2004.
2. E. Biham and R. Chen. *New Results on SHA-0 and SHA-1*. In Crypto'04 Rump Session, August 2004.
3. E. Biham, R. Chen, A. Joux, P. Carribault, W. Jalby and C. Lemuet. *Collisions in SHA-0 and Reduced SHA-1*. In Advances in Cryptology – Eurocrypt'05 , Springer-Verlag, May 2005.

4. J. Coron, Y. Dodis, C. Malinaud, and P Puniya *Merkle-Damgrd Revisited : How to Construct a Hash Function* In Advances in Cryptology – Crypto'05, Springer-Verlag, 2005.
5. F. Chabaud and A. Joux. *Differential Collisions in SHA-0*. In Advances in Cryptology – Crypto'98, Springer-Verlag, August 1998.
6. I. Damgård. *A Design Principle for Hash Functions*, In Advances in Cryptology – Crypto'89, Springer-Verlag, 1990.
7. M. Daum and S. Lucks. *The Story of Alice and her Boss* In Rump session of Eurocrypt'05. http://www.cits.rub.de/MD5Collisions/.
8. H. Handschuh and H. Gilbert *Security Analysis of SHA-256 and Sisters*. Proceedings of the Workshop on Selected Areas in Cryptography - SAC'03, Springer-Verlag, 2003.
9. S. Halevi and H. Krawczyk *Strengthening Digital Signatures via Randomized Hashing*, Internet-Draft, May 12, 2005. http://www.ietf.org/internet-drafts/draft-irtf-cfrg-rhash-00.txt.
10. P. Hawkes and M. Paddon and G. Rose. *On Corrective Patterns for the SHA-2 Family*. http://eprint.iacr.org/2004/207
11. A. Joux. *Collisions for SHA-0*. In Rump session of Crypto'04, August 2004.
12. C. Jutla and A. Patthak *A Simple and Provably Good Code for SHA Message Expansion*, IACR Eprint archive, Report 2005/247, http://eprint.iacr.org/2005/247.
13. V. Klima: *Finding MD5 Collisions on a Notebook PC Using Multi-message Modifications*, IACR Eprint archive, Report 2005/102, http://eprint.iacr.org/2005/102.
14. A. Lenstra and X. Wang and B. de Weger. *Colliding X.509 Certificates*, IACR Eprint archive, Report 2005/067. http://eprint.iacr.org/.
15. R. Merkle. *One Way hash Functions and DES*, In Advances in Cryptology – Crypto'89, Springer-Verlag, 1990.
16. K. Matusiewicz and J. Pieprzyk. *Finding Good Differential Patterns for Attacks on SHA-1*. IACR Eprint archive, December 2004.
17. Y. Naito and Y. Sasaki and N. Kunihiro and K. Ohta. *Improved Collision Attack on MD4* IACR Eprint archive, Report 2005/151.
18. V. Rijmen and E. Oswald. *Update on SHA-1*. In Topics in Cryptology – CT-RSA 2005, Springer-Verlag, 2005.
19. NIST. *Secure hash standard*. Federal Information Processing Standard, FIPS 180, May 1993.
20. NIST. *Secure hash standard*. Federal Information Processing Standard, FIPS 180-1, April 1995.
21. NIST. *Secure hash standard*. Federal Information Processing Standard, FIPS 180-2, August 2002.
22. X. Wang, F. Guo, X. Lai, and H. Yu. *Collisions for Hash Functions MD4, MD5, HAVAL-128 and RIPEMD*. In Rump session of Crypto'04 and IACR Eprint archive, August 2004.
23. X. Wang, X. Lai, F. Guo, H. Chen, X. Yu. *Cryptanalysis for Hash Functions MD4 and RIPEMD*. In Advances in Cryptology – Eurocrypt'05, Springer-Verlag, May 2005.
24. X. Wang and Y.L. Yin and H. Yu. *Finding Collisions in the full SHA-1*. In Advances in Cryptology – Crypto'05, Springer-Verlag, 2005.
25. X. Wang and H. Yu. *How to Break MD5 and Other Hash Functions*. In Advances in Cryptology – Eurocrypt'05, Springer-Verlag, May 2005.
26. X. Wang and H. Yu and Y.L. Yin. *Efficient Collision Search Attacks on SHA-0*. In Advances in Cryptology – Crypto'05, Springer-Verlag, 2005.
27. X. Wang, A. Yao, and F. Yao, *New Collision search for SHA-1*, Rump Session Crypto'05.

RFID-Tags for Anti-counterfeiting*

Pim Tuyls[1] and Lejla Batina[2]

[1] Philips Research Laboratories,
Prof. Holstlaan 4, 5656 AA, Eindhoven, The Netherlands
[2] Katholieke Universiteit Leuven, ESAT/COSIC,
Kasteelpark Arenberg 10, B-3001 Leuven-Heverlee, Belgium
{Pim.Tuyls, Lejla.Batina}@esat.kuleuven.ac.be,
Pim.Tuyls@philips.com

Abstract. RFID-tags are becoming very popular tools for identification of products. As they have a small microchip on board, they offer functionality that can be used for security purposes. This chip functionality makes it possible to verify the authenticity of a product and hence to detect and prevent counterfeiting. In order to be successful for these security purposes too, RFID-tags have to be resistant against many attacks, in particular against cloning of the tag. In this paper, we investigate how an RFID-tag can be made unclonable by linking it inseparably to a Physical Unclonable Function (PUF). We present the security protocols that are needed for the detection of the authenticity of a product when it is equipped with such a system. We focus on off-line authentication because it is very attractive from a practical point of view. We show that a PUF based solution for RFID-tags is feasible in the off-line case.

Keywords: RFID, counterfeiting, authentication, ECC, Physical Unclonable Function (PUF).

1 Introduction

RFID-tags are low-cost pervasive devices that target to provide identification of goods. They consist of an antenna connected to a microchip. Because of the presence of this microchip, they can be considered as a next generation of bar codes with added functionality. In supply chain management they allow for tracking of a product in several stages and locations. Several applications are being developed that can process the data obtained from the tags for their own purposes, such as automated inventory management, automated quality control, access control, payment systems and general security applications. Clearly, one of the main success factors for a large deployment of RFID-tag based systems is the price of the tags. Currently the prices range from a few cents up to 1\$. Very cheap tags do not carry a battery but obtain their power from the electromagnetic field generated by the reader querying the tag.

* Lejla Batina is funded by a research grant of the Katholieke Universiteit Leuven, Belgium. This work was supported by Concerted Research Actions GOA-Mefisto 2000/06 and GOA-Ambiorix 2005/11 of the Flemish Government and by the FWO projects (G.0141.03) and (G.0450.04).

An emerging application that goes beyond identification, is the use of RFID-tags for anti-counterfeiting purposes [1]. By locating an RFID-tag with specific product and reference information on a product, one aims to verify the authenticity of the product. Loosely speaking the verification is performed as follows. When a product passes a reader, the reader checks whether the necessary and authentic product and reference information is present on the tag. For this purpose it runs a protocol with the tag. If the necessary information is there and verified to be authentic, the product is declared to be genuine and otherwise not. However, by capturing the necessary authentication information (obtained *e.g.* by eavesdropping the protocol between the tag and the reader), and by storing it in a new chip, the attacker has effectively made a clone of the original tag that cannot be distinguished from an original tag by a reader. In order to make this cloning of the tag infeasible, it should not be possible to derive the tag secrets by active or passive attacks. Recently a lightweight version of such a protocol was developed in [1].

We stress however that it is rather easy to physically clone a tag. This means that an attacker can capture the RFID-tag, investigate it, read out its memory (with reasonable effort) and in particular its security related data (identification number, reference information, keys, etc). Then she produces a new tag with exactly the same data in its memory. When this tag is embedded into a product, it is impossible for a reader to distinguish an authentic product from a fake one. In order to protect an RFID-tag against this type of cloning attack, one can of course attempt to prevent read out its memory by using several protective measures [23, 18]. However these measures will increase the price of the tag so much that it will become unacceptably high for its main application. In order to thwart the physical cloning attacks we propose to use Physical Unclonable structures (so-called PUFs) for storing secret key material in the tag. PUFs have been proposed as a cost-effective mean to produce unclonable tokens for identification [20, 21]. They are realized as a physical system such that the function is easy to evaluate but hard to clone.

The contributions of this paper are:

1. We identify the technological components of anti-counterfeiting technology and give a general protocol for verifying the authenticity of an item.
2. We propose a solution for anti-counterfeiting based on RFID-tags and PUFs [25, 22, 19, 20]. Our solution withstands physical cloning attacks as well as active and passive attacks on the verification protocols. In particular, we present a solution based on PUFs that are inseparably bound to an IC.
3. We present protocols for the off-line situation (as far as we are aware this is the first time that the off-line case has been considered). Our construction for the off-line case is designed in such a way that it inherits its security from the underlying cryptographic algorithms (signature and secure identification scheme) used.
4. We show that the construction that we propose is feasible on a constrained device such as an RFID-tag. In order to minimize the area constraints of a tag, we sacrifice slightly the efficiency of the involved cryptographic algorithms. The obtained performance is still sufficient for our application.

The paper is organized as follows. In Sect. 2 we identify the required technological components for anti-counterfeiting technology. Additionally, a general protocol for the verification of the authenticity of a product is given. Section 3 mentions related work. An overview of PUFs and associated key-extraction algorithms is given in Sect. 4. In Sect. 5 we introduce unclonable RFID-tags with an Integrated PUF on board. Furthermore, we present verification protocols for off-line authentication. Finally, in Sect. 6 we investigate the efficiency of the off-line verification protocol in detail.

2 Model

In order to protect a product against cloning (counterfeiting) a detection mark is embedded into the product or its packaging. This detection mark consists of a physical and a digital part. The mark is put there by a legitimate authority. The attacker (counterfeiter) has access to all components of this detection mark; *i.e.* she can read it, remove it from the product and investigate it. Based on the information that she obtained from investigating the legal detection mark, she produces a fake detection mark. The goal of the attacker is to produce a fake detection mark that can only with small probability be distinguished from an authentic one.

2.1 Components of Anti-counterfeiting Technology

In order to protect a product against counterfeiting, technological means are needed to verify whether the product is authentic or not. In order to make an item unclonable, the following two components are needed.

1. *Physical protection.* This is obtained by using unclonable physical structures embedded in the package (removal of the structure leads to its destruction). One or more unique *fingerprints* derived from the physical structure will be printed on the product for the verification of the authenticity of the product.
2. *Cryptographic protection* serving two goals. Firstly, cryptography provides techniques (digital signatures) to detect and prevent tampering with data (fingerprints) derived from a physical object. Secondly, it provides secure identification protocols to identify a product. Those protocols do not leak any necessary identification information to an eavesdropper attacking (actively or passively) the communication channel.

Good candidates for unclonable physical structures, that can be used for physical protection purposes, are so-called Physical Unclonable Functions (PUFs) [25].

2.2 A General Anti-counterfeiting Protocol

We give intuition for protocols that can be used to check the authenticity of a product based on embedding a PUF in the product in combination with the use of cryptographic techniques.

First there is an enrollment phase, which is performed by some trusted authority. During this phase the following steps are performed.

1. Several fingerprints are derived from the PUF by challenging it with multiple challenges and recording the responses. These responses are then turned into binary fingerprints (and some auxiliary data are derived for use during the verification phase).
2. These challenges, fingerprints and auxiliary data are then signed with the secret key sk of the issuer of the product (the issuer is assumed to be trustworthy).
3. The signatures, the challenges (corresponding to the fingerprints) and maybe some auxiliary data (needed to perform processing during the authentication phase) are also printed on the product (and/or stored in a database).

During the verification phase, the authenticity is checked by running the following protocol.

1. The verification device reads the challenges and auxiliary data.
2. The verification device challenges the physical structure with one of the challenges printed on the product. After having measured the responses, it derives the fingerprint from the response based on the auxiliary data.
3. Then, using the fingerprint derived in step 2., the verification device checks the signature to verify that the fingerprint, challenges and auxiliary data were printed on the product by a legitimate authority. If the signature is not correct, the product is not authentic.

We briefly analyze the security of this protocol. An attacker who wants to counterfeit the product has to embed a fake physical structure on the product that produces correct fingerprints to the challenges (with correct signatures). Under the assumption that the physical structure is unclonable, she cannot produce a clone of the originally embedded physical structure. More precisely, we assume that given some challenges $c_1, \ldots, c_n$ and corresponding fingerprints $s_1, \ldots, s_n$ she cannot produce a (fake) physical structure that produces the same fingerprints $s_1, \ldots, s_n$ given the original challenges $c_1, \ldots, c_n$. On the other hand she can produce another structure and create challenges, auxiliary data and fingerprints $s'_1, \ldots, s'_n$ according to the procedures used during enrollment. However, since she does not know the secret key sk and the responses of her fake structure will be different with very high probability, she will not be able to put the correct signatures on these data. The verification device will detect that the signatures are not correct and reject this as a fake product.

We note that the number of fingerprints that can be verified during a verification session is very limited by time and space constraints. Furthermore, the attacker can easily capture the required fingerprints (by measuring the responses according to the challenges printed on the product). Therefore the production of a clone only requires the fabrication of a physical structure (PUF) producing the same fingerprints for a limited number of challenges.

2.3   RFID Systems

The PUF based solution for preventing counterfeiting of goods that was presented above can be improved with active components, that are inseparably linked with a PUF. An example consists of an RFID-tag equipped with a microchip that is inseparably bound to a PUF. The precise construction is explained in Sect. 4. Because of the presence of a microchip a secure identification protocol can be run without revealing any information on the fingerprint of the PUF. Additionally, by inseparably linking the chip and the PUF, it becomes possible to prevent leakage of the PUF measurement to the outside world.

Typical RFID systems consist of the following two components: the *RFID-tag* and a *reader*. The reader will perform the verification to detect whether a tag is authentic or not. The RFID-tag consists of an antenna connected to a microchip that can store and read data and has possibly some dedicated hardware to perform a small amount of computations. Typically, the power for performing operations is obtained from the RF-field (by inductive coupling). A reader can read and write data from/on a tag. The reader is often linked with some system that can perform computations on the data that it receives from tags.

In order to use RFID-tags for anti-counterfeiting purposes, we proceed as follows. An RFID-tag containing reference information is embedded in a product. The (identification) data stored in the memory of the tag is signed with the secret key sk of the legitimate issuer. The tag communicates with a reader for verification purposes over a public channel. The ROM memory of the tag is accessible to the attacker. The reader has a certified public key pk corresponding the issuer's secret key for verification of the digital signatures.

3   Related Work

The two most related papers to ours are [1] and [12]. Both deal with the cloning problem of RFID-tags and hence with the problem of using RFID-tags for anti-counterfeiting purposes. The focus of these papers is on efficient protocols for authenticating these tags. In these papers, one focuses on authentication of RFID-tags in the on-line situation; *i.e.* the reader shares a secret with the RFID-tag that is being authenticated. Clearly, when RFID-tags will become widely used, this is not a reasonable assumption.

4   Physical Unclonable Functions

For the sake of clarity we start with a definition of a PUF [9].

Definition 1. *A Physical Unclonable Function is a function that maps challenges to responses and that is embodied in a physical object. It satisfies the following properties:*

1. *Easy to evaluate: the physical object can be evaluated in a short amount of time.*
2. *Hard to characterize: from a number of measurements performed in polynomial time, an attacker who no longer has the device and who only has a limited (polynomial) amount of resources can only obtain a negligible amount of knowledge about the response to a challenge that is chosen uniformly at random.*

More formally the PUF model is as follows. We denote the PUF response to a challenge C during the enrollment phase by $X \in \mathbb{R}^n$ and during the verification phase by $Y \in \mathbb{R}^n$ (the pair (C, X) is called a Challenge-Response pair or CRP). The PUF response according to a fake PUF is denoted by Z. The responses X, Y, Z are modeled as random variables with probability distribution $\mathbb{P}_{X,Y,Z}$.

Definition 2. *Let $\delta, \epsilon_a, \epsilon_e \geq 0$. A joint distribution $\mathbb{P}_{X,Y,Z}$ on $(\mathbb{R}^n)^3$ is called $(\delta, \epsilon_a, \epsilon_e)$-reliable if it satisfies i) $Prob(d(Y, X) > \delta) \leq \epsilon_a$ and ii) $Prob(d(Z, X) \leq \delta) \leq \epsilon_e$; here the probabilities are over the joint distribution $\mathbb{P}_{X,Y,Z}$.*

This definition implies that if the enrollment and authentication measurements (according to the same challenge C) are performed on the same PUF, then these responses are with high probability very close to each other. When on the other hand the measurements are performed on different PUFs (modeling the fact that the PUF used during authentication might be fake), the responses are with high probability far apart.

We propose to equip the microchip on an RFID-tag with a PUF that is inseparably linked to the chip. More precisely we define this as follows.

Definition 3. *An Integrated Physical Unclonable Function (I-PUF) is a PUF that additionally satisfies the following properties.*

1. *The I-PUF is inseparably bound to a chip which means that any attempt to remove the PUF from the chip leads to the destruction of the PUF and the chip.*
2. *It is impossible to tamper with the communication (measurement data) between the chip and the PUF.*
3. *The output of the PUF is inaccessible to an attacker.*

In the remainder of the paper we will only use I-PUFs, while we will often use just the abbreviation PUF.

The two best known examples of such I-PUFs are silicon PUFs [8] and coating PUFs [19]. For coating PUFs it is expected that the additional measurement circuit requires less than 1000 gates.

4.1 Key Extraction

In this paper, the term key extraction always refers to key extraction from noisy data. Generally speaking a key extraction algorithm is built on a Secret Extraction Code [24][1]. For the sake of simplicity we describe the algorithm in terms of a *shielding function* [14] or (G, W)-*pair* [26], which generates a special set of Secret Extraction Codes, while having all the necessary properties.

A function $G(.,.) : \mathbb{R}^n \times \mathcal{W} \to \{0, 1\}^k$ is called δ-contracting if for all X there exists *helper data* $W \in \mathcal{W}$ such that for all X' that lie within a sphere of radius

---

[1] This construction can be applied to discrete and continuous data. An equivalent construction for the discrete case, called Fuzzy Extractors, was developed by Dodis *et al.* in [7].

δ of X ($\|X' - X\| \leq \delta$) $G(X', W) = G(X, W)$ ($\mathcal{W}$ denotes the space of helper data. At this point it has to be considered as some abstract space.). We use δ-contracting functions to extract keys $S = G(X, W)$ from noisy data X using *helper data W*. A function $G(.,.)$ is called ϵ-revealing if the helper data W leaks less than ϵ bits on S (in the information theoretic sense), *i.e.* $\mathbf{I}(W; S) \leq \epsilon$. An (ϵ, δ)-shielding function $G : \mathbb{R}^n \times \mathcal{W} \rightarrow \{0, 1\}^k$ is a function that is δ-contracting and ϵ-revealing. It is used to extract a secret of length k from the PUF response as follows.

- **Enrollment Phase:** The PUF is subjected to a challenge C and the response X is measured. Then a random key S is chosen from $\{0, 1\}^k$ and helper data W is computed by solving $G(X, W) = S$ for W. The quadruplet $(\text{ID}_{\text{PUF}}, C, W, S)$ is then stored in a CRP database.
- **Verification Phase:** When the PUF is inserted into the reader the PUF's identity is sent to the verifier. The verifier chooses a random challenge C from his database and sends it to the PUF together with the corresponding helper data W. Then the PUF is subjected to the challenge C and its response X' is measured. A key S' is then computed as $S' = G(X', W)$.

Notice that if $G(.,.)$ is δ-robust and if $\mathbb{P}_{X,Y,Z}$ is $(\delta, \epsilon_a, \epsilon_e)$-reliable, then we obtain $\text{Prob}(G(Y, W) = S) \geq 1 - \epsilon_a$ and $\text{Prob}(G(Z, W) = \bot) \geq 1 - \epsilon_e$, which expresses that FRR (False Rejection Rate) and FAR (False Acceptance Rate) are at most ϵ_a and ϵ_e respectively. In the case of a passive attacker, the extracted key S can then be used securely since $\mathbf{I}(W; S) \leq \epsilon$. Note that by adding a privacy amplification this can be guaranteed (if the Réniy entropy is sufficiently large). Also note that this procedure can be used to set up a shared secret key between an I-PUF and a verifier (reader).

Since the PUF responses are often analog data[2], the helper data typically consists of three parts. The first part W_1 allows to quantise the signal into a binary representation while the second part W_2 implements the error correction and the random key choice on the binary data. The third part is used for privacy amplification. For a detailed example, we refer the reader to [22] for the case of optical PUFs.

4.2   Example

We present a brief example of key extraction from noisy (binary) data. It shows that the required processing at the side of the RFID-tag is low. Assume for the sake of simplicity that the responses X are uniformly random binary strings of length k, *i.e.* $X \in \{0, 1\}^k$. Furthermore, we assume that the authentication measurement performed during the verification phase can be modeled as a noisy observation over a binary symmetric channel with cross-over probability p. Let $\mathcal{C}$ be an error correcting code, with l codewords. Then, for a key $s \in_R \{0, \ldots, l - 1\}$ the helper

---

[2] In the case of an optical PUF the PUF response is a speckle pattern which can be seen as an analog picture. In the case of a coating PUF the responses are given by capacitance values which are analog signals.

data $w(x, s) = x \oplus c_s$ is generated during the enrollment phase (where $c_s \in C$). During the verification phase, the tag measures y and computes $G(y, w(x; s)) = \mathsf{Dec}(y \oplus w(x; s))$ (Dec denotes the decoding algorithm of the error-correcting code C). Clearly, if y corresponds to the same challenge (and the same PUF), s is obtained after decoding while otherwise a random code-word is obtained or a decoding error. Hence, the tag has to perform an XOR operation and a decoding operation. On a tag with some S-RAM (Static RAM) available (which most tags have), the decoding costs less than 1000 gates[3].

5   Unclonable RFID-Tags

5.1   Set-Up

In order to make unclonable RFID-tags, we introduce RFID-tags whose microchips are equipped with an I-PUF.

In our construction, the PUF is used as a secure memory for storing secret keys. The secret key s which is usually stored in (protected) ROM or EEPROM is derived from the PUF, when needed. In order to enable the generation of the secret key s during authentication, helper data w is stored in (publicly accessible) ROM (EEPROM). The key s is derived from the response X of the PUF by means of a key extraction algorithm (Fuzzy Extractor and the helper data w are used here). It was mentioned in Sect. 4.1 that the public helper data w reveals only a negligible amount of information on the key s. Given our assumption on I-PUFs in Def. 3, it follows that the key s is securely stored in the PUF.

5.2   Off-Line Authentication

We introduce our PUF-Certificate-Identity-based Identification scheme (PUF-Cert-IBI) by following the definition of Certificate-based IBI in [3]. Let $SI = (K_g, P, V)$ denote a standard identification scheme (SI-scheme) where K_g denotes the key generation algorithm, and P, V denote the interactive protocols run by the prover and verifier respectively. Let $SS = (\mathrm{SK}_g, \mathrm{Sign}, V_f)$ be a standard signature scheme (SS-scheme) [6] with SK_g denoting the key generation algorithm, Sign denoting the signing algorithm and V_f the verification algorithm run by a verifier. We assign to each tag an identity I (this might be the serial number or EPC-code of the tag or the serial number of the product in which it has been embedded). To the PUF, the SI, the SS scheme and the identity I an Identity-Based Identification scheme $(\mathrm{MK}_g, \mathrm{UK}_g, \hat{P}, \hat{V})$ is associated as follows.

During **enrollment** the issuer uses SK_g as the master-key generation algorithm MK_g. This means that the master key msk is used for generating signatures and the corresponding public key mpk for verification of the signatures. The user key generation algorithm UK_g consists of the following steps. For each RFID-tag, having identity I, the issuer then creates a public-secret key pair

---

[3] In the case of coating PUFs the codewords are relatively short (200 bits) and the information rate is high. In that case BCH codes are efficient in use.

(pk, sk) using the algorithm K_g on input 1^k. The couple (pk, sk) is the public-secret key pair for the SI-scheme. The issuer runs the following protocol with the tag.

- It requests the tag to challenge its PUF with a challenge c and to measure the response $x(c)$.
- The tag sends $x(c)$ to the issuer.
- Based on the knowledge of $x(c)$ and sk, the issuer determines the helper data w such that $sk = G(x, w)$.
- The helper data w are written into the ROM (EEPROM) memory of the tag.

Finally, the issuer creates the following certificate that is also stored in the ROM of the tag Cert $\leftarrow (pk, \mathrm{Sign}(msk, pk\|I))$. The usk is then put to $usk \leftarrow$ (PUF, Cert).

During **authentication**, the tag (in the role of the prover) runs the following steps with a verifier.

- The tag runs the protocol $\hat{P}$ which consists of the following steps.
  - It challenges the PUF with c, measures the response $y(c)$ and computes $sk \leftarrow G(y(c), w)$.
  - Initialisation of the prover protocol P of the $\mathcal{SI}$ scheme with sk.
  - It includes the certificate Cert in the first step of the algorithm P.
- The verifier uses (mpk, I) as input for the verification algorithm V.
- When the verifier receives Cert from the tag, it first verifies Cert by running $V_f(mpk, pk\|I, \mathrm{Sign}(msk, pk\|I))$.
- If the certificate Cert is invalid the protocol is aborted.
- If Cert is valid, the verifier initializes V with pk and runs it.
- If V accepts, then the verifier accepts.

The security of our PUF-Certificate-Identity-based identification scheme follows from the following theorem. This theorem is very similar to theorem 4.2 in [3]. The proof of the theorem presented there, can be applied here with minor modifications and is therefore omitted.

Theorem 1. *Let $\mathcal{SI}$ be an SI-scheme and SS a uf-cma [4] secure SS-scheme. Let PUF-Cert-IBI be the corresponding PUF-Certificate-Identity based Identification scheme presented above. If the scheme $\mathcal{SI}$ is impersonation-atk secure then PUF-Cert-IBI is impersonation-atk secure for atk $\in \{pa, aa, ca\}$ (pa: passive attack, aa: active attack, ca: concurrent attack).*

It follows from this theorem, that by choosing an appropriate SI-scheme (withstanding a *pa*, *aa* or *ca*) the PUF-Cert-IBI inherits the same property. If only resistance against passive attacks is needed, the Schnorr Identification scheme can be used. It is known that this scheme is secure against passive attacks under the discrete logarithm assumption. It is also secure against active attacks under the one-more-discrete-logarithm assumption. An alternative is to use Okamoto's identification scheme [16], which is secure against passive, active and concurrent attacks under the discrete logarithm assumption.

---

[4] *uf-cma*: existential unforgeability under chosen message attack.

5.3   Storage Requirements

In order to minimize the size of the ROM memory of the tag as small as possible, we propose to use Elliptic Curve Discrete Log based secure identification schemes. This makes an implementation on an Elliptic Curve (EC) possible. For the signature algorithm $\mathcal{SS}$ we take then the ECDSA approach. This makes the size of the signatures no larger than 326 bits. The identification protocol investigated in detail is the Schnorr identification protocol. For the sake of completeness the ECC version of the protocol is given in Appendix. The total storage requirement for the public information (sP, Cert) is in total at most 500 bits.

6   Implementation

In this section, we discuss implementation issues, *i.e.* efficiency and size of the hardware if the off-line RFID identification protocol is implemented on an RFID-tag. As an example we take the Schnorr identification protocol, which allows a user to prove knowledge of x given the public information g^x in a group where the discrete log problem is difficult. For the sake of efficiency, we investigate the efficiency of this protocol on an elliptic curve over $\text{GF}(2^{163})$.

6.1   Elliptic Curves over $\text{GF}(2^n)$

Elliptic Curve Cryptography (ECC) relies on a group structure induced on an elliptic curve. The set of points on an elliptic curve (with one special point added, the so-called point at infinity $\mathcal{O}$) together with point addition as a binary operation has the structure of an abelian group. Here we consider a finite field of characteristic 2, *i.e.* $\text{GF}(2^n)$. A non-supersingular elliptic curve E over $\text{GF}(2^n)$ is defined as the set of solutions $(x, y) \in \text{GF}(2^n) \times \text{GF}(2^n)$ of the equation:

$$y^2 + xy = x^3 + ax^2 + b, \tag{1}$$

where $a, b \in \text{GF}(2^n)$, $b \neq 0$, together with $\mathcal{O}$.

The point or scalar multiplication is the basic operation for cryptographic protocols based on ECDLP; it is easily performed via repeated group operations. One can visualize these operations in a hierarchical structure. Point multiplication is at the top level. At the next (lower) level are the point operations, which are closely related to the coordinates used to represent the points. The lowest level consists of finite field operations such as addition, subtraction, multiplication and inversion required to perform the group operations.

The easiest way to calculate the point or scalar multiplication is by means of the basic double-and-add algorithm [16].

The point addition in affine coordinates is performed according to the formulae in [5]. In either case, the computation requires one field inversion (I), two field multiplications (M) and one squaring (S), or $1I + 2M + 1S$. As we are interested in hardware implementations, we count squarings and multiplications together as they are both executed on the same multiplier.

The inversion operation is very costly in hardware and can be avoided by choosing one of many options for projective coordinates. However, the number of multiplications is increased in this case, which makes the choice of a multiplier even more crucial for an efficient implementation. To summarize, we consider squaring as a special case of multiplication and inversion is ignored. The addition of two field elements requires the modulo 2 addition of the coefficients of the elements. In hardware, a bit-parallel adder requires n XOR gates and the sum can be computed in one clock cycle.

Another option for scalar or point multiplication is to use the so-called "Montgomery ladder"[11]. According to López and Dahab [15], the Montgomery representation requires less memory and offers a better protection against side-channel attacks. These both facts are very useful in this case as memory $i.e.$ registers are very "expensive" in hardware implementations. Also, side-channel attacks are an issue on RFID tags and also some cheap protection $i.e.$ by means of balanced implementations is desirable.

The idea of Montgomery dealt with speeding up the calculation of only the x-coordinate of the result. More precisely, to add two points their difference is used as an input parameter while the y-coordinate is not used in the algorithm. This fact is justified by cryptographic applications that rarely use the y-coordinate. The algorithm for scalar multiplication is a variant of the binary method and was considered by López and Dahab [15]. They have also introduced an option for recovering the y-coordinate.

We introduce the following notation: $P_4 = (x_4, y_4) = P_2 - P_1$, $P_5 = (x_5, y_5) = 2P_1$ and $P_3 = P_1 + P_2$. The point P_4 is included because the method for point multiplication, as introduced by Montgomery, is defined by the fact that to add two points their difference should be known (while y-coordinate is not needed).

For point operations (addition and doubling) we consider the formulae of López and Dahab in $GF(2^n)$. The operation count is $A : D = 5M : 6M$ (2). Here, A and D are the point operations and M is a field multiplication. We remind the reader that field addition in hardware for $GF(2^n)$ is just a simple bit-wise XOR operation and therefore is not taken into account. We use the formulae for point operations in the case of simple projective coordinates $i.e.$ $x_i = (X_i/Z_i), i = 1, 2$. The results of point doubling and point addition, $i.e.$ $X_5 = X(P_5)$ and $X_3 = X(P_3) = X(P_1 + P_2)$ respectively, are calculated as:

$$X_5 = X_1{}^4 + b \cdot Z_1{}^4$$
$$Z_5 = X_1{}^2 \cdot Z_1{}^2 . \tag{2}$$

$$X_3 = x_4 \cdot Z_3 + (X_1 \cdot Z_2) \cdot (Z_1 \cdot X_2),$$
$$Z_3 = (X_1 \cdot Z_2 + X_2 \cdot Z_1)^2 .$$

6.2 ECC Operations

In this section we describe ECC operations at each level by following the top-down approach.

Point Multiplication: For the point multiplication we chose the method of Montgomery that maintains the relationship $P_2 - P_1$ as invariant [17]. It uses a representation where computations are performed on the x-coordinate only.

Point Addition and Doubling: We start from Eqs. (2), but the goal is to save some registers, as it is known that this part is usually the largest portion of the total area. As the previous formulae require 3 intermediate registers (2 for addition and 1 for doubling) [15], we eliminate 2 intermediate registers by introducing a few additional steps (cf. Algorithm 1). Therefore, we get the sequences of operations that require only one intermediate variable (T). Moreover, this value is manipulated only twice for addition and it could be even stored in some RAM. In this way we made a trade-off between speed and area as point operations require now 7 and 8 multiplications for addition and doubling (instead of 5 and 6 M respectively). Furthermore, point operation can be also easily balanced to achieve some simple side-channel protection such as in [2].

Algorithm 1. EC point addition and doubling

Require: $X_1, Z_1, X_2, Z_2, x_4 = x(P_2 - P_1)$
Ensure: $X(P_1 + P_2) = X(P_3) = X_3, Z_3$
1. $Z_3 \leftarrow X_2 \cdot Z_1$
2. $X_3 \leftarrow X_1 \cdot Z_2$
3. $Z_3 \leftarrow X_3 + Z_3$
4. $Z_3 \leftarrow Z_3{}^2$
5. $X_3 \leftarrow X_1 \cdot Z_2$
6. $X_3 \leftarrow X_3 \cdot X_2$
7. $X_3 \leftarrow X_3 \cdot Z_1$
8. $T \leftarrow x_4 \cdot Z_3$
9. $X_3 \leftarrow X_3 + T$

Require: $b \in \mathrm{GF}(2^n), X_1, Z_1$
Ensure: $X(2P_1) = X(P_5) = X_5, Z_5$
1. $Z_5 \leftarrow Z_1{}^2$
2. $Z_5 \leftarrow Z_5{}^2$
3. $Z_5 \leftarrow b \cdot Z_5$
4. $X_5 \leftarrow X_1{}^2$
5. $X_5 \leftarrow X_5{}^2$
6. $X_5 \leftarrow X_5 + Z_5$
7. $Z_5 \leftarrow X_1{}^2$
8. $Z_5 \leftarrow Z_5 \cdot Z_1$
9. $Z_5 \leftarrow Z_5 \cdot Z_1$

An Algorithm for Field Multiplication: The standard way to compute the product $c(x) = a(x) \cdot b(x) \bmod f(x)$ is the one that uses convolution and to which we refer to as the classical algorithm [4].

The most compact architecture for this multiplication is the classical bit-serial multiplier (the MSB or the LSB multiplier) [4].

6.3 A Prototype Elliptic Curve Processor

The Elliptic Curve Processor (ECP) is shown in Fig. 1. The operation blocks are as follows:

- Control Unit(CU)
- Arithmetic Unit (ALU)
- Registers
- Memory: RAM

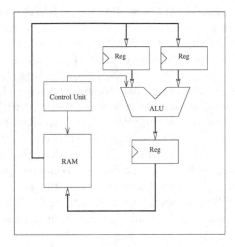

Fig. 1. Architecture of the elliptic curve processor

The Control Unit takes care of scalar multiplication, point operations and all conversions to suitable representation. It also commands the ALU which performs field multiplication, addition and inversion.

The largest part of the ALU is finite field multiplier, which is the MSB bit-serial multiplier [4]. The inversion operation is also performed by the multiplier using Fermat's theorem.

6.4   Estimated Results

Here we estimate the performance of the ECC processor for the field GF(2^{163}). The irreducible polynomial is the pentanomial $f(x) = x^{163} + x^7 + x^6 + x^3 + 1$. One point multiplication takes $163 \cdot 15M = 2445M$. Conversion of coordinates $A \to P$ and $P \to A$ takes respectively $2M$ and $I + 2M$. Assuming that inversion is done by means of Fermat the total for conversion is around $300M$. This all together results in approximately $3000M$. One field multiplication (M) takes 163 cycles, which results in 489000 cycles for point multiplication. With a clock frequency of even $1MHz$ one point multiplication would take less than half a second, which is reasonably fast.

The estimated area complexity for the bit-serial multiplier is around $16n$, so for $n = 163$ we get around 2.6 kgates. Modular addition takes 163 XOR gates, so it sums up to around 3 kgates. The complexity of the FSMs used is hard to estimate, but as those are only some control logic it should not be too large. However, the registers that are required might take quite large area as 1FF is at least 6 NANDs. This is the most crucial aspect of the design. However, as 3 registers are absolutely necessary for ALU, we believe that this hardware component can be of the order of 5 kgates, depending on technology. We assume that EC parameters as well as other pre-calculated input values can be stored in memory blocks. It may further slow-down the performance but there is certainly enough margin for that according to the RFID specifications [1]. This also follows from the fact that the operating frequency for RFID tags is actually $13.56MHz$ according to the ISO 18000-3 standard while our estimates were made assuming the operating frequency of $1MHz$. Another option to minimize hardware complexity would be to decrease the field size. Namely, 163 bit long key sizes correspond to RSA keys that are much longer than 1024 bits [13]. More precisely, one could achieve that level of security with around 130 bits long ECC keys. Consequently, scaling down ECC parameters would result in a roughly linear decrease of hardware complexity. The fact that ECC is a suitable technology for RFIDs was also concluded in the work of Wolkerstorfer [27]. That work is the first complete ECC low-power and compact implementation that meets the constraints imposed by the EPC standard. Yet, our solution can be even smaller as our off-line authentication do not require full ECDSA algorithm to be executed on a single tag. That allows for further optimization with respect to area.

7   Concluding Remarks

In this paper we have shown that by equipping RFID-tags with I-PUFs, the tags become unclonable and hence suitable for anti-counterfeiting purposes. Using our protocols, both the physical cloning attack as well as the cloning attack based on (actively or passively) attacking the protocol between the tag and the reader can be prevented. It has been shown that the required protocols are feasible on an RFID-tag in the off-line situation.

Acknowledgement. The authors thank Gregory Neven for his comments on an earlier version on this manuscript and for the nice and constructive discussions on this topic.

References

1. S. A. Weis A. Juels. Authenticating pervasive devices with human protocols. In V. Shoup, editor, *Advances in Cryptology: Proceedings of CRYPTO 2005*, volume 3621 of *Lecture Notes in Computer Science*, pages 293–308. Springer-Verlag, 2005.
2. L. Batina, N. Mentens, B. Preneel, and I. Verbauwhede. Side-channel aware design: Algorithms and architectures for elliptic curve cryptography over $GF(2^n)$. In *Proceedings of the IEEE International Conference on Application-Specific Systems, Architectures, and Processors (ASAP'05)*, Samos, Greece, July 23-15 2005. IEEE Computer Society Press.
3. Mihir Bellare, Chanathip Namprempre, and Gregory Neven. Security proofs for identity-based identification and signature schemes. In C. Cachin and J. Camenisch, editors, *Proceedings of Eurocrypt 2004*, volume 3027 of *Lecture Notes in Computer Science*, pages 268–286. Springer-Verlag, 2004.
4. T. Beth and D. Gollmann. Algorithm engineering for public key algorithm. *IEEE Journal on Selected Areas in Communications*, 7(4):458–465, May 1989.

5. I. Blake, G. Seroussi, and N. P. Smart. *Elliptic Curves in Cryptography*. London Mathematical Society Lecture Note Series. Cambridge University Press, 1999.

6. Y. Dodis, J. Katz, S. Xu, and M. Yung. Strong Key-Insulated Signature Schemes. In Y. Desmedt, editor, *Proceedings of 6th International Workshop on Practice and Theory in Public Key Cryptosystems (PKC 2003)*, number 2567 in LNCS, pages 130–144. Springer-Verlag, 2003.

7. Y. Dodis, M. Reyzin, and A. Smith. Fuzzy extractors: How to generate strong keys from biometrics and other noisy data. In C. Cachin and J. Camenisch, editors, *Proceedings of Eurocrypt 2004*, volume 3027 of *Lecture Notes in Computer Science*, pages 523–540. Springer-Verlag, 2004.

8. B. Gassend et al. Silicon physical unknown functions. *Proc. 9th ACM Conference on Computer and Communications Security*, November 2002.

9. B. Gassend, D. Clarke, M. van Dijk, and Srinivas Devadas. Controlled physical random functions. *In Proceedings of the 18th Annual Computer Security Conference*, December 2002.

10. D. Johnson and A. Menezes. The elliptic curve digital signature algorithm (ECDSA). Technical Report CORR 99-34, Department of Combinatorics & Optimization, University of Waterloo, Canada, February 24 2000. http://www.cacr.math.uwaterloo.ca.

11. M. Joye and S.-M. Yen. The montgomery powering ladder. In B. S. Kaliski Jr., Ç. K. Koç, and C. Paar, editors, *Proceedings of 4th International Workshop on Cryptographic Hardware and Embedded Systems (CHES)*, number 2523 in Lecture Notes in Computer Science, pages 291–302. Springer-Verlag, 2002.

12. A. Juels. Strengthening EPC Tags against Cloning. March 2005. manuscript.

13. A. Lenstra and E. Verheul. Selecting cryptographic key sizes. In H. Imai and Y. Zheng, editors, *Proceedings of Third International Workshop on Practice and Theory in Public Key Cryptography (PKC 2000)*, number 1751 in Lecture Notes in Computer Science, pages 446–465. Springer-Verlag, 2000.

14. J.P. Linnartz and P. Tuyls. New shielding functions to enhance privacy and prevent misuse of biometric templates. In J. Kittler and M. Nixon, editors, *Proc. of the 3rd Conference on Audio and Video Based Person Authentication*, volume 2688 of *Lecture Notes in Computer Science*, pages 238–250. Springer-Verlag, 2003.

15. J. López and R. Dahab. Fast multiplication on elliptic curves over $GF(2^m)$. In Ç. K. Koç and C. Paar, editors, *Proceedings of 1st International Workshop on Cryptographic Hardware and Embedded Systems (CHES)*, volume 1717 of *Lecture Notes in Computer Science*, pages 316–327. Springer-Verlag, 1999.

16. A. Menezes, P. van Oorschot, and S. Vanstone. *Handbook of Applied Cryptography*. CRC Press, 1997.

17. P. Montgomery. Speeding the pollard and elliptic curve methods of factorization. *Mathematics of Computation*, Vol. 48:243–264, 1987.

18. Michael Neve, Eric Peeters, David Samyde, and Jean-Jacques Quisquater. Memories: a Survey of their Secure Uses in Smart Cards. In *2nd International IEEE Security In Storage Workshop (IEEE SISW 2003)*, pages 62–72, Washington DC, USA, 2003.

19. B. Skoric P. Tuyls. Secret key generation from classical physics. *Philips Research Book Series*, September 2005.

20. R. Pappu. Physical one-way functions. *Science*, 297(6):2026, 2002.

21. G. J. Simmons. Identification of data, devices, documents and individuals. In *Proc. 25th Ann. Intern. Carnahan Conference on Security Technology*, pages 197–218, Taipei, Taiwan, ROC, October 1–3, 1991. IEEE.

22. B. Skoric, P. Tuyls, and W. Ophey. Robust key extraction from physical unclonable functions. In J. Ionnidis, A.D. Keromytis, and M. Yung, editors, *Proceedings of the Applied Cryptography and Network Security Conference 2005*, volume 3531 of *Lecture Notes in Computer Science*, pages 407–422. Springer-Verlag, 2005.
23. S. P. Skorobogatov and R. J. Anderson. Optical fault induction attacks. In B. S. Kaliski Jr., Ç. K. Koç, and C. Paar, editors, *Proceedings of the 4th International Workshop on Cryptographic Hardware and Embedded Systems (CHES)*, volume 2523 of *Lecture Notes in Computer Science*, pages 2–12. Springer-Verlag, 2002.
24. P. Tuyls and J. Goseling. Capacity and examples of template protecting biometric authentication systems. In D. Maltoni and A.K. Jain, editors, *Proceedings of Biometric Authentication Workshop*, volume 3087 of *Lecture Notes in Computer Science*, pages 158–170. Springer-Verlag, 2004.
25. P. Tuyls, B. Skoric, S. Stallinga, A.H.M. Akkermans, and W. Ophey. Information theoretical security analysis of physical unclonable functions. In A.S. Patrick and M. Yung, editors, *Proceedings of 9th Financial Cryptography and Data Security Conference*, volume 3570 of *Lecture Notes in Computer Science*, pages 141–155. Springer-Verlag, 2005.
26. M van Dijk and P. Tuyls. Robustness, reliability and security of biometric key distillation in the information theoretic setting. In N. Cerf and J. Cardinal, editors, *Proceedings of the 26th Benelux Symposium on Information Theory*, volume 26 of *Proceedings of the WIC*, 2005.
27. J. Wolkerstorfer. Scaling ECC Hardware to a Minimum. In ECRYPT workshop - Cryptographic Advances in Secure Hardware - CRASH 2005, September 6-7 2005. invited talk.

Appendix

Schnorr Identification Protocol Based on ECDLP

Here we specify the Schnorr identification protocol based on ECDLP that could be performed in the case of off-line authentication. In this case a tag proves its identity to a reader in a 3-pass protocol.

1. **Common Input:** The set of system parameters in this case consists of: (q, FR, a, b, P, n, h). Here, q specifies the finite field, FR is a field representation, a, b, define an elliptic curve, P is a point on the curve of order n and h is the cofactor [10]. In this case of a tag authentication, most of these parameters are assumed to be fixed.
2. **Prover-Tag Input:** The prover's secret a such that $Z = -aP$.
3. **Protocol:** The protocol involves exchange of the following messages:

$$
\begin{array}{lll}
\underline{\text{Prover } P} & & \underline{\text{Verifier } V} \\
r \in_R \mathbb{Z}_n & & \\
X \leftarrow rP & \xrightarrow{\quad X \quad} & \\
& \xleftarrow{\quad e \quad} & e \in_R \mathbb{Z}_{2^t} \\
y = ae + r & \xrightarrow{\quad y \quad} & \\
& & \text{If } yP + eZ = x \\
& & \text{then accept else reject}
\end{array}
$$

More precisely, steps of the protocol are:

- *Commitment by a Prover-Tag:* The tag picks $r \in_R \{0, \ldots, n-1\}$, and sends $x = rP$ to the reader.
- *Challenge from a Verifier-Reader:* The reader picks a number $e \in [1, 2^t]$ and sends it to the tag.
- *Response from a Tag:* The tag computes $y = ae + r$ and sends it to the reader.
- The verifier checks that $yP + eZ$ equals x. Check: $yP + eZ = (ae+r)P + eZ = aeP + rP + (-eaP) = rP = x$

A "Medium-Field" Multivariate Public-Key Encryption Scheme

Lih-Chung Wang[1,*], Bo-Yin Yang[2,3,**], Yuh-Hua Hu[4,*], and Feipei Lai[4]

[1] Department of Applied Mathematics, National Donghua University,
Hualien, Taiwan
lcwang@math.ndhu.edu.tw
[2] Department of Mathematics, Tamkang University,
Tamsui, Taiwan
[3] Taiwan Information Security Center, Taipei
by@moscito.org
[4] Department of Computer Science and Engineering,
National Taiwan University,
Taipei, Taiwan
{d92015, flai}@csie.ntu.edu.tw

Abstract. Electronic commerce fundamentally requires two different public-key cryptographical primitives, for key agreement and authentication. We present the new encryption scheme MFE, and provide a performance and security review. MFE belongs to the $\mathcal{MQ}$ class, an alternative class of PKCs also termed Polynomial-Based, or multivariate. They depend on multivariate quadratic systems being unsolvable.

The classical trapdoors central to PKC's are modular exponentiation for RSA and discrete logarithms for ElGamal/DSA/ECC. But they are relatively slow and will be obsoleted by the arrival of QC (Quantum Computers). The argument for $\mathcal{MQ}$-schemes is that they are usually faster, and there are no known QC-assisted attacks on them.

There are several $\mathcal{MQ}$ digital signature schemes being investigated today. But encryption (or key exchange schemes) are another story — in fact, only two other $\mathcal{MQ}$-encryption schemes remain unbroken. They are both built along "big-field" lines. In contrast MFE uses medium-sized field extensions, which makes it faster. For security and efficiency, MFE employs an iteratively triangular decryption process which involves rational functions (called by some "tractable rational maps") and taking square roots. We discuss how MFE avoids previously known pitfalls of this genre while addressing its security concerns.

Keywords: multivariate ($\mathcal{MQ}$) public key cryptosystem, Galois field, extended triangular form, tame-like map, tractable rational map, MFE.

* Sponsored in part by National Science Council under Grant NSC-94-2115-M-259-002.
** Correspondence should be addressed to this author, who is partially sponsored by the Taiwan Information Security Center (TWISC) project as well as the National Science Council under Grant NSC-94-2115-M-032-010. A version of this paper is available as a TWISC tech report and will be placed on the IACR ePrint archive.

D. Pointcheval (Ed.): CT-RSA 2006, LNCS 3860, pp. 132–149, 2006.
© Springer-Verlag Berlin Heidelberg 2006

1   Introduction

Electronic commerce requires at least the following fundamental cryptological primitives: one digital signature scheme, one public-key encryption or key exchange scheme, one hash function, and one symmetric cipher. The first two involve public-key cryptosystems which are based on computationally difficult problems. Currently deployed PKCs most often involve the integer factoring problem (RSA, ESign, Rabin) or the discrete log problem (ECC, ElGamal/DSA).

We aim to introduce a new public-key encryption scheme that may be used for key exchange. This is one of many schemes based on the difficulty of solving a system of polynomial equations. PKCs of this class are usually described as $\mathcal{MQ}$, multivariate, or polynomial-based schemes.

Before we proceed, let us first answer the inevitable question of why researching alternative schemes at all when RSA does just fine. One reason is for diversity. When quantum computers that can handle 4000 quantum bits becomes reality, *Shor's Algorithm* [Sho94] can break all the abovementioned classical cryptosystems very quickly. $\mathcal{MQ}$-schemes are among the alternative PKCs that are weakened by quantum computers (via Grover's Algorithm [Gro96]), but not fatally wounded. Another is for better efficiency in resource (time, power or chip area) usage. This can let us do public-key cryptography in low-resource environments, or make do with cheaper components where we already use PKI.

We will introduce $\mathcal{MQ}$-schemes (Sec. 2), and discuss their state of the art in Sec. 3. We construct the central map of our schemes in Sec. 4, and explain our idea based on an overlay of two stepwise triangular systems. Further details are given in Sec. 5 and the Appendices. Aspects of security are sketched in Sec. 6–8, and we conclude with a discussion of performance and future possibilities.

2   About Multivariate or $\mathcal{MQ}$-Schemes

An $\mathcal{MQ}$-scheme is a cryptosystem whose security depend on this problem:

> Solve the system $p_1 = p_2 = \cdots = p_m = 0$, where each p_i is a quadratic polynomial in $x_1, \ldots, x_n$. All coefficients and variables are in $\mathbb{K} = \mathrm{GF}(q)$.

This problem is called $\mathcal{MQ}$ (for multivariate quadratics). The complexity clearly depend on q, the size of the finite field $\mathbb{K}$ (usually called the *base field*). [GJ79] proved $\mathcal{MQ}$ to be generically NP-hard even over the smallest field, i.e., when $q = 2$. Of course, that does not necessarily imply $\mathcal{MQ}$ to be difficult *on average*, but prevailing expert opinion does expect it to be exponential-time.

The public map $\mathcal{P}$ is the set of quadratics $(p_1, \ldots, p_m)$. Of course, we need a trapdoor to build a public-key cryptosystem. In every practical $\mathcal{MQ}$-schemes, this is accomplished by having a $\mathcal{P}$ that is composed of three maps as in $\mathcal{P} = T \circ \mathcal{Q} \circ S$. $\mathcal{Q}$ is the *central map* and it is quadratic. S and T are linear (affine) maps. We can write them as $S : \mathbf{x} \mapsto \mathbf{x}' = \mathrm{M}_S \mathbf{x} + \mathbf{c}_S$, $T : \mathbf{y}' \mapsto \mathbf{y} = \mathrm{M}_T \mathbf{y}' + \mathbf{c}_T$. Some authors also represent this as $\mathcal{P} : \mathbf{x} \in \mathbb{K}^n \overset{S}{\mapsto} \mathbf{x}' \overset{\mathcal{Q}}{\mapsto} \mathbf{y}' \overset{T}{\mapsto} \mathbf{y} \in \mathbb{K}^m$. We may set $\mathcal{P}(0) = 0$. The public key is then the $mn(n+3)/2$ nonconstant coefficients of

$\mathcal{P}$ (we assume $q > 2$). The secret key comprise the $n(n+1) + m(m+1)$ entries of $(M_S^{-1}, M_T^{-1}, c_S, c_T)$, plus parameters in $\mathcal{Q}$ needed for taking its inverse.

If $\mathcal{Q}$ is a random quadratic, then $\mathcal{P}$ would be equally random and infeasible to decompose. But that is impossible, since we need to invert $\mathcal{Q}$ efficiently. Thus *the security of an $\mathcal{MQ}$-scheme depends on the infeasibility of decomposing maps in addition to that of solving large systems.*

3   Current $\mathcal{MQ}$-Schemes and Taxonomy

[WP05] is a good reference on the nomenclature and state-of-the-art on $\mathcal{MQ}$-schemes today. According to its classification, known $\mathcal{MQ}$-schemes (extant and broken) are a handful of modifiers applied to four different basic trapdoors (all must be modified in practice) and two combinations:

C^* **or MIA:** C^* (and HFE below) can be used both for digital signatures and for encryption. Proposed by Matsumoto-Imai [MI88], broken and revamped by Patarin into the signature scheme C^{*-} or MIA$-$ [Pat95, PGC98]. NESSIE-recommended SFLASH [Nessie, PCG01a] is an instance. Ding proposed PMI+ (MIAi+) for encryption [DG05] (a modification from its precursor MIAi [Din04] after the cryptanalysis of Fouque *et al* [FGS05]).

HFE: The basic scheme of Patarin's Hidden Field Equations [Pat96] is broken ([CDF03, FJ03]), but HFE$-$ or HFE$v-$ (or QUARTZ, [PCG01]) for signatures and HFEi (or IPHFE, [DS05]) for encryption are not. This genre of schemes are burdened by its slow private map.

UOV: Unbalanced-Oil-and-Vinegar by Kipnis *et al* [KPG99], a modification of the earlier and broken [KS98] Oil-and-Vinegar. Useful for signing only and secure only for some awkward choices of parameters [BWP05], hence usually appears in combination with STS.

STS: Stepwise Triangular System. Variables are solved one by one in domino fashion. This basic trapdoor and its $\pm$ modifications is broken by techniques of [GC00] (also [CSV93, WBP04, YC05]). Two useful combinations follow:

STS-UOV: A better name might be LuOV (*Layered unbalanced Oil-and-Vinegar*). In segments, vinegar variables are added and linear systems are solved *a la* UOV. All extant examples are very new signature schemes. The first is enTTS (Enhanced Tame Transformation Signature, [YC05]), which is a *sparse variant* just like TRMS (Tractable Rational Map Signature, [WHL+05]). In the slightly later Rainbow [DS05a], Ding *et al* decide to omit the sparsity.

STS-R: Stepwise Triangular System Repeated. Iteration of triangular runs are made to cover inevitable rank vulnerabilities in a triangular system.

[WP05] calls MIA/HFE *mixed-field* ([YC05] terms them *big-field*). These are run mostly over one single large field even though the public map is over a smaller field. STS/UOV is in contrast called *single-field* or *true* [WP05, YC05] because we actually work with the field units. Note that in TRMS several field extensions are used at once. But if we expand all products, we see that they only serve to create an efficiently-invertible sparse central map.

4   A Central Map for an $\mathcal{MQ}$-Scheme, on Medium Fields

We now describe our idea, a central map $\mathcal{Q}$ with a very different flavor:

1. We use throughout the private map one particular field extension $\mathbb{L}$ above the base field $\mathbb{K}$. But this field extension does not cover (nearly) all the variables. Hence, we are not dealing with a *big-field* $\mathcal{MQ}$-scheme like in MIA or HFE. Hence the title "Medium Field" and the name MFE.

2. We are solving for the variables in stepwise fashion, *without using vine-gar variables*. Indeed, our scheme might be said to descend spiritually from TTM [Moh99], which pioneered the STS-R approach, to combine two tri-angular maps to cover the critical small end of the triangle. However, all implementations of TTM had fatal flaws, and our techniques and ideas are radically different, involving what have been called "tractable rational maps" [WC04].

3. Our intended as practical example will have a base field of $\mathbb{K} = \mathrm{GF}(2^{16})$. While it is not unknown for multivariates to have such a base field, it is usually the result of scaling up for security reasons. Here we are designing from the ground up to use such a big field.

4. Our currently favored example scheme is also *tame-like* [YC05]. This makes the key generation process more efficient.

4.1   The Central Map

We define $\mathcal{Q} : \mathbb{L}^{12} \to \mathbb{L}^{15}$ as follows:

$$
\begin{cases}
Y_1 = X_1 + X_5X_8 + X_6X_7 + Q_1; \\
Y_2 = X_2 + X_9X_{12} + X_{10}X_{11} + Q_2; \\
Y_3 = X_3 + X_1X_4 + X_2X_3 + Q_3; \\
Y_4 = X_1X_5 + X_2X_7; \qquad\qquad Y_5 = X_1X_6 + X_2X_8; \\
Y_6 = X_3X_5 + X_4X_7; \qquad\qquad Y_7 = X_3X_6 + X_4X_8; \qquad (1) \\
Y_8 = X_1X_9 + X_2X_{11}; \qquad\quad\; Y_9 = X_1X_{10} + X_2X_{12}; \\
Y_{10} = X_3X_9 + X_4X_{11}; \qquad\quad Y_{11} = X_3X_{10} + X_4X_{12}; \\
Y_{12} = X_5X_7 + X_2X_{11}; \qquad\quad Y_{13} = X_5X_{10} + X_7X_{12}; \\
Y_{14} = X_6X_9 + X_8X_{11}; \qquad\quad Y_{15} = X_6X_{10} + X_8X_{12}.
\end{cases}
$$

Here each X_i and Y_i is in $\mathbb{L} = \mathbb{K}^k$. Since $\mathbb{L} = \mathbb{K}^k$. We split $X_1, X_2, X_3, Q_1, Q_2, Q_3$ into components in $\mathbb{K}^k$, such that $q_1 = 0$, $q_2 = (x_1')^2$ and for $i = 3 \cdots 3k$, q_i is a more or less a random quadratic in variables $(x_1', \ldots, x_{i-1}')$.

$$
X_1 = \begin{bmatrix} x_1' \\ x_2' \\ \vdots \\ x_k' \end{bmatrix}, \; X_2 = \begin{bmatrix} x_{k+1}' \\ x_{k+2}' \\ \vdots \\ x_{2k}' \end{bmatrix}, \; X_3 = \begin{bmatrix} x_{2k+1}' \\ x_{2k+2}' \\ \vdots \\ x_{3k}' \end{bmatrix}; \; Q_1 = \begin{bmatrix} q_1 \\ q_2 \\ \vdots \\ q_k \end{bmatrix}, \; Q_2 = \begin{bmatrix} q_{k+1} \\ q_{k+2} \\ \vdots \\ q_{2k} \end{bmatrix}, \; Q_3 = \begin{bmatrix} q_{2k+1} \\ q_{2k+2} \\ \vdots \\ q_{3k} \end{bmatrix}.
$$

4.2 An Inverse to the Central Map

Idea: We may arrange $X_1, X_2, \ldots, X_{12}, Y_4, Y_5, \ldots, Y_{15} \in \mathbb{L}$, into 2×2 matrices:

$$M_1 = \begin{bmatrix} X_1 & X_2 \\ X_3 & X_4 \end{bmatrix}, \qquad M_2 = \begin{bmatrix} X_5 & X_6 \\ X_7 & X_8 \end{bmatrix}, \qquad M_3 = \begin{bmatrix} X_9 & X_{10} \\ X_{11} & X_{12} \end{bmatrix};$$

$$M_1 M_2 = \begin{bmatrix} Y_4 & Y_5 \\ Y_6 & Y_7 \end{bmatrix}, \; M_1 M_3 = \begin{bmatrix} Y_8 & Y_9 \\ Y_{10} & Y_{11} \end{bmatrix}, \; M_2^T M_3 = \begin{bmatrix} Y_{12} & Y_{13} \\ Y_{14} & Y_{15} \end{bmatrix}.$$

(2)

$\mathcal{Q}$ is inverted in three triangular steps, simple linear algebra gives that

$$Y_4 Y_7 - Y_5 Y_6 = \det(M_1 M_2) = \det M_1 \det M_2,$$

and similarly,

$$Y_8 Y_{11} - Y_9 Y_{10} = \det M_1 \det M_3, \; Y_{12} Y_{15} - Y_{13} Y_{14} = \det M_2 \det M_3.$$

Thus, knowing $Y_4, \ldots, Y_{15}$, we can find $\det M_1$, $\det M_2$, and $\det M_3$, provided that none of them is zero (we will need a square-root taking operation, which is one-to-one and onto, and not very hard – as we shall show – in a char $= 2$ field, and for appropriately chosen k). Further

$$Y_1 = X_1 + \det M_2 + Q_1, \; Y_2 = X_2 + \det M_1 + Q_2, \; Y_3 = X_3 + \det M_3 + Q_3.$$

Therefore, having found $\det M_1, \det M_2, \det M_3$, we reduce the components of Y_1, Y_2, Y_3 to a triangular form in the x_i:

$$X_1 + Q_1 = Y_1 + \sqrt{(Y_4 \times Y_7 + Y_5 \times Y_6)(Y_8 \times Y_{11} + Y_9 \times Y_{10})(Y_{12} \times Y_{15} + Y_{13} \times Y_{14})^{-1}}$$
$$X_2 + Q_2 = Y_2 + \sqrt{(Y_4 \times Y_7 + Y_5 \times Y_6)(Y_8 \times Y_{11} + Y_9 \times Y_{10})^{-1}(Y_{12} \times Y_{15} + Y_{13} \times Y_{14})}$$
$$X_3 + Q_3 = Y_3 + \sqrt{(Y_4 \times Y_7 + Y_5 \times Y_6)^{-1}(Y_8 \times Y_{11} + Y_9 \times Y_{10})(Y_{12} \times Y_{15} + Y_{13} \times Y_{14})}$$

then we apply a second triangular step to compute X_1, X_2, and X_3 component by component. If $X_1 \neq 0$, from $\det M_1$ we can also find X_4. We can now obtain the rest of the variables. **However, it is not necessary to have $X_1 \neq 0$.** This will make a difference in the security analysis — we omit the details, please see Appendix B.

It remains to flesh out our skeletal description, explain our design decisions, and try to show how our approach avoids the mistakes made in earlier designs.

5 Sample Implementations Using a "Tower" Approach

We start by taking $\mathbb{L} = \mathbb{K}^4$. We will use (for simplicity) $q_i = c_i x_{i-1} x_{i-2}$ for $i = 3 \cdots 11$. $\mathcal{S}$ and $\mathcal{T}$ are respectively affine maps of $\mathbb{L}^{12}$ and $\mathbb{L}^{15}$, selected according to Eq. 5 such that the resultant public map $\mathcal{P} = \mathcal{T} \circ \mathcal{Q} \circ \mathcal{S}$ does not have a constant term.

5.1   The Sample Scheme MFE-1

We will take $\mathbb{K} = \mathrm{GF}(2^{16})$. $\mathbb{K}$ is implemented as a degree-two extension of $\mathrm{GF}(2^8)$, and $\mathbb{L}$ as a degree four extension of $\mathbb{K}$ that is a composition of two degree-two extensions. We multiply in $\mathrm{GF}(2^8)$ via a 256×256 (64kB) table. We may alternatively use log-and-exp tables, which are somewhat slower.

From $\mathrm{GF}(2^8)$ to $\mathbb{L}$ we need to do three degree-two extensions. This can be aided by the following observation: Let $F' = F[x]/(x^2 + x + \alpha)$ be a valid deg-2 extension of the char-2 finite field F, then so is $F'' = F'[y]/(y^2 + y + \alpha x)$ a valid extension of F'. The arithmetic of the "tower" extensions may then be done efficiently using Karatsuba multiplication/inversion ([KO63]):

$$(ax + b)(cx + d) = [(a + b)(c + d) + bd] + [\alpha ac + bd], \qquad (3)$$

$$\text{similarly,} \qquad (ax + b)^{-1} = [b(a + b) + \alpha a^2]^{-1} [ax + (a + b)]; \qquad (4)$$

where we are operating in a field extension from F to $F[x]/(x^2 + x + \alpha)$. Multiplication by α is a lot easier than a regular multiplication because it is fixed. Squaring is also easier than a regular multiplication. Indeed, we find a normal basis of $\mathrm{GF}(2^8)$, i.e., an x such that $(x, x^2, x^4, x^8, x^{16}, x^{32}, x^{64}, x^{128})$ forms a basis of $\mathrm{GF}(2^8)$, and represent by a byte $\sum_{i=0}^{7} b_i 2^i$ the element $\sum_{i=0}^{7} b_i x^{2^i} \in \mathrm{GF}(2^8)$. Build a multiplication (or log-exp) table accordingly. Squaring then become no more than a byte rotation. It is consistent with our own implementations that *with every extra stage in the tower, a multiplication or division takes a little more than* $3\times$ *time. A squaring always cost less than* $1/10$ *of a multiplication.*

Keys and Generation: The private key is the coefficients in S, T and the c_i's for a total of 5,904 elements of $\mathbb{K}$ or about 12kB. The public key comprise the $60 \times 48 \times 51/2 = 73,440$ coefficients of $\mathcal{P}$ or about 147kB.

While $\mathcal{MQ}$-schemes typically use interpolation or a similar technique (cf. [Wol04]) to generate the key, **a faster method [YC05, YCCh04] applies here because our scheme is tame-like** [YC05]. Let $\mathcal{P}$ be given by the quadratics

$$y_k = \sum_i P_{ik} x_i + \sum_i Q_{ik} x_i^2 + \sum_{i>j} R_{ijk} x_i x_j, \; k = 1 \cdots m.$$

Expand each central equation in Y_i into its 4 y_i' components. Each central equation in y_γ' has less than 20 cross-terms $\pi_{\alpha\beta} x_\alpha' x_\beta'$, where $\pi_{\alpha\beta}$'s are constants of the system or one of the c_i's (we keep these metadata in a precomputed table).

$$P_{ik} = \sum_{\gamma=1}^{m} \left[(\mathrm{M}_T)_{k,\gamma} \left([\gamma \le 12](\mathrm{M}_S)_{\gamma i} + \sum_{\pi_{\alpha\beta} \, x_\alpha' x_\beta' \, \mathrm{in} \, y_\gamma'} \pi_{\alpha\beta} \left((\mathrm{M}_S)_{\alpha i}(\mathbf{c}_S)_\beta + (\mathbf{c}_S)_\alpha (\mathrm{M}_S)_{\beta i} \right) \right) \right]$$

$$Q_{ik} = \sum_{\gamma=1}^{m} \left[(\mathrm{M}_T)_{k,\gamma} \left(\sum_{\pi_{\alpha\beta} \, x_\alpha' x_\beta' \, \mathrm{in} \, y_\gamma'} \pi_{\alpha\beta} \, (\mathrm{M}_S)_{\alpha i} (\mathrm{M}_S)_{\beta i} \right) \right]$$

$$R_{ijk} = \sum_{\gamma=1}^{m} \left[(M_T)_{k,\gamma} \left(\sum_{\pi_{\alpha\beta} \, x'_\alpha x'_\beta \text{ in } y'_\gamma} \pi_{\alpha\beta} \left((M_S)_{\alpha i}(M_S)_{\beta j} + (M_S)_{\alpha j}(M_S)_{\beta i} \right) \right) \right]$$

In the formula for P_{ik}, the notation $[\gamma \le 12]$ means a term that is only present if $\gamma \le 12$. For every pair $i < j$, we first find

$$\overline{R}_{ijk} = \sum_{\pi x_\alpha x_\beta \text{ is a term of } y_k} \left[\pi \left((M_S)_{\alpha i}(M_S)_{\beta j} + (M_S)_{\alpha j}(M_S)_{\beta i} \right) \right]$$

for every k, then multiply the vector by the matrix M_T to find all R_{ijk} at once ([YC05, YCCh04]). We are then able to compute the entire key in less than 5 million $\mathbb{K}$-multiplications, Finally, the constant part $\mathbf{c}_T$ of $\mathcal{T}$ is computed thus:

$$(\mathbf{c}_T)_k = \sum_{p=1}^{n} \left[(M_T)_{k,\ell} \left((\mathbf{c}_S)_\ell + \sum_{\pi \, x_\alpha x_\beta \text{ in } y_\ell} \pi \, (\mathbf{c}_S)_\alpha (\mathbf{c}_S)_\beta \right) \right]. \tag{5}$$

Details of encryption and decryption operations may be filled in as above.

5.2   Other Sample Schemes

Aside from the "regular" scheme with $\mathbb{K} = GF(2^{16})$ and $k = 4$. We will also present contrasting data for other instances of MFE.

MFE-1′: Here we use $k = 5$ ($\mathbb{L} = GF(2^{80})$) instead of 4. The computations are significantly more complex and time-consuming.

MFE-0: In what we shall call the "mini"-implementation, we use $\mathbb{K} = GF(2^8)$ and $\mathbb{L} = GF(2^{32})$. Everything else is as in the above section.

MFE-0⁺: Run like MFE-0, but we use redundancy to make failure to decrypt less likely. When encrypting, always treat two blocks (B, B') by sending $\mathcal{P}(B), \mathcal{P}(B'), \mathcal{P}(B' \boxplus B^{-1})$, where B^{-1} means to take the patched inverse in $\mathbb{K}$ of every component, and $\boxplus$ means addition modulo $|\mathbb{K}|$.

$GF(2^8)$ multiplications are three times faster than $GF(2^{16})$ multiplications. Therefore, running the encryption function of MFE-0 three times is still faster than one run of the encryption function of MFE-1, and let us have the smaller key sizes of MFE-0. Other implications are given below.

6   Security: A Basic Overview

Security analysis for $\mathcal{MQ}$-schemes are hampered by the lack of "provable security". As far as we know, the only attempt in this area is due to [Cou03], which is not followed up actively. As a result, while it is easy to show that an $\mathcal{MQ}$-scheme is insecure by presenting a cryptanalysis. So far a cryptologist can only show that current attacks don't work and are not likely to work. We try to do the best we can under the circumstances. Currently known attacks on polynomial-based PKCs can be roughly classified into four kinds:

Correlation/Statistical Attacks: A common systemic attacks agains symmetric ciphers, but usually not applied against PKCs. We will describe how known attacks do not apply.

Linear Algebra-Based (Rank) Attacks: There are several attacks that are quite generic against when the target scheme is not of the big-field type. We just give numbers below, please refer to Appendix.

Algebraic Attacks: Today this means any attack whose functionality comes down to solving a system of equations, usually distilled out of the structure of the system. We will summarize what is out there.

Very Specialized Attacks: Obviously, an attacks has to focus on some aspect or structural element of the cryptosystem. Some do have wide applicability to a whole class of schemes. Others do not. While this does not detract from the sheer intellectual worth or ingenuity of such cryptanalytic work, many ideas (e.g., the Gilbert-Minier attack [GM02] on the original SFLASH) simply do not work on other schemes.

We repeat that showing our scheme to resist known attacks does not guarantee security. It is an "original sin" for $\mathcal{MQ}$-schemes today. We only do what we can.

6.1 Cryptanalysis Using Rank Attacks

"Rank Attacks" encompasses the High Rank, Low Rank, and Separation of Oil-and-Vinegar attacks. These are basic attacks against the STS or UOV based trapdoors in $\mathcal{MQ}$-schemes. *Summary: Rank Attacks do not work on any of our sample encryption schemes.*

Separation of Oil and Vinegar (UOV) Attacks: Security level for the "mini" version (MFE-0) is about 2^{100}; the "full" version (MFE-1) is about 2^{140}.

High Rank (Dual Rank) Attack: Security level is about 2^{128} for the "mini" version (MFE-0) and 2^{181} for the full-version (MFE-1).

Low Rank (Rank or MinRank) Attack: Security level is about 2^{128} for the "mini" version (MFE-0) and 2^{172} for the full-version (MFE-1).

See Appendix A and references, particularly [GC00, YC05] for more details.

6.2 Cryptanalysis Under Specialized Attacks

We tested our scheme not to succumb to any earlier known special attacks, and will henceforth ignore specialized attacks without wider applicability.

One specialized attack that is specifically designed around is the Patarin Relations, which can also be considered an algebraic attack or a linear-algebra attack. It can defeat many systems that can be described as of intrinsic rank 2.

Eq. 2 specifically had its matrix products arranged $M_1 M_2$, $M_1 M_3$ and $M_2^T M_3$. This take full advantage of the incommutativity of matrix multiplication. Other arrangements will create lots of Patarin relations. For example, if we have $N = M_1 M_2$ and $N' = M_2 M_3$ as matrices with components linear in the Y_i's, then we will have the relations corresponding to $N M_3 = M_1 N'$. As the central equations Eq. 2 are written, no such Patarin relations can be found. We should only need to test this over GF(2); we actually tested it over GF(4).

7   Algebraic Cryptanalysis

Basically, an algebraic attack refers to any technique that ends with a system-solving exercise. There may be guessing, or there may not be. The system may be linear, as in Patarin relations vs. C^* [Pat95], or non-linear, as in the Courtois/Faugère-Joux attack on HFE [CDF03, FJ03].

As characterized above, eventually an algebraic attack comes down to solving a system. At the moment, the state of the art is represented by the $\mathbf{F_5}$ Gröbner Bases algorithm of J.-C. Faugère [Fau02] while the best in commercially available software is a version of its predecessor $\mathbf{F_4}$ [Fau99]. Few know how to program $\mathbf{F_5}$ correctly, and certainly the only known implementation of $\mathbf{F_5}$ is the one used by Dr. Faugère to break HFE challenge 1.

The alternative methods of XL/FXL by Courtois $et\ al$ [CKP+00] has been analyzed in some depth [Die04, YC04]. Today XL is usually considered to be a poor relative of $\mathbf{F_4}$-$\mathbf{F_5}$ [AFS+04]. The asymptotic behavior of the Gröbner Bases-XL family is described by [BFS04, BFS+05, YCCo04].

7.1   On Extraction of More Tractable Systems

An important remark is that the attack against HFE challenge 1 involves an algebraic extraction of the actual system to be solved, one that is significantly more overdetermined than the original.

Such an extraction method does not yet exist in general. At the moment, we have no way of distinguishing our encryption scheme from random quadratics. This conclusion is supported by some experiments that we ran, trying to solve systems directly very miniaturized version of MFE, using $GF(4)$ as the base field, and the tool is MAGMA and the version of $\mathbf{F_4}$ built therein.

Finally, we cannot rule out the possibility that a method of extracting some solvable system exists, as for example in [JKM+05]. However, we have checked for it and as far as we can tell, no known method of such extraction works.

Let us describe the [JKM+05] attack briefly. This is an attack on the encryption scheme TRMCv2 of Wang and Chang [WC04]. The scheme in question has many variables and equations, but there is a subsystem of 7 variables and 11 equations. This was part of the trapdoor in TRMCv2. The weakness is that by running a simplified version of XL, the attacker can essentially isolate this central subsystem. It transpires that the algorithm terminates at the degree that is required to solve the central subsystem, instead of the much higher degree that would be required of a generic system with as many equations and variables.

In addition to trying to execute the attack of [JKM+05] and determining that it does not work at a sufficiently low degree, we ran a clique-finding algorithm on our central polynomials and found no such central subsystem.

7.2   On the Speed of Equation-Solving

It seems that the most effective XL-derived method is FXL [YCCo04], and the same idea applies to $\mathbf{F_5}$ at least in the generic case. There are also cases [YC04] where XL will not and $\mathbf{F_5}$ may not work.

Indeed, our formulation satisfies a lemma from [YC04] which says that XL will not work and FXL will take longer. $\mathbf{F_5}$ is also expected to take longer. However, in the following we still assume that FXL and $\mathbf{F_5}$ will function as if the polynomials are generic.

Assuming generic equations, $\mathbf{F_4}$-$\mathbf{F_5}$ works at the smallest degree D where the coefficient of t^D in $(1-t)^{m-n}(1+t)^m$ is negative ([BFS$^+$05, Die04, YC04]). The dominant term of the time complexity is given by $E\left(\binom{n+D-1}{D}\right)$, where $E(N)$ is the cost of elimination on a $N \times N$ matrix equation. Here we have 48 variables and 60 equations. Assuming a field size of $q = 256$ (our "mini" version MFE-0), and $E(N) = N^2(4 + \lg N/4)$ in cycles (a very optimistic assumption that dense-matrix elimination can work asymptotically like sparse-matrix system solving that we take from [YCCo04]), we get about 2^{97} cycles or about a 2^{93} multiplications security level.

The conclusion is, then, that even discounting the possibility that such methods don't function at all, Gröbner Bases and related methods should be more effective than Rank Attacks against our schemes, but does not reduce them down below 2^{80}, with a lot of safety margin.

8 Correlation or Statistical Cryptanalysis and Defenses

A correlation or statistical attack works by finding imbalances of some kind in the ciphertext. Not many attacks on public-key cryptosystems use correlation or statistical artifacts. We comment on only two particular items specifically.

One is the attack [FGS05] on the scheme PMI. Fouque $et\ al$ used differential cryptanalysis with a one-sided statistical distinguisher. While ingenious, this does not apply to our scheme and this is confirmed by some empirical tests.

The other is much more relevant. Our system requires $X_1X_4 - X_2X_3$, $X_5X_8 - X_6X_7$, and $X_9X_{12} - X_{10}X_{11}$ all to be non-zero to get a successful decryption. Thus a single block will fail to decrypt with the probability $3/|\mathbb{L}|$.

For the "mini" sample scheme MFE-0, this chance of failure is about 2^{-30} and for the "regular" sample scheme MFE-1, it is about 2^{-62}. This results in a possible attack by guessing at decryption failures. However, there are no ways an attacker can easily check that two decryption faults correspond to a zero in the same determinant. Nor can we generate easily from two samples where $X_1X_4 = X_2X_3$ another such point to make use of the algebraic variety. We can not guarantee that this takes care of all correlation attacks based on decryption faults. However, we can expect any such attack to be significantly more difficult than just finding one such inscrutable ciphertext. Therefore, we can presume that our "regular" schemes are quite secure enough under such attacks.

8.1 Possible Cryptanalysis Via Correlation and Timings

Note that $X_1 = 0$ is much more useful to an attacker than any of the determinants being zero, since it is an affine formula. An adversary that can distinguish whether $X_1 = 0$ can execute the following attack:

Cryptanalysis: encrypt random blocks (vectors) and send to a decryption device (or oracle) for decryption continuously; register the blocks B_i whose corresponding ciphertext result in decryption failures (or timing "tells"). Every time a block B_i is registered, send $aB_i + (1-a)B_j$ for decryption for a few random a's and for each $j < i$ to find out if it correspond to a vector where $X_1 = 0$ (because such vectors form an affine subspace). Collect 12 such blocks in an expected $12|\mathbb{L}|$ attempts. With high probability we have found the affine subspace $X_1 = 0$. Restrict the polynomials to this affine subspace, and we can perform a MinRank attack on the reduced equations corresponding to Y_3. Since each evaluation takes about $mn(n+3)/2$ multiplications, total time used is $12|\mathbb{L}| \cdot [mn(n+3)/2\mathbb{K}$-multiplications + decryption time$]$.

This idea of using decryption failures seems to have been proposed by Proos *et al* [HNP+03]. The elegant cryptanalysis proposed in that paper resulted in a revision in the current version of NTRU Encryption. The idea of using Min-Rank after some other reduction may have been invented by J. Ding against the predecessor version to enTTS [DY04].

The analysis above shows that the "mini" scheme MFE-0 cannot be used if a Proos-like attack can execute. However, all is not lost. According to Appendix B we can decrypt even when $X_1 = 0$ without an appreciable speed difference, and the same straightforward attack cannot function with $X_1X_4 - X_2X_3 = 0$, say, because it is not an affine relation. Therefore, the method in Appendix B guarantees that such a cryptanalysis will not operate.

Further, for MFE-1, we have $|\mathbb{L}| = 2^{64}$, $n = 48$, $m = 60$, we find the complexity of the cryptanalysis to be about 2^{85} multiplications in $\mathbb{L}$, which is just about barely enough even if a good distinguisher exists for a Proos-like attack. For MFE-1$'$, where $\mathbb{L} = \mathbb{K}^5$ rather than $\mathbb{L} = \mathbb{K}^4$, the system runs quite a bit slower but the Proos-style attack will still have a cryptanalytic complexity above 2^{90} even if it works. Finally, Even if we can find an alternative way to execute the Proos attack on MFE-0, we can still use MFE-0$^+$.

9   Performance Data

Having shown that our schemes safe under known attacks, it then becomes meaningful to test performances. We compare our first implementations with the Crypto++ library (benchmarks at `www.eskimo.com/~weidai/benchmarks.html`).

We wrote our programs in plain C. The Crypto++ libraries are of course highly optimized binaries. We think that the data above shows that the schemes we propose are competitive with RSA and ECC. Of course, Crypto++ is not nearly as well optimized for ECC as for RSA.

For comparison's sake, we recompiled our programs in C51 and tested for performance on a 8051. One block decryption of MFE-1 ($\mathbb{K} = GF(2^{16})$, $\mathbb{L} = \mathbb{K}^4$) can run on the following smart card development platform (24kB EEPROM, including private key and code; 256 byte **idata**, 10 MHz basic Intel 8052, no extra RAM) in 0.28s. MFE-0 or MFE-0$^+$, if applicable, will be even faster.

Table 1. Our Schemes on a 1.6GHz Opteron compared with Crypto++ library

Scheme	BlockLen	PublKey	SecrKey	Genkey	SecrMap	PublMap
RSA-1024	1024 bits	128 B	320 B	0.86 sec	4.75 ms	0.18 ms
RSA-2048	2048 bits	256 B	640 B	2.71 sec	28.13 ms	0.45 ms
ECIES-155	310 bits	40 B	20 B	0.02 ms	7.91 ms	12.09 ms
MFE-1	512 bits	12 kB	147 kB	9.90 ms	32 μs	0.86 ms
MFE-1$'$	640 bits	18 kB	283 kB	23.40 ms	48 μs	1.79 ms
MFE-0	256 bits	6 kB	73 kB	2.21 ms	2.3 μs	0.12 ms
MFE-0$^+$	512 bits	6 kB	73 kB	2.21 ms	7.0 μs	0.39 ms

Given that RSA-1024 on an Infineon SLE-66X64-2P (a costly card with 208 kB ROM, 5052 bytes RAM, 64 kB EEPROM, and 1100-bit Advanced Crypto Engine) takes 0.4s at the same clock, this shows that the idea is of particular interest for situations where resources are scarce. This continues the trend of [ACD+03, YCCh04], that is, multivariates are worth investigating for low-resource and pervasive cryptography even without the interest of diversity.

We also note that usually decrypting is centralized at the servers while encrypting is done by those the masses sending data to the servers. So decrypting is more likely to be resource-intensive. Conversely, someone who uses a smart card to verify his or her identity is more likely to want to receive sensitive data, so a smart card (a low resource item) is more likely to want to do decryption.

A lot of optimizations remains to be done for a new scheme, of course, in particular the degree-five extension may be implemented better. We will of course pursue this direction in the future.

10   Discussions and Summary

We make a few comments about history and speculate on the future.

10.1   A Little History

This is the triangular (or tame, or *de Jonquiere*) map of algebraic geometry:

$$y_1 = x_1, \; y_2 = x_2 + f_2(x_1), \; y_3 = x_3 + f_3(x_1, x_2), \ldots, y_n = x_n + f_n(x_1, \ldots, x_{n-1}).$$

A PKC based on a triangular central map is known to be weak early on. The idea of using a composition of triangular maps to cover the vulnerability (Segmentwise Triangular System Repeated, STS-R) is pioneered by TTM [Moh99]. However, the execution was faulty and no rank-safe instances are available [GC00, YC05].

10.2   An Issue of Terminology

The seminal idea for our scheme is invented by L.-C. Wang [WC05]. The basic approach is not limited to that of the original TTM. It is more versatile, e.g., in a char $= 2$ field Eq. 2 may start like this without being really different:

$$y_1' := [Y_1]_1 = (x_1')^2 + [\det M_2]_1.$$

This type of map is called the "Tractable Rational Map" according to Wang *et al* [WC04, WC05], who term PKCs that uses compositions of "tractable rational maps" as "tractable rational map cryptosystems" (TRMC) [WC04, WC05]. Of course, there is a fine line between being nicely general and overly broad. For example, the central maps of HFE and C^* are both "tractable rational" maps (just as they are "tame transformation" maps, which themselves are a subsets of "tractable rational" maps; it must be added that the authors of TRMC do not claim HFE as a TRMC). How to demarcate clearly between various cryptosystems has not been agreed by all scholars of the $\mathcal{MQ}$ genre. An interested reader can look up what is claimed as "Tractable Rational Map Cryptosystem" in [WC04, WC05] as well as the eponymous pending patent application.

11   Conclusion

No other current instances of multivariates with the STS-R structure, such as TTM (Tame Transformation Method), are being employed today. We think we have shown that while the devil is in the details, there is some merit to the idea of STS-R, in particularly. However, to make it useful, we have to generalize by using more general operations than TTM. In particular, we need introduce rational operations and square roots. Therefore it is not TTM any more.

The introduction of new tricks naturally may introduce vulnerabilities. We showed how the failure of decryption may be enough of a discrepancy for cryptanalysis as Proos *et al* did for the previous version of NTRU. We also show our attempts at avoiding a similar fate. We leave to future historians to judge what and how our scheme will be considered.

As is lamented in many discussions (e.g., [WP05]) about $\mathcal{MQ}$-schemes, some measure of provable security seems to be hard to come by. Many scholars are studying this topic. But even so, we hope to have shown that there is some life in $\mathcal{MQ}$-schemes, in particular non-big-field types.

References

[ACD+03]   M. Akkar, N. Courtois, R. Duteuil, and L. Goubin, *A Fast and Secure Implementation of SFLASH*, PKC 2003, LNCS v. 2567, p. 267–278.

[AFS+04]   G. Ars, J.-C. Faugère, M. Sugita, M. Kawazoe, and H. Imai, *Comparison of XL and Gröbner Bases Algorithms over Finite Fields*. Asiacrypt 2004, LNCS v. 3329, p. 323–337.

[BFS04]   M. Bardet, J.-C. Faugère, and B. Salvy, *Complexity of Gröbner Basis Computations for Regular Overdetermined Systems*, INRIA report RR-5049, and presentation at the ICSPP conference honoring Daniel Lazard.

[BFS+05]   M. Bardet, J.-C. Faugère, B. Salvy, and B.-Y. Yang, *Asymptotic Behaviour of the Degree of Regularity of Semi-Regular Polynomial Systems*, presentation at the MEGA 2005 conference and a chapter of Ph.D. thesis by M. Bardet, 2004.

[BWP05]   A. Braeken, C. Wolf, and B. Preneel, *A Study of the Security of Unbalanced Oil and Vinegar Signature Schemes*, CT-RSA 2005, LNCS v. 3376, p. 29-43.

[CSV93]   D. Coppersmith, J. Stern, and S. Vaudenay, *Attacks on the Birational Permutation Signature Schemes*, Crypto 1993, LNCS v. 773, p. 435–443.

[Cou03]   N. Courtois, *Generic Attacks and the Security of Quartz*, PKC 2003, LNCS v. 2567, p. 351–364. Also see E-Print Archive 2004/143.

[CDF03]   N. Courtois, M. Daum, and P. Felke, *On the Security of HFE, HFEv-, and Quartz*, PKC 2003, LNCS v. 2567, p. 337–350.

[CKP+00] N. Courtois, A. Klimov, J. Patarin, and A. Shamir, *Efficient Algorithms for Solving Overdefined Systems of Multivariate Polynomial Equations*, Eurocrypt 2000, LNCS v. 1807, p. 392–407.

[Die04]   C. Diem, *The XL-algorithm and a conjecture from commutative algebra*, Asiacrypt 2004, LNCS v. 3329, p. 338–353.

[Din04]   J. Ding, *A New Variant of the Matsumoto-Imai Cryptosystem through Perturbation*, PKC 2004, LNCS v. 2947, p. 305–318.

[DG05]   J. Ding and J. Gower. *Inoculating Multivariate Schemes Against Differential Attacks*, private communication and manuscript, E-Print Archive, 2005/255.

[DS05]   J. Ding and D. Schmidt, *Cryptanalysis of HFEv and Internal Perturbation of HFE*, PKC 2005, LNCS v. 3386, p. 288–301.

[DS05a]   J. Ding and D. Schmidt, *Rainbow, a new Digitial Multivariate Signature Scheme*, ACNS 2005, LNCS v. 3531, p. 164–175.

[DY04]   J. Ding and Y. Yin, *Cryptanalysis of a TTS Implementation*, presentation at the IWAP 2004 conference.

[Fau99]   J.-C. Faugère, *A New Efficient Algorithm for Computing Gröbner Bases (F4)*, Journal of Pure and Applied Algebra, 139 (1999), p. 61–88.

[Fau02]   J.-C. Faugère, *A New Efficient Algorithm for Computing Gröbner Bases without Reduction to Zero (F5)*, Proc. ISSAC, ACM Press, 2002.

[FJ03]   J.-C. Faugère and A. Joux, *Algebraic Cryptanalysis of Hidden Field Equations (HFE) Cryptosystems Using Gröbner Bases*, Crypto 2003, LNCS v. 2729, p. 44-60.

[FGS05]   P.-A. Fouque, L. Granboulan, and J. Stern, *Differential Cryptanalysis for Multivariate Schemes*, Eurocrypt 2005, LNCS v. 3494, p. 341–353.

[GJ79]   M. Garey and D. Johnson, *Computers and Intractability, A Guide to the Theory of NP-completeness*, Freeman and Co., 1979, p. 251.

[GM02]   H. Gilbert and M. Minier, *Cryptanalysis of SFLASH*, Eurocrypt 2002, LNCS v. 2332, pp. 288–298.

[GC00]   L. Goubin and N. Courtois, *Cryptanalysis of the TTM Cryptosystem*, Asiacrypt 2000, LNCS v. 1976, p. 44–57.

[Gro96]   L. K. Grover, *A fast quantum mechanical algorithm for database search*, Proc. 28th Annual ACM Symposium on Theory of Computing (1996), p. 212–220.

[HNP+03] N. Howgrave-Graham, P. Nguyen, D. Pointcheval, J. Proos, J. Silverman, A. Singer, and W. Whyte, *The Impact of Decryption Failures on the Security of NTRU decryption*, Crypto 2003, LNCS v. 2729, p. 226–246.

[JKM+05] A. Joux, S. Kunz-Jacques, F. Muller, P.-M. Ricordel, *Cryptanalysis of the Tractable Rational Map Cryptosystem*, PKC 2005, LNCS v. 3386, pp. 258–274.

[KO63]   A. Karatsuba and Yu. Ofman, *Multiplication of Many-Digital Numbers by Automatic Computers*, Doklady Akad. Nauk SSSR 145(1962), p. 293-294. Translation in Physics-Doklady 7(1963), p. 595-596.

[KPG99]   A. Kipnis, J. Patarin, and L. Goubin, *Unbalanced Oil and Vinegar Signature Schemes*, Crypto 1999, LNCS v. 1592, p. 206–222.

[KS98]   A. Kipnis and A. Shamir, *Cryptanalysis of the Oil and Vinegar Signature Scheme*, Crypto 1998, LNCS v. 1462, p. 257–266.

[MI88]   T. Matsumoto and H. Imai, *Public Quadratic Polynomial-Tuples for Efficient Signature-Verification and Message-Encryption*, Eurocrypt 1988, LNCS v. 330, p. 419–453.

[Moh99]   T. Moh, *A Public Key System with Signature and Master Key Functions*, Communications in Algebra, 27 (1999), pp. 2207–2222.

[Nessie]   NESSIE project homepage: http://www.cryptonessie.org.

[Pat95]   J. Patarin, *Cryptanalysis of the Matsumoto and Imai Public Key Scheme of Eurocrypt'88*, Crypto 1995, LNCS v. 963, p. 248–261.

[Pat96]   J. Patarin, *Hidden Fields Equations (HFE) and Isomorphisms of Polynomials (IP): Two New Families of Asymmetric Algorithms*, Eurocrypt 1996, LNCS v. 1070, p. 33–48.

[PGC98]   J. Patarin, L. Goubin, and N. Courtois, C^*_{-+} *and HM: Variations Around Two Schemes of T. Matsumoto and H. Imai*, Asiacrypt 1998, LNCS v. 1514, p. 35–49.

[PCG01]   J. Patarin, N. Courtois, and L. Goubin, *QUARTZ, 128-Bit Long Digital Signatures*, CT-RSA'01, LNCS v. 2020, p. 282–297. Update available at [Nessie].

[PCG01a]   J. Patarin, N. Courtois, and L. Goubin, *FLASH, a Fast Multivariate Signature Algorithm*, CT-RSA 2001, LNCS v. 2020, p. 298–307. Update available at [Nessie].

[Sho94]   P. W. Shor, *Algorithms for quantum computation: Discrete logarithms and factoring*, Proc. 35th Annual Symposium on Foundations of Computer Science (S. Goldwasser, ed.), IEEE Computer Society Press (1994), p. 124-134.

[WC04]   L.-C. Wang, and F.-H. Chang, *Tractable Rational Map Cryptosystem*, manuscript, E-Print Archive 2004/046.

[WC05]   L.-C. Wang, and F.-H. Chang, *Revision of Tractable Rational Map Cryptosystem*, manuscript, on the E-Print Archive.

[WHL$^+$05]   L.-C. Wang, Y.-H. Hu, F.-P. Lai, C.-Y. Chou, and B.-Y. Yang, *Tractable Rational Map Signature*, PKC 2005, LNCS v. 3386, p. 244–257.

[Wol04]   C. Wolf, *Efficient Public Key Generation for Multivariate Cryptosystems*, Proc. ERACOM Conference and Workshop on Cryptographic Algorithms and their Uses, July 5-6, 2004, also see E-Print Archive 2003/089.

[WBP04]   C. Wolf, A. Braeken, and B. Preneel, *Efficient Cryptanalysis of RSE(2)PKC and RSSE(2)PKC*, SCN '04, LNCS v. 3352, p. 294–309.

[WP05]   C. Wolf and B. Preneel, *Taxonomy of Public-Key Schemes based on the Problem of Multivariate Quadratic Equations*, manuscript, E-Print Archive 2005/077.

[WP05a]   C. Wolf and B. Preneel, *Superfluous keys in Multivariate Quadratic asymmetric systems*, PKC 2005, LNCS v. 3386, p. 275–287. Extended version at E-Print Archive 2004/361.

[YC04]   B.-Y. Yang and J.-M. Chen, *All in the XL Family: Theory and Practice*, ICISC 2004, LNCS v. 3506, p. 67–86.

[YC05]   B.-Y. Yang and J.-M. Chen, *Rank Attacks and Defence in Tame-Like Multivariate PKC's*, ACISP 2005, LNCS v. 3574, p. 518–531. Older version at E-Print Archive 2004/061.

[YCCh04]   B.-Y. Yang, J.-M. Chen, and Y.-H. Chen, *TTS: High-Speed Signatures from Low-End Smartcards,*, CHES 2004, LNCS v. 3156, p. 371-385.

[YCCo04]   B.-Y. Yang, J.-M. Chen, and N. Courtois, *On Asymptotic Security Estimates in XL and Gröbner Bases-Related Algebraic Cryptanalysis*, ICICS 2004, LNCS v. 3269, p. 401-413.

A   A Brief Description of Rank Attacks

We describe the linear algebra based attacks briefly.

Separation of Oil-and-Vinegar: Consider a set of polynomials $p_i(x_1, \ldots, x_n)$ where the set of variables $\{x_1, \ldots, x_n\}$ can be partitioned into disjoint portions $\mathcal{V} \uplus \mathcal{O}$, such that no quadratic term has both factors in the *oil set* $\mathcal{O}$. If we specify each variable in the *vinegar set* $\mathcal{V}$, we can solve for variables in $\mathcal{O}$ as a linear system. This is called a UOV structure.

Kipnis *et al* attacked UOV structures by distilling the *oil subspace* [KPG99, KS98]. If the size of the minimal vinegar set in the central equations is v, then we can find the subspace spanned by the oil variables in $q^{2v-n-1}(n-v)^4$ field multiplications.

We may have to manuever further, but such a distillation usually leads to a cryptanalysis of the scheme. A program to find maximal cliques can verify that if we ignore the q_i terms and work with $\mathbb{L}$, then $m = 15$, $n = 12$, $v = 9$. So even for $\mathbb{L} = \mathrm{GF}(2^{32})$ of the "mini" version (cf. Sec. 5), the security level way above 2^{100}, high enough. This seems reasonable because the Kipnis attack seems more inclined toward signature schemes.

High Rank Attack: We can associate with every quadratic polynomial a symmetric matrix. To be exact $p = \sum_{i \leq j} a_{ij} x_i x_j + \sum_i b_i x_i$ corresponds (for char $= 2$) to $\mathrm{M}_p := [A_{ij}]$, where $A_{ij} = a_{ij}$ if $i < j$, a_{ji} if $i > j$, and 0 if $i = j$. Usually $r = \mathrm{rank}\,\mathrm{M}_p$, if and only if we may write $\sum_{i \leq j} a_{ij} x_i x_j = \sum l_a l_b$ for a minimum independent set of linear forms $l_1, \ldots, l_r$.

Equations in the public key tend to be full rank as are most of their linear combinations. However, when a variable x_i does not appear in a polynomial p, the associated matrix will be singular, i.e., $\mathrm{rank}\,\mathrm{M}_p < n$. Thus, if some variable appears in only one central equation, for most pairs of public polynomials (p_i, p_j), we can find a linear combination $p_i + \lambda_{ij} p_j$ that is less than full rank. The same goes for linear combinations of $(u + 1)$-tuples of public polynomials if a variable appears only in u central equation. A simple and a more algebraic (and complete) implementation of this idea is given by [GC00] and [CSV93] respectively.

All told, this attack costs around $\left(un^2 + \frac{n^3}{6}\right) q^u$ multiplications if all goes correctly. Here, each X_i appears at least in 4 equations, so even for $\mathbb{L} = \mathrm{GF}(2^{32})$ in MFE-0 (our "mini" scheme) in Sec. 5, we have a security level above 2^{128}. It is quite a bit higher for the "regular" scheme MFE-1.

Low Rank Attack: This is approximately dual to the previous attack.

If p has rank r, then a random vector $\mathbf{x}$ satisfy $\mathrm{M}_p \mathbf{x} = 0$ with probability q^{-r}. We guess at $\mathbf{x}$ and try to solve for the linear combination that is M_p. For encryption schemes $m > n$, so there are too many matrices spanned by those corresponding to the public polynomials. In this case we must guess at $x_1, \ldots, x_k$, where $k = \lceil m/n \rceil$. This makes the linear system $\sum \lambda_i M_{p_i} x_j = 0$, $j = 1 \cdots k$ in the λ_i overdetermined. If there is a unique linear combination with the minimum rank r, we expect to find it within q^{kr} tries.

This is also known as the MinRank kernel attack. When there are more than one kernel of the same minimal rank that are mostly disjoint, we can do better [YC05]. If there are c such kernels, then we expect to find one within $q^{kr} kmn(m+n)/c$ field multiplications.

Here, $Y_3 = X_3 + \det M_3 + X_1 X_2 + [\vdots]$, or rather its first component, corresponds to a single equation with the smallest rank where $k = 2$, $r = 2$, $q = 2^{32}$ (for the "mini" version MFE-0). Thus the formulas of [GC00, YC05] both gives more than 2^{128} as the security level.

Please refer to [GC00] for details on High and Low Rank attacks, [KPG99, BWP05] on the unbalanced Oil and Vinegar scheme, and [YC05] for a recent summary.

B   Inverting $\mathcal{Q}$ and Circumventing $X_1 = 0$

Here is the last complete algorithm we implemented.

1. First find X_1, X_2, X_3 in a triangular manner from

$$X_1 + Q_1 = Y_1 + \sqrt{(Y_4 \times Y_7 + Y_5 \times Y_6)(Y_8 \times Y_{11} + Y_9 \times Y_{10})(Y_{12} \times Y_{15} + Y_{13} \times Y_{14})^{-1}}$$

$$X_2 + Q_2 = Y_2 + \sqrt{(Y_4 \times Y_7 + Y_5 \times Y_6)(Y_8 \times Y_{11} + Y_9 \times Y_{10})^{-1}(Y_{12} \times Y_{15} + Y_{13} \times Y_{14})}$$

$$X_3 + Q_3 = Y_3 + \sqrt{(Y_4 \times Y_7 + Y_5 \times Y_6)^{-1}(Y_8 \times Y_{11} + Y_9 \times Y_{10})(Y_{12} \times Y_{15} + Y_{13} \times Y_{14})}$$

   The actual pre-computations are:
   (a) Calculate $\det(M_1 M_2) = Y_4 \times Y_7 + Y_5 \times Y_6$.
   (b) Calculate $\det(M_1 M_3) = Y_8 \times Y_{11} + Y_9 \times Y_{10}$.
   (c) Calculate $\det(M_2^T M_3) = Y_{12} \times Y_{15} + Y_{13} \times Y_{14}$.
   (d) Calculate $\det M_1 = \sqrt{\det(M_1 M_2) \det(M_1 M_3)/\det(M_2^T M_3)}$.
   (e) Calculate $\det M_2 = \det(M_1 M_2)/\det M_1$, $\det M_3 = \det(M_1 M_3)/\det M_1$.
   (f) Calculate $Y_1 + Q_1$, $Y_2 + Q_2$, $Y_3 + Q_3$ and the triangular substitutions.

2. if $X_1 \neq 0$ compute M_1^{-1} and thereby M_2 and M_3, and we are done.
3. if $X_1 = 0$, we let $B = (\det(M_2^T M_3))^{-1}$ and compute $A = X_2^{-1}$, then

$$X_7 = Y_4 A$$
$$X_8 = Y_6 A$$
$$X_{11} = Y_8 A$$
$$X_{12} = Y_9 A$$

$$X_9 = \det(M_3)\, B\, (Y_{12}X_8 + Y_{14}X_7)$$
$$X_{10} = \det(M_3)\, B\, (Y_{13}X_8 + Y_{15}X_7)$$
$$X_5 = \det(M_2)\, B\, (Y_{12}X_{12} + Y_{13}X_{11})$$
$$X_6 = \det(M_2)\, B\, (Y_{14}X_{12} + Y_{15}X_{11})$$
$$X_4 = \det(M_3)\, B\, (Y_6 X_6 + Y_7 X_5)$$

Note that this avoids trouble if any other variable vanishes! We can also see that this case takes 1 fewer multiplication and 4 fewer additions after a careful count, and should pad the time upwards with some delaying action.

A New Security Proof for Damgård's ElGamal

Kristian Gjøsteen*

Department of Telematics,
Norwegian University of Science and Technology, 7491 Trondheim, Norway
kristian.gjosteen@item.ntnu.no

Abstract. We provide a new security proof for a variant of ElGamal
proposed by Damgård, showing that it is secure against non-adaptive
chosen ciphertext attack. Unlike previous security proofs for this cryptosystem, which rely on somewhat problematic assumptions, our underlying problem is similar to accepted problems such the Gap and Decision
Diffie-Hellman problems.

1 Introduction

Damgård [7] defined a variant of the ElGamal public key cryptosystem, and proposed a proof of security against non-adaptive chosen ciphertext (CCA1) attacks
based on an assumption now commonly known as the knowledge-of-exponent
assumption [10, 2, 3]. Unfortunately, the knowledge-of-exponent assumption is
a somewhat strange and impractical assumption, and it would be better if we
could do without it.

In [3], one security proof is given for Damgård's cryptosystem and one for
the so-called CS-lite scheme [6], the latter modeled on the former. We propose a
security proof for Damgård's cryptosystem modeled upon the real security proof
for CS-lite.

CS-lite essentially consists of two parts, an ϵ-smooth and an ϵ-universal hash
proof system (HPS). The ϵ-smooth HPS is used to hide the message, and the
ϵ-universal HPS ensures that an adversary cannot create valid ciphertexts with
a certain property.

Damgård's cryptosystem does not include the ϵ-universal HPS, but instead
verifies the property for each ciphertext directly, discarding discarding such ciphertexts. Except for this, the two systems are essentially identical. This suggest
that the ϵ-universal hash proof system in CS-lite points out exactly what we need
in order to prove Damgård's scheme secure without the knowledge-of-exponent
assumption.

We therefore propose a new problem similar to conventional problems such
as the Gap Diffie-Hellman problem. If the new problem is hard, we are able to
show that Damgård's cryptosystem is semantically secure against a non-adaptive
chosen ciphertext attack.

Our new assumption is defined in Sect. 2 and the new security proof is in
Sect. 3. But first we need to recall the definition of hash proof systems in Sect. 1.1.

---

* Supported by the Norwegian Research Council.

D. Pointcheval (Ed.): CT-RSA 2006, LNCS 3860, pp. 150–158, 2006.

We actually define a family of cryptosystems. Note that for the instantiation based on Paillier's Composite Residuosity, the knowledge of exponent assumption does not hold for the relevant subgroup, so the approach of [3] would not work.

While all of the instantiations described are homomorphic, the scheme based on the Decision Composite Residuosity problem is additively homomorphic (multiplying ciphertexts corresponds to adding messages), which is a very useful property.

1.1 Oracle Smooth Projective Hash Proof Systems

We refer the reader to [5] for more information about projective hash families and hash proof systems.

Let G be a set and let H be a subset of G. We say that a set W is a *witness set* for H if there is an easily computable bijection $\rho : W \to H$. This bijection allows one to prove that an element $x \in G$ really is in H by presenting an element $w \in W$ such that $\rho(w) = x$. This obviously assumes that it is easy to recognize elements of W.

For two sets S, S', denote by $\mathrm{Map}(S, S')$ the set of maps from S to S'. Let L be a finite, abelian group. We are interested in looking at maps from G to L. There is a natural map $\mathrm{Map}(G, L) \to \mathrm{Map}(H, L)$ given by restriction. From ρ we get a bijection $\rho^* : \mathrm{Map}(H, L) \to \mathrm{Map}(W, L)$. We also denote the natural map $\mathrm{Map}(G, L) \to \mathrm{Map}(W, L)$ by ρ^*.

A *projective hash family* is a tuple $(G, H, L, L', W, \rho, M)$, where G is a set, H is a subset of G, L is a group, L' is a subgroup of L, W is a witness set for H with isomorphism ρ, M is a subset of $\mathrm{Map}(G, L)$.

Let x be sampled randomly from $G \setminus H$, let f be sampled uniformly at random from M, and let y be sampled uniformly at random from L'. We say that the projective hash family is (t, ϵ)-*oracle-smooth* if for any algorithm A with runtime less than t that accepts as input triples from $(G \setminus H) \times \mathrm{Map}(W, L) \times L$ and outputs a bit b, we have that

$$\mathrm{Adv}_A^{SHF(G,H,L,L',W,\rho,M)} =$$
$$|\Pr[A(x, \rho^*(f), f(x)) = 1] - \Pr[A(x, \rho^*(f), f(x)y) = 1]| < \epsilon.$$

The algorithm A is given access to an oracle that evaluates f at any point in H, but refuses to evaluate f on points in $G \setminus H$.

We note that the definitions given in [5] are statistical, but our definition is computational. For statistical results, we omit the time bound t.

As usual, a *hash proof system* P for (G, H) is a projective hash family $(G, H, L, L', W, \rho, M)$ along with efficient algorithms for sampling M and W, and evaluating the functions $f \in M$ and $\rho^*(f) \in \mathrm{Map}(W, L)$ (given descriptions of the functions). The algorithm for evaluating f is called the *private evaluation algorithm*, and the algorithm for evaluation $\rho^*(f)$ is called the *public evaluation algorithm*.

The sampling algorithms sample from some distribution that is δ-close to the uniform distribution (that is, the statistical distance between the distributions is at most δ). We say that δ is the hash proof system's *approximation error*.

We shall provide constructions for smooth hash proof systems in the next section.

2 Oracle Subgroup Membership Problems

Let G be a finite, abelian group. Let H be a non-trivial, proper subgroup of G, and suppose that there is a subgroup J of G such that $H \cap J$ is trivial and $HJ = G$. Then $H \times J \simeq G$.

There is an isomorphism $H \times J \to G$ given by $(x, y) \mapsto xy$. Let $\sigma : G \to H \times J$ be the inverse of this isomorphism. The *splitting* problem for (G, H, J) is to compute this map for a given random $z \in G$.

The *subgroup membership* problem for (G, H, J) is to decide if a *challenge element* $x \in G$ is sampled uniformly from H or from $G \setminus H$. (A stronger assumption [9] samples the challenge element (x, y) uniformly either from $H \times J$ or from $(G \times G) \setminus (H \times J)$.)

The *gap splitting* problem for (G, H, J) is the same as the splitting problem, but any solver is given access to an oracle that for any element $x \in G$ returns 1 if $x \in H$, and 0 otherwise. This oracle is called the *gap oracle*.

We propose the following new problem.

Definition 1. *The* oracle subgroup membership *problem for* (G, H, J) *is the following problem. The adversary receives* (G, H, J) *and has free access to a gap oracle for* (G, H, J). *After some computation, he requests the challenge element* $x \in G$. *He must then decide if* $x \in H$ *or if* $x \in G \setminus H$, *but he may no longer query the gap oracle.*

The adversary must be prevented from using the gap oracle after he has received the challenge element to keep the problem from being trivial.

A *decision function* for (G, H, J) is a function $d : G \to \{0, 1\}$ such that $d(x) = 1$ if and only if $x \in H$. We say that (G, H, J) is a *trapdoor (oracle) subgroup membership problem* if there exists an efficiently computable decision function.

We let $\mathrm{Adv}_A^{OSM(G,H,J)}$ denote the advantage of the adversary A against the (trapdoor) oracle subgroup membership problem. If E is the event that the adversary A decides correctly,

$$\mathrm{Adv}_A^{OSM(G,H,J)} = 2|\mathrm{Pr}[E] - 1/2|.$$

Note that the gap splitting problem and the oracle subgroup membership problem are interactive problems. Given the splitting and subgroup membership problems, the gap and oracle problems follow immediately by adding the appropriate oracles.

2.1   Diffie-Hellman Problems

Let L be a cyclic group of order p generated by g. We write the group multiplicatively.

The *discrete logarithm to base g* is the group isomorphism $\log_g : L \to \mathbb{Z}_p$ given by $\log_g g = 1$.

The *Computational Diffie-Hellman (CDH)* problem [8] in L is, for random $x, y \in L$, find z such that $\log_g z = \log_g x \log_g y$.

The *Decision Diffie-Hellman (DDH)* problem [4] in L is, for random $x, y \in L$ and z, decide if z is random or if $\log_g z = \log_g x \log_g y$.

The *Strong Diffie-Hellman (SDH)* problem [1] in L is the same as the CDH problem, except that the problem solver is given access to a DDH oracle (on input of $(y', z') \in L \times L$, it decides if $\log_g z' = \log_g x \log_g y'$).

Let $G = L \times L$, let H be the subgroup generated by (g, x) (where x is a random, non-trivial element), and let J be the subgroup generated by $(1, g)$. Then CDH is the splitting problem (G, H, J), DDH is the subgroup membership problem (G, H, J), and SDH is the gap splitting problem (G, H, J).

If $a = \log_g x$, then $(y, z) \in H$ if and only if $z = y^a$. This means that a decision function exists.

Remark 1. To get a proper equivalence of problems, we need to consider the group H as sampled from some distribution.

Remark 2. There is a stronger version of SDH known as the *Gap Diffie-Hellman* problem [11], where the gap oracle answers queries for arbitrary x.

Hash Proof System. We describe a projective hash family $(G, H, L, L', W, \rho, M)$.

Let $L' = L$. Let (g_1, g_2) be a generator for H. Let $W = \mathbb{Z}_p$, and $\rho(w) = (g_1, g_2)^w$. Finally, let M be the set homomorphisms $G \to L$ of the form $(x, y) \mapsto x^{k_1} y^{k_2}$, where $(k_1, k_2) \in \mathbb{Z}_p \times \mathbb{Z}_p$.

We note that for any $f(x, y) = x^{k_1} y^{k_2}$ in M, the element $f(g_1, g_2) \in L$ is sufficient to allow efficient computation of $\rho^*(f)$, since $\rho^*(f)(w) = (g_1^w)^{k_1} (g_2^w)^{k_2} = (g_1^{k_1} g_2^{k_2})^w = f(g_1, g_2)^w$.

This family is statistically 0-smooth [5]. Any sampling can be done uniformly at random by sampling uniformly from $\{0, \ldots, p-1\}$, so the approximation error is 0.

2.2   Composite Residuosity Problems

Let $n = pq$ be an RSA modulus where p and q are safe primes and $\gcd(n, \phi(n)) = 1$. Let G be the subgroup of quadratic residues in $\mathbb{Z}_{n^2}^*$, let H be the subgroup isomorphic to the subgroup of quadratic residues in $\mathbb{Z}_n^*$, and let J be the subgroup generated by residue class containing $1 + n$.

We get a splitting problem (G, H, J) called the *Computational Composite Residuosity (CCR)* problem, a subgroup membership problem (G, H, J) called the *Decision Composite Residuosity (DCR)* problem. These were first proposed by Paillier [12].

A decision function exists, since $z \in H$ if and only if $z^{|H|} = 1$.

Remark 3. The knowledge-of-exponent assumption does not hold for the subgroup H, since we have the map $x \mapsto x^n$ taking elements of G to elements of H.

Hash Proof System. We describe a projective hash family $(G, H, L, L', W, \rho, M)$.

Let g be a generator for H. Let $L = G$, $L' = J$, $M = \mathrm{Hom}(G, G)$, and $W = \mathbb{Z}_{\phi(n)}$. Let $[w] \in \mathbb{Z}_n$ be the residue class represented by $w \in \mathbb{Z}$. Then $\rho([w]) = g^w$.

Any $f \in \mathrm{Hom}(G, G)$ is of the form $x \mapsto x^k$. Therefore, the elements in the set $\{0, \ldots, n\phi(n)/4 - 1\}$ are useful descriptions of the homomorphisms. Again, $f(g)$ is a useful description of $\rho^*(f)$, since $\rho^*(f)([w]) = f(g^w) = (g^k)^w = f(g)^w$.

Assume $p < q$. As was shown in [5], this hash-family is $1/p$-smooth.

Without p and q, we cannot sample elements uniformly from $\mathrm{Hom}(G, G) \simeq \mathbb{Z}_{\phi(n)}$, but we can sample $4/(p - 1)$-close to uniformly by sampling uniformly from the set $\{0, \ldots, \lfloor n/4 \rfloor - 1\}$.

Given only n, we cannot sample uniformly from W, but by sampling uniformly from $\{0, \ldots, \lfloor n/4 \rfloor - 1\}$, we get a sample distribution on W that is $4/p$-close to uniform. Therefore, the approximation error is $4/p$.

2.3   Symmetric Subgroup Membership Problems

Let $n = pq$ be an RSA modulus such that $p' = 2n + 1$ is a prime. Let G be the subgroup of quadratic residues in $\mathbb{F}_{p'}^*$, let H be the subgroup of order p and J be the subgroup of order q.

Alternatively, let a, b, c, d be primes such that $p = 2ab + 1$ and $q = 2cd + 1$ are prime. Set $n = pq$. Let G be the subgroup of $\mathbb{Z}_n^*$ with Jacobi symbol 1, let H be the subgroup of order $2ac$ and J be the subgroup of order bd.

A decision function exists, since $z \in H$ if and only if $z^{|H|} = 1$.

These problems are further discussed in [9].

Hash Proof System. We describe a projective hash family $(G, H, L, L', W, \rho, M)$.

Let g be a generator for H. Let $L = G = L'$, $W = \mathbb{Z}_{|H|}$, $\rho([w]) = g^w$, where $[w]$ is the residue class represented by $w \in \mathbb{Z}$, and $M = \mathrm{Hom}(G, G)$.

Any $f \in \mathrm{Hom}(G, G)$ is of the form $x \mapsto x^k$. Therefore, the elements in the set $\{0, \ldots, |G| - 1\}$ are useful descriptions of the homomorphisms. Again, $f(g)$ is a useful description of $\rho^*(f)$, since $\rho^*(f)(w) = f(g^w) = (g^k)^w = f(g)^w$.

Let ℓ be the smallest prime dividing $|J|$. As was shown in [9], this hash-family is $(t, 1/\ell + \epsilon')$-smooth, where ϵ' is the advantage of any algorithm with run-time at most t against the oracle subgroup membership problem (G, J, H).

In the finite field case we know $|G|$ and can sample elements uniformly from $\mathrm{Hom}(G, G) \simeq \mathbb{Z}_{|G|}$. In the ring case $n/4$ is a usable approximation of $|G|$. If we know that $|H| < 2^t$, we can sample 2^{-t_0}-close to uniform from W by sampling from $\{0, \ldots, 2^{t+t_0} - 1\}$.

3   The Cryptosystem

First we describe the cryptosystem based on a trapdoor oracle subgroup membership problem (G, H, J) with a decision function d, and a hash proof system P with projective hash family $(G, H, L, L', W, \rho, M)$.

Key generation. The input is the hash proof system. A function f from M is sampled using the sampling algorithms of P. The public key is $pk = (G, L, L', W, \rho, \rho^*(f))$, the private key is $sk = (G, H, L, L', f, d)$.

Encryption. The input is the public key and a message $m \in L'$. An element w is sampled uniformly at random from W and $\rho * (f)(w)$ is computed using the public evaluation algorithm. The ciphertext is $(\rho(w), \rho^*(f)(w)m)$.

Decryption. The input is the secret key and a ciphertext $(x, y) \in G \times L$. The algorithm first verifies that $x \in H$ by computing $d(x)$. If $x \notin H$, failure is reported and the ciphertext is discarded. Otherwise, the message $yf(x)^{-1} \in L'$ is returned.

Before we give the new security proof, we show that Damgård's cryptosystem is really the same as our cryptosystem instantiated with the Diffie-Hellman group structure.

In this case, the key generation algorithm does as follows. It selects a group L of order p with a generator g_1, samples a uniformly from $\{0, \ldots, p-1\}$ and sets $g_2 = g_1^a$. The subgroup H is generated by (g_1, g_2). Then the key generation algorithm samples k_1 and k_2 uniformly from $\{0, \ldots, p-1\}$ and computes $s = g_1^{k_1} g_2^{k_2}$. The public key is then (L, g_1, g_2, s), the private key is (L, a, k_1, k_2).

The encryption algorithm samples w uniformly from $\{0, \ldots, p-1\}$, computes the ciphertext as $(x_1, x_2, y) = (g_1^w, g_2^w, s^w m)$.

The decryption algorithm checks that $(x_1, x_2) \in H$ by checking that $x_1^a = x_2$. If it is, it returns the message $yx_1^{-k_1}x_2^{-k_2}$.

In Damgård's original scheme, the public key also consists of (L, g_1, g_2, s), but s is computed as g_1^b for a random b. Our key generation algorithm yields the exact same key distribution as Damgård's scheme. Indeed, if $b \equiv k_1 + ak_2$ (mod p) the public keys would be equal.

Theorem 1. *Let (G, H, J) be a trapdoor oracle subgroup membership problem and let P be a hash proof system with projective hash family $(G, H, L, L', W, \rho, M)$ and approximation error δ. Let A be a non-adaptive chosen ciphertext adversary against the semantic security of the cryptosystem based on P. Then*

$$\mathrm{Adv}_A \leq \mathrm{Adv}_{A'}^{OSM(G,H,J)} + \mathrm{Adv}_{A''}^{SHF(G,H,L,L',W,\rho,M)} + 2\delta,$$

where the algorithms A' and A'' have essentially the same run-time as A.

Proof. We use the standard techniques of game-hopping.

Game 0. The initial game is the usual non-adaptive chosen ciphertext attack against semantic security, which proceeds as follows: A simulator runs the key generation algorithm with the hash proof system as input and receives a public and a private key. It gives the public key to the adversary, and responds to the adversary's decryption queries by running the decryption algorithm with the secret key.

When the adversary outputs his chosen messages, the simulator runs the encryption algorithm on one message chosen at random and gives the resulting challenge ciphertext to the adversary. The adversary then tries to guess which message was encrypted.

If the adversary guesses correctly, we say that he wins the game.

Game 1. The first modification we make is to sample the element of W used for creating the challenge ciphertext from the uniform distribution, not via P's sampling algorithm. The difference in game behaviour is bounded by the approximation error δ.

Game 2. Next, instead of applying $\rho^*(f)$ to an element of W in the encryption, we sample from the uniform distribution on H and apply f. This is a purely conceptual change, and the game behaviour does not change.

Game 3. Now we sample not from H, but from $G \backslash H$ when creating the challenge ciphertext. We claim that there is a adversary A' against the trapdoor oracle subgroup membership problem (G, H, J) whose advantage is equal to the change in behaviour.

The algorithm A' takes (G, H, J) as input. To simulate the key generation, the hash proof system's sampling algorithms are used to construct the public and private keys. Obviously, A' does not know d. Therefore, the private key will be deficient and the decryption algorithm must be changed.

When the adversary requests decryptions, the gap oracle is used to check that the group element really is in H. If it is, f is used to decrypt the message.

When the adversary submits its messages, the algorithm requests its challenge element x and computes the challenge ciphertext as $(x, f(x)m)$. If the adversary guesses correctly, we output 1, otherwise 0.

If $x \in H$, everything proceeds as in Game 2. If $x \in G \backslash H$, everything proceeds as in Game 3.

Game 4. To prepare for the adversary against the hash proof system, we sample the function f from the uniform distribution on M. The difference in behaviour is bounded by the approximation error δ.

Game 5. In this game we compute the ciphertext as $f(x)ym$, where y is sampled uniformly from L'. We claim that there is an adversary A'' against the hash proof system whose advantage is equal to the change in behaviour.

The algorithm takes $(G, H, L, L', W, \rho, M)$, $x \in G \setminus H$, $\rho^*(f)$ and $z \in G$ as input. It constructs the public key from its input. To answer decryption queries (x', y'), it passes x' onto its evaluation oracle. If the oracle refuses to answer, then $x' \notin H$ and the ciphertext does not decrypt. If the oracle replies with z, the decryption $y'z^{-1}$ is returned.

When the adversary submits its messages, the challenge ciphertext is (x, zm). If the adversary guesses correctly, we output 1, otherwise 0.

If $z = f(x)$, then everything proceeds as in Game 4. If $z = f(x)y$, where y is a random element of L', then everything proceeds as in Game 5.

To conclude the proof, we simply note that in Game 5, the distribution of the ciphertext is independent of the message, the adversary gets no information about the message, and therefore he has no advantage. $\qquad\square$

It is worthwhile to note that all of the instantiations of this scheme are homomorphic, so they are not secure against adaptive chosen ciphertext attacks. When instantiated with the Decision Composite Residuosity problem, the scheme can easily be made additively homomorphic, a property that is very useful.

4   Concluding Remarks

We have given a security proof for a generalized variant of Damgård's ElGamal. The underlying assumption is a new problem called (trapdoor) oracle subgroup membership problem. The new problem seems much more credible than the knowledge of exponent assumption that has previously been used to prove the security. The security proof also highlights the relationship between Damgård's ElGamal and CS-lite nicely, where the subgroup membership check in the former performs the same function as the verification of ϵ-universal hash in the latter.

While the new problem is similar to established problems such as the gap splitting problems, it is in certain respects very different. The gap splitting problems is hard if the splitting problem does not reduce to the subgroup membership problem. This means that studying the gap problem is studying whether efficient reductions between to related problems exists. There is no similar way to look at the oracle subgroup membership problem. It still remains to be seen if the oracle subgroup membership problem can find other applications in cryptography.

Acknowledgments

The author would like to thank Alex Dent for very useful discussions, and the anonymous referees for helpful comments.

References

1. Michel Abdalla, Mihir Bellare, and Phillip Rogaway. The Oracle Diffie-Hellman assumptions and an analysis of DHIES. In David Naccache, editor, *Proceedings of CT-RSA 2001*, volume 2020 of *LNCS*, pages 143–158. Springer-Verlag, 2001.
2. Mihir Bellare and Adriana Palacio. The knowledge-of-exponent assumptions and 3-round zero-knowledge protocols. In Matthew K. Franklin, editor, *Proceedings of CRYPTO 2003*, volume 3152 of *LNCS*, pages 273–289. Springer-Verlag, 2004.
3. Mihir Bellare and Adriana Palacio. Towards plaintext-aware public-key encryption without random oracles. In Pil Joong Lee, editor, *Proceedings of ASIACRYPT 2004*, volume 3329 of *LNCS*, pages 48–62. Springer-Verlag, 2004.
4. D. Boneh. The Decision Diffie-Hellman problem. In *Proceedings of the Third Algorithmic Number Theory Symposium*, volume 1423 of *LNCS*, pages 48–63. Springer-Verlag, 1998.
5. Ronald Cramer and Victor Shoup. Universal hash proofs and a paradigm for adaptive chosen ciphertext secure public-key encryption. In Lars R. Knudsen, editor, *Proceedings of EUROCRYPT 2002*, volume 2332 of *LNCS*, pages 45–64. Springer-Verlag, 2002.

6. Ronald Cramer and Victor Shoup. Design and analysis of practical public-key encryption schemes secure against adaptive chosen ciphertext attack. *SIAM Journal on Computing*, 33(1):167–226, 2003.
7. Ivan Damgård. Towards practical public key systems secure against chosen ciphertext attacks. In Joan Feigenbaum, editor, *Proceedings of CRYPTO '91*, volume 576 of *LNCS*, pages 445–456. Springer-Verlag, 1992.
8. W. Diffie and M. E. Hellman. New directions in cryptography. *IEEE Transactions on Information Theory*, 22:644–654, 1976.
9. Kristian Gjøsteen. Symmetric subgroup membership problems. In Serge Vaudenay, editor, *Proceedings of Public Key Cryptography 2005*, volume 3386 of *LNCS*, pages 104–119. Springer-Verlag, 2005.
10. Satoshi Hada and Toshiaki Tanaka. On the existence of 3-Round zero-knowledge protocols. In Hugo Krawczyk, editor, *Proceedings of CRYPTO '98*, volume 1462 of *LNCS*, pages 408–423. Springer-Verlag, 1998.
11. Tatsuaki Okamoto and David Pointcheval. The gap-problems: A new class of problems for the security of cryptographic schemes. In Kwangjo Kim, editor, *Proceedings of Public Key Cryptography 2001*, volume 1992 of *LNCS*, pages 104–118. Springer-Verlag, 2001.
12. P. Paillier. Public-key cryptosystems based on composite degree residue classes. In Jacques Stern, editor, *Proceedings of EUROCRYPT '99*, volume 1592 of *LNCS*, pages 223–238. Springer-Verlag, 1999.

Stand-Alone and Setup-Free Verifiably Committed Signatures

Huafei Zhu and Feng Bao

Department of Information Security, I²R, A-Star, Singapore 119613
{huafei, baofeng}@i2r.a-star.edu.sg

Abstract. In this paper, a novel construction of stand-alone and setup-free verifiably committed signatures from RSA − an open problem advertised by Dodis and Reyzin in their speech [16] is presented. The methodology used in this paper is reminiscent of the concept of verifiably encrypted signatures introduced by Asokan et al [1, 2]. We suggest to encrypt only a random salt used to generate a virtual commitment that will be embedded into Cramer-Shoup's signature scheme and to prove the validity of the signature with respect to this encrypted value. Our construction is provably secure assuming that the underlying Cramer-Shoup's signature scheme is secure against adaptive chosen-message attack, and Paillier's encryption is one-way. We thus provide an efficient solution to Dodis-Reyzin's open problem.

Keywords: Off-line fair-exchange, Setup-free, Stand-alone property, Verifiably committed signature.

1 Introduction

The research of fair exchange protocols has a rich history due to its fundamental importance. In the following, we only briefly mention the body of research most relevant to our results, and refer the reader to [1, 2, 3, 4, 8, 11, 13, 17, 23, 24, 26] for general references.

A fair-exchange protocol typically consists of three participants: a client (a primary signer), a merchant (a verifier) and a trusted third party (TTP). TTP can be on-line at the expense of the TTP becoming a potential bottleneck, or off-line, meaning that it only gets involved when something goes wrong. Off-line fair-exchange protocols can be classified into two categories: with or without initial-key-setup procedures. An off-line fair-exchange protocol is called *setup-free* if no initial-key-setup procedure between a primary signer and its TTP is involved except for one requirement that the primary signer can obtain and verify TTP's certificate and vice versa. An off-line fair-exchange protocol is called *setup-driven* if an initial-key-setup protocol between a primary signer and its TTP must be involved such that at the end of the key setup protocol, the primary signer and its TTP share prior auxiliary information. This shared auxiliary information enables TTP to convert any valid partial signature into the corresponding full signature if a confliction occurs between the primary signer and its verifier (and thus the fairness of protocols can be achieved inherently).

D. Pointcheval (Ed.): CT-RSA 2006, LNCS 3860, pp. 159–173, 2006.

Dodis and Reyzin [16] have already formalized a unified model for off-line fair-exchange protocols as a new cryptographic primitive called verifiably committed signatures. Verifiably committed signatures are the following things: a primary signer Alice can produce a partial signature to her verifier Bob; upon receiving what she needs from Bob, she can convert it to a full signature. If she refuses, a trusted third party Charlie (arbitrator) can do it for her upon the receipt of a partial signature and proper verification that Bob fulfilled his obligation to Alice.

A sibling notion of off-line fair-exchange protocols is verifiably encrypted signatures which was introduced by Asokan et al [1, 2] in 1988. Verifiably encrypted signatures are the following things: a primary signer Alice wants to show her verifier Bob that she has signed a message, but does not want to Bob to possess her signature of that message, and Alice will give her signature to Bob only when she gets what she needs from Bob. Alice can achieve this by encrypting her signature using the public key of a TTP, and sending this to Bob along with a proof that she has given him a valid encryption of her signature. Bob can verify that Alice has signed the message, but cannot deduced any information about her signature. Later, in the protocol, if Alice is unable to or unwilling to reveal her signature, Bob can ask the trusted third party Charlie (arbitrator/TTP) to reveal Alice's signature.

1.1   Previous Works Within Setup-Driven Model

Notice that the existence of off-line fair-exchange protocol within setup-driven model is obvious in the standard complexity model assuming that two underlying signatures are secure in the sense of [21] in the standard complexity model. That is, suppose a primary signer has public/secret key (pk_1, sk_1) for the first signature scheme, and at the same time the prime signer and its TTP share another public/secret key (pk_2, sk_2) of the second signature scheme. By $PK = (pk_1, pk_2)$ we denote the public key of the entire signature scheme, and by $SK = (sk_1, sk_2)$, we denote the correspondent secret keys. Now given a message m, the primary signer produces its partial signature σ_1 on the message m. A full signature of the message m is defined as $\sigma = (\sigma_1, \sigma_2)$, where σ_2 is the signature of m correspondent the public/secret key pair (pk_2, sk_2). It is easy to verify that the resulted signature is a verifiably committed signature scheme.

The main criticism of the above two-signature based construction is its loose of the stand-alone property. We say a fair-exchange protocol (equivalently, a verifiably committed signature scheme) is stand-alone if any final signature (or any full signature) produced according to the corresponding off-line fair exchange protocol is the same as it were produced by an ordinary signature scheme only, i.e., given a valid partial signature scheme σ', the primary signer de-randomizes the mask of the random salt. The output is an ordinary signature of message m. There are a collection of efficient constructions of fair-exchange protocols with stand-alone property within setup-driven model which are sketched below:

The off-line fair-exchange protocol proposed by Boyd and Foo is setup-driven [10] which is constructed from convertible signatures – a cryptographic primitive first introduced by Boyar, Chaum, Damgård and Pedersen [6]. The general idea of

Boyd and Foo's work is that − a primary signer first registers its public key in its own certificate authority (CA). Then the primary signer and its TTP run a key setup algorithm together for generating public/secret key pair for their specified resolution algorithm. Finally, a partial signature of the transaction information is generated in such a way that the correctness of this partial signature can be verified only by a designated verifier Bob, and at the same time the designed verifier is convinced that the trust third party is able to convert it into a normal signature which anyone could verify.

Alternative attractive work is due to Park, Chong and Siegel [26]. In PODC 2003, Park, Chong and Siegel proposed a light-weight fair-exchange protocol by distributing the computation of RSA signature [26]. The proposed scheme requires an initial-key-setup procedure. Although Park, Chong and Siegel's idea for constructing off-line fair-exchange protocol is very interesting, this protocol is totally breakable in its registration phase. A remedy scheme constructed from the Gap Diffie-Hellman problem has been proposed by Dodis and Reyzin within the setup-driven model [16].

1.2 Previous Works Within Setup-Free Model

The first construction of off-line fair exchange protocols within this category seems Stadler[28]. Since then a collection of nice works are presented (e.g., [1, 2, 3, 4, 11, 24] and so on...). The ideas behind these works are that: a primary signer Alice encrypts her signature using the public key of a TTP, and sends this to her verifier Bob along with a proof that she has given him a valid encryption of her signature. We remark that an encrypted signature together with the validity proof in essence is a partial signature scheme. The revealed Alice's signature is the full signature. Thus, the existence of verifiably encrypted signatures implies the existence of verifiably committed signatures. As a result, the concept of verifiably committed signatures is a general notion of verifiably encrypted signatures [1, 2].

Very recently, Boneh, Gentry, Lynn and Shacham (BGLS) [9] proposed an attractive construction of verifiably encrypted signatures from bilinear aggregate signature schemes. This is a novel construction without any initial-key-setup procedure involved since a primary signer Alice simply needs to register her public key with TTP rather than perform a complex initial key setup protocol. The significant features of the BGLS protocol are its efficiency and its low communication overhead. Since the BGLS scheme is two-signature based, the stand-alone property of the BGLS protocol is completely NOT satisfied ($e(\omega, g_2)= e(h, v)$ $e(\mu, v'))$.

1.3 Problem Statement

We have discussed a collection of off-line fair exchange protocols and showed that each protocol mentioned above has one of shortcomings below:

− an initial key setup procedure is involved (e.g, [10] and [16]);
− stand-alone property is violated (e.g., [29] and [9]).

This leaves an interesting research problem: how to construct stand-alone and setup-free verifiably committed signature with low communication and computation complexity? In fact, Dodis and Reyzin have already advertised a open problem in their speech − constructing stand-alone and setup-free verifiably committed signatures from the RSA problem [16].

1.4   Our Contribution

We propose an efficient construction of stand-alone and setup-free verifiably committed signatures from RSA − an open problem advertised by Dodis and Reyzin in their speech [16]. The methodology used in this paper is reminiscent of the concept of verifiably encrypted signatures introduced by Asokan et al [1, 2]. We here allow a primary signer Alice to encrypt only the random salt used for generating of a probabilistic signatures and to prove the validity of the signature with respect to this encrypted value rather than to encrypt a signature using the public key of a TTP, and to send this to her verifier Bob along with a proof that she has given him a valid encryption of her signature. Our approach is exemplified by virtual commitment-based signatures[1] and Paillier's encryption scheme. In essence, we propose an efficient construction of stand-alone and setup-free verifiably committed signature scheme from Cramer-Shoup's trapdoor hash signature scheme [12] since Cramer-Shoup's trapdoor hash signature scheme provides us an ad hoc approach to construct setup-free verifiably committed signature schemes where a group G can be chosen independently with RSA modulus. We are able to show that our construction is provably secure within our model assuming that the strong RSA problem is hard, and Paillier's encryption is one-way.

2   Virtual Commitment-Based Signatures

2.1   Statistically Hiding and Computationally Binding Commitments

A commitment scheme has a probabilistic polynomial time key generation KG, which on input k, where k stands for security parameter of the commitment scheme, outputs a public key PK, the common reference string. Associated with this public key are a message space m_{PK}, a commitment space c_{PK} and two polynomial time algorithms c_{PK} and d_{PK}. To commit to a message $m \in m_{PK}$, we choose a string r uniformly at random. We then given m, r as input to c_{PK}. The resulting outputs is (c, d), where $c \in c_{PK}$, while d is the auxiliary information needed to open the commitment. Typically $d = (m, r)$. To open a

---

[1] A virtual commitment not a real commitment since the value $H(m)$ is a publicly known datum. Nevertheless, it has the same structure of commonly used commitment schemes such as Pedersen's commitment scheme [27] and Fujisaki-Okamoto commitment scheme [19]. Virtual commitment-based signatures are those derived from virtual commitments, e.g., Cramer-Shoup's signature scheme [12] and Zhu's signature scheme [29], we refer the reader to Section 2 for more details.

message the sender sends d to the receiver. The receiver computes $d_{PK}(c, d)$. When the commitment is constructed as above the output of this computation is m. If something is wrong, e.g, $c \notin c_{PK}$ or d is not a valid opening of the commitment, the output of the decommitment algorithm is $\bot$. There are two kinds of commitments extensively used in security community: 1) Pedersen's commitment scheme and 2) Fujisaki-Okamoto commitment scheme (both commitment schemes are statistically hiding and computationally binding):

1) Pedersen's commitment scheme [27]: Let $p = 2q + 1$ and $G \in Z_p^*$ be a cyclic group. Let g_1 and g_2 be two generators of G of order q such that both discrete logarithm of g_1 in base g_2 and the discrete logarithm of g_2 in base g_1 are unknown by P and V. We denote $C(x, r) = g_1^x g_2^r \bmod p$ a commitment to x in base (g_1, g_2), where $r \in Z_q$ is randomly selected.

2) Fujisaki-Okamoto commitment scheme (this commitment scheme first appeared in [19] and reconsidered by Damgård and Fujisaki [15]): Let s be a security parameter. The public key is a k_2-bit RSA modulus, where P, Q are two large safe primes. We assume that neither P nor V knows factorization N. Let g_1 be a generator of QR_N and g_2 be an element of large order of the group generated by g_1 such that both discrete logarithm of g_1 in base g_2 and the discrete logarithm of g_2 in base g_1 are unknown by P and V. We denote $C(x, r) = g_1^x g_2^r \bmod N$ a commitment to x in base (g_1, g_2), where r is randomly selected over $\{0, 2^s N\}$.

2.2   Derived Signatures

Given an instance of Pedersen's commitment scheme, a secure signature scheme can be derived with minor modification (specifying the security parameters so that the security of the derived signature can be rigorously proved) based on the technique of Cramer and Shoup as follows:

- Key generation algorithm: Let p, q be two large safe primes such that $p - 1 = 2p'$ and $q - 1 = 2q'$, where p', q' are two l'-bit primes. Let $n = pq$ and QR_n be the quadratic residue of Z_n^*. Let x, h be two generators of QR_n. Also chosen are a group G of order s, where s is $(l + 1)$-bit prime, and two random generators g_1, g_2 of G. The public key is (n, h, x, g_1, g_2, H) along with an appropriate description of G including s. The private key is (p, q).
- Signature algorithm: To sign a message m, a $(l+1)$-bit prime e and a string $t \in Z_s$ is chosen at random. The equation $y^e = x h^{H(g_1^t g_2^{H(m)})} \bmod n$ is solved for y. The corresponding signature of the message m is (e, t, y), where the variable $g_1^t g_2^{H(m)}$ is a virtual commitment.
- Verification algorithm: given a putative triple (e, t, y), the verifier first checks that e is an odd $(l + 1)$-bit number. Second it checks the validation that $x = y^e h^{-H(g_1^t g_2^{H(m)})} \bmod n$. If the equation is valid, then the verifier accepts, otherwise, it rejects.

Notice that a virtual commitment embedded into Cramer-Shoup's trapdoor hash signature can be Pedersen's commitment scheme [27] and Fujisaki-Okamoto commitment scheme [19]. The resulting Cramer-Shoup's trapdoor hash signature is immune to adaptive chosen-message attack under joint assumptions of the strong RSA problem as well as the existence of collision free hash function.

We remark the Cramer-Shoup's trapdoor hash signature can be simplified if Fujisaki-Okamoto commitment scheme [19] is defined over G which is described as follows [29].

- Key generation algorithm: Let p, q be two large safe primes (i.e., $p - 1 = 2p'$ and $q - 1 = 2q'$, where p', q' are two primes with length $(l' + 1)$). Let $n = pq$ and QR_n be the quadratic residue of Z_n^*. Let $X, g, h \in QR_n$ be three generators chosen uniformly at random. The public key is (n, g, h, X, H), where H is a collision free hash function with output length l. The private key is (p, q).
- Signature algorithm: To sign a message m, a $(l + 1)$-bit prime e and a string $t \in \{0, 1\}^l$ are chosen at random. The equation $y^e = Xg^t h^{H(m)} \bmod n$ is solved for y. The corresponding signature of the message m is (e, t, y).
- Verification algorithm: Given a putative triple (e, t, y), the verifier checks that e is an $(l + 1)$-bit odd number. Then it checks the validity of $X = y^e g^{-t} h^{-H(m)} \bmod n$. If the equation is valid, then the signature is valid. Otherwise, it is rejected.

Since the group G defined in Cramer-Shoup's trapdoor hash signature scheme, can be chosen arbitrarily. Thus, to construct stand-alone and setup-free verifiably committed signature scheme, it is enough for us to consider an instance of Cramer-Shoup's trapdoor hash signatures only.

3   Syntax and Security Definitions

3.1   Syntax

Definition: A setup-free verifiably committed signature scheme involves a primary signer Alice, a verifier Bob and an arbitrator (or TTP) Charlie, and is given by the following efficient procedures:

- Key generator KG: On input k_1, a primary signer Alice generates (PK, SK), and proves to her CA that these values are correctly generated and then obtains a valid certificate $Cert_A$ from her CA. On input k_2, an arbitrator (co-signer) generates (APK, ASK) and proves to his CA that these values are correctly generated and then obtains a valid certificate $Cert_C$ from his CA. For simplicity, we also assume that Alice and Charlie both obtain a pair of valid certificates $(Cert_A, Cert_C)$ from public bulletin boards.
- Fully signing algorithm Sig and its correspondent verification algorithm Ver: These are conventional signing and verification algorithms. $Sig(m, SK)$ run by the primary signer, outputs a full signature σ on m, while $Ver(m, \sigma, PK)$ run by any verifier, outputs 1 (accept) or 0 (reject);
- Partially signing algorithm $PSig$ and the correspondent verification algorithm $PVer$: These are partial signing and verification algorithms, which are similar to ordinary signing and verification algorithms, except they can

depend on the public arbitration key APK. $PSig(m, SK, PK, APK)$, run
by the primary signer, outputs a partial signature σ', while $PVer(m, \sigma',$
$PK, APK)$, run by any verifier, outputs 1 (accept) or 0 (reject);
- Resolution algorithm Res: This is a resolution algorithm run by the arbitrator in case the primary signer refuses to open her signature σ to the verifier,
  who in turn possesses a valid partial signature σ' on m and a proof that he
  fulfilled his obligation to the primary signer. In this case, Res $(m, \sigma', ASK,$
  $PK)$ should output a valid full signature of m.

Correctness of verifiably committed signatures states that:

- $Ver(m, Sig(m, SK), PK) = 1$;
- $PVer(m, PSig(m, SK, PK, APK), PK, APK) = 1$;
- $Ver(m, Res(PSig(m, SK, PK, APK), ASK, APK, PK), PK) = 1$.

Stand-alone property of verifiably committed signatures states that: given a
valid partial signature scheme $\sigma'(m)$, the primary signer de-randomizes the mask
of the random salt. The output is an ordinary signature of message m respect
to the public key PK.

3.2   The Definition of Security

Recall that the concept of verifiably committed signatures is formalized the same
thing as off-line fair-exchange protocols. Thus the security definition of verifiably
committed signatures should consist of ensuring three aspects: security against a
primary signer Alice, security against a verifier Bob, and security against a arbi-
trator/TTP Charlie. The security definitions of verifiably committed signatures
follows from Dodis and Reyzin [16] and we thus refer the reader to Appendix A
for more details.

4   Building Blocks

In this section, we briefly describe cryptographic primitives that are used to
construct our stand-alone and setup-free verifiably committed signatures.

4.1   Paillier's Public Key Encryption Scheme

Paillier investigated a novel computational problem, called Composite Residuosity Class Problem, and its applications to public key cryptography in [25]. Our
construction will heavily rely on this probabilistic encryption scheme which is
sketched below.

  The public key is a k_1-bit RSA modulus $n = pq$, where p, q are two large safe
primes. The plain-text space is Z_n and the cipher-text space is $Z_{n^2}^*$. To encrypt
$\alpha \in Z_n$, one chooses $r \in Z_n^*$ uniformly at random and computes the cipher-text
as $E_{PK}(a, r) = g^a r^n \bmod n^2$, where $g = (1 + n)$ has order n in $Z_{n^2}^*$. The private
key is (p, q).

The encryption function is homomorphic, i.e., $E_{PK}(a_1, r_1) \times E_{PK}(a_2, r_2)$ mod $n^2 = E_{PK}(a_1 + a_2 \bmod n, r_1 \times r_2 \bmod n)$.

Another interesting result of Paillier's public key encryption scheme is that it can be viewed as a commitment scheme as well since given a cipher-text $c := g^a r^n \bmod n^2$, we first compute $a \in Z_n$ from the following equation $\frac{L(c^\lambda \bmod n^2)}{L((1+n)^\lambda \bmod n^2)} \bmod n$ and then compute r from the equation $r \in Z_n^* = c'^{n^{-1} \bmod \lambda}$ mod n, where $\lambda = lcm(p-1, q-1)$, $n = pq$.

4.2 Proof of Knowledge of Encryptions

Given a $u = g^x$ defined over a cyclic group G, and also given a cipher-text $c = Enc(x)$ computed from Paillier's encryption scheme, the prover should provide a proof that she knows u contains the same value of the encryption, and it lies in a given interval I specified in the protocol. An efficient protocol presented by Damgård and Jurik [14] can be tailored for our purpose.

- Let T be the maximum bit length of x. The prover chooses at random w, an integer of length $T + 2l$, where l is a security parameter. He sends $a = Enc(w)$ and $b = g^w$ to the verifier. Here we assume that the security parameter k_1 of Paillier's scheme is larger than $T + 2l$;
- The verifier chooses a l-bit challenge e;
- The prover opens the encryption $a(Enc(x)^e) \bmod N^2$ and the commitment bu^e, to reveal in both cases the number $z = w + ex$ defined over integer $\mathcal{Z}$. The verifier checks the opening were correct.
- The prover then uses Boudot's protocol [5] to prove that the hiding value lies in I.

The protocol can be made non-interactive in the standard way using a hash function $\mathcal{RO}$ and the Fiat-Shamir paradigm. It is also statistically zero-knowledge in the random oracle model. Our construction will also heavily rely on this non-interactive proof of knowledge protocol.

5 Construction and Proof of Security

In this section, we will propose our construction and prove its security in the sense of the security definition in the Appendix A. Basically we propose a secure verifiably committed signature scheme from Cramer-Shoup's (CS) trapdoor hash signature scheme [12].

5.1 Our Construction

We now transfer the CS signature scheme into a stand-alone, setup-free verifiably committed signature scheme below.

- Primary signer's key generation algorithm: on input k_1, a PPT primary signer Alice generates two large safe primes p and q such that $p - 1 = 2p'$ and $q - 1 = 2q'$, where p', q' are two l'-bit primes. Alice also outputs x, h, two random elements of QR_n, where $n = pq$ and QR_n the quadratic residue of Z_n^*. Finally, Alice outputs a description of a group G of order s, where s is $(l+1)$-bit prime, and two random elements g_1 and g_2 of G. The public key of a primary signer is (n, h, x, g_1, g_2, H), along with an appropriate description of G including s. The private key is (p, q). Alice also proves to her CA that all values are correctly generated and then obtains her certificate $Cert_A$ from her CA;

- Arbitrator/TTP's key generation algorithm: on input k', a PPT arbitrator/TTP Charlie generates a k'-bit RSA modulus $N = p_c q_c$, where p_c, q_c are two large safe primes. It outputs $g = (1 + N)$ that has order N in $Z_{N^2}^*$. The public key is $APK = (g, N)$. The private key is $ASK = (p_c, q_c)$. Charlie also proves to his CA that all values are correctly generated and then obtains his certificate $Cert_B$ from his CA;

- Fully signing algorithm Sig and its correspondent verification algorithm Ver: To sign a message m, a $(l + 1)$-bit prime e and a string $t \in Z_s$ is chosen at random. The equation $y^e = xh^{H(g_1^t g_2^{H(m)})} \bmod n$ is solved for y. The corresponding signature of the message m is (e, t, y). Given a putative triple (e, t, y), the verifier first checks that e is an odd $(l + 1)$-bit number. Second it checks the validation of the equation $x = y^e h^{-H(g_1^t g_2^{H(m)})} \bmod n$. If the equation is valid, then the verifier accepts, otherwise, it rejects.

- Partially signing algorithm $PSig$: On input a message m, $(l+1)$-bit prime e and a string $t \in Z_s$ is chosen at random. The equation $y^e = xh^{H(g_1^t g_2^{H(m)})} \bmod n$ is solved for y. Then Alice hides t by computing $u = g_1^t$ and $Enc(APK, t)$ together with a proof pr that she knows that u contains the same number as the encryption, and then uses Boudot's protocol to prove that the encrypted value $t \in I$ in a non-interactive way by using a hash function $\mathcal{RO}$ and the Fiat-Shamir's paradigm. The partial signature is defined by $\sigma' = (e, y, u, c, pr)$.

- The correspondent partial signature verification algorithm $PVer$: Given a putative signature $\sigma' = (e, y, u, c, pr)$, the verifier first checks that e is an odd $(l + 1)$-bit number. Second it checks the validity of the equation $x = y^e h^{-H(u g_2^{H(m)})} \bmod n$. If the equation is valid, then the verifier further check the validity of proof pr that u contains the same number as the encryption, and then uses Boudot's protocol to verify that the encrypted value $t \in I$. If it is valid then the verifier accepts, otherwise, it rejects.

- Resolution algorithm Res: Given $\sigma' = (e, y, u, c, pr)$ and a proof that Bob fulfilled his obligation to the primary signer. If the verification is passed, then Charlie outputs a valid full signature of (e, y, t) using his decryption key, otherwise, it rejects.

5.2   The Proof of Security

In this section, we will show that the stand-alone, setup-free verifiably committed signature scheme constructed above is secure assuming that the underlying the Cramer-Shoup's signature scheme is secure against adaptive chosen-message attack, and Paillier's encryption is semantically secure.

Lemma 1: Our construction is secure against malicious primary signer Alice.

Proof: Suppose Alice is able to provide a valid partial signature $\sigma' = (e, y, u, c, pr)$ correspondent to a message m, where pr means that she knows that u contains the same number as the encryption and the encrypted value $t \in I$. Since σ' is valid from the viewpoints of its verifier and TTP, by rewinding Alice, both verifier and cosigner can extract $t \in I$ such that $u = g_1^t$, $c = \text{Enc}(APK, t)$ and $y^e = xh^{H(g_1^t g_2^{H(m)})}$. It follows that the designated TTP can always transform any valid partial signature scheme into the correspondent valid signature $\sigma = (e, y, t)$.

Lemma 2: Our construction is secure against malicious verifier Bob assuming that the underlying Paillier's encryption scheme is one-way.

Proof: We convert any attacker $\mathcal{B}$ that attacks our verifiably committed signature scheme into an inverter $\mathcal{B}'$ of the underlying encryption scheme. That is, given a random cipher-text c, $\mathcal{B}'$ will obtain the correspondent plain-text m_c with non-negligible probability with the help of the attacker $\mathcal{B}$ (if this event happens with at least non-negligible probability in the following simulation, we then arrive at the contradiction of the security assumption that Paillier's encryption scheme is one-way). $\mathcal{B}'$ now describe our simulation of the environment for attacker $\mathcal{B}$ as follows:

- $\mathcal{B}'$ runs Alice to generate the primary signer's public/secret key (PK, SK) as that in the real verifiably committed signature scheme described above. $\mathcal{B}'$ then obtains PK and SK from Alice.
- $\mathcal{B}'$ then runs Charlie to generate the arbitrator's public/secret key (APK, ASK) as that in the real verifiably committed signature scheme described above. $\mathcal{B}'$ then obtains APK from Charlie.

Given the target cipher-text c, we first describe a simulator for the oracle P as follows: let q_{Psig} be the total number of P queries made by $\mathcal{B}$, and let λ be a random number chosen from $\{1, q_{Psig}\}$ by $\mathcal{B}'$. We further consider the following two cases of queries to the oracle P:

- If $i \in \{1, q_{Psig}\}$ and $i \neq \lambda$, then $\mathcal{B}'$ runs the partial signing oracle as the real partial signature scheme;
- If $i \in \{1, q_{Psig}\}$ and $i = \lambda$, for the given target cipher-text c, $\mathcal{B}'$ chooses a random string f, z and u in the correct interval specified in the real protocol and computes the encryption $\text{Enc}(APK, z)$ of z. Then it computes c' from the equation $\text{Enc}(APK, z) = c' \, c^f$. At the same time, it computes u' from the equation $g^z \bmod s = u' \, u^f$.
- Given u, $\mathcal{B}'$ computes (e, y) from the equation $y^e = xh^{H(ug_2^{H(m_\lambda)})}$, this is possible since $\mathcal{B}'$ knows SK;
- Finally $\mathcal{B}'$ assigns f be the hash value of the random oracle $\mathcal{RO}$ specified in the non-interactive proof system. Similarly, for a given u, one can simulate view for the proof of knowledge $\log_{g_1}(u)$ that lies in the correct interval using Boudot's technique [5].

$\mathcal{B}'$ simulates R oracle queries as follows:

- If (m_j, σ'_j) that is in the partial signature query list and if $j \neq \lambda$, then R outputs t_i;
- If (m_j, σ'_j) that is in the partial signature query list and if $j = \lambda$, then R outputs $\perp$;
- If (m_j, σ'_j) that is NOT in the partial signature query list, then R outputs $\perp$.

Notice that the probability that the simulator outputs $\perp$ is $1 - 1/q_{Psig}$ for the queries whose partial signatures are listed in the P oracle query. Thus when the adversary outputs a valid full signature (m^*, σ^*) whose partial signature is the list of P oracle query, the probability that $\mathcal{B}'$ can invert the target cipher-text c with probability at least ϵ/q_{Psig}, where ϵ stands for the probability that $\mathcal{B}$ can break our verifiably committed signature scheme.

Lemma 3: Our construction is secure against malicious arbitrator Charlie assuming that the underlying CS scheme is secure against adaptive chosen-message attack. In other words, our construction is secure against malicious arbitrator under joint assumptions that the hardness of the strong-RSA problem and the existence of collision free hash functions.

Proof: Suppose Charlie is able to forgery partial signature σ' with non-negligible probability, then by rewinding Charlie, we can extract t from the valid proof pr. It follows that Charlie is able to output a valid forgery signature from the Cramer-Shoup's signature scheme with non-negligible probability. Since the underlying Cramer-Shoup's trapdoor hash signature has already proved to be secure against adaptive chosen-message attack under joint assumptions of the strong RSA problem as well as the existence of collision free hash function. It follows that our construction is secure against malicious arbitrator under joint assumptions that the hardness of the strong-RSA problem and the existence of collision free hash functions.

In summary, we have proved the main result below:

Theorem: The stand-alone, setup-free verifiably committed signature scheme constructed above is provably secure assuming that the underlying CS signature scheme is secure against adaptive chosen-message attack, and Paillier's encryption is one-way.

6   Conclusion

In this paper, we have proposed an efficient construction of stand-alone and setup-free verifiably committed signature scheme. We have shown that our construction is provably secure assuming that the underlying the Cramer-Shoup's signature scheme is secure against adaptive chosen-message attack, and Paillier's encryption is one-way.

Acknowledgment

The first author thanks Professor Yevgeniy Dodis and Professor Leonid Reyzin for their continuous encouragement.

References

1. N. Asokan, M. Schunter, M. Waidner: Optimistic Protocols for Fair Exchange. ACM Conference on Computer and Communications Security 1997: 7 - 17.
2. N. Asokan, V. Shoup, M. Waidner: Optimistic Fair Exchange of Digital Signatures (Extended Abstract). EUROCRYPT 1998: 591 - 606.
3. Feng Bao: An Efficient Verifiable Encryption Scheme for Encryption of Discrete Logarithms. CARDIS 1998: 213 - 220.
4. F. Bao, R. Deng, and W. Mao, Efficient and practical fair exchange protocols with off-line TTP, IEEE Symposium on Security and Privacy, IEEE Computer Society Press, 1998, page 77- 85.
5. Fabrice Boudot: Efficient Proofs that a Committed Number Lies in an Interval. Proc. of EUROCRYPT 2000: 431 - 444, Springer Verlag.
6. Joan Boyar, David Chaum, Ivan Damgård, Torben P. Pedersen: Convertible Undeniable Signatures. CRYPTO 1990: 189 - 205
7. A. Boldyreva. Efficient threshold signatures, multisignatures and blind signatures based on the Gap Diffie Helman group signature scheme. PKC 2003, LNCS 2567.
8. Michael Ben-Or, Oded Goldreich, Silvio Micali, Ronald L. Rivest: A Fair Protocol for Signing Contracts (Extended Abstract). ICALP 1985: 43 - 52.
9. Dan Boneh, Craig Gentry, Ben Lynn, Hovav Shacham: Aggregate and Verifiably Encrypted Signatures from Bilinear Maps. EUROCRYPT 2003: 416 - 432
10. C. Boyd, E. Foo: Off-Line Fair Payment Protocols Using Convertible Signatures. ASIACRYPT 1998: 271 - 285
11. Jan Camenisch, Victor Shoup: Practical Verifiable Encryption and Decryption of Discrete Logarithms. CRYPTO 2003: 126 - 144
12. R. Cramer and V. Shoup. Signature scheme based on the Strong RAS assumption. 6th ACM Conference on Computer and Communication Security, Singapore, ACM Press, November 1999.
13. Ivan Damgård: Practical and Provably Secure Release of a Secret and Exchange of Signatures. EUROCRYPT 1993: 200 - 217
14. Ivan Damgård, Mads Jurik: Client/Server Tradeoffs for Online Elections. Proc. of Public Key Cryptography 2002: 125 - 140. Springer Verlag.
15. Ivan Damgård, Eiichiro Fujisaki: A Statistically-Hiding Integer Commitment Scheme Based on Groups with Hidden Order. Proc. of ASIACRYPT 2002: 125 - 142, Springer Verlag.
16. Y.Dodis, L. Reyzin. Breaking and Repairing Optimistic Fair Exchange from PODC 2003, ACM Workshop on Digital Rights Management (DRM), October 2003.
17. Juan A. Garay, Markus Jakobsson, Philip D. MacKenzie: Abuse-Free Optimistic Contract Signing. CRYPTO 1999: 449 - 466
18. Shafi Goldwasser, Silvio Micali, Ronald L. Rivest: A Digital Signature Scheme Secure Against Adaptive Chosen-Message Attacks. SIAM J. Comput. 17(2): 281 - 308 (1988).

19. E. Fujisaki, T. Okamoto. Statistically zero knowledge protocols to prove modular polynomial relations. Crypto'97. 16 - 30, 1997.
20. E. Fujisaki, T. Okamoto. Statistical zero-knowledge protocols to prove modular polynomial relations. Crypto'97, LNCS 1294, Springer-verlag, 1997.
21. S. Goldwasser, S. Micali, R. Rivest: A Digital Signature Scheme Secure Against Adaptive Chosen-Message Attacks. SIAM J. Comput. 17(2): 281 - 308, 1988.
22. L. Guillou, J. Quisquater. A practical zero-knowledge protocol fitted to security microprocessors minimizing both transmission and memory. Eurocrypt'88, 123 - 128, 1988.
23. Wenbo Mao: Verifiable Escrowed Signature. ACISP 1997: 240 - 248
24. S. Micali: Simple and fast optimistic protocols for fair electronic exchange. PODC 2003: 12 - 19.
25. Pascal Paillier: Public-Key Cryptosystems Based on Composite Degree Residuosity Classes. Proc. of EUROCRYPT 1999: 223 - 238, Springer Verlag.
26. J. Park, P.Chong, H. Siegel: Constructing fair-exchange protocols for E-commerce via distributed computation of RSA signatures. PODC 2003: 172 - 181
27. T. Pedersen. Non-interactive and information-theoretic secure verifiable secret sharing. Proc of CRYPTO91, Springer LNCS 576, page 129-140.
28. Markus Stadler: Publicly Verifiable Secret Sharing. EUROCRYPT 1996: 190 - 199
29. Huafei Zhu: Constructing Committed Signatures from Strong-RSA Assumption in the Standard Complexity Model. Public Key Cryptography 2004: 101 - 114

Appendix A: The Definition of Security Due to Dodis and Reyzin [16]

Security against malicious primary signer Alice: Intuitively, a primary signer Alice should not provide a partial signature which is valid both from the viewpoints of a verifier and an arbitrator but which will not be opened into the primary signer's full signature by the honest arbitrator[2]. More formally, Let P be an oracle simulating the partial signing procedure $PSig$, and R be an oracle simulating the resolution procedure Res. Let k be system security parameter. We require that any probabilistic polynomial time Adv succeeds with at most negligible probability in the following experiment.

Experiment 1 (security against malicious primary signer Alice):

- Key generation: $(SK^*, PK, ASK, APK) \leftarrow KG^*(1^k)$, where KG^* denotes the run of key generator KG with the dishonest primary signer by the adversary, and SK^* denotes the adversary's states.
- Res oracle query: In this phase, for each adaptively chosen message m_j, the adversary computes its partial signature σ_j' for m_j. Finally the adversary forward σ_j' to the oracle R to obtain the full signature σ_j of message m_j, where $1 \leq j \leq p(k)$, and $p(\cdot)$ is a polynomial. At the end of R oracle query, the adversary produces a message and its full signature pair (m, σ), i.e.,

---

[2] The security preventing a malicious third party from forging valid partial signatures is stated as security against an malicious arbitrator below as a malicious arbitrator is the most powerful adversary in the security model.

$(m, \sigma') \leftarrow Adv^R(SK^*, PK, APK)$, $\sigma \leftarrow Adv(m, \sigma', SK^*, APK, PK)$, where $m \neq m_j$, $1 \leq j \leq p(k)$.

- Success of $Adv := [PVer(m, \sigma', APK, PK) = 1 \wedge Ver(m, \sigma, PK) = 0]$.

Definition 1. A verifiably committed signature is secure against malicious primary signer Alice, if any probabilistic polynomial time adversary Adv associated with Resolution oracle, succeeds with at most negligible probability, where the probability takes over coin tosses in $KG(\cdot)$, $PSig(\cdot)$ and $R(\cdot)$.

Security against malicious verifier Bob. We consider the following scenario: suppose a primary signer Alice and a verifier Bob are trying to exchange signature in a fair way. Alice wants to commit to the transaction by providing her partial signature. Of course, it should be computationally infeasible for Bob to compute the correspondent full signature from any partial signature. More formally, we require that any probabilistic polynomial time adversary Adv succeeds with at most negligible probability in the following experiment:

Experiment 2 (security against malicious verifier Bob):

- Key generation: $(SK, PK, ASK, APK) \leftarrow KG(1^k)$, where KG is run by the honest primary signer and honest arbitrator/TTP Charlie. Adversary Adv are admitted to make queries to the two orales P and R.
- P and R oracle query: For each adaptively chosen message m_j, the adversary obtains the partial signature σ_j' of message m_j by querying the partial signing oracle P. Then the adversary forward σ_j' to the resolution oracle R to obtain the full signature σ_j of message m_j, where $1 \leq j \leq p(k)$, and $p(\cdot)$ is a polynomial. At the end of oracle both P and R queries, the adversary produces a message-full signature pair $(m, \sigma) \leftarrow Adv^{P,R}(PK, APK)$.
- Success of adversary $Adv := [Ver(m, \sigma, PK) = 1 \wedge m \notin Query(Adv, R)]$, where $Query(Adv, R)$ is the set of valid queries the adversary Adv asked to the resolution oracle R, i.e., (m, σ') such that $PVer(m, \sigma') = 1$.

Definition 2. A verifiably committed signature is secure against any malicious verifier Bob, if any probabilistic polynomial time adversary Adv associated with partial signing oracle P and the resolution oracle R, succeeds with at most negligible probability, where the probability takes over coin tosses in $KG(\cdot)$, $P(\cdot)$ and $R(\cdot)$.

Security against malicious arbitrator Charlie. Even though the arbitrator is semi-trusted, the primary signer does not want this arbitrator to produce a valid signature which the primary signer did not intend on producing. To achieve this goal, we require that any probabilistic polynomial time adversary Adv associated with partial signing oracle P, succeeds with at most negligible probability in the following experiment:

Experiment 3 (security against malicious arbitrator Charlie):

- Key generation: $(SK, PK, ASK^*, APK) \leftarrow KG^*(1^k)$, where $KG^*(1^k)$ is run by the dishonest cosigner or arbitrator. Adversary Adv are admitted to make queries to the partial signing oracle P.

- P oracle query: For each adaptively chosen message m_j, the adversary obtains the partial signature σ_j' for m_j from the oracle P, where $1 \le j \le p(k)$, and $p(\cdot)$ is a polynomial. At the end of the partial partial signing oracle query, the adversary produces a message-full signature pair (m, σ), i.e., $(m, \sigma) \leftarrow Adv^P(ASK^*, PK, APK)$.
- Success of adversary $Adv := [Ver(m, \sigma, PK) = 1 \wedge m \notin Query(Adv, P)]$, where $Query(Adv, P)$ is the set of valid queries Adv asked to the partial oracle P, i.e., (m, σ') such that $PVer(m, \sigma') = 1$.

Definition 3. A verifiably committed signature is secure against malicious arbitrator Charlie, if any probabilistic polynomial time adversary Adv associated with partial signing oracle P, succeeds with at most negligible probability, where the probability takes over coin tosses in $KG(\cdot)$, $P(\cdot)$.

Definition 4. A verifiably committed signature is secure if it is secure against malicious primary signer Alice, malicious verifier Bob and malicious arbitrator Charlie.

Toward the Fair Anonymous Signatures: Deniable Ring Signatures

Yuichi Komano[1], Kazuo Ohta[2], Atsushi Shimbo[1], and Shinichi Kawamura[1]

[1] Toshiba Corporation,
1, Komukai Toshiba-cho, Saiwai-ku,
Kawasaki 212-8582, Japan
{yuichi1.komano, atsushi.shimbo, shinichi2.kawamura}@toshiba.co.jp
[2] The University of Electro-Communications,
Chofugaoka 1-5-1, Chofu-shi, Tokyo 182-8585, Japan
ota@ice.uec.ac.jp

Abstract. Ring signature scheme, proposed by Rivest et al., allows a signer to sign a message anonymously. In the ring signature scheme, the signer who wants to sign a document anonymously first chooses some public keys of entities (signers) and then generates a signature which ensures that one of the signer or entities signs the document. In some situations, however, this scheme allows the signer to shift the blame to victims because of the anonymity. The group signature scheme may be a solution for the problem; however, it needs a group manager (electronic big brother) who can violate the signer anonymity without notification, and a complicated key setting.

This paper introduces a new concept of a signature scheme with signer anonymity, *a deniable ring signature scheme* ($\mathcal{DRS}$), in which no group manager exists, and the signer should be involved in opening the signer anonymity. We also propose a concrete scheme proven to be secure under the assumption of the DDH (decision Diffie Hellman) problem in the random oracle model.

Keywords: Group signatures, Ring signatures, DLP, CDH, DDH, Random oracle model.

1 Introduction

1.1 Background

Rivest et al. [17] proposed a ring signature scheme to give a solution for leaking a secret (classified information). In the ring signature scheme, no group manager exists and signer's anonymity is perfectly ensured. The basic and promising methodology was invented by Abe et al. [2]. In some situations, however, the ring signature scheme allows the signer to shift the blame to entities[1] (victims) because of its anonymity.

---

[1] Hereafter, we call a signer whose public key is utilized in generating a signature *an entity*.

D. Pointcheval (Ed.): CT-RSA 2006, LNCS 3860, pp. 174–191, 2006.

In these situations, a group signature scheme may solve the problem. Group signature scheme, proposed by Chaum and van Heyst [11], also allows a signer to sign a document anonymously. The group signature scheme consists of three entities: a group manager GM, signers S's, and a verifier V. In the group signature scheme, each signer joins the group and generates a signing key (secret key) by interacting with GM. The signature signed with the secret key is checked by the group public key published by GM for the group. Note that V only ensures that the signature is signed by one of the signers in the group or not; on the other hand, GM can easily identify the signer with the group secret key which corresponds to the group public key without the interaction with the signer. Much research have been done to obtain a secure and practical scheme [3, 4, 6].

In some situations, however, the following problems arise concerning the group signature scheme: (G1) the cost of interaction in join and key generation phases is very high, (G2) GM should be well-protected because the secret information (a group secret key, user privacy, etc.) should be concentrated in GM, and (G3) signers are anxious that their anonymity will be or has been violated by GM without notification. Namely, if we utilize the group signature scheme in order to solve the problem in which the signer can shift the blame, another problems arise.

1.2 Prime Causes of Problems from (G1) to (G3)

Before we discuss the $\mathcal{DRS}$, we first discern the truth of the problems from (G1) to (G3) in the group signature scheme.

Let us review the functionalities of GM in the group signature scheme. There are three functionalities; (F1) of issuing a credential to each signer which ensures that the corresponding signer belongs to the group, (F2) of interacting with each signer so that the signer generates her secret key (signing key), and (F3) of opening a signer anonymity (*i.e.*, of canceling the signer anonymity) when some problem arises[2].

Let us consider the relation between the problems from (G1) to (G3) and the functionalities from (F1) to (F3). In this setting, the functionalities (F2) and (F3) seem to cause the problems from (G1) to (G3). Indeed, (F2) leads to (G1) and (G2), and (F3) raises (G3). Therefore, the goal of this paper is to construct a scheme in which no electronic big brother who realizes (F2) and (F3) exists; namely, each signer generates her key pair by herself, and the signer must be involved in opening the signer anonymity. This involution alerts the signer to the violation of her anonymity, except in the case in which all entities who have not signed disavow the signature.

Now, let us return to the group signature scheme of Chaum and van Heyst [11]. The reference [11] proposes four group signature schemes. We found that, surprisingly, GM in the fourth scheme of [11] is only required to have the functionality of (F1). This means that the fourth scheme seems to be our goal; to our regret, however, the scheme is not publicly verifiable. Namely, the verification of a signature is performed by an interaction between the signer and the

---

[2] Bellare et al. [6] divides the group manager into two entities; an *issuer* who realizes (F1) and (F2), and an *opener* who realizes (F3).

verifier anonymously. Note that this interaction not only decreases the performance but also requires an anonymous channel in order to ensure the signer anonymity.

In 2002, Naor [13] proposed a deniable ring authentication, in which a prover can confirms a verifier that the prover does or does not authenticate a message anonymously. The deniable ring authentication needs only the third party (PKI); however, its verification is not publicly verifiable, too. Namely, the deniable ring authentication requires an iteration between the prover and the verifier over an anonymous channel.

1.3   Our Contribution

This paper first proposes a new concept of *a deniable ring signature scheme* ($\mathcal{DRS}$). The $\mathcal{DRS}$ consists of two entities in the PKI setting (no group manager[3] GM): signers $\mathcal{S}$'s and a verifier $\mathcal{V}$. $\mathcal{S}$ publishes her public key to the PKI with a (setup free) key generation algorithm, and to sign a document anonymously, she chooses some public keys of respective entities and generates the signature with the message, public keys, and her public and secret keys. $\mathcal{V}$ can verify the signature with public keys of the signer and entities, like the verification of the ring signature scheme. The difference from the ring signature is that the $\mathcal{DRS}$ allows $\mathcal{V}$ to interact with the signer and entities in order to confirm that the signer/entity generates the signature or not with a zero knowledge interactive proof (ZKIP).

The $\mathcal{DRS}$ solves the problem in which the signer can shift the blame to entities, *i.e.*, the $\mathcal{DRS}$ allows entities to claim the false charge by interacting with the verifier, even if the signer tries to shift the blame to her. Moreover, since the $\mathcal{DRS}$ has no GM and its key generation algorithm is setup free, it solves the problems from (G1) to (G3) concerned with the group signature scheme. Furthermore, since the deniable ring signature is publicly verifiable, no anonymous interaction (no anonymous channel) is required between the signer and the verifier in the verification.

Second, this paper discusses a security model of the $\mathcal{DRS}$ by comparing it with the models of the group signature scheme [11, 4, 6] and the ring signature scheme[4] [17, 2]. We show that it is enough for us to take care of the anonymity, traceability, and non-frameability, as well as the case of the dynamic group signature scheme. Note that the definition of these security requirements for the $\mathcal{DRS}$ slightly differ from those for the dynamic group signature scheme [6]. Since the $\mathcal{DRS}$ has no the group manager, we do not need to take some oracles discussed in [6] into consideration (SendToI$(\cdot, \cdot)$, SendToU$(\cdot, \cdot)$, RReg$(\cdot)$, and WReg$(\cdot, \cdot)$). As for the definition of anonymity, we follow the definition described in [8] in which an attacker is restricted so as not to corrupt signers (entities) who are in the group of signers corresponding to the target signature.

---

[3] In the $\mathcal{DRS}$, a signer selects a group of signers dynamically in the same manner as that of the ring signature scheme. Note that we can construct the scheme by assuming a GM who has a functionality of only (F1).

[4] Cramer et al. [12] introduced a concept of *one-out-of-n* signature scheme.

This paper also constructs a concrete instantiation of the $\mathcal{DRS}$ based on the undeniable signature scheme [9, 15, 14] which is proven to be secure in the random oracle model [5]. The anonymity is proven under the assumption of the decision Diffie-Hellman (DDH) problem; on the other hand, the non-frameability is ensured under the assumption of the computational Diffie-Hellman (CDH) problem. Since the DDH problem is reduced to the CDH problem, the proposed scheme is proven to be secure under the assumption of the DDH problem.

Note that the $\mathcal{DRS}$ can be regarded as *a group signature scheme without a group manager*. The $\mathcal{DRS}$ does not realize an original purpose of the ring signature scheme, *e.g.*, leaking a secret (classified information) to a third party; because the third party can identify (search for) the signer who cannot deny the signature with (sometimes terrible) endeavors.

2   Assumptions

Referring to [7], we first review the computational Diffie-Hellman (CDH) and decision Diffie-Hellman (DDH) problems. Our concrete instantiation is proven to be anonymous, and non-frameable under the assumption of the CDH problem and the DDH problem, respectively.

Definition 1. *Let p and q be primes such that $q|p - 1$. $\mathcal{G}$ denotes sub-group of $\mathbf{Z}_p^*$ generated by an element g whose order is q. We write $\mathcal{I}$ for an algorithm to solve the following problems.*

> *CDH) We define that $\mathcal{I}$ (τ, ϵ)-solves the CDH problem if $\mathcal{I}$ whose running time is bounded by τ, given g, g^a, and g^b (a and b are randomly chosen from $\mathbf{Z}_q$), outputs $g^{ab} \bmod p$ with success probability more than ϵ.*
> *DDH) We define that $\mathcal{I}$ (τ, ϵ)-solves the DDH problem if $\mathcal{I}$ whose running time is bounded by τ, given g, h, z_1, and z_2, decides whether $\log_g z_1 = \log_h z_2$ or not with advantage more than ϵ. Here, the advantage of $\mathcal{I}$ is defined as $\Pr[\mathcal{I}(g, h, z_1, z_2) = 1|(g, h, z_1, z_2) \in \mathbf{D}] - \Pr[\mathcal{I}(g, h, z_1, z_2) = 1|(g, h, z_1, z_2) \in \mathbf{R}]$, where $\mathbf{D}$ and $\mathbf{R}$ denote the sets $\{(g, h, z_1, z_2)| \log_g z_1 = \log_h z_2\}$ and $\{(g, h, z_1, z_2)|g, h, z_1, z_2 \overset{R}{\leftarrow} \mathcal{G}\}$, respectively.*

We define that the CDH problem (the DDH problem, respectively) is (τ, ϵ)-hard if there is no algorithm $\mathcal{I}$ which can (τ, ϵ)-solve the CDH problem (the DDH problems, respectively).

3   Deniable Ring Signature Scheme

This section introduces a new concept of *a deniable ring signature scheme* ($\mathcal{DRS}$) in which an entity (victim) whose public key is utilized in signing can claim the false charge. The $\mathcal{DRS}$ can be regarded as *a group signature scheme without a group manager*. We denote the number of signers published in the system by L. In the following definition, we assume that a signer P_{i_k} dynamically selects the

$L'-1$ entities $P_{i_1}, \cdots, P_{i_{k-1}}, P_{i_{k+1}}, \cdots, P_{i_{L'}}$ from the list of signers List published by the PKI and runs the signing algorithm with her own secret key and public keys of the entities.

Definition 2 (Deniable Ring Signature Scheme ($\mathcal{DRS}$)). *The deniable ring signature scheme $\mathcal{DRS}$ consists of the following algorithms.*

1) *A probabilistic key generation algorithm $\mathcal{K}$, given a security parameter k, outputs a pair of public and secret keys, $(\mathsf{pk}_{i_j}, \mathsf{sk}_{i_j})$, for each signer P_{i_j}.*
2) *A probabilistic signing algorithm $\mathcal{S}$, given a message M, a secret key sk_{i_k} of signer P_{i_k}, and the public keys $(\mathsf{pk}_{i_1}, \cdots, \mathsf{pk}_{i_{L'}})$ as an input, outputs a pair of message, signature, and public keys (or signers' ID), $(M, \sigma, \mathsf{pk}_{i_1}, \cdots, \mathsf{pk}_{i_{L'}})$.*
3) *A deterministic verification algorithm $\mathcal{V}$, given $(M, \sigma, \mathsf{pk}_{i_1}, \cdots, \mathsf{pk}_{i_{L'}})$ as an input, determines whether σ is valid for M and $\mathsf{pk}_{i_1}, \cdots, \mathsf{pk}_{i_{L'}}$ or not.*
4) *Probabilistic confirmation and disavowal algorithms $\mathcal{C}/\mathcal{D}$, performed by the interaction between a signer P_{i_j} and a verifier V with input $(M, \sigma, \mathsf{pk}_{i_1}, \cdots, \mathsf{pk}_{i_{L'}})$, convinces V that P_{i_j} does or does not generate σ.*

4   Correctness and Security Requirements

Intuitively, the security notions required for the group signature are also required for the $\mathcal{DRS}$. Namely, the unforgeability, anonymity, un-linkability, traceability, exculpability, coalition resistance, and framing should be convincing. For the group signature scheme with dynamic group setting, Bellare et al. [6] introduced three security requirements; anonymity, traceability, and non-frameability.

Taking their discussion into consideration, we fix the security requirements of the $\mathcal{DRS}$. As in the group signature scheme, we can see that the anonymity, traceability, and non-frameability are sufficient requirements for the $\mathcal{DRS}$ instead of the above seven requirements. Note that the definition of three security requirements for the $\mathcal{DRS}$ differ from those for the dynamic group signature scheme [6]. Since the $\mathcal{DRS}$ has no the group manager, we do not need to take some oracles discussed in [6] into consideration ($\mathsf{SendToI}(\cdot, \cdot)$, $\mathsf{SendToU}(\cdot, \cdot)$, $\mathsf{RReg}(\cdot)$, and $\mathsf{WReg}(\cdot, \cdot)$).

4.1   Oracles

In this subsection, let us review the oracles utilized in the security considerations (following subsections). Let List be a list issued by PKI. MList is a list of malicious signers who are corrupted or registered by an attacker. We call a signer who does not belong to MList *an honest signer*. GSet is a list of message-signature pairs which is generated through a challenge oracle query $\mathsf{Ch}_b(\cdot, \cdot, \cdot)$: Note that an attacker cannot query the pair included in GSet to a confirmation/disavowal oracle $\mathcal{C}/\mathcal{D}(\cdot, \cdot, \cdot)$.

$\mathsf{Add}(i)$: *An add user oracle* is invoked to add an honest signer with identity i to List. If a signer with identity i already exists, then the oracle returns ϵ. Otherwise, the oracle runs the key generation algorithm and adds the signer (and her public key) to List. Finally, the oracle returns pk_i.

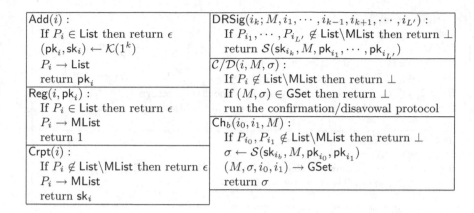

Fig. 1. Description of Oracles

$\mathsf{Reg}(i, \mathsf{pk}_i)$: With *a signer register oracle*, an attacker can register a new signer with public key pk_i in List. The oracle also adds the signer to MList.

$\mathsf{Crpt}(i)$: *A corrupt oracle* is utilized to corrupt the signer whose identity is i. An attacker can draw the secret key sk_i of signer P_i from the oracle.

$\mathsf{DRSig}(i_k; M, i_1, \cdots, i_{k-1}, i_{k+1}, \cdots, i_{L'})$: *A signing oracle* is given the identity of an honest signer P_{i_k}, message M, and identities of entities $P_{i_1}, \cdots, P_{i_{k-1}}, P_{i_{k+1}}$, $\cdots, P_{i_{L'}}$ to output a signature σ of the honest signer and entities.

$\mathsf{Ch}_b(i_0, i_1, M)$: *A challenge oracle* is utilized in the definition of the anonymity. For a challenge bit $b \in \{0, 1\}$, the oracle, given the identities (i_0, i_1) and message M, returns a target signature $\mathcal{S}(\mathsf{sk}_{i_b}, M, \mathsf{pk}_{i_0}, \mathsf{pk}_{i_1})$ of P_{i_0} and P_{i_1} for M. Note that an attacker cannot only corrupt signers P_{i_0} and P_{i_1} (see [8]); moreover, the attacker cannot run the confirmation/disavowal protocol for the target signature with P_{i_0} and P_{i_1}. In the case of the confirmation/disavowal oracle query for the target signature, the challenge oracle adds the target signature to GSet.

$\mathcal{C}/\mathcal{D}(i, M, \sigma)$: *A confirmation/disavowal oracle*, given the identity i and message-signature pair (M, σ), runs the confirmation/disavowal protocol (interacts with an attacker) if P_i is an honest signer. This oracle does not run the confirmation/disavowal algorithm if the attacker inputs the target signature output by the challenge oracle.

4.2 Correctness

We call that the $\mathcal{DRS}$ is correct if; the signature generated from the signing algorithm properly is accepted by the verification algorithm, the signer of the signature is identified by confirmation/disavowal algorithm. We formalize the correctness with an experiment $\mathsf{Exp}_{\mathcal{DRS}, \mathcal{A}}^{\mathsf{corr}}(k)$ for the $\mathcal{DRS}$, an adversary $\mathcal{A}$, and security parameter k, described in Figure 2. An advantage $\mathsf{Adv}_{\mathcal{DRS}, \mathcal{A}}^{\mathsf{corr}}(k)$ is defined as follows:

$$\mathsf{Adv}_{\mathcal{DRS}, \mathcal{A}}^{\mathsf{corr}}(k) = \Pr[\mathsf{Exp}_{\mathcal{DRS}, \mathcal{A}}^{\mathsf{corr}}(k) = 1]$$

Experiment $\mathsf{Exp}^{\mathsf{corr}}_{\mathcal{DRS},\mathcal{A}}(k)$

  List $\leftarrow \phi$; MList $\leftarrow \phi$; GSet $\leftarrow \phi$

  $(i_k; M, i_1, \cdots, i_{L'}) \leftarrow \mathcal{A}(1^k, \mathsf{Add}(\cdot), \mathsf{Reg}(\cdot, \cdot), \mathsf{Crpt}(\cdot), \mathcal{C}/\mathcal{D}(\cdot, \cdot, \cdot), \mathsf{DRSig}(\cdot; \cdot, \cdot))$

  If $P_{i_1}, \cdots, P_{i_{L'}} \notin \mathsf{List} \backslash \mathsf{MList}$ then return 0

  $\sigma \leftarrow \mathcal{S}(\mathsf{sk}_{i_k}, M, \mathsf{pk}_{i_1}, \cdots, \mathsf{pk}_{i_{L'}})$; If $\mathcal{V}(M, \sigma, \mathsf{pk}_{i_1}, \cdots, \mathsf{pk}_{i_{L'}}) = 0$ return 1

  If $\mathcal{C}/\mathcal{D}(i_k, M, \sigma)$ succeeds in disavowal protocol then return 1

  If $\mathcal{C}/\mathcal{D}(i_j \neq i_k, M, \sigma)$ succeeds protocol in confirmation then return 1

  return 0

Experiment $\mathsf{Exp}^{\mathsf{anon}-b}_{\mathcal{DRS},\mathcal{A}}(k)$

  List $\leftarrow \phi$; MList $\leftarrow \phi$; GSet $\leftarrow \phi$

  $d \leftarrow \mathcal{A}(1^k, \mathsf{Ch}_b(\cdot, \cdot, \cdot), \mathsf{Add}(\cdot), \mathsf{Reg}(\cdot, \cdot), \mathsf{Crpt}(\cdot), \mathcal{C}/\mathcal{D}(\cdot, \cdot, \cdot), \mathsf{DRSig}(\cdot; \cdot, \cdot))$

  If P_i in GSet is in MList then return 0

  return d

Experiment $\mathsf{Exp}^{\mathsf{trace}}_{\mathcal{DRS},\mathcal{A}}(k)$

  List $\leftarrow \phi$; MList $\leftarrow \phi$; GSet $\leftarrow \phi$

  $(M, \sigma, \mathsf{pk}_{i_1}, \cdots, \mathsf{pk}_{i_{L'}}) \leftarrow \mathcal{A}(1^k, \mathsf{Add}(\cdot), \mathsf{Reg}(\cdot, \cdot), \mathsf{Crpt}(\cdot), \mathcal{C}/\mathcal{D}(\cdot, \cdot, \cdot), \mathsf{DRSig}(\cdot; \cdot, \cdot))$

  If $\mathcal{V}(M, \sigma, \mathsf{pk}_{i_1}, \cdots, \mathsf{pk}_{i_{L'}}) = 0$ then return 0

  If $P_{i_1}, \cdots, P_{i_{L'}}$ can disavow (M, σ) then return 1 else return 0

Experiment $\mathsf{Exp}^{\mathsf{nf}}_{\mathcal{DRS},\mathcal{A}}(k)$

  List $\leftarrow \phi$; MList $\leftarrow \phi$; GSet $\leftarrow \phi$

  $(M, \sigma, \mathsf{pk}_{i_1}, \cdots, \mathsf{pk}_{i_{L'}}) \leftarrow \mathcal{A}(1^k, \mathsf{Add}(\cdot), \mathsf{Reg}(\cdot, \cdot), \mathsf{Crpt}(\cdot), \mathcal{C}/\mathcal{D}(\cdot, \cdot, \cdot), \mathsf{DRSig}(\cdot; \cdot, \cdot))$

  If $\mathcal{V}(M, \sigma, \mathsf{pk}_{i_1}, \cdots, \mathsf{pk}_{i_{L'}}) = 0$ then return 0

  If the followings are satisfied then return 1 else return 0:

    - For some $t \in [1, L']$, P_{i_t} cannot disavow (M, σ)

    - $\mathcal{A}$ did not query $\mathsf{Crpt}(i_t)$ or $\mathsf{DRSig}(i_t; M, i_1, \cdots, i_{t-1}, i_{t+1}, \cdots, i_{L'})$

Fig. 2. Experiment of Correctness, Anonymity, traceability, and non-frameability

We say that the $\mathcal{DRS}$ is *correct* if $\mathsf{Adv}^{\mathsf{corr}}_{\mathcal{DRS},\mathcal{A}}(k)$ is negligible for any probabilistic polynomial-time adversary $\mathcal{A}$ and security parameter k.

4.3   Anonymity

Formal Definition: For the $\mathcal{DRS}$, any adversary $\mathcal{A}$, a bit $b \in \{0, 1\}$, and security parameter k, we define the experiment $\mathsf{Exp}^{\mathsf{anon}-b}_{\mathcal{DRS},\mathcal{A}}(k)$ described in Figure 2. An advantage $\mathsf{Adv}^{\mathsf{anon}}_{\mathcal{DRS},\mathcal{A}}(k)$ is defined as follows:

$$\mathsf{Adv}^{\mathsf{anon}}_{\mathcal{DRS},\mathcal{A}}(k) = |\Pr[\mathsf{Exp}^{\mathsf{anon}-1}_{\mathcal{DRS},\mathcal{A}}(k) = 1] - \Pr[\mathsf{Exp}^{\mathsf{anon}-0}_{\mathcal{DRS},\mathcal{A}}(k) = 1]|$$

$$= |2 \Pr[\mathsf{Exp}^{\mathsf{anon}-b}_{\mathcal{DRS},\mathcal{A}}(k) = b] - 1|$$

We say that the $\mathcal{DRS}$ is *anonymous* in $(\tau, q_{Ch}, q_H, q_S, q_C, \epsilon)$ if the advantage is less than ϵ for any adversary $\mathcal{A}$, with time bound τ, who is allowed to access the challenge oracle, hash oracle, signing oracle, confirmation/disavowal oracles q_{Ch}, q_H, q_S, q_C times, respectively. Note that if the system includes L signers, $\mathcal{A}$ can query the signer register oracle and corrupt oracle at most $L - 2$ times in total.

Discussion: The definition of anonymity for the $\mathcal{DRS}$ is based on the indistinguishability, which is similar to that of the group signature scheme [6, 1, 8]. The difference between them is: in the $\mathcal{DRS}$, since an entity who wants to open the signer anonymity adaptively selects a signer and runs the confirmation/disavowal protocols with the signer, we replace the open oracle $\mathbf{Open}(gmsk, \cdot, \cdot)$ of [6, 8] with the confirmation/disavowal oracle $\mathcal{C}/\mathcal{D}(\cdot, \cdot, \cdot)$. Note that our definition of the anonymity is restricted compared to that of [6]: in restricting an attacker so as not to corrupt the signers whom the attacker queries to the challenge oracle, the similar discussion is done in the group signature scheme [8]. This restriction causes that the number of entities $\mathcal{A}$ can corrupt is bounded by (at most) $L - 2$. Note that the attacker can adaptively output queries $\mathsf{Add}(\cdot)$, $\mathsf{Reg}(\cdot, \cdot)$, $\mathsf{Crpt}(\cdot)$, and $\mathsf{DRSig}(\cdot; \cdot, \cdot)$ to each oracle, respectively. The anonymity suggests the unlinkability in addition to the anonymity.

4.4 Traceability

Formal Definition: For the $\mathcal{DRS}$, any adversary $\mathcal{A}$ and security parameter k, we define the experiment $\mathsf{Exp}_{\mathcal{DRS},\mathcal{A}}^{\mathrm{trace}}(k)$ described in Figure 2. An advantage $\mathsf{Adv}_{\mathcal{DRS},\mathcal{A}}^{\mathrm{trace}}(k)$ is defined as follows:

$$\mathsf{Adv}_{\mathcal{DRS},\mathcal{A}}^{\mathrm{trace}}(k) = \Pr[\mathsf{Exp}_{\mathcal{DRS},\mathcal{A}}^{\mathrm{trace}}(k) = 1]$$

We say that the $\mathcal{DRS}$ is *traceable* in $(\tau, q_H, q_S, q_C, \epsilon)$ if the advantage is less than ϵ for any adversary $\mathcal{A}$, with time bound τ, who is allowed to access the hash oracle, signing oracle, and confirmation/disavowal oracles q_H, q_S, and q_C times, respectively. Note that if the system includes L signers, $\mathcal{A}$ can query the signer register oracle and corrupt oracle at most $L - 1$ times in total.

Discussion: Traceability is a property such that no attacker can produce a forgery from which an entity is detected as the signer by the confirmation/disavowal protocol. With regard to the traceability of the $\mathcal{DRS}$, we should take the case into consideration where the group of malicious signers generate an untraceable signature. Namely, we allow the attacker $\mathcal{A}$ to register/corrupt all L signers (entities) by himself. Moreover, $\mathcal{A}$ can adaptively output queries $\mathsf{Add}(\cdot)$ and $\mathsf{DRSig}(\cdot; \cdot, \cdot)$ to each oracle, respectively. The traceability suggests the unforgeability, exculpability, coalition resistance, in addition to the traceability.

4.5 Non-frameability

Formal Definition: For the $\mathcal{DRS}$, any adversary $\mathcal{A}$ and security parameter k, we define the experiment $\mathsf{Exp}_{\mathcal{DRS},\mathcal{A}}^{\mathrm{nf}}(k)$ described in Figure 2. An advantage $\mathsf{Adv}_{\mathcal{DRS},\mathcal{A}}^{\mathrm{nf}}(k)$ is defined as follows:

$$\mathsf{Adv}_{\mathcal{DRS},\mathcal{A}}^{\mathrm{nf}}(k) = \Pr[\mathsf{Exp}_{\mathcal{DRS},\mathcal{A}}^{\mathrm{nf}}(k) = 1]$$

We say that the $\mathcal{DRS}$ is *non-frameable* in $(\tau, q_H, q_S, q_C, \epsilon)$ if the advantage is less than ϵ for any adversary $\mathcal{A}$, with time bound τ, who is allowed to access the hash oracle, signing oracle, and confirmation/disavowal oracles q_H, q_S, and q_C times, respectively. Note that if the system includes L signers, $\mathcal{A}$ can query the signer register oracle and corrupt oracle at most $L - 1$ times in total.

Discussion: Non-frameability is a property such that no attacker can produce a forgery from which an honest entity is detected as the signer by the confirmation/disavowal protocol. The definition of non-frameability of the $\mathcal{DRS}$ is almost the same as that of the dynamic group signature scheme. We allow the attacker $\mathcal{A}$ to output queries $\mathsf{Add}(\cdot)$, $\mathsf{Reg}(\cdot,\cdot)$, $\mathsf{Crpt}(\cdot)$, $\mathsf{DRSig}(\cdot;\cdot,\cdot)$, and $\mathcal{C}/\mathcal{D}(\cdot,\cdot,\cdot)$ to each oracle, respectively; however, $\mathcal{A}$ is restricted not to corrupt the entity (victim) P_t who cannot disavow the forgery and not to query the message M^* (corresponding to the forgery) to $\mathsf{DRSig}(t; M^*, \cdot)$. This restriction causes that the number of signers (entities) $\mathcal{A}$ can corrupt is bounded by (at most) $L - 1$. The non-frameability suggests the unforgeability, exculpability, and coalition resistance, in addition to the framing.

5   Concrete Scheme

In this section, we propose a concrete deniable ring signature scheme[5] and give the security considerations.

5.1   Protocol

The concrete scheme is based on the zero-knowledge undeniable signature scheme [9]. More precisely, in order to ensure the provable security, we utilize the zero-knowledge undeniable signature scheme with the probabilistic full domain hash function.

Protocol 1 (Deniable Ring Signature Scheme). Let p be an l bit prime, and q an l_1 bit prime such that $q|p-1$. $\mathcal{G}$ denotes sub-group of $\mathbf{Z}_p^*$ generated by an element g whose order is q. We write $H: \{0,1\}^* \to \mathbf{Z}_q$, $G: \{0,1\}^* \to \mathcal{G}$, and $F: \{0,1\}^* \to \{0,1\}^{l_2}$ for the one-way and collision-resistant hash functions.

Key Generation: A signer P_{i_j} chooses $s_{i_j} \overset{R}{\leftarrow} \mathbf{Z}_q$ as a secret key, and then P_{i_j} computes her public key $k_{i_j} = g^{s_{i_j}} \bmod p$ and sends it to PKI. PKI authorizes the signer P_{i_j}, adds P_{i_j} to List, and returns the credential for the registration to List.

Signing: For a message m, the signer decides the number of the group, L', and selects $k \overset{R}{\leftarrow} [1, L']$. Hereafter, we denote the signer as P_{i_k}. Next, P_{i_k} randomly chooses $L' - 1$ entities (signers) from List, $P_{i_1}, \cdots, P_{i_{k-1}}, P_{i_{k+1}}, \cdots, P_{i_{L'}}$ whose keys are denoted as $k_{i_1}, \cdots, k_{i_{k-1}}, k_{i_{k+1}}, \cdots, k_{i_{L'}}$, respectively. P_{i_k} then chooses $R \overset{R}{\leftarrow} \{0,1\}^{l_3}$, and computes $w = G(M, R)$ and $S = w^{s_{i_k}} \bmod p$ which is also utilized by the confirmation and disavowal protocols. For $j = 1, \cdots, i - 1, i + 1, \cdots, L'$; P_{i_k} chooses $c_{i_j}, d_{i_j} \overset{R}{\leftarrow} \mathbf{Z}_q$, and computes $a_{i_j} = g^{c_{i_j}} k_{i_j}^{d_{i_j}} \bmod p, b_{i_j} = w^{c_{i_j}} S^{d_{i_j}} \bmod p$, respectively. P_{i_k} then chooses $\tilde{e} \overset{R}{\leftarrow} \mathbf{Z}_q$, and calculates $a_{i_k} = g^{\tilde{e}} \bmod p, b_{i_k} = w^{\tilde{e}} \bmod p$. P_{i_k} computes $d = H(M, R, w, S, k_{i_1}, \cdots, k_{i_{L'}}, a_{i_1}, b_{i_1},$

---

[5] Chaum and Pedersen [10] utilized a similar technique to construct blind signature schemes for wallet databeses.

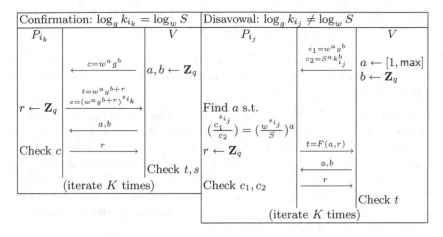

Fig. 3. Confirmation and Disavowal Protocols [9, 15]

$\cdots, a_{i_{L'}}, b_{i_{L'}})$, sets $d_{i_k} = d - (d_{i_1} + \cdots + d_{i_k-1} + d_{i_k+1} + \cdots + d_{i_{L'}}) \bmod q$, $c_{i_k} = \tilde{e} - d_{i_k} s_{i_k} \bmod q$, and outputs $(M, R, S, c_{i_1}, d_{i_1}, \cdots, c_{i_{L'}}, d_{i_{L'}}, k_{i_1}, \cdots, k_{i_{L'}})$ as a message-signature pair.

Verification: For the message-signature pair $(M, R, S, c_{i_1}, d_{i_1}, \cdots, c_{i_{L'}}, d_{i_{L'}}, k_{i_1}, \cdots, k_{i_{L'}})$, a verifier V first computes $w = G(M, R)$. For $j = 1, \cdots, L'$; V computes $a_{i_j} = g^{c_{i_j}} k_{i_j}^{d_{i_j}} \bmod p$, $b_{i_j} = w^{c_{i_j}} S^{d_{i_j}} \bmod p$, respectively. V computes $d = H(M, R, w, S, y, k_{i_1}, \cdots, k_{i_{L'}}, a_{i_1}, b_{i_1}, \cdots, a_{i_{L'}}, b_{i_{L'}})$. V checks if $d \stackrel{?}{=} d_{i_1} + \cdots + d_{i_{L'}} \bmod q$. If the equality holds, V accepts the signature; otherwise rejects.

Confirmation and Disavowal [9, 15]: (Figure 3) The confirmation and disavowal protocols are executed by the interaction between a verifier V and a signer P_{i_j}.

Note 1. The concrete deniable ring signature scheme is not compact [4] since its signature length increases in proportion to the number of entities (signers) the signer adaptively selects in the first step of signing algorithm. In order to elude the tracing, the signer should select plenty of signers. There is a trade-off between the anonymity, and both the computation cost and the signature length.

5.2 Security Considerations

Let us consider the correctness and the security of the concrete scheme. It is easy to see that the scheme is correct; we omit the proof. With regard to security requirements, as we will claim, the non-frameability is guaranteed by the CDH assumption; however, since the anonymity is ensured by the DDH assumption, the security of the concrete scheme is based on the DDH assumption.

Theorem 1 (Anonymity). *If the DDH problem is (τ', ϵ')-hard, then the concrete scheme containing L signers is anonymous in $(\tau, q_{Ch}, q_H, q_G, q_S, q_C, \epsilon)$. Here,*

$$\begin{cases} \tau' \leq \tau + (q_G + q_H + q_S + q_C)\mathsf{T}_H + q_S\mathsf{T}_S + 2q_C\mathsf{T}_C \\ \epsilon' \geq \frac{1}{Lq_{Ch}}\left(1 - \frac{(q_H+q_S+1)(q_S+1)}{2^{l_3}}\right)\left(1 - \frac{q_C}{q}\right)\left(1 - \frac{1}{q}\right)\epsilon \end{cases}$$

hold. In the above inequalities, T_H, T_S, and T_C denote the time needed in sim-ulation of a query to the random oracles, signing oracles, and confirmation and disavowal oracles.

Proof of Theorem 1: Let $\mathbf{P} = (g, h, z_1, z_2)$ be an instance of the DDH problem. We construct an inverter $\mathcal{I}$ which utilizes an attacker $\mathcal{A}$ against the anonymity of the concrete scheme to decide whether $\log_g z_1 = \log_h z_2$ or not. $\mathcal{I}$ selects $t \in [1, L]$ and set the public key $k_t = z_1$ (it reduces the success probability into $\frac{1}{L}\epsilon$). Note that $\mathcal{A}$ can adaptively corrupt or register at most $L - 2$ signers: If $P_t \in \mathsf{MList}$ then $\mathcal{I}$ aborts. Especially, if $\mathcal{A}$ outputs a corrupt query for t then $\mathcal{I}$ cannot answer $\mathsf{sk}_t = \log_g z_1$. This is because we adopt a restricted definition [8] for the anonymity. See Appendix A for detail. ∎

Theorem 2 (Traceability). *There exists no attacker $\mathcal{A}$ against the concrete scheme who can output an untraceable forgery with non-negligible probability.*

Proof of Theorem 2: Theorem 2 is proven by the contradiction (not the reduction between the hardness of breaking the traceability and the hardness of mathematical problem, *e.g.*, the discrete logarithm problem). See Appendix B for detail. ∎

Theorem 3 (Non-Frameability). *If the CDH problem is (τ', ϵ')-hard, then the concrete scheme containing L signers is traceable in $(\tau, q_H, q_G, q_S, q_C, \epsilon)$. Here,*

$$\begin{cases} \tau' \leq \tau + (q_G + q_H + q_S + q_C)\mathsf{T}_H + q_S\mathsf{T}_S + 2q_C\mathsf{T}_C \\ \epsilon' \geq \frac{1}{L}\left(\epsilon - \frac{(q_H+q_G+2q_S+q_C)q_S}{2^{l_3}} - \frac{q_C+1}{q}\right)\epsilon \end{cases}$$

hold. (Notations are the same as those in Theorem 1)

Proof of Theorem 3: Due to space limitation, we omit the proof. The strategy of the proof is similar to that for the undeniable signature scheme given in [14]. We will give it in the full paper. ∎

Note that, under the assumption of the GDH problem (gap Diffie Hellman prob-lem, [15]), we can easily prove Theorem 3; however, it is not meaningful to do so. This is because, under the assumption of the GDH problem, we assume that there is an oracle which answers the DDH problem. Namely the non-frameability under the assumption of GDH problem makes sense when the anonymity is breakable. Therefore, we give a somewhat complicated proof under the assumption of the CDH problem.

6   Discussion

In this section, we compare the deniable ring signature scheme with the group and ring signature schemes.

We first discuss the group setting. Group signature requests each signer to interact with GM in joining the group and generating her secret key (signing key). Both the static and dynamic group signature schemes, only GM decides (adds and revokes) the member of the group. On the other hand, the ring and deniable ring signature schemes allow each signer dynamically to select the group of signers for the message being signed.

Next, let us consider the opening of the signer anonymity. In the group signature scheme, GM can open the signer anonymity by himself with the group secret key held by GM. As for the ring signature scheme, no one can detect the signer at all. With regard to the deniable ring signature scheme, anyone cannot detect the signer by himself: If one can cooperate with the signer who wants to confirm or disavow the signature, he can detect the signer or winnow the candidates by one. On the contrary, for the signer who generates the deniable ring signature, the signer can notice the violation of the anonymity, except for the case where all entities who do not generate the signature disavow it. Therefore, the confidence of anonymity is improved by selecting plenty of entities for the signature.

Finally, we compare the computation cost. As for the group signature, generally speaking, the cost in signing and verification is independent of the number of signers of the group. With the ring and deniable ring signature schemes (the concrete scheme), however, the cost increases in proportion to the number of signers of the group.

7   Conclusion and Future Works

This paper first introduced a new notion of *the deniable ring signature scheme* (≈ *the group signature scheme without a group manager*). We then discuss the security requirements for the deniable ring signature scheme. Moreover, we proposed the concrete scheme proven to be secure under the assumption of the DDH in the random oracle model. The proposed scheme, based on the undeniable signature scheme [9], solves the problems concerned with the ring signature (problem that the signer can shift the blame to other entities) and the group signature scheme (problems from (G1) to (G3)); however, the signing and verification require higher computation cost than the group signature. Construction of a more efficient deniable ring signature scheme remains an open problem.

References

1. M. Abdalla and B. Warinschi. On the minimal assumptions of group signature schemes. In L. Lopez, S. Qing, and E. Okamoto, editors, *The 2004 International Conference on Information and Communications Security – ICICS 2004*, volume 3269 of *Lecture Notes in Computer Science*, pages 1–13, Berlin, Heidelberg, New York, 2004. Springer-Verlag.
2. M. Abe, M. Ohkubo, and K. Suzuki. 1-out-of-n signatures from a variety of keys. *IEICE Transactions Fundamentals of Electronics, Communications and Computer Sciences*, E87–A(1):131–140, 2004.

3. G. Ateniese, J. Camenisch, M. Joye, and G. Tsudik. A practical and provably secure coalition-resistant group signature scheme. In M. Bellare, editor, *Advances in Cryptology — CRYPTO 2000*, volume 1880 of *Lecture Notes in Computer Science*, pages 255–270, Berlin, Heidelberg, New York, 2000. Springer-Verlag.

4. M. Bellare, D. Micciancio, and B. Warinschi. Foundations of group signatures: Formal definitions, simplified requirements, and a construction based on general assumptions. In E. Biham, editor, *Advances in Cryptology — EUROCRYPT 2003*, volume 2656 of *Lecture Notes in Computer Science*, pages 614–629, Berlin, Heidelberg, New York, 2003. Springer-Verlag.

5. M. Bellare and P. Rogaway. Random oracles are practical: A paradigm for designing efficient protocols,. In *Proc. of the First ACM Conference on Computer and Communications Security*, pages 62–73. ACM Press, 1993.

6. M. Bellare, H. Shi, and C. Zhang. Foundations of group signatures: The case of dynamic groups. In A. Menezes, editor, *Topics in Cryptography – CT-RSA 2005*, volume 3376 of *Lecture Notes in Computer Science*, pages 136–153, Berlin, Heidelberg, New York, 2005. Springer-Verlag.

7. D. Boneh. The decision diffie-hellman problem. In J. Buhler, editor, *Algorithmic Number Theory, Third International Symposium — ANTS-III,*, volume 1423 of *Lecture Notes in Computer Science*, pages 48–63, Berlin, Heidelberg, New York, 1998. Springer-Verlag.

8. J. Camenisch and J. Groth. Group signatures: Better efficiency and new theoretical aspect. In C. Blundo and S. Cimato, editors, *SCN '04*, volume 3352 of *Lecture Notes in Computer Science*, pages 120–133, Berlin, Heidelberg, New York, 2005. Springer-Verlag.

9. D. Chaum. Zero-knowledge undeniable signatures. In I. Damgård, editor, *Advances in Cryptology — EUROCRYPT'90*, volume 473 of *Lecture Notes in Computer Science*, pages 458–464, Berlin, Heidelberg, New York, 1991. Springer-Verlag.

10. D. Chaum and P. Pedersen. Wallet databeses with observers. In E. F. Brickell, editor, *Advances in Cryptology — CRYPTO '92*, volume 740 of *Lecture Notes in Computer Science*, pages 89–105, Berlin, Heidelberg, New York, 1993. Springer-Verlag.

11. D. Chaum and E. van Heyst. Group signatures. In D. W. Davies, editor, *Advances in Cryptology - EuroCrypt '91*, pages 257–265, Berlin, 1991. Springer-Verlag. Lecture Notes in Computer Science Volume 547.

12. R. Cramer, I. Damgård, and B. Schoenmakers. Proofs of partial knowledge and simplified design of witness hiding protocols. In Y. Desmedt, editor, *Advances in Cryptology — CRYPTO '94*, volume 893 of *Lecture Notes in Computer Science*, pages 174–187, Berlin, Heidelberg, New York, 1995. Springer-Verlag.

13. M. Naor. Deniable ring authentication. In M. Yung, editor, *Advances in Cryptology — CRYPTO 2002*, volume 2442 of *Lecture Notes in Computer Science*, pages 481–498, Berlin, Heidelberg, New York, 2002. Springer-Verlag.

14. W. Ogata, K. Kurosawa, and S.-H. Heng. The security of the fdh variant of chaum's undeniable signature scheme. In S. Vaudenay, editor, *Public Key Cryptography – PKC 2005*, volume 3386 of *Lecture Notes in Computer Science*, pages 328–345, Berlin, Heidelberg, New York, 2005. Springer-Verlag.

15. T. Okamoto and D. Pointcheval. The gap-problems: A new class of problems for the security of cryptographic schemes. In K. Kim, editor, *Public Key Cryptography — PKC 2001*, volume 1992 of *Lecture Notes in Computer Science*, pages 104–118, Berlin, Heidelberg, New York, 2001. Springer-Verlag.

16. D. Pointcheval and J. Stern. Security arguments for digital signatures and blind signatures. *Journal of Cryptology*, 13(3):361–396, 2000.

17. R. L. Rivest, A. Shamir, and Y. Tauman. How to leak a secret. In C. Boyd, editor, *Advances in Cryptology - ASIACRYPT 2001*, volume 2248 of *Lecture Notes in Computer Science*, pages 526–565, Berlin, Heidelberg, New York, 2001. Springer-Verlag.

A   Proof of Theorem 1 (Anonymity)

We give the of proof of Theorem 1. In order to prove the theorem, we construct an inverter $\mathcal{I}$ which utilizes an adversary $\mathcal{A}$ against the anonymity of the democratic group signature as an oracle to solve a DDH instance $\mathbf{P} = (g, h, z_1, z_2)$.

We first claim the following lemma with regard to the number of challenge oracle queries. The proof is similar to that of [6].

Lemma 1. *If there exists an adversary $\mathcal{A}$ who breaks the anonymity of the democratic group signature in $(\tau, q_{Ch}, q_H, q_G, q_S, q_C, \epsilon)$, then there exits an adversary $\mathcal{B}$ who breaks the anonymity of the democratic group signature in $(\tau, 1, q_H, q_G, q_S, q_C, \epsilon_B)$. Here, $\epsilon_B = \frac{\epsilon}{q_{Ch}}$ holds.*

Hereafter, let us construct $\mathcal{I}$ which utilizes $\mathcal{B}$ as an oracle to solve $\mathbf{P} = (g, h, z_1, z_2)$.

Strategy: $\mathcal{I}$ first selects $t \xleftarrow{R} [1, L]$. $\mathcal{I}$ runs $\mathcal{B}$ by simulating answers for the queries which $\mathcal{B}$ outputs to each oracle. For the add user oracle query for t, especially, $\mathcal{I}$ sets (simulates) $k_t = z_1$. As for the challenge oracle query for (i_0, i_1, M^*), $\mathcal{I}$ first chooses $b \xleftarrow{R} \{0, 1\}$. If $i_b = t$ then $\mathcal{I}$ simulates the answer as follows. $\mathcal{I}$ chooses $R^* \xleftarrow{R} \{0, 1\}^{l_3}$, and sets $w^* = G(M^*, R^*) := h$ and $S^* = z_2$. For $j \in \{0, 1\}$, $\mathcal{I}$ selects $c_{i_j}^*, d_{i_j}^* \xleftarrow{R} \mathbf{Z}_q$, and computes $a_j^* = g^{c_j^*} k_j^{d_j^*} \bmod p$ and $b_j^* = w^{*c_j^*} S^{*d_j^*} \bmod p$. $\mathcal{I}$ calculates $d^* = d_{i_0}^* + d_{i_1}^* \bmod q$, sets (simulates) $d^* := H(M^*, R^*, w^*, S^*, k_{i_0}, k_{i_1}, a_{i_0}^*, b_{i_0}^*, a_{i_1}^*, b_{i_1}^*)$, and returns $(M^*, R^*, S^*, c_{i_0}^*, d_{i_0}^*, c_{i_1}^*, d_{i_1}^*, k_{i_0}, k_{i_1})$. Finally, $\mathcal{I}$ receives b' output by $\mathcal{B}$ and returns 1 if $b = b'$; otherwise $\mathcal{I}$ returns 0. In this strategy, the following inequality holds between the advantage of $\mathcal{B}$ (ϵ_B) and that of $\mathcal{I}$ (ϵ').

$$\epsilon' = |\Pr[\mathcal{I}(\mathbf{P}) = 1 | \mathbf{P} \in \mathbf{D}] - \Pr[\mathcal{I}(\mathbf{P}) = 1 | \mathbf{P} \in \mathbf{R}]|$$

$$= \left| \epsilon_B + \frac{1}{2} - \frac{\Pr[\mathcal{I}(\mathbf{P}) = 1 \wedge \mathbf{P} \in \mathbf{R}]}{\Pr[\mathbf{P} \in \mathbf{R}]} \right|$$

$$= \left| \epsilon_B + \frac{1}{2} - \left\{ \Pr[\mathcal{I}(\mathbf{P}) = 1 \wedge \mathbf{P} \in \mathbf{D}] + \Pr[\mathcal{I}(\mathbf{P}) = 1 \wedge \mathbf{P} \in \mathbf{R} \backslash \mathbf{D}] \right\} \right|$$

$$= \left| \epsilon_B + \frac{1}{2} - \left\{ \frac{\Pr[\mathcal{I}(\mathbf{P}) = 1 | \mathbf{P} \in \mathbf{D}]}{q} + \Pr[\mathcal{I}(\mathbf{P}) = 1 | \mathbf{P} \in \mathbf{R} \backslash \mathbf{D}] \Pr[\mathbf{P} \in \mathbf{R} \backslash \mathbf{D}] \right\} \right|$$

$$\geq \left| \epsilon_B + \frac{1}{2} - \left\{ \frac{1}{q} \left(\epsilon_B + \frac{1}{2} \right) + \frac{1}{2} \left(1 - \frac{1}{q} \right) \right\} \right|$$

$$= \left(1 - \frac{1}{q} \right) \epsilon_B$$

If we denote the probability for the case where $\mathcal{I}$ fails to simulate the answers for the queries which $\mathcal{B}$ outputs by $\Pr[\mathsf{Fail}]$, we finally have that $\epsilon' \geq \Pr[\neg\mathsf{Fail}](1 - \frac{1}{q})\epsilon_{\mathcal{B}} = \Pr[\neg\mathsf{Fail}](1 - \frac{1}{q})\frac{\epsilon}{q_{Ch}}$ (by Lemma 1).

The followings describe the way for simulating the answers for the queries output by $\mathcal{B}$. For simplicity, we assume that the identities of signers are $i = 1, 2, \cdots, L$.

Answer to the add user oracle query for i: If $P_i \in \mathsf{List}$ then $\mathcal{I}$ return ϵ. If $i = t$ then $\mathcal{I}$ sets $k_t = z_1$, adds P_t into List, and returns k_t. Otherwise, $\mathcal{I}$ sets $k_i = g^{s_i} \bmod p$ for $s_i \xleftarrow{R} \mathbf{Z}_q$, adds P_i into List, and returns k_i.

Answer to the register oracle query for (i, k_i): If $P_i \in \mathsf{List}$ then $\mathcal{I}$ return ϵ. If $i = t$ then $\mathcal{I}$ aborts. Otherwise, $\mathcal{I}$ sets k_i as a public key of P_i, adds P_i into both List and MList, respectively, and returns 1.

Answer to the corrupt oracle query for i: If $P_i \notin \mathsf{List}\backslash\mathsf{MList}$ then $\mathcal{I}$ return ϵ. If $i = t$ then $\mathcal{I}$ aborts. Otherwise, $\mathcal{I}$ adds P_i into MList and returns s_i.

Answer to G-oracle query for (M, R): If $(M, R, r, w, S) \in \mathsf{G\text{-}List}$ then $\mathcal{I}$ returns w. Otherwise, $\mathcal{I}$ chooses $r \xleftarrow{R} \mathbf{Z}_q$, and calculates $w = g^r \bmod p$ and $S = z_1^r \bmod p$. $\mathcal{I}$ then sets (simulates) $G(M, R) = w$, adds (M, R, r, w, S) into $\mathsf{G\text{-}List}$, and returns w.

Answer to H-oracle query for $(M, R, w, S, k_{i_1}, \cdots, k_{i_{L'}}, a_{i_1}, b_{i_1}, \cdots, a_{i_{L'}}, b_{i_{L'}})$: If $(M, R, w, S, k_{i_1}, \cdots, k_{i_{L'}}, a_{i_1}, b_{i_1}, \cdots, a_{i_{L'}}, b_{i_{L'}}, d) \in \mathsf{H\text{-}List}$ then $\mathcal{I}$ returns d. Otherwise, $\mathcal{I}$ chooses $d \xleftarrow{R} \mathbf{Z}_q$, sets (simulates) $H(M, R, w, S, k_{i_1}, \cdots, k_{i_{L'}}, a_{i_1}, b_{i_1}, \cdots, a_{i_{L'}}, b_{i_{L'}}) = d$, adds $(M, R, w, S, k_{i_1}, \cdots, k_{i_{L'}}, a_{i_1}, b_{i_1}, \cdots, a_{i_{L'}}, b_{i_{L'}}, d)$ into $\mathsf{H\text{-}List}$, and returns d.

Answer to the signing oracle query for $(i_k \neq t; M, i_1, \cdots, i_{L'})$: For $j \in \{1, \cdots, L'\}$, if $P_{i_j} \notin \mathsf{List}$ or $P_{i_j} \in \mathsf{MList}$ then $\mathcal{I}$ returns $\bot$. Otherwise, $\mathcal{I}$ simulates G and H in the same manner as we have described, generates a signature with the signing key s_{i_k} and randomness $R \in \{0, 1\}^{l_3}$, and returns the signature.

Answer to the signing oracle query for $(i_k = t; M, i_1, \cdots, i_{L'})$: $\mathcal{I}$ chooses $R \xleftarrow{R} \{0, 1\}^{l_3}$, simulates $G(M, R) = w = g^r \bmod p$, and sets $S = k_t^r \bmod p$. For $j \in \{1, \cdots, L'\}$, $\mathcal{I}$ chooses $c_{i_j}, d_{i_j} \xleftarrow{R} \mathbf{Z}_q$, calculates $a_{i_j} = g^{c_{i_j}} k_{i_j}^{d_{i_j}} \bmod p$ and $b_{i_j} = w^{c_{i_j}} S^{d_{i_j}} \bmod p$, and sets (simulates) $d = d_{i_1} + \cdots + d_{i_L} \bmod q$. If $(M, R, w, S, k_{i_1}, \cdots, k_{i_{L'}}, a_{i_1}, b_{i_1}, \cdots, a_{i_{L'}}, b_{i_{L'}}, *) \in \mathsf{H\text{-}List}$ then $\mathcal{I}$ aborts (with probability at most $\frac{q_H + q_S + 1}{2^{l_3}}$). Otherwise, $\mathcal{I}$ adds $(M, R, w, S, k_{i_1}, \cdots, k_{i_{L'}}, a_{i_1}, b_{i_1}, \cdots, a_{i_{L'}}, b_{i_{L'}}, d)$ into $\mathsf{H\text{-}List}$, and returns $(M, R, S, c_{i_1}, d_{i_1}, \cdots, c_{i_{L'}}, d_{i_{L'}}, k_{i_1}, \cdots, k_{i_{L'}})$.

Answer to the confirmation/disavowal query for $(i \neq t, M, R, S)$: If $P_i \notin \mathsf{List}\backslash\mathsf{MList}$ then $\mathcal{I}$ returns $\bot$. Otherwise, $\mathcal{I}$ gets $w = G(M, R)$ (by simulating G if necessary), and tests if $S = w^{s_i} \bmod p$ holds or not. If it holds then $\mathcal{I}$ runs the confirmation protocol (interacts with $\mathcal{B}$) with the secret key s_i. Otherwise, $\mathcal{I}$ runs the disavowal protocol (interacts with $\mathcal{B}$) with s_i.

Answer to the confirmation/disavowal oracle query for $(i = t, M, R, S)$: $\mathcal{I}$ first checks if $(M, R, *, S, *) \in \mathsf{G\text{-}List}$ holds or not. If it holds then $\mathcal{I}$ runs the confirmation protocol by rewinding $\mathcal{B}$ (at most twice). Otherwise, from the fact that w is equal to $S^{1/s_t} \bmod p$ by accident without querying M, R to G, $\mathcal{I}$ runs the disavowal protocol by rewinding $\mathcal{B}$ (at most twice). Here, if $w = S^{1/s_t} \bmod p$ holds by accident, $\mathcal{I}$ fails to simulate $\mathcal{B}$'s view (with probability $\frac{1}{q}$).

Answer to the challenge oracle query for (i_0, i_1, M^*)**:** $\mathcal{I}$ first chooses $b \xleftarrow{R} \{0,1\}$. If $i_b \neq t$ then $\mathcal{I}$ aborts. Otherwise, $\mathcal{I}$ simulates the answer as follows. $\mathcal{I}$ chooses $R^* \xleftarrow{R} \{0,1\}^{l_3}$, and sets $w^* = G(M^*, R^*) := h$ and $S^* = z_2$. In this setting, if $(M^*, R^*, *, *, *) \in$ G-List then $\mathcal{I}$ aborts (with probability at most $\frac{q_G + q_S + q_C + 1}{2^{l_3}}$). For $j \in \{0, 1\}$, $\mathcal{I}$ then selects $c^*_{i_j}, d^*_{i_j} \xleftarrow{R} \mathbf{Z}_q$ and computes $a^*_j = g^{c^*_j} k_j^{d^*_j} \bmod p, b^*_j = w^{*c^*_j} S^{*d^*_j} \bmod p$. $\mathcal{I}$ calculates $d^* = d^*_{i_0} + d^*_{i_1} \bmod q$ and sets (simulates) $d^* := H(M^*, R^*, w^*, S^*, k_{i_0}, k_{i_1}, a^*_{i_0}, b^*_{i_0}, a^*_{i_1}, b^*_{i_1})$. If $(M^*, R^*, *, *, *, *, *)$ has already been in H-List then $\mathcal{I}$ aborts (with probability at most $\frac{q_H + q_S + 1}{2^{l_3}}$). Otherwise $\mathcal{I}$ returns $M^*, R^*, S^*, c^*_{i_0}, d^*_{i_0}, c^*_{i_1}, d^*_{i_1}, k_{i_0}$, and k_{i_1}.

Analysis: Let us estimate the probability $\Pr[\neg\mathsf{Fail}]$. As for the challenge oracle query, $\mathcal{I}$ succeeds in guessing $t \in [1, L]$ successfully with probability $\frac{1}{L}$, and the probability for the case where $\mathcal{I}$ succeeds in simulating the challenge signature is estimated as $(1 - \frac{q_H + q_S + 1}{2^{l_3}})(1 - \frac{q_G + q_S + q_C + 1}{2^{l_3}})$. The probability for the case where $\mathcal{I}$ succeeds in answering the $\mathsf{DRSig}(t; \cdot, \cdot)$ queries is evaluated by $(1 - \frac{q_H + q_S + 1}{2^{l_3}})^{q_S}$. On the other hand, the probability for the case where $\mathcal{I}$ succeeds in answering the $\mathcal{C}/\mathcal{D}(t, \cdot)$ queries is estimated by $(1 - \frac{1}{q})^{q_C} \geq 1 - \frac{q_C}{q}$. Therefore, we have

$$\Pr[\neg\mathsf{Fail}] \geq \frac{1}{L}\left(1 - \frac{(q_H + q_S + 1)}{2^{l_3}}\right)^{q_S + 1}\left(1 - \frac{q_C}{q}\right) \geq \frac{1}{L}\left(1 - \frac{(q_H + q_S + 1)(q_S + 1)}{2^{l_3}}\right)\left(1 - \frac{q_C}{q}\right):$$

this inequality and $\epsilon' \geq \Pr[\neg\mathsf{Fail}]\left(1 - \frac{1}{q}\right)\frac{\epsilon}{q_{Ch}}$ give Theorem 1. ∎

B   Proof of Theorem 2 (Traceability)

In this section, we show that there exists no attacker $\mathcal{A}$ who outputs an untraceable forgery $(M^*, R^*, S^*, c^*_{i_1}, d^*_{i_1}, \cdots, c^*_{i_{L'}}, d^*_{i_{L'}}, k_{i_1}^*, \cdots, k_{i_{L'}}^*)$ where $\log_g k_{i_j} \neq \log_{w^*} S^*$ for all $j \in \{1, \cdots, L'\}$ with non-negligible probability. Let us review the splitting lemma [16].

Lemma 2 (Splitting Lemma). *Assume that X and Y be sets. For $A \subseteq X \times Y$, if $\Pr_{(x,y) \in X \times Y}[A(x,y)] \geq \epsilon$ holds, then there exists $Z \subseteq X$ (heavy row) such that;*

1. $\Pr_{x \in X}[x \in Z] \geq \frac{\epsilon}{2}$,

2. if $a \in Z$ then $\Pr_{y \in Y}[A(a, y)] \geq \frac{\epsilon}{2}$, *and*

3. $\Pr_{x \in X}[x \in Z | \exists y \text{ s.t., } A(x, y)] \geq \frac{1}{2}$.

Assume that $\mathcal{A}$ can output an untraceable forgery $(M^*, R^*, S^*, c^*_{i_1}, d^*_{i_1}, \cdots, c^*_{i_{L'}}, d^*_{i_{L'}}, k_{i_1}^*, \cdots, k_{i_{L'}}^*)$ where $\log_g k_{i_j} \neq \log_{w^*} S^*$ for all $j \in \{1, \cdots, L'\}$ with non-negligible probability. From Lemma 2, by rewinding $\mathcal{A}$, we can get two

untraceable forgeries $(M^{*(1)}, R^{*(1)}, S^{*(1)}, c_{i_1^*(1)}, d_{i_1^*(1)}, \cdots, c_{i_{L'}^*(1)}, d_{i_{L'}^*(1)}, k_{i_1^*(1)}, \cdots,$
$k_{i_{L'}^*(1)})$ and $(M^{*(2)}, R^{*(1)}, S^{*(2)}, c_{i_1^*(2)}, d_{i_1^*(2)}, \cdots, c_{i_{L''}^*(2)}, d_{i_{L''}^*(2)}, k_{i_1^*(2)}, \cdots, k_{i_{L''}^*(2)}),$
where $M^{*(1)} = M^{*(2)}$, $R^{*(1)} = R^{*(2)}$, $w^{*(1)} = w^{*(2)} = w^*$, $S^{*(1)} = S^{*(2)} = S^*$,
$L' = L''$, $k_{i_j^*(1)} = k_{i_j^*(2)}$, $d_{i_j^*(1)} \neq d_{i_j^*(2)}, a_{i_j^*(1)} = a_{i_j^*(2)},$ and $b_{i_j^*(1)} = b_{i_j^*(2)}$ for some
i_j (see the simulation below).

In this case, since $a_{i_j^*(1)} = a_{i_j^*(2)}$ and $d_{i_j^*(1)} \neq d_{i_j^*(2)}$ hold, we have $c_{i_j^*(1)} +$
$d_{i_j^*(1)} \log_g k_{i_j} = c_{i_j^*(2)} + d_{i_j^*(2)} \log_g k_{i_j}$; hence we can calculate $\log_g k_{i_j} = \frac{c_{i_j^*(2)} - c_{i_j^*(1)}}{d_{i_j^*(1)} - d_{i_j^*(2)}}$
$\mathrm{mod} q$.

On the other hand, since $b_{i_j^*(1)} = b_{i_j^*(2)}$ and $d_{i_j^*(1)} \neq d_{i_j^*(2)}$ hold, we have
$c_{i_j^*(1)} + d_{i_j^*(1)} \log_{w^*} S^* = c_{i_j^*(2)} + d_{i_j^*(2)} \log_{w^*} S^*$; hence we can calculate $\log_{w^*} S^* = \frac{c_{i_j^*(2)} - c_{i_j^*(1)}}{d_{i_j^*(1)} - d_{i_j^*(2)}} \mathrm{mod}\ q$.

The above two equalities give $\log_{w^*} S^* = \log_g k_{i_j}$ for some i_j; however, it
contradicts the assumption, $\log_g k_{i_j} \neq \log_{w^*} S^*$ for all $j \in \{1, \cdots, L'\}$. Therefore,
we can conclude that there exists no attacker $\mathcal{A}$ who can output an untraceable
forgery with non-negligible probability.

First Run: $\mathcal{I}$ runs repeatedly $\mathcal{A}$ with random tape Ω to get a forgery. If $\mathcal{A}$
fails to output a forgery $1/\epsilon_1$ times then $\mathcal{I}$ aborts. Here, ϵ_1 denotes the success
probability of $\mathcal{A}$ for the first run (in the simulation) and is estimated by $\epsilon_1 \geq \epsilon - \frac{1}{q}$.
If $\mathcal{A}$ outputs a forgery then $\mathcal{I}$ runs $\mathcal{A}$ with the same tape for the second run.

Simulating the Answers $H(\cdot), \mathrm{DRSig}(\cdot; \cdot, \cdot), \mathcal{C}/\mathcal{D}(\cdot, \cdot)$: Almost the same as
those described in Appendix A, respectively. Note that the format of the elements
in G-List is slightly modified, and we only deal with the case where $i \neq t$ for the
signing and confirmation/disavowal oracle queries $(\cdot; \cdot, \cdot)$ and $(\cdot, \cdot)$, respectively.

Answer to the add user oracle query for i: If $P_i \in$ List then $\mathcal{I}$ return ϵ.
Otherwise, $\mathcal{I}$ sets $k_i = g^{s_i} \mathrm{mod}\ p$ for $s_i \xleftarrow{R} \mathbf{Z}_q$, adds P_i into List, and returns k_i.

Answer to the register oracle query for (i, k_i): If $P_i \in$ List then $\mathcal{I}$ return
ϵ. Otherwise, $\mathcal{I}$ sets k_i as a public key of P_i, adds P_i into both List and MList,
respectively, and returns 1.

Answer to the corrupt oracle query for i: If $P_i \notin$ List\MList then $\mathcal{I}$ return
ϵ. Otherwise, $\mathcal{I}$ adds P_i into MList and returns s_i.

Answer to G-oracle query for (M, R): If $(M, R, r, w) \in$ G-List then $\mathcal{I}$ returns
w. Otherwise, $\mathcal{I}$ chooses $r \xleftarrow{R} \mathbf{Z}_q$, and calculates $w = g^r \mathrm{mod}\ p$. $\mathcal{I}$ then sets
(simulates) $G(M, R) = w$, adds (M, R, r, w) into G-List, and returns w.

End of the First Run: $\mathcal{I}$ receives a forgery $(M^{*(1)}, R^{*(1)}, S^{*(1)}, c_{i_1^*(1)}, d_{i_1^*(1)}, \cdots,$
$c_{i_{L'}^*(1)}, d_{i_{L'}^*(1)}, k_{i_1^*(1)}, \cdots, k_{i_{L'}^*(1)})$ output by $\mathcal{A}$. $\mathcal{I}$ then searches $(M^{*(1)}, R^{*(1)}, r^{*(1)},$
$w^{*(1)})$ for G-List. If it is not in G-List then $\mathcal{I}$ aborts before the n-the query for
H-oracle (see below, with probability $\frac{1}{q}$). $\mathcal{I}$ then searches $(M^{*(1)}, R^{*(1)}, w^{*(1)},$
$S^{*(1)}, k_{i_1^*(1)}, \cdots, k_{i_{L'}^*(1)}, a_{i_1^*(1)}, b_{i_1^*(1)}, \cdots, a_{i_{L'}^*(1)}, b_{i_{L'}^*(1)}, d^{*(1)})$ for H-List. If it is not

in H-List, then $\mathcal{I}$ aborts (with probability $\frac{1}{q}$). Assume that it is n-th query to H-oracle.

Second Run: $\mathcal{I}$ repeatedly runs $\mathcal{A}$ with the tape Ω with which $\mathcal{A}$ outputs a forgery at the first run. If $\mathcal{A}$ fails to output a forgery $1/\epsilon_2$ times then $\mathcal{I}$ aborts. Here, ϵ_2 denotes the success probability of $\mathcal{A}$ for the second run (in the simulation) and is estimated by $\epsilon_2 \geq \frac{1}{q_H}(\frac{\epsilon}{2} - \frac{1}{q})$. $\mathcal{I}$ runs $\mathcal{A}$ with the same input as that utilized in the first run until the n-th query to H-oracle. If n-th query to H-oracle happens then $\mathcal{I}$ answers $d^{*(2)} \neq d^{*(1)}$; then simulates the answers to the queries output by $\mathcal{A}$ to each oracle in the same manner.

End of the Second Run: $\mathcal{I}$ receives a forgery $(M^{*(2)}, R^{*(2)}, S^{*(2)}, c_{i_1^{*(2)}}, d_{i_1^{*(2)}},$
$\cdots, c_{i_{L''}^{*(2)}}, d_{i_{L''}^{*(2)}}, k_{i_1^{*(2)}}, \cdots, k_{i_{L''}^{*(2)}})$ output by $\mathcal{A}$. $\mathcal{I}$ then searches $(M^{*(2)}, R^{*(2)},$
$w^{*(2)}, S^{*(2)}, k_{i_1^{*(2)}}, \cdots, k_{i_{L''}^{*(2)}}, a_{i_1^{*(2)}}, b_{i_1^{*(2)}}, \cdots, a_{i_{L''}^{*(2)}}, b_{i_{L''}^{*(2)}}, d^{*(2)})$ for H-List. If it is not in H-List, then $\mathcal{I}$ aborts (with probability $\frac{1}{q}$). Assume that it is n'-th query to H-oracle. Moreover, in $n \neq n'$ then $\mathcal{I}$ aborts. Note that $n = n'$ holds with probability $\frac{1}{q_H}$. Since $d^{*(2)} \neq d^{*(1)}$, for some $j \in \{1, \cdots, L'\}$, $d_{i_j^{*(1)}} = d_{i_j^{*(2)}}$ holds. ∎

Practical Second-Order DPA Attacks for Masked Smart Card Implementations of Block Ciphers*

Elisabeth Oswald, Stefan Mangard, Christoph Herbst, and Stefan Tillich

Institute for Applied Information Processing and Communcations (IAIK),
Graz University of Technology, Inffeldgasse 16a, A–8010 Graz, Austria
{elisabeth.oswald, stefan.mangard, christoph.herbst,
stefan.tillich}@iaik.tugraz.at

Abstract. In this article we describe an improved concept for second-order differential-power analysis (DPA) attacks on masked smart card implementations of block ciphers. Our concept allows to mount second-order DPA attacks in a rather simple way: a second-order DPA attack consists of a pre-processing step and a DPA step. Therefore, our way of performing second-order DPA attacks allows to easily assess the number of traces that are needed for a successful attack. We give evidence on the effectiveness of our methodology by showing practical attacks on a masked AES smart card implementation. In these attacks we target inputs and outputs of the SubBytes operation in the first encryption round.

1 Introduction

Higher-order DPA attacks were already mentioned in Kocher *et al.*'s pioneering article [KJJ99]: "Of particular importance are high-order DPA functions that combine multiple samples from within a trace." Subsequently, several researchers have tried to implement attacks based on this very brief sketch. Messerges was the first researcher to successfully report on a second-order DPA attack in [Mes00].

Since the publication of these two articles little progress has been made. Only recently, the topic was picked up again, see [ABG04], [WW04], [PSDQ05], [SPQ05] and [JPS05]. However, none of these articles have tackled the practical issues that arise when performing higher-order DPA attacks on software implementations on smart cards.

In this article we present a way to formulate second-order DPA attacks that are practical for smart card implementations. Our attacks are simple to mount, it is easy to assess their complexity and they can be applied to any implementation that uses additive masking as DPA countermeasure. Our results are compelling: we can attack a masked AES implementation on an 8-bit micro controller with

---

* The work described in this paper has been supported in part by the European Commission through the IST Programme under Contract IST-2002-507270 SCARD and through the Austrian Science Fund (FWF) under grant number P16952.

D. Pointcheval (Ed.): CT-RSA 2006, LNCS 3860, pp. 192–207, 2006.

no more than 400 traces. The exact moments of time when intermediate values are being manipulated do not need to be known.

This article is organized as follows. In Sect. 2 we review Messerges' original second-order DPA attack and survey related work. In Sect. 3, we explain our concept of second-order attacks, assess the complexity and formulate various attack scenarios that are relevant for masked implementations of block ciphers. In Sect. 4 we show the results of our new attacks on a masked AES implementation. We conclude this article in Sect. 5. There are two appendices to this paper. In App. A we briefly discuss an alternative pre-processing method. In App. B we provide a graphical description of the AES implementation that is targeted in Sect. 4.

2   Second-Order DPA Attacks

The attack described in [Mes00], targets the exclusive-or (short: XOR) operation of a byte of the key and a byte of masked data. It is assumed that in the implementation under attack, the mask is generated and subsequently exclusive-ored with the data prior to the exclusive-or operation that involves the key byte:

```
t=1: m = rand()  (generate mask-byte)
t=2: x = p ⊕ m  (XOR mask with plaintext-byte)
t=3: y = x ⊕ k  (XOR masked plaintext with key-byte)
```

Fig. 1. A code sample of a typical masked key addition

In the attack, the point in the power trace $s_j[t = 1]$ that corresponds to the time when the mask is generated (line 1 in Fig. 1) is subtracted from the point in the power trace $s_j[t = 3]$ that corresponds to the time when the masked data is XORed with the key byte (line 3 in Fig. 1). The joint distribution of these two power samples allows to derive the key-byte bit by bit. For every bit in the plaintext byte the adversary calculates the mean values $\overline{S_0} = \sum_j |s_j[t = 1] - s_j[t = 3]|$ (if the plaintext bit is 0) and $\overline{S_1} = \sum_j |s_j[t = 1] - s_j[t = 3]|$ (if the plaintext bit is 1). If $\overline{S_0} - \overline{S_1} > 0$ then the key bit is 1, otherwise it is zero. A proof for the soundness of the attack is given in [Mes00].

In the attack, it is mandatory to use the absolute value of the differences, because otherwise the difference of means is 0 in both cases. In addition, it is necessary that the mean value of the power traces are roughly the same, otherwise the difference of means also does not lead to conclusive results. This property can be achieved by using the distance-of-mean (short: DOM) test as described in statistic textbooks or by using the Pearson correlation coefficient.

Several questions arise when studying this methodology. An important one for practical attacks is how to identify the interesting points in the power trace? Other questions are how many traces are required for reliable statistics and whether the approach can be improved by using different statistics. In his article, Messerges tries to answer the last of these questions. He concludes that using the absolute value of the difference is a sound approach.

2.1 Related Work

Waddle *et al.* [WW04] were the next to investigate this topic in detail. They investigated how higher-order attacks can be mounted in a way to minimize the additional effort compared to standard DPA attacks. More precisely, their goal was to formulate higher-order attacks as standard DPA attacks with a pre-processing step. Our idea will follow this line of thought.

In their article, Waddle *et al.* suggest to multiply the appropriate points in the power trace in order to produce a DPA peak (this approach was already mentioned by Chari *et al.* in [CJRR99]). Waddle *et al.* also tackle some of the questions that we raised at the end of the previous section. That is, they deal with the issue of finding the interesting points in the power trace. They suggest two methods to find the points of interest. Firstly, they propose the so-called zero-offset 2DPA which works if the masked value and the mask are processed at the same time. If this is the case, there is only one point of interest and the power traces can simply be squared. Secondly, they propose the so-called FFT 2DPA which essentially is a DPA on the FFT (fast fourier transform) of the power traces. For both proposals they investigate how the number of samples needs to be increased for reliable statistics. They conclude for both cases that a significant increase is to be expected due to the pre-processing step.

Peeters *et al.* [PSDQ05] have implemented an attack similar to the zero-offset 2DPA on an FPGA. They have concluded that the zero-offset 2DPA idea works but requires significantly more traces than a standard DPA.

Joye *et al.* [JPS05] have analyzed how the height of the DPA peak is related to the number of samples and the power consumption model under a certain definition of a signal-to-noise ratio.

Summarizing the related work it turns out that so far the arising questions have only been answered in part. Joye *et al.*'s article gives theoretical foundations and allows to assess the efficiency of higher-order attacks in theory in a certain model. Peeters *et al.* have confirmed that some of the ideas of Waddle *et al.* can be applied to FPGA implementations. Messerges has shown that if the points of interest can somehow be found, software implementations on smart cards can be attacked.

We aim to develop an attack strategy for software implementations on smart cards that is versatile, simple to implement, and easy to analyze.

3 Practical Second-Order DPA Attacks

In this section we outline our strategy for second-order DPA attacks. We first explain the assumptions that we make, then we explain the idea of our strategy and last we develop different attack scenarios.

In the following we assume that the instantaneous power consumption of the device under attack depends linearly on the Hamming-weight (short: HW) of the processed data.

Assumption. Let a be a value $\in \{0,1\}^n$ and $C(a)$ denote the power consumption of the value a. Then the power consumption C of the device at the time when a is processed is proportional to the Hamming-weight of the value a : $C(a) \approx HW(a)$.

In this paper, we focus on implementations where $n = 8$, *i.e.* we study 8-bit micro controllers. We use the following simple observation to explain a large class of second-order DPA attacks.

Observation. Let a and b be values $\in \{0,1\}$, let $\oplus$ denote the exclusive-or operation, and let $HW(x)$ denote the Hamming-weight of x. Then the following relation holds with probability one:

$$HW(a \oplus b) = |HW(a) - HW(b)|. \tag{1}$$

Consequently, we can correctly predict $|C(a) - C(b)|$ with $HW(a \oplus b)$ if $a, b \in \{0, 1\}$.

We can use this observation to mount second-order DPA attacks: In the first step, the adversary chooses a point in a power trace, subtracts it from the rest of the trace and takes the absolute value of the result. In the second step, the adversary tests for all keys whether the Hamming-weight of the exclusive-or of the two intermediate values under attack correlates to the pre-processed power traces. Only for the correct key and for the correct point, a peak will occur in the power trace. If there is no peak for any key then the attacker chooses another point. We can formalize our approach as follows.

Second-Order DPA Attack. Let $\mathcal{T}$ be the set of power traces that were acquired during the execution of a known algorithm using a set of known texts P_i, using a set of unknown masks M_i and using an unknown key K. We define a standard DPA attack to be a first-order DPA attack that is based on the Pearson correlation coefficient. Let $F_v(P_i)$ denote an intermediate value that is computed by the algorithm with input P_i and with a part of the unknown key K. We attack two intermediate values $F_1(P_i) \oplus M_i$ and $F_2(P_i) \oplus M_i$.

1st Step: We fix an interval I of length l for all power traces $T \in \mathcal{T}$. This interval is determined by an educated guess for the time frame in which $F_1(P_i) \oplus M_i$ and $F_2(P_i) \oplus M_i$ are processed. For each trace T we do the following. We calculate a pre-processed trace that contains all values $|I_a - I_b|$ $\forall I_a, I_b \in I \subseteq T$.

2nd Step: We make a standard DPA attack on the pre-processed power traces. In this attack, we guess a part of the key K to predict the value $HW(F_1(P_i) \oplus F_2(P_i))$.

The value $|C(F_1(P_i) \oplus M_i) - C(F_2(P_i) \oplus M_i)|$ occurs in the pre-processed traces. This value is due to the two attacked intermediate results $F_1(P_i) \oplus M_i$ and $F_2(P_i) \oplus M_i$. In the DPA attack on the pre-processed traces there occur peaks at these positions if the key guess is correct.

Remark. It is important to notice that we have given a description that is more general than Messerges' original approach. Whereas his predictions are always made for individual bits of one intermediate result, we allow to predict several bits. When using several bits in the attack, then Observation 1 does not hold in general; it only holds for some values.

Remark. The result of the pre-processing step are traces of length $\frac{l(l-1)}{2}$. This is because $|I_a - I_b| = |I_b - I_a|$.

In the following section we look at the complexity of a second-order DPA attack and then, we formulate various second-order DPA attack scenarios that are relevant for masked implementations of block ciphers.

3.1   Complexity of Our Second-Order DPA Attack

The complexity of a DPA attack is typically determined by the number of traces that have to be acquired for a successful attack. Another factor that is relevant for a practical application is the length of the traces. We discuss both complexity aspects.

Number of traces. Our approach of performing second-order DPA attacks consists of two steps. The first step consists of pre-processing the acquired power traces. The second step consists of performing a standard DPA attack. Remember that we have pointed out that we perform this standard DPA attack by predicting several bits of an intermediate value. These predictions do not always coincide with what happens inside the device because Observation 1 does not hold if $a, b \in \{0,1\}^n, n > 1$. We have to compensate these errors by increasing the number of measurements. The pre-processing step potentially increases the noise in the measurements. The effect of pre-processing has been studied in [CCD00] and [Man04]. In these articles the effect of the increase of uncorrelated noise has been investigated. However, in many micro controllers the noise that occurs in subsequent clock cycles is highly correlated. Thus, the increase of the noise due to pre-processing is not necessarily severe.

We conclude that the complexity of the second-order DPA attack is mainly determined by the number of predictions that do not match the internal value that is being processed. The influence of the pre-processing step is small and depends on the device under attack. In the attacks that we have performed in practise it turned out that the pre-processing step has virtually no influence on the number of samples.

Length of traces. In our approach we work with traces of length $\frac{l(l-1)}{2}$. In comparison to a standard DPA attack on an interval of length l, the complexity is squared.

3.2   Attacking One Masked Table Look-Up

One way to protect an implementation of a block cipher against (first-order) DPA attacks is to mask the intermediate values that occur during the computation. This is typically achieved by adding (exclusive-oring) a random value to the plaintext. The description of the block cipher is modified such that it maintains the masking.

Assumptions. Assume that the table S of the original cipher is replaced by a masked table S' such that $S'(X \oplus M) = S(X) \oplus M$ for a fixed mask M. There are two possibilities to attack such a table look-up; either one attacks the first or one attacks the last encryption round. Because both attacks follow the same principle we explain the attack on the first round only.

Attack on the first round. We use the input of the table look-up $P \oplus K \oplus M$ and the output of the table look-up $S'(P \oplus K \oplus M) = S(P \oplus K) \oplus M$ in the first encryption round for our attack. We assume that we have recorded the power trace of the first round of the algorithm.

In the first step of the attack, we locate the sequence of table look-up operations. We make an educated guess for the time frame when $S(P \oplus K) \oplus M$ and $P \oplus K \oplus M$ are computed and perform the pre-processing step. In the second step, we predict $|C(S(P \oplus K) \oplus M) - C(P \oplus K \oplus M)|$ with $HW(S(P \oplus K) \oplus (P \oplus K))$ and perform a standard DPA attack. Therefore we need to know one byte of the plaintext byte and guess one byte of the key.

Number of traces needed. In previous work we have shown that the number of traces in a standard DPA is determined by the correlation ρ between the correct predictions and the traces, see [Man04]. Based on this correlation coefficient, the number of traces can be calculated as follows

$$ N = 3 + 8 \left(\frac{Z_\alpha}{\ln \frac{1+\rho}{1-\rho}} \right)^2 . \tag{2} $$

In order to assess the correlation coefficient for this scenario in practice, we study the correlation in the idealized model where $C(a) = HW(a)$ and a has 8 bits. We use the AES S-box for the table S in our calculations. Then, we calculate the correlation between $|HW(S(P \oplus K) \oplus M) - HW(P \oplus K \oplus M)|$ and $HW(S(P \oplus K) \oplus (P \oplus K))$. This can be done easily with a computer. It turns out that the correlation coefficient for a second-order DPA attack on one masked 8-bit table look-up is 0.2405.

Setting $Z_{0.9999} = 3.7190$ and $\rho = 0.2405$, and evaluating (2) shows that $N = 462$ is an upper bound for the number of traces.

The immediate question that arises is whether we could do better by either using less bits in our predictions or by applying a simple but non-linear function (for instance raising to the power β) to our pre-processed data as suggested by [JPS05]. The answer can be easily obtained by calculating the correlation between $|HW(S(P \oplus K) \oplus M) - HW(P \oplus K \oplus M)|^\beta$ and $HW(S(P \oplus K) \oplus (P \oplus K))$. We have calculated this correlation for different values of β and for attacks on different numbers of bits of $S(P \oplus K) \oplus (P \oplus K)$. It turns out that attacking a full byte is the best choice[1] and that varying β does not lead to a significant improvement of the correlation coefficient (see Tab. 1).

---

[1] In App. A we show that using multiplication for pre-processing, such as suggested in [WW04] and [CJRR99], leads to smaller correlations.

Table 1. Exact correlation values for the scenario described in Sect. 3.2. The correlation increases when more bits are used in the prediction. The correlation increases slightly for $\beta = \{2,3\}$ but decreases for higher values of β.

β	1	2	3	4	5	6
1 Bit	0.0861	0.0985	0.0950	0.0869	0.0775	0.0685
2 Bits	0.1119	0.1315	0.1283	0.1189	0.1080	0.0972
3 Bits	0.1415	0.1652	0.1604	0.1482	0.1341	0.1203
4 Bits	0.1723	0.1914	0.1834	0.1674	0.1496	0.1327
5 Bits	0.1936	0.2100	0.2003	0.1822	0.1623	0.1435
6 Bits	0.2092	0.2291	0.2186	0.1987	0.1767	0.1559
7 Bits	0.2278	0.2460	0.2341	0.2125	0.1887	0.1661
8 Bits	0.2405	0.2622	0.2501	0.2273	0.2021	0.1782

3.3   Attacking Two Masked Table Look-Ups

Assumptions. Assume that the table S of the original cipher is replaced by another masked table S' such that $S'(X \oplus M) = S(X) \oplus M'$ for fixed masks M and M'. The outputs of two table look-ups are then $S(X_1) \oplus M'$ and $S(X_2) \oplus M'$. There are two possibilities to attack the table look-up outputs of such an implementation. One possibility is to target two different table look-up outputs in the first encryption round. The second option is to attack one table look-up output in the first and one table-lookup output in the last encryption round.

Attack on the first round. Assume that we attack the outputs of two table look-ups in the first encryption round. Hence we use $S(P_1 \oplus K_1) \oplus M'$ and $S(P_2 \oplus K_2) \oplus M'$ in our attack. In the first step we locate the sequence of table look-ups and make an educated guess for the time frame when $S(P_1 \oplus K_1) \oplus M'$ and $S(P_2 \oplus K_2) \oplus M'$ are computed and perform the pre-processing. In the second step we predict $|C(S(P_1 \oplus K_1) \oplus M') - C(S(P_2 \oplus K_2) \oplus M')|$ with $HW(S(P_1 \oplus K_1) \oplus S(P_2 \oplus K_2))$ and perform a standard DPA attack. We need to know two bytes of plaintext and guess two bytes of the key for this attack.

Attack on the first and the last round. Assume that we use the output of one S-box in the first encryption round and the output of one S-box in the last encryption round. Hence, we use $S(P \oplus K_1) \oplus M'$ and $S(C \oplus K_2) \oplus M'$ in our attack. In the first step, we locate the first and the last encryption round and perform the pre-processing step. In the second step, we predict $|C(S(P \oplus K_1) \oplus M') - C(S(C \oplus K_2) \oplus M')|$ with $HW(S(P \oplus K_1) \oplus S(C \oplus K_2))$ and perform a standard DPA attack. Therefore, we need to know one byte of the plaintext and one byte of the ciphertext and we need to guess one byte of the key of the first round and one byte of the key in the last round.

Number of traces needed. Because the key guess in this scenario is based on 16 bits, it gets impractical to calculate the correlation coefficient exactly. Therefore, we have decided to estimate it based on 100000 plaintexts. The estimation of the correlation between $|HW(S(P \oplus K_1) \oplus M') - HW(S(C \oplus K_2) \oplus M')|^\beta$ and $HW(S(P \oplus K_1) \oplus S(C \oplus K_2))$ leads to the values shown in Tab. 2.

Table 2. Simulated correlation values for the scenario described in Sect. 3.3. The correlation increases when more bits are used in the prediction. The correlation increases slightly for $\beta = \{2,3\}$ but decreases for higher values of β.

β	1	2	3	4	5	6
1 Bit	0.0851	0.0894	0.0944	0.0788	0.0698	0.0587
8 Bits	0.2322	0.2563	0.2517	0.2265	0.2043	0.1755

Because we only want to illustrate that the correlation coefficients are approximately the same as in the previous attack, we only give the numbers for attacking 1 bit and for attacking 8 bits.

3.4   Attacking a Masked Key Addition

This scenario is the same as the one described by Messerges in [Mes00].

Assumptions. The plaintext P is concealed with a random mask M: $P \oplus M$. During the key addition, the masked plaintext is exclusive-ored with the key: $P \oplus M \oplus K$. The manipulation of M and the computation of $P \oplus M \oplus K$ occur somewhen during the (initial) phase of the algorithm. We assume for the attack that we have recorded the power trace of the initial phase of the algorithm.

Attack on the key addition. We use the value of the mask M and and the value $P \oplus M \oplus K$ of the key addition in our attack. In the first step, we locate the sequence of key addition operations. We make an educated guess for the time frame when M and $P \oplus M \oplus K$ are computed. In the second step, we predict $|C(M) - C(P \oplus M \oplus K)|$ with $HW(P \oplus K)$ and perform a standard DPA attack. For the prediction we need to know one bit of the plaintext and we need to guess one bit of the key.

Number of traces. In this scenario we can only attack one bit of an intermediate result at a time. We have calculated the correlation between $HW(P \oplus K)$ and $|HW(M) - HW(P \oplus M \oplus K)|^{\beta}$ for different values of β. Tab. 3 shows the results for different values of β.

Table 3. Exact correlation values for the scenario described in Sect. 3.4. The correlation increases only slightly for $\beta = \{2,3\}$ but decreases for higher values of β.

β	1	2	3	4	5	6
1 Bit	0.0846	0.0912	0.0879	0.0806	0.0717	0.0626

4   Attacking a Masked AES Smart Card Implementation

In order to verify the theoretical discussions presented in Sect. 3, we have performed these attacks in practice. The target of these attacks was a masked AES smart card implementation. This implementation is described in Sect. 4.1. For the attacks, we have executed this implementation of AES on a micro controller whose power consumption is proportional to the Hamming weight of the data it processes. The results of two second-order DPA attacks on this implementation are reported in Sect. 4.2 and 4.3.

4.1   Masking AES in Software for a Smart Card Implementation

In our masked software implementation of AES the inputs and outputs of each operation are masked additively. In the following paragraphs we briefly sketch how the masked versions of the four AES operations have been implemented. A graphical description can be found in App. B.

Masked AddRoundKey: The AddRoundKey operation does not change the mask and therefore it does not require special attention in our masked implementation. Essentially, we use the same AddRoundKey operation as in an unmasked implementation.

Masked SubBytes: We mask the SubBytes operation S with values M and M' (the masks). Therefore, we have to derive a new masked S-box S' with the property that $S'(X \oplus M) = S(X) + M'$.

Masked ShiftRows: The ShiftRows operation is done in combination with the SubBytes operation by reading and writing the state bytes in a specific order. Therefore, no separate masking effort is required for ShiftRows.

Masked MixColumns: As MixColumns is a linear operation it is sufficient to calculate the MixColumns operation with the used masks in addition to the normal calculation with the masked AES state. In order to minimize the overhead for calculation, we make sure that the state before MixColumns is always masked with the same four values. Hence, the output of MixColumns is also masked with the same four masks. The four output masks only need to be calculated once per AES encryption or decryption. The reason for using four different masks for a column is that the four bytes of the column are combined with each other during the MixColumns operation. If the same mask would be used on each byte of the column, then intermediate values of MixColumns could be processed unmasked if the masks cancel each other out.

4.2   Attacking Two S-Box Outputs in the First Encryption Round

In the scenario described in Section 3.3, we predict the power consumption by calculating the Hamming-weight of the exclusive-or of two outputs of the masked SubBytes operation: $HW(S(P_1 \oplus K_1) \oplus S(P_2 \oplus K_2))$. Using this technique, we have targeted the first two key bytes of our masked AES implementation during the first encryption round.

In order to reduce the computational effort that is needed for the attack, we have first made an educated guess for the time frame when $S(P_1 \oplus K_1) \oplus M'$ and $S(P_1 \oplus K_1) \oplus M'$ are computed. For this purpose, we have measured the power

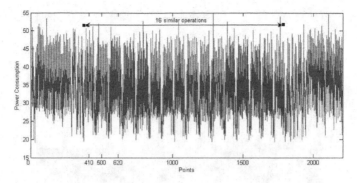

Fig. 2. A part of the first encryption round. A sequence of 16 similar operations is clearly visible in the power trace.

consumption during one execution of AES. We have inspected the power trace and we have deduced when the first round takes place. Within the first round (see Fig. 2), there is a distinct part where 16 similar operations take place. This part corresponds to the operations AddRoundKey and SubBytes. These operations are executed in combination for each byte of the AES state. Since we decided to target the outputs of the first two SubBytes operations, we selected the interval 410 to 620 of the power trace (see Fig. 2) for our attack.

After having selected this interval, we have made 3000 measurements of the power consumption of the micro controller while it was performing AES encryptions of random plaintexts. Subsequently, we have performed the pre-processing operation described in Section 3. This means that for each of the 3000 power traces we have calculated the absolute value of the difference of all pairs of points in the interval 410 to 620. We have done this computation by first subtracting the points 411 to 620 from the point 410. We refer to the absolute value of these differences as a segment. The segment that has been calculated based on point 410 consists of 210 values.

The next segment was calculated based on point 411. This segment contained the absolute value of the difference between the points 412 to 620 and 411. It consists of 209 values. Following this strategy, we have calculated corresponding segments based on all remaining points in the interval from 412 to 620. This lead to 210 segments in total, where the largest segment consisted of 210 values and the smallest one consisted of just one value.

For the attack, we have concatenated the 210 segments of each power trace. After the pre-processing step, we therefore had 3000 traces, where each trace consisted of 210 segments. Based on these traces, a standard DPA attack making hypotheses about the value of $HW(S(P_1 \oplus K_1) \oplus S(P_2 \oplus K_2))$ has been performed. Since the intermediate result $HW(S(P_1 \oplus K_1) \oplus S(P_2 \oplus K_2))$ depends on two key bytes, 65536 key guesses were necessary.

The result of this attack for the correct key guess is shown in Fig. 3. Several peaks are clearly visible in this figure. There is more than one peak in the result because the targeted micro controller manipulates the two attacked intermediate

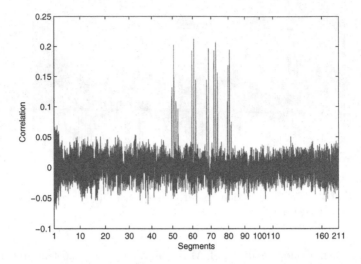

Fig. 3. Result of a second-order DPA attack on the interval 410 to 620 of the original power traces

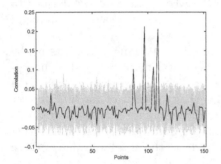

Fig. 4. The result of all 65536 key guesses in an attack on segment 61

Fig. 5. Correlation coefficients for all 65536 keys depending on the number of power traces that are used in the attack

results in more than two clock cycles. The highest peak that is shown in Fig. 3 is located in segment 61 and has the value 0.21. This segment contains the result of the second-order DPA attack mounted based on the absolute values of the differences between the points 471 to 620 and 470 in the original traces. The segment consists of 150 values.

In order to show that only the correct key produces a peak in this second-order attack, we have attacked this segment based on all 65536 key hypotheses. The results of this attack are shown in Fig. 4. The results for the 65535 incorrect keys are plotted in gray. The result for the correct key is plotted in black. It can be observed that indeed only the correct key leads to significant peaks.

We have also analyzed how many samples are needed to obtain a significant peak in segment 61. Figure 5 shows how the correlation coefficients evolve depending on the number of used power traces. The correlation coefficient for the correct key guess is shown in black. The correlation coefficients for the incorrect key guesses are shown in gray. It can be observed in Fig. 5 that roughly 400 traces are needed to perform a second-order DPA attack on the output of two S-box operations. This confirms our theoretical estimate for the correlation coefficient and the number of samples given in Sect. 3.2.

For our attack we have used a standard PC with 2 GB of RAM and a standard digital oscilloscope. It took us roughly one hour to make the 3000 measurements. We compressed the power traces by integrating the absolute values of each clock cycle. The compression step required about 23 minutes. The pre-processing step for the second-order DPA attack (*i.e.*, the calculation and concatenation of the segments) took about 5 minutes. Attacking segment 61 based on 65536 key hypotheses took less than two minutes. An attack on all 210 segments can be performed within a few hours. Hence, this type of attack can be easily performed in practice. The time that is needed for the attack is mainly determined by the transfer speed of the hard disk. On a standard PC not all power traces and hypotheses can be kept in memory simultaneously.

4.3   Attacking an S-Box in the First Encryption Round

In the scenario that is described in Sect. 3.2, the power consumption of the attacked device is predicted based on the Hamming-weight of the exclusive-or of the input and output of a masked SubBytes operation: $(P_1 \oplus K_1) \oplus S(P_1 \oplus K_1)$.

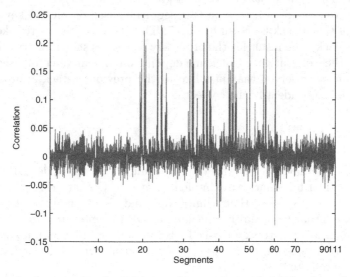

Fig. 6. Result of a second-order DPA attack on an interval of 111 points of the original power traces

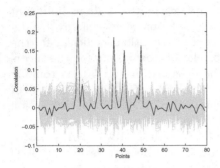
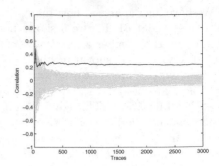

Fig. 7. The result of all 256 key guesses in an attack on segment 33

Fig. 8. Correlation coefficients for all 256 key guesses depending on the number of power traces that are used in the attack

We have targeted the first encryption round in our analysis and we have again acquired 3000 power traces of the attacked device.

Based on visually inspecting the power traces, we have made an educated guess for the time frame when the input and output of the attacked S-box is processed. We have considered an interval of 111 points for the attack. Just like in the previous section, we have performed a pre-processing step to calculate the absolute value of the difference between all possible pairs of points in this interval.

Based on each power trace, 110 segments have been calculated and concatenated. The resulting traces have been used as input for a standard DPA attack predicting the Hamming-weight of $(P_1 \oplus K_1) \oplus S(P_1 \oplus K_1)$. The result for the correct key guess is shown in Fig. 6. As expected, there are again several peaks visible in this trace. The highest peak occurs in segment 33 and has the value 0.24.

We have performed an attack on this segment based on all 256 key guesses. The result of this attack is shown in Fig. 7. The result for the correct key guess is shown in black. The result for the other key guesses is shown in gray. Figure 8 shows how the correlation coefficients depend on the number of used power traces. It can be observed that just like in the previous attack approximately 400 traces suffice to identify the correct key.

5   Conclusion

In this article, we have presented a way to formulate second-order DPA attacks that are practical for smart card implementations. Our attacks are simple to mount, it is easy to assess their complexity and they can be applied to any implementation that uses additive masking as DPA countermeasure. Our results are compelling: we can attack a masked AES implementation with no more than 400 traces without needing to know the exact moments when intermediate values are being manipulated.

Our work clearly shows that second-order DPA attacks are a practical threat for masked software implementations. Consequently, masking by itself is insufficient to protect masked smart card implementation of block ciphers against power-analysis attacks.

References

[ABG04]   Mehdi-Laurent Akkar, Régis Bevan, and Louis Goubin. Two Power Analysis Attacks against One-Mask Methods. In Bimal K. Roy and Willi Meier, editors, *Fast Software Encryption, 11th International Workshop, FSE 2004, Delhi, India, February 5-7, 2004, Revised Papers*, volume 3017 of *Lecture Notes in Computer Science*, pages 332–347. Springer, 2004.

[CCD00]   Christophe Clavier, Jean-Sébastien Coron, and Nora Dabbous. Differential Power Analysis in the Presence of Hardware Countermeasures. In Çetin Kaya Koç and Christof Paar, editors, *Cryptographic Hardware and Embedded Systems – CHES 2000, Second International Workshop, Worcester, MA, USA, August 17-18, 2000, Proceedings*, volume 1965 of *Lecture Notes in Computer Science*, pages 252–263. Springer, 2000.

[CJRR99]   Suresh Chari, Charanjit S. Jutla, Josyula R. Rao, and Pankaj Rohatgi. Towards Sound Approaches to Counteract Power-Analysis Attacks. In Michael J. Wiener, editor, *Advances in Cryptology - CRYPTO '99, 19th Annual International Cryptology Conference, Santa Barbara, California, USA, August 15-19, 1999, Proceedings*, volume 1666 of *Lecture Notes in Computer Science*, pages 398–412. Springer, 1999.

[JPS05]   Marc Joye, Pascal Paillier, and Berry Schoenmakers. On Second-Order Differential Power Analysis. In *Cryptographic Hardware and Embedded Systems - Proceedings of CHES 2005*. Springer, 2005.

[KJJ99]   Paul C. Kocher, Joshua Jaffe, and Benjamin Jun. Differential Power Analysis. In Michael Wiener, editor, *Advances in Cryptology - CRYPTO '99, 19th Annual International Cryptology Conference, Santa Barbara, California, USA, August 15-19, 1999, Proceedings*, volume 1666 of *Lecture Notes in Computer Science*, pages 388–397. Springer, 1999.

[Man04]   Stefan Mangard. Hardware Countermeasures against DPA – A Statistical Analysis of Their Effectiveness. In Tatsuaki Okamoto, editor, *Topics in Cryptology - CT-RSA 2004, The Cryptographers' Track at the RSA Conference 2004, San Francisco, CA, USA, February 23-27, 2004, Proceedings*, volume 2964 of *Lecture Notes in Computer Science*, pages 222–235. Springer, 2004.

[Mes00]   Thomas S. Messerges. Using Second-Order Power Analysis to Attack DPA Resistant Software. In Çetin Kaya Koç and Christof Paar, editors, *Cryptographic Hardware and Embedded Systems – CHES 2000, Second International Workshop, Worcester, MA, USA, August 17-18, 2000, Proceedings*, volume 1965 of *Lecture Notes in Computer Science*, pages 238–251. Springer, 2000.

[PSDQ05]   Eric Peeters, Francois-Xavier Standaert, Nicolas Donckers, and Jean-Jacques Quisquater. Improved Higher Order Side-Channel Attacks with FPGA experiments. In *Cryptographic Hardware and Embedded Systems - Proceedings of CHES 2005*. Springer, 2005.

[SPQ05]   Francois-Xavier Standaert, Eric Peeters, and Jean-Jacques Quisquater. On the Masking Countermeasure and Higher-Order Power Analysis Attacks. In *ITCC 2005*, 2005.

[WW04]   Jason Waddle and David Wagner. Towards Efficient Second-Order Power Analysis. In Marc Joye and Jean-Jacques Quisquater, editors, *Cryptographic Hardware and Embedded Systems – CHES 2004, 6th International Workshop, Cambridge, MA, USA, August 11-13, 2004, Proceedings*, volume 3156 of *Lecture Notes in Computer Science*, pages 1–15. Springer, 2004.

A   Using Multiplication as Pre-processing Method

Waddle *et al.* [WW04] and Chari *et al.* [CJRR99] have suggested to use multiplication as a pre-processing method. In order to illustrate that this is inferior to using the absolute difference, we study this technique for the scenario described in Sect. 3.2.

With multiplication as pre-processing method, we use $HW(S(P \oplus K) \oplus (P \oplus K))$ to predict $(HW(S(P \oplus K) \oplus M) * HW(P \oplus K \oplus M))^{\beta}$. It is easy to exactly calculate the correlation between $HW(S(P \oplus K) \oplus (P \oplus K))$ and $(HW(S(P \oplus K) \oplus M) * HW(P \oplus K \oplus M))^{\beta}$ with a computer. We performed this calculation for different values of β and for attacks on different numbers of bits of $S(P \oplus K) \oplus (P \oplus K)$. The results are given in Tab. 4. The correlation coefficients are clearly much lower than the ones in Tab. 1 which means that this pre-processing technique is less effective.

Table 4. Exact correlation values for the scenario described in Sect. 3.2 when multiplication is used as pre-processing method. The correlation increases when more bits are used in the prediction. The correlation increases slightly for $\beta = \{2, 3, 4\}$ but decreases for higher values of β.

β	1	2	3	4	5	6
1 Bit	-0.0327	-0.0531	-0.0627	-0.0644	-0.0610	-0.0551
2 Bits	-0.0437	-0.0706	-0.0830	-0.0846	-0.0797	-0.0717
3 Bits	-0.0548	-0.0888	-0.1045	-0.1069	-0.1010	-0.0911
4 Bits	-0.0636	-0.1032	-0.1223	-0.1261	-0.1202	-0.1096
5 Bits	-0.0698	-0.1134	-0.1346	-0.1391	-0.1330	-0.1215
6 Bits	-0.0761	-0.1236	-0.1468	-0.1517	-0.1449	-0.1323
7 Bits	-0.0817	-0.1328	-0.1579	-0.1634	-0.1563	-0.1429
8 Bits	-0.0871	-0.1415	-0.1681	-0.1737	-0.1660	-0.1515

B   Graphical Description of the Masked AES Implementation

Figure 9 shows, in correspondence with the notation used in Sect. 4, how our masked AES implementation works.

We use six masks in total. The masks M and M' are the input and output masks for SubBytes, respectively. At the start of an AES encryption, the masked S-box is calculated for M and M'. The masks M1, M2, M3, and M4 are the input masks for MixColumns. The corresponding output masks M1', M2', M3', and

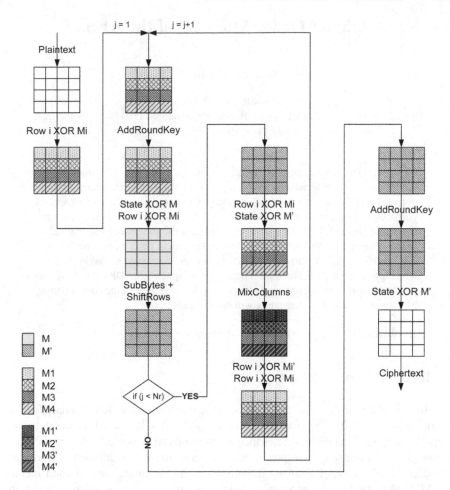

Fig. 9. Graphical description of the masked AES implementation

M4' are determined by applying MixColumns to the input masks. Fig. 9 shows how all AES operations are masked and how the masks are changed between the operations.

We have verified that our masked AES implementation is secure against first-order DPA attacks by attacking all intermediate values.

Higher Order Masking of the AES

Kai Schramm and Christof Paar

Horst Görtz Institute for IT Security (HGI),
Universitätsstr. 150, Ruhr University Bochum, Germany, 44780 Bochum, Germany
{schramm, cpaar}@crypto.ruhr-uni-bochum.de

Abstract. The development of masking schemes to secure AES implementations against side channel attacks is a topic of ongoing research. Many different approaches focus on the AES S-box and have been discussed in the previous years. Unfortunately, to our knowledge most of these countermeasures only address first-order DPA. In this article, we discuss the theoretical background of higher order DPA. We give the expected measurement costs an adversary has to deal with for different hardware models. Moreover, we present a masking scheme which protects an AES implementation against higher order DPA. We have implemented this masking scheme for various orders and present the corresponding performance details implementors will have to expect.

Keywords: AES, Higher Order DPA, Masking Countermeasure.

1 Introduction

The *Advanced Encryption Standard* (AES) is the worldwide de-facto standard for symmetric encryption [10]. Therefore, it is very likely that it will be used for many different purposes ranging from high-performance applications such as video stream encryption to low-cost (low memory, low power consumption) implementations on smart cards. Especially in the case of software implementations for smart cards limited memory (ROM, RAM, XRAM) poses a challenging constraint for implementors. Even worse, side channel attacks based on differential power analysis (DPA) [13, 17] and its various branches such as higher order differential power analysis (HODPA) [16, 2] require considerable effort to come up with efficient yet secure implementations which do not succumb to these attacks.

A lot of effort has been devoted in the past years to the development of efficient countermeasures for AES implementations against first-order DPA [7]. Especially the so-called random masking technique has been suggested many times [15, 8, 1, 11] but also algebraic techniques to protect AES implementations against side channel attacks have been proposed in various publications [4, 18, 9, 19]. Unfortunately, HODPA attacks which have been discussed in several articles [16, 21, 12] are often capable to break these first-order countermeasures.

In this article, we would like to present a masking scheme which protects AES implementations against HODPA attacks. In Section 2, we start with a theoretical discussion of HODPA attacks and derive the correlation coefficients of HODPA attacks for various orders and hardware architectures. In Section 3, we propose an AES masking scheme which uses multiple masks and is secure against HODPA attacks. In Section 4, we

D. Pointcheval (Ed.): CT-RSA 2006, LNCS 3860, pp. 208–225, 2006.
© Springer-Verlag Berlin Heidelberg 2006

present performance-related details of various HODPA-resistant AES implementations which we have programmed on a test device. We conclude this article in Section 5.

2   HODPA: Theoretical Issues

We assume an adversary encrypts N plaintexts X_j and measures the corresponding power traces $P_j(t)$. As discussed in [7, 16, 2, 21], we define a DPA of order d as the correlation of the product of d power signals $P_j(t_1), ..., P_j(t_d)$ with a selected function f of the known plaintext X_j and a key hypothesis K_h.

$$\rho(\prod_{i=1}^{d} P(t_i), f(X_j, K_h)) = \frac{COV[\prod_{i=1}^{d} P(t_i), f(X_j, K_h))]}{\sqrt{V[\prod_{i=1}^{d} P(t_i)]}\sqrt{V[f(X_j, K_h))]}} \quad (1)$$

Since the adversary is generally only able to measure a finite number N of power traces $P_j(t)$, the estimated correlation coefficient $\hat{\rho}(N)$ is computed using the approximated covariance and variances.

$$COV[\prod_{i=1}^{d} P(t_i), f(X_j, K_h))] = \frac{1}{N} \sum_{j=0}^{N-1} \prod_{i=1}^{d} P_j(t_i) f(X_j, K_h)$$

$$- \left(\frac{1}{N} \sum_{j=0}^{N-1} \prod_{i=1}^{d} P_j(t_i)\right)\left(\frac{1}{N} \sum_{j=0}^{N-1} f(X_j, K_h)\right) \quad (2)$$

$$V[\prod_{i=1}^{d} P(t_i)] = \frac{1}{N} \sum_{j=0}^{N-1} \left(\prod_{i=1}^{d} P_j(t_i) - \frac{1}{N} \sum_{j=0}^{N-1} \prod_{i=1}^{d} P_j(t_i)\right)^2 \quad (3)$$

$$V[f(X, K_h))] = \frac{1}{N} \sum_{j=0}^{N-1} \left(f(X_j, K_h) - \frac{1}{N} \sum_{j=0}^{N-1} f(X_j, K_h)\right)^2. \quad (4)$$

2.1   Second Order DPA Against the AES S-Box Input (HW-Model)

We suppose an adversary performs a second-order DPA based on an l-bit key hypothesis K_h against the masked input of an AES S-box in round one with $1 \leq l \leq n = 8$. Furthermore, we assume that a random mask M leaks at time t_1, the masked S-box input $X \oplus K \oplus M$ leaks at time t_2 and that the adversary is able to measure the corresponding power signals $P(t_1)$ and $P(t_2)$. If the power contribution ϵ of all bits is equal[1] and coupling effects among the bits are neglected, the power signals $P(t_1)$ and $P(t_2)$ can be modelled as

$$P(t_1) = \epsilon \sum_{i=0}^{n-1} M[i] + N_1 \quad \text{and} \quad P(t_2) = \epsilon \sum_{i=0}^{n-1} (X \oplus K \oplus M)[i] + N_2$$

---

[1] Hamming weight model.

where the additive noise terms N_1 and N_2 are assumed to be independent Gaussian random variables with $(0, \sigma^2)$. As shown in Appendix A, the correlation coefficient is

$$\rho(P(t_1)P(t_2), W(X \oplus K_h)) = \frac{\epsilon^2 \frac{1}{4}(\frac{l}{2} - u)}{\sqrt{(\epsilon^4 \frac{n^2}{16}(3 + 2(n-1)) + \epsilon^2 \frac{n}{2}(\sigma^2 + \sigma^2 n) + \sigma^4)} \sqrt{\frac{l}{4}}} \tag{5}$$

where $W(X \oplus K_h) \leq l$ denotes the Hamming weight of the guessed lower l bits of the unmasked S-box input and u denotes the number of correctly guessed key bits, $0 \leq u \leq l$. The expression can be simplified, if we assume that the power signals only depend on the Hamming weights ($\epsilon = 1$) and that the uncorrelated noise terms N_1 and N_2 are neglected ($\sigma = 0$).

$$\rho(P(t_1)P(t_2), W(X \oplus K_h)) = \frac{\frac{1}{4}(\frac{l}{2} - u)}{\sqrt{(\frac{n^2}{16}(3 + 2(n-1))} \sqrt{\frac{l}{4}}} \tag{6}$$

The AES S-box input in the first round is a linear function of a plaintext byte X and a key byte K. As a result, the resulting correlation coefficient shows a linear characteristic. It is proportional to the number of correctly guessed key bits and reaches its minimum (maximum)[2], if all key bits are guessed correctly (incorrectly). Moreover, wrong key guesses which are "close" to the correct key guess (e.g. $l-1$ bits are guessed correctly and only one bit incorrectly) will result in a correlation coefficient which is "close" to its minimum. Therefore, DPA attacks usually focus on the output of non-linear functions, such as the AES S-boxes, because wrong key guesses will result in correlation coefficients which are clearly distinguishable from the correct key guess. Please note, however, that in [5] it was shown that S-boxes which are not perfectly non-linear (e.g. the DES S-boxes) may result in "ghost peaks".

2.2 Multi-bit HODPA Against the AES S-Box Output (HW-Model)

We suppose an adversary performs a DPA of order d against the S-box output, i.e., the adversary correlates the product of d power signals with a selected function of the unmasked S-box output based on the key hypothesis K_h. Moreover, we presume that the leakage of a variable is equal to its Hamming weight (i.e. $\epsilon = 1$ and $\sigma = 0$) and that the adversary knows that the Hamming weights of $d-1$ random masks $M_1, ..., M_{(d-1)}$ leak at times $t_1, ..., t_{(d-1)}$ and that the masked S-box output $S(X \oplus K) \oplus M_1 \oplus ... \oplus M_{(d-1)}$ leaks at time t_d. As shown in the Appendix B, the correlation coefficient ρ of the product $\prod_{i=1}^d P(t_i)$ and the Hamming weight of the correctly predicted S-box output $W(S(X \oplus K_h))$ for $K_h = K$ is

$$\rho\left(\prod_{i=1}^d P(t_i), W(S(X \oplus K_h))\right) = \frac{2^{-(d+1)}(n^{(d+1)} - n) + n2^{-d}(d \bmod 2) - (\frac{n}{2})^{(d+1)}}{\sqrt{\left((\frac{n}{4} + \frac{n^2}{4})^d - (\frac{n}{2})^{(2d)}\right)\left(\frac{n}{4}\right)}} \quad \text{if } K_h = K \tag{7}$$

---

[2] The minimum and maximum have an equal magnitude, i.e. the correlation coefficient is a symmetric function.

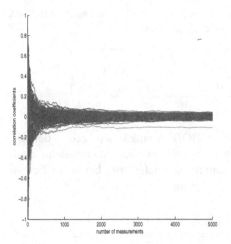

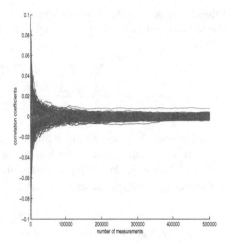

Fig. 1. Correlation plot of a simulated second-order DPA against the AES S-box output according to the HW-model with no noise

Fig. 2. Correlation plot of a simulated third-order DPA against the AES S-box output according to the HW-model with no noise

where n denotes the bit length of all intermediate variables. In the case of AES, $n = 8$, i.e. 8 bits of K_h must be guessed to predict the S-box output. Thus, the correlation coefficient reduces to

$$\rho\left(\prod_{i=1}^{d} P(t_i), W(S(X \oplus K))\right) = \frac{2^{(3-d)}(\mathrm{d}\bmod 2) - 2^{(-d+2)}}{\sqrt{(18^d - 2^{(4d)})}\sqrt{2}} \quad \text{if} \quad K_h = K \quad (8)$$

while wrong key hypotheses, i.e. $K_h \neq K$, result in correlation coefficients which converge to zero for an increasing number of measurements N due to the non-linear characteristics of the AES S-box. In Figures 1 and 2, the results of a simulated second-order and third-order DPA based on the Hamming weight model are shown. In both cases the correlation coefficients corresponding to the correct key hypotheses are clearly visible, however, significantly more measurements ($\approx$ a factor of 10^2) are required to successfully perform the third-order DPA[3].

In Table 1, the correlation coefficients of correct key hypotheses for DPA attacks of orders $d = 1, ..., 7$ are listed. Please note that the correlation coefficients approximately decrease by a factor of 10 with order d and, moreover, feature alternating signs. In order

Table 1. Correlation coefficients of a successful HODPA for a given order

DPA Order d	1	2	3	4	5	6	7
Corr. Coeff. ρ	1	-0.0857	0.0085	$-8.901 \cdot 10^{-4}$	$9.638 \cdot 10^{-5}$	$-1.064 \cdot 10^{-5}$	$1.191 \cdot 10^{-6}$

---

[3] Also note that the magnitude of the correlation coefficients in the third-order plot has approx. decreased by a factor of 10.

Table 2. Number of measurements N required to achieve an $|SNR|$ of ≥ 5 in simulated HODPA attacks (averaged over 100 simulated DPA attacks for each order d)

DPA Order d	1	2	3	4
N	31.14 $\approx 3.13 \cdot 10^1$	3941.67 $\approx 3.94 \cdot 10^3$	415513.67 $\approx 4.16 \cdot 10^5$	44383112.11 $\approx 4.44 \cdot 10^7$

to define some kind of quality rating regarding a HODPA attack, we need to define a signal-to-noise ratio (SNR) which expresses how much the estimated correlation coefficient of the correct key hypothesis deviates from the estimated correlation coefficients of the wrong key hypotheses for a certain number of measurements N.

$$SNR(N) = \frac{\hat{\rho}\left(\prod_{i=1}^{d} P(t_i), W(S(X \oplus K_h))|K_h = K\right)}{\sqrt{V[\hat{\rho}\left(\prod_{i=1}^{d} P(t_i), W(S(X \oplus K_h))\right)|K_h \neq K]}} \tag{9}$$

Experimental results showed that an $|SNR|$ of ≥ 5 is a reasonable threshold, i.e. it results in satisfactory HODPA attacks for which the correct key guess is clearly distinguishable from wrong key guesses. Table 2 lists the average number of measurements N required to achieve an $|SNR|$ of ≥ 5. These numbers were derived from statistical simulations, i.e. for a given order d 100 simulated HODPA attacks were performed. The numbers given in Table 2 clearly show that the measurement costs grow exponentially with DPA order d (see [7]). However, it must be pointed out that practical HODPA attacks will most certainly require more measurements. For example, the assumption that only 31 measurements are required to perform a first-order DPA is extremely optimistic and usually not achieveable in a noisy measurement environment. In order to give a better estimation regarding the measurement costs we analyzed an 8051-based microcontroller whose power consumption behaviour matches surprisingly well the Hamming weight model. For this architecture the power leakage of some 8-bit variable X at time t_X can be modelled as

$$P(t_X) = offset + \epsilon \cdot W(X) + \sigma \cdot N.$$

In an experiment we analyzed $256 \cdot 1000 = 256000$ power traces and determined an average $offset = 10$ mA, current gain $\epsilon = 3.72$ mA, and Gaussian noise with a standard deviation $\sigma = 1.9636$ mA and $N \sim N(0,1)$. Using these parameters we simulated[4] 100 DPA attacks for each order $d = 1, ..., 4$ in order to determine the average measurement costs required to achieve an $|SNR| \geq 5$. The results are listed in Table 3.

As a result of the additive Gaussian noise $\sigma \cdot N$, the measurement costs roughly increase by a factor of ≈ 10. Hence, the use of a noise generator as an add-on countermeasure is certainly reasonable to make HODPA attacks more difficult. In Figures 3 and 4

---

[4] Real-world HODPA attacks against the 8051 microcontroller would have been possible, but with regard to the high measurement costs for orders $d > 2$ we decided to simulate these attacks.

Table 3. Number of measurements N required to achieve an $|SNR|$ of ≥ 5 in simulated HODPA based on the HW-model (parameters: $Offset = 10$ mA, $\epsilon = 3.72$ mA, $\sigma = 1.9636$ mA) attacks (averaged over 100 simulated DPA attacks for each order d)

DPA Order d	1	2	3
N	225.85 $\approx 2.26 \cdot 10^2$	12539.1 $\approx 1.25 \cdot 10^4$	3527564 $3.52 \approx \cdot 10^6$

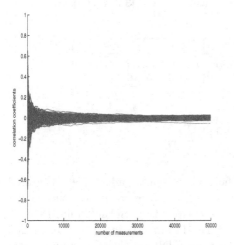

Fig. 3. Correlation plot of a simulated second-order DPA against the AES S-box output according to the HW-model (parameters: $Offset = 10$ mA, $\epsilon = 3.72$ mA, $\sigma = 1.9636$ mA)

Fig. 4. Correlation plot of a simulated third-order DPA against the AES S-box output according to the HW-model (parameters: $Offset = 10$ mA, $\epsilon = 3.72$ mA, $\sigma = 1.9636$ mA)

two correlation plots of a simulated second and third-order DPA are shown for the aforementioned parameters.

2.3   Single-Bit HODPA Against the AES S-Box Output (General Model)

In the previous sections, we proposed theoretical results of HODPA attacks against hardware architectures with regard to the Hamming weight model. However, as discussed in [6, 3, 14], this model is of limited use in real-world attacks. In [16], a more general model was presented which focuses on a single bit and comprehends all remaining noise sources[5] as a Gaussian distributed random variable. According to this model the two possible probability distributions of a power signal $P(t_i)$ are defined as

$$f(P(t_i)|b = 0) \sim N(-\epsilon, \sigma^2) \quad and \quad f(P(t_i)|b = 1) \sim N(\epsilon, \sigma^2) \qquad (10)$$

depending on the state of some bit b, e.g. an S-box output bit, which leaks at time t_i. As derived in Appendix C, if the key K is guessed correctly, the correlation coefficient of the product $\prod_{i=1}^{d} P(t_i)$ and an S-box output bit $b = S(X \oplus K_h)[j]$ with $0 \leq j \leq 7$ is

---

[5] i.e. both arithmetic noise and measurement noise.

Table 4. Correlation coefficients of a successful single-bit HODPA for various orders d with parameters $\epsilon = 3.1838$ mA and $\sigma = 16.9143$ mA according to the general model

DPA Order d	1	2	3	4	5	6	7
Corr. Coeff. ρ	0.185	-0.0342	$6.30 \cdot 10^{-3}$	$-1.20 \cdot 10^{-3}$	$2.17 \cdot 10^{-4}$	$-4.01 \cdot 10^{-5}$	$7.41 \cdot 10^{-6}$

Table 5. Number of measurements N required to achieve an $|SNR|$ of ≥ 5 in simulated single-bit HODPA attacks with parameters $\epsilon = 3.1838$ mA and $\sigma = 16.9143$ mA (averaged over 100 simulated DPA attacks for each order d)

DPA Order o	1	2	3	4
N	801.8 $\approx 8.02 \cdot 10^2$	22614.37 $\approx 2.26 \cdot 10^4$	1291118.02 $\approx 1.29 \cdot 10^6$	17705001.01 $\approx 1.77 \cdot 10^7$

$$\rho\left(\prod_{i=1}^{d} P(t_i), S(X \oplus K_h)[j]\right) = \frac{(-1)^{(d+1)}\epsilon^d}{\sqrt{(\epsilon^2 + \sigma^2)^d}} \qquad \text{if} \quad K_h = K \qquad (11)$$

In order to estimate the correlation coefficient for various orders d we measured 1000 power traces from a test device[6]. Using this set of measurements we analyzed the power consumption caused by S-box 0 output bit 0 in round one and determined a mean $\epsilon = 3.1838$ mA and a standard deviation $\sigma = 16.9143$ mA. Using these parameters, we were able to estimate the correlation coefficients of DPA attacks for various orders d. These numbers are listed in Table 4. As in the previous section, we also performed simulated DPA attacks for various orders d in order to determine the average number of measurements required to extract the correct key, i.e. to achieve an $|SNR| \geq 5$. These numbers are given in Table 5 and again show an exponential increase. Finally, in Figures 5 and 6 two correlation plots of a simulated second and third-order DPA are shown.

3   Secure HODPA AES Masking Scheme

In this section, we propose an AES masking scheme which is secure against HODPA attacks. We assume that the adversary knows the exact points in time when any occuring intermediate variable leaks in the side channel trace and that she/he is able to measure the corrsponding power signal.

Definition. *A masking scheme which applies $(d - 1)$ independent, random masks to blind a subkey-dependent intermediate variable is considered secure, if an adversary must perform an DPA attack of order d, i.e. she/he must correlate at least d power signals with a selected function of the subkey-hypothesis, in order to successfully determine the secret subkey.*

---

[6] A smart card which is based on the AVR architecture and runs a software implementation of AES. From our measurements we derived that this architecture does not agree very well with the Hamming weight model.

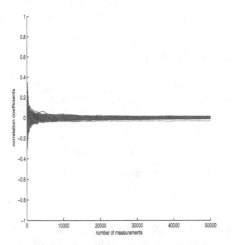 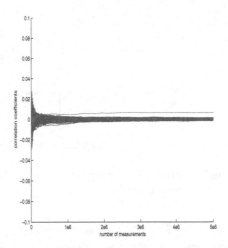

Fig. 5. Correlation plot of a simulated single-bit second-order DPA against the AES S-box output according to the general model (parameters: $\epsilon = 3.1838$ mA and $\sigma = 16.9143$ mA)

Fig. 6. Correlation plot of a simulated single-bit third-order DPA against the AES S-box output according to the HW-model (parameters: $\epsilon = 3.1838$ mA and $\sigma = 16.9143$ mA)

Let us consider a very simple and naive masking scheme based on a modified S-box S^* which is shown in Figure 7. We assume that the same set of $d - 1$ input and output masks $M_1, ..., M_{d-1}$ are used to thwart DPA attacks up to order $d - 1$. Unfortunately, this scheme has several vulnerabilities. First, note that the $d - 1$ masks at the S-box output can be regarded as a single x-or mask $M = M_1 \oplus ... \oplus M_{d-1}$. While the x-or sum M may never leak by itself as an intermediate variable in the side channel trace, the variables $K \oplus M$, $X \oplus K \oplus M$ and $S(X \oplus K) \oplus M$ do occur and thus cause a leakage. We observed that this gives rise to the following two counterintuitive second-order attacks even if $d - 1 > 1$ masks M_i are used. The two correlation coefficients

$$\rho\Big(W(S(X \oplus K) \oplus M) \cdot W(K \oplus M), W(S(X \oplus K_h) \oplus K_h)\Big)$$

and

$$\rho\Big(W(S(X \oplus K) \oplus M) \cdot W(X \oplus K \oplus M), W(S(X \oplus K_h) \oplus X \oplus K_h)\Big)$$

will result in distinct peaks, if the correct key hypothesis is guessed. A simple way to thwart both attacks is to use a different set of input and output masks for an S-box. Furthermore, let us assume that different input and output masks are used for an S-box, however, the same two sets of $d - 1$ input masks M_i and output masks N_i are used for all S-boxes. As suggested in [11], this leads to the following second-order attack

$$\rho\Big(W(S(X \oplus K_X) \oplus N)W(S(Y \oplus K_Y) \oplus N), W(S(X \oplus K_{HX}))W(S(Y \oplus K_{HY}))\Big)$$

where $N = N_1 \oplus ... \oplus N_{d-1}$ denotes the x-or sum of the output masks, X, Y denote two arbitrary plaintext bytes, K_X, K_Y the two corresponding key bytes in the first round

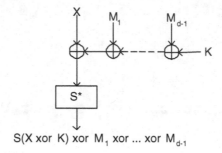

$$S(X \text{ xor } K) \text{ xor } M_1 \text{ xor } ... \text{ xor } M_{d-1}$$

Fig. 7. Insecure AES masking scheme using the same $d - 1$ input and output masks to thwart DPA attacks of order d

and K_{HX}, K_{HY} the two corresponding key hypotheses guessed by the adversary. Thus, the hypothesis space is increased to 16 key bits which is still feasible. An insufficient measure to counteract this second-order attack would be the random permutation[7] of the 16 S-boxes in each round, since this would merely increase the measurement costs by a factor of $16 \cdot 15 \cdot \frac{1}{2} = 120$. A better countermeasure is the usage of different input and output masks for each S-box.

Design Rule. *Every AES S-box S_j^* with $1 \leq j \leq 16$ should use a different set of $d - 1$ input masks $M_{(j,1)}, ..., M_{(j,d-1)}$ and output masks $N_{(j,1)}, ..., N_{(j,d-1)}$ for each round to thwart DPA attacks of orders $< d$.*

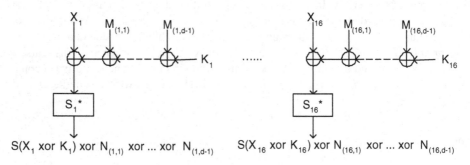

$$S(X_1 \text{ xor } K_1) \text{ xor } N_{(1,1)} \text{ xor } ... \text{ xor } N_{(1,d-1)} \qquad S(X_{16} \text{ xor } K_{16}) \text{ xor } N_{(16,1)} \text{ xor } ... \text{ xor } N_{(16,d-1)}$$

Fig. 8. Secure AES masking scheme which uses $d - 1$ different input and output masks for each S-box to thwart DPA attacks of order d

3.1 S-Box Recomputation

In the case of AES, 8-bit x-or masks are used to blind elements in $GF(2^8)$. As stated in [16, 1, 11], the only transformation in AES which requires special attention with regard to masking is the non-linear S-box, which performs an inversion in $GF(2^8)$ followed by an affine bitwise transformation. For this reason, an x-or mask M will not propagate unchanged through the S-box.

---

[7] i.e. a temporal desynchronisation of the power traces.

$$s((X \oplus K) \oplus M) = s(X \oplus K) \oplus R \neq s(X \oplus K) \oplus M \quad \text{for any } X \oplus K, M \neq \{0\}$$

The S-box must be modified in such a way that $s((X \oplus K) \oplus M) = s(X \oplus K) \oplus M$ for $\forall\, X \oplus K$. This can be achieved twofold: either by simple recomputation of the S-box [15] or by algebraic methods [4, 18, 9, 19]. The disadvantage of algebraic methods is that they are usually not very efficient when implemented in software and generally do not address higher order masking. In [15], a very simple recomputation algorithm was proposed which blinds the index of a table S with mask M, the output with mask N and stores it as a new table S'.

```
For (i=0; i <= 255; i = i + 1)
    A = S(i)
    S'(i x-or M) = A x-or N
End
```

This algorithm requires 256 read and write instructions and 512 bytes RAM[8] for tables S and S'. In [20], the following "split and swap" algorithm was suggested.

```
For (i=0; i <= 255; i = i + 1)
    S(i) = S(i) x-or N  //Apply the output mask N
End
For (j=0; j <= 7; j = j + 1)
    If (M[j] = 1) Then
        (1) Split S into succeeding blocks of 2^j elements
        (2) Swap pairwise the (2n)th and (2n+1)th,... block
    End
End
```

where $M[j]$ denotes bit j of input mask M. The initial step which applies the output mask N requires 256 read and write instructions. If bit $M[j]$ is set, $2^{8-(j+1)}2^{j+1} = 256$ read and write instructions are required for each split-and-swap operation. This results in an average total of $256 + 4 \cdot 256 = 1280$ read and write instructions. As an advantage, only 256 bytes of RAM are required to recompute the S-box. We propose the following S-box recomputation algorithm which requires 256 read and write instructions and only 256 bytes of RAM.

```
Determine the most significant bit for which M[j]=1, j=7,...,0
//For example, if M=0x1C, then j=4

For (i=0; i <= 255; i = i + 2^(j+1) )
    For (l=0; l < 2^j; l = l + 1)
        A = S(i x-or l)
        B = S(i x-or l x-or M)
        S(i x-or l) = B x-or N
        S(i x-or l x-or M) = A x-or N
    End
End
```

---

[8] We will see later that in the case of higher order masking it is inefficient to store table S in ROM and only S' in RAM even though it makes sense to do so in the case of first-order masking.

Let us assume we would want to apply $d - 1$ input masks $M_1, ..., M_{d-1}$ and $d - 1$ output masks $N_1, ..., N_{d-1}$ to the original S-box S which has been copied from ROM into RAM. Since the x-or sums $M = M_1 \oplus ... \oplus M_{d-1}$ and $N = N_1 \oplus ... \oplus N_{d-1}$ shall never leak during the execution of the cipher, one possibility is to recompute the S-box $d - 1$ times:

$$recompute(S, M_1, N_1) \rightarrow ... \rightarrow recompute(S, M_{d-1}, N_{d-1})$$

The order of the recomputation steps is arbitrary. Fortunately, it is only necessary to perform these $d - 1$ recomputations for the very first S-box in round one. Once the first S-box is masked with M and N, it is easy to derive a new S-box with input masks $U_1, ..., U_{d-1}$ and output masks $V_1, ..., V_{d-1}$ by using the *chain of masks* U' and V'.

$$U' = U_1 \oplus M_1 \oplus ... \oplus U_{d-1} \oplus M_{d-1}$$
$$V' = V_1 \oplus N_1 \oplus ... \oplus V_{d-1} \oplus N_{d-1}$$

Thus, the x-or sum U' removes the previous input masks M_i and adds the new input masks U_i, while the x-or sum V' removes the previous output masks N_i and adds the new output masks V_i in one step. It is important that the previous and new masks are stacked up in the alternating order given above to avoid any possible side channel vulnerabilities. As a result, only a single recomputation step $recompute(S, U', V')$ is required to derive the new S-box independent of the number of masks $d - 1$.

3.2 Mask Propagation and the MixColumn Transformation

For AES implementations which must be secure against first-order DPA, only, it is sufficient to use a single 8-bit mask M for the entire algorithm. As a matter of fact, the mask M will simply propagate through the MixColumn transformation and no attention must be paid to correct the mask after it has propagated through the MixColumn transformation[9].

$$MixCol \begin{pmatrix} S(X_1 \oplus K_1) \oplus M \\ S(X_2 \oplus K_2) \oplus M \\ S(X_3 \oplus K_3) \oplus M \\ S(X_4 \oplus K_4) \oplus M \end{pmatrix} = MixCol \begin{pmatrix} S(X_1 \oplus K_1) \\ S(X_2 \oplus K_2) \\ S(X_3 \oplus K_3) \\ S(X_4 \oplus K_4) \end{pmatrix} \oplus \begin{pmatrix} M \\ M \\ M \\ M \end{pmatrix}$$

With regard to an AES implementation resistant against a DPA attack of order d let us assume that $d - 1$ different input masks $M_{(j,1)}, ..., M_{(j,d-1)}$ and $d - 1$ different output masks $N_{(j,1)}, ..., N_{(j,d-1)}$ are used for each S-box j in the first round with $M_j = M_{(j,1)} \oplus ... \oplus M_{(j,d-1)}$ and $N_j = N_{(j,1)} \oplus ... \oplus N_{(j,d-1)}$ and $1 \leq j \leq 16$. In this case, the masks do change after they have propagated through the MixColumn transformation.

$$MixCol \begin{pmatrix} S(X_1 \oplus K_1) \oplus N_1 \\ S(X_2 \oplus K_2) \oplus N_2 \\ S(X_3 \oplus K_3) \oplus N_3 \\ S(X_4 \oplus K_4) \oplus N_4 \end{pmatrix} = MixCol \begin{pmatrix} S(X_1 \oplus K_1) \\ S(X_2 \oplus K_2) \\ S(X_3 \oplus K_3) \\ S(X_4 \oplus K_4) \end{pmatrix} \oplus \begin{pmatrix} N'_1 \\ N'_2 \\ N'_3 \\ N'_4 \end{pmatrix}, \begin{pmatrix} N'_1 \\ N'_2 \\ N'_3 \\ N'_4 \end{pmatrix} \neq \begin{pmatrix} N_1 \\ N_2 \\ N_3 \\ N_4 \end{pmatrix}$$

---

[9] Please note that special care has to be taken when a single mask M is used in connection with the MixColumn transformation. The computation of the MixColumn transformation must be performed in a carefully chosen order so that the mask M is never cancelled out at any time.

In order to follow the propagation of the output masks $N_{(j,1)}, ..., N_{(j,d-1)}$, the Mix-Colum transformation must be executed an additional $d - 1$ times for each column.

$$\begin{pmatrix} N'_{(1,1)} \\ N'_{(2,1)} \\ N'_{(3,1)} \\ N'_{(4,1)} \end{pmatrix} = MixCol \begin{pmatrix} N_{(1,1)} \\ N_{(2,1)} \\ N_{(3,1)} \\ N_{(4,1)} \end{pmatrix}, ..., \begin{pmatrix} N'_{(13,d-1)} \\ N'_{(14,d-1)} \\ N'_{(15,d-1)} \\ N'_{(16,d-1)} \end{pmatrix} = MixCol \begin{pmatrix} N_{(13,d-1)} \\ N_{(14,d-1)} \\ N_{(15,d-1)} \\ N_{(16,d-1)} \end{pmatrix}$$

For example, in an implemenation secure against second-order DPA attacks, the MixColumn transformation must be executed an additional $4 \cdot 2 = 8$ times.

4   HODPA-Resistant AES Implementations

We implemented the following AES implementations on an AVR-based smard card in assembly: an unmasked AES implementation, a first-order DPA-resistant AES implementation using a single mask for the entire AES and, finally, implementations resistant to second, third and fourth-order DPA using different input and output masks for all S-boxes. The details such as code sizes and data sizes of these implementations are given in Table 6. Moreover, the number of cycles required for an encryption and the corresponding execution times[10] are given, as well.

Table 6. Details of various HODPA resistant AES implementations

DPA resistance	S-box algo.	code size [bytes/ROM]	data size [bytes/RAM]	cycles	time [ms]
unprotected	-	1078	16	4625	0.925
1st order resistant	[15]	2422	264	8701	1.74
2nd order resistant	[15]	2798	592	193199	38.6
3rd order resistant	[15]	3350	624	197263	39.5
4th order resistant	[15]	3962	656	201255	40.2
2nd order resistant	see 3.1	2614	336	243581	48.7
3rd order resistant	see 3.1	3164	368	247573	49.5
4th order resistant	see 3.1	4174	400	260229	52.0

Due to the diffusion characteristics of the MixColumn transformation [10], an S-box output in round two depends on 32 key bits and in round three already on 128 key bits. Because of performance issues, we only masked the first three and the last three rounds in our AES implementations. Furthermore, we developed two sets of AES implementations resistant to HODPA. The first set listed in Table 6 uses the simple S-box recomputation alogrithm suggested in [15] which requires 512 bytes of RAM but is quick. The second set listed in Table 6 uses our proposed S-box recomputation alogrithm which requires only 256 bytes of RAM but could not be implemented as efficiently in assembly as the simple S-box recomputation algorithm due to pointer arithmetic issues.

---

[10] Under the assumption that the device is clocked at 5 MHz.

5   Conclusion

In this article we investigated the theoretical background of HODPA attacks and proposed several ideas how to protect an AES software implementation against such attacks. From our simulated experiments based on different hardware architectures it became clear that HODPA requires a huge number of measurements which exponentially increases with the order of the attack. Hence, a very simple way to protect a device against HODPA would be the use of a protocol which bounds the number of possible encryptions for a secret key. Moreover, we showed that the use of a noise generator as an add-on countermeasure does also increase the measurement costs considerably. We have presented details of various HODPA-resistant AES implementations which were programmed in assembly. In our benchmark tests it became clear that the permanent S-box recomputation is the major bottleneck and slows down the HODPA-resistant implementations, however this should not be an issue, if AES is used in challenge response-based protocols. In theory it would be possible to store all S-boxes in ROM, however, this would require $256 \cdot 256 \cdot 256 = 16$ MB , which is not feasible with currently available smart card microcontrollers.

Acknowledgements

We would like to thank Kerstin Lemke and Ahmad Sadeghi for the helpful discussions. Furthermore, we would like to thank Robert Szerwinski for implementing the various HODPA-resistant AES versions in assembly.

References

1. M.-L. Akkar and C. Giraud. An Implementation of DES and AES Secure against Some Attacks. In Ç. K. Koç, D. Naccache, and C. Paar, editors, *Cryptographic Hardware and Embedded Systems — CHES 2001*, volume LNCS 2162, pages 309–318. Springer-Verlag, 2001.
2. M.-L. Akkar and L. Goubin. A Generic Protection against High-Order Differential Power Analysis. In T. Johansson, editor, *Fast Software Encryption — FSE 2003*, volume 2887, pages 192–205. Springer-Verlag, 2003.
3. Mehdi-Laurent Akkar, Régis Bevan, Paul Dischamp, and Didier Moyart. Power Analysis, What Is Now Possible... In Tatsuaki Okamoto, editor, *Advances in Cryptology - ASIACRYPT 2000*, volume LNCS 1976, pages 489–502. Springer, 2000.
4. J. Blömer, J. Guajardo, and V. Krummel. Provably Secure Masking of AES. In H. Handschuh and M. Anwar Hasan, editors, *Selected Areas in Cryptography — SAC 2004*, volume 3357, pages 69–83. Springer-Verlag, August 2004.
5. E. Brier, C. Clavier, and F. Olivier. Correlation Power Analysis with a Leakage Model. In M. Joye and J.-J. Quisquater, editors, *Cryptographic Hardware and Embedded Systems — CHES 2004*, volume 3156, pages 16–29. Springer-Verlag, 2004.
6. S. Chari, C. S. Jutla, J. R. Rao, , and P. Rohatgi. A Cautionary Note Regarding the Evaluation of AES Candidates on Smart Cards. In *Proceedings: Second AES Candidate Conference (AES2)*, Rome, Italy, March 1999.

7. S. Chari, C. S. Jutla, J. R. Rao, , and P. Rohatgi. Towards Sound Approaches to Counteract Power-Analysis Attacks. In *Advances in Cryptology — CRYPTO '99*, volume LNCS 1666, pages 398 – 412. Springer-Verlag, August 1999.

8. C. Clavier and J.-S. Coron. On Boolean and Arithmetic Masking against Differential Power Analysis. In Ç. K. Koç and C. Paar, editors, *Cryptographic Hardware and Embedded Systems — CHES 2000*, volume LNCS 1965, pages 231 – 237. Springer-Verlag, 2000.

9. N. T. Courtois and L. Goubin. An Algebraic Masking Method to Protect AES Against Power Attacks. http://eprint.iacr.org/2005/204.pdf, 2005. Cryptology ePrint Archive: Report 2005/204.

10. J. Daemen and V. Rijmen. *The Design of Rijndael*. Springer Verlag, Berlin, 2002.

11. J. D. Golic and C. Tymen. Multiplicative Masking and Power Analysis of AES. In B.S. Kaliski, Ç. K. Koç, and C. Paar, editors, *Cryptographic Hardware and Embedded Systems — CHES 2002*, volume 2523, pages 198–212. Springer-Verlag, 2002.

12. M. Joye, P. Paillier, and B. Schoenmakers. On Second-Order Differential Power Analysis. In *accepted to Cryptographic Hardware and Embedded Systems — CHES 2005*. Springer-Verlag, 2005.

13. P. Kocher, J. Jaffe, and B. Jun. Differential Power Analysis: Leaking Secrets. In *Advances in Cryptology — CRYPTO '99*, volume LNCS 1666, pages 388–397. Springer-Verlag, 1999.

14. K. Lemke, K. Schramm, and C. Paar. DPA on n-Bit Sized Boolean and Arithmetic Operations and Its Application to IDEA, RC6 and the HMAC-Construction. In M. Joye and J.-J. Quisquater, editors, *Cryptographic Hardware and Embedded Systems — CHES 2004*, volume 3156, pages 205–219. Springer-Verlag, August 2004.

15. T. S. Messerges. Securing the AES Finalists Against Power Analysis Attacks. In B. Schneier, editor, *Fast Software Encryption — FSE 2000*, volume LNCS 1978, pages 150 – 164. Springer-Verlag, 2000.

16. T. S. Messerges. Using Second-Order Power Analysis to Attack DPA Resistant Software. In Ç. K. Koç and C. Paar, editors, *Cryptographic Hardware and Embedded Systems — CHES 2000*, volume LNCS 1965, pages 238 – 251. Springer-Verlag, 2000.

17. T. S. Messerges, E. A. Dabbish, and R. H. Sloan. Investigations of Power Analysis Attacks on Smartcards. In *USENIX Workshop on Smartcard Technology*, pages 151–162, 1999.

18. E. Oswald and K. Schramm. An Efficient Masking Scheme for AES Software Implementations. In *Workshop on Information Security Applications — WISA 2005*. Springer-Verlag, 2005.

19. A. G. Rostovtsev and O.V. Shemyakina. AES Side Channel Attack Protection Using Random Isomorphisms. http://eprint.iacr.org/2005/087.pdf, 2005. Cryptology ePrint Archive: Report 2005/087.

20. E. Trichina, D.S. Seta, and L. Germani. Simplified Adaptive Multiplicative Masking for AES. In B.S. Kaliski, Ç. K. Koç, and C. Paar, editors, *Cryptographic Hardware and Embedded Systems — CHES 2002*, volume 2523, pages 187–197. Springer-Verlag, 2002.

21. J. Waddle and D. Wagner. Towards Efficient Second-Order Power Analysis. In M. Joye and J.-J. Quisquater, editors, *Cryptographic Hardware and Embedded Systems — CHES 2004*, volume 3156, pages 1–15. Springer-Verlag, 2004.

A   Second Order DPA Against the AES S-Box Input (HW Model)

Given are two power signals $P(t_1)$ and $P(t_2)$ according to the HW-model

$$P(t_1) = \epsilon \sum_{i=0}^{n-1} M[i] + N_1 \quad \text{and} \quad P(t_2) = \epsilon \sum_{i=0}^{n-1} (X \oplus K \oplus M)[i] + N_2$$

with $N_1, N_2 \sim N(0, \sigma^2)$ and with n equal to the bit length of all intermediate variables, i.e. in the case of AES $n = 8$. Then, the covariance of the product $P(t_1) \cdot P(t_2)$ and the Hamming weight of the hypothesized S-box input $W(X \oplus K_h)$ is

$$COV[P(t_1) \cdot P(t_2), W(X \oplus K_h)] = COV[P(t_1) \cdot P(t_2), W(X \oplus K_h) - \frac{l}{2}]$$

$$= E[P(t_1)P(t_2) \cdot (W(X \oplus K_h) - \frac{l}{2})] - E[P(t_1)P(t_2)] \cdot \underbrace{E[(W(X \oplus K_h) - \frac{l}{2})]}_{=0}$$

$$= E[\epsilon^2 \underbrace{\sum_{i=0}^{n-1} \sum_{h=0, h \neq i}^{n-1} \sum_{j=0}^{l-1} M[i](M \oplus X \oplus K)[h](X \oplus K_h)[j]]}_{\epsilon^2 \cdot n \cdot (n-1) \cdot l \cdot \frac{1}{8}}$$

$$- \frac{l}{2} E[\epsilon^2 \underbrace{\sum_{i=0}^{n-1} \sum_{h=0, h \neq i}^{n-1} M[i](M \oplus X \oplus K)[h]]}_{\frac{l}{2} \cdot \epsilon^2 \cdot n \cdot (n-1) \cdot \frac{1}{4}}$$

$$+ E[\epsilon^2 \underbrace{\sum_{i=l}^{n-1} \sum_{j=0}^{l-1} M[i](M \oplus X \oplus K)[i](X \oplus K_h)[j]])}_{\epsilon^2 \cdot (n-l) \cdot \frac{1}{4} \cdot l \cdot \frac{1}{2}} - \frac{l}{2} E[\epsilon^2 \underbrace{\sum_{i=l}^{n-1} M[i](M \oplus X \oplus K)[i]]}_{\frac{l}{2} \cdot \epsilon^2 \cdot (n-l) \cdot \frac{1}{4}}$$

$$+ E[\epsilon^2 \underbrace{\sum_{i=0}^{l-1} \sum_{j=0}^{l-1} M[i](M \oplus X \oplus K)[i](X \oplus K_h)[j]])}_{\epsilon^2 \cdot (l-u) \cdot \frac{1}{4} + \epsilon^2 \cdot l \cdot (l-1) \cdot \frac{1}{4} \cdot \frac{1}{2}} - \frac{l}{2} E[\epsilon^2 \underbrace{\sum_{i=0}^{l-1} M[i](M \oplus X \oplus K)[i]]}_{\frac{l^2}{2} \cdot \epsilon^2 \cdot \frac{1}{4}}$$

$$+ E[N_2 \cdot \epsilon \cdot \underbrace{\sum_{i=0}^{n-1} M[i] \cdot (\sum_{i=0}^{l-1} X \oplus K_h[i] - \frac{l}{2})]}_{=0}$$

$$+ E[N_1 \cdot \epsilon \cdot \underbrace{\sum_{i=0}^{n-1} (M \oplus X \oplus K)[i] \cdot (\sum_{i=0}^{l-1} X \oplus K_h[i] - \frac{l}{2})]}_{=0}$$

$$+ E[N_1 \cdot N_2 \cdot \underbrace{(\sum_{i=0}^{l-1} X \oplus K_h[i] - \frac{l}{2})]}_{=0} = \frac{1}{4} \cdot \epsilon^2 \cdot (\frac{l}{2} - u)$$

where u denotes the number of correctly guessed key bits, $0 \leq u \leq l$. The variance of the Hamming weight of the hypothesized S-box input is

$$V[W(X \oplus K_h)] = V[W(X \oplus K_h) - \frac{l}{2}] = \frac{l}{4}$$

The variance of the product $P(t_1) \cdot P(t_2)$ can be expressed as

$$V[P(t_1) \cdot P(t_2)] = \underbrace{V[\epsilon^2 \sum_{i=0}^{n-1} \sum_{j=0}^{n-1} M[i](M \oplus X \oplus K)[j]]}_{\epsilon^4 \cdot n^2 \cdot \frac{1}{16} \cdot (3 + 2(n-1))} + \underbrace{V[N_2 \cdot \epsilon \cdot \sum_{i=0}^{n-1} M[i]]}_{\sigma^2 \cdot \epsilon^2 \cdot \frac{n}{4} + \sigma^2 \cdot \epsilon^2 \cdot \frac{n^2}{4}}$$

$$+ \underbrace{V[N_1 \cdot \epsilon \cdot \sum_{i=0}^{n-1} M \oplus X \oplus K[i]]}_{\sigma^2 \cdot \epsilon^2 \cdot \frac{n}{4} + \sigma^2 \cdot \epsilon^2 \cdot \frac{n^2}{4}} + \underbrace{V[N_1 \cdot N_2]}_{\sigma^4}$$

$$= \epsilon^4 \cdot n^2 \cdot \frac{1}{16} \cdot (3 + 2(n-1)) + \epsilon^2 \frac{n}{2}[(1+n)\sigma^2] + \sigma^4$$

This results in the correlation coefficient

$$\rho(P(t_1) \cdot P(t_2), W(X \oplus K_h)) = \frac{COV[P(t_1) \cdot P(t_2), W(X \oplus K_h)]}{\sqrt{V[P(t_1) \cdot P(t_2)]} \cdot \sqrt{V[W(X \oplus K_h)]}}$$

$$= \frac{COV[P(t_1) \cdot P(t_2), W(X \oplus K_h) - \frac{l}{2}]}{\sqrt{V[P(t_1) \cdot P(t_2)]} \cdot \sqrt{V[W(X \oplus K_h) - \frac{l}{2}]}}$$

$$= \frac{\frac{1}{4}\epsilon^2(\frac{l}{2} - u)}{\sqrt{(\epsilon^4 \frac{n^2}{16}(3 + 2(n-1)) + \epsilon^2 \frac{n}{2}(\sigma^2 + \sigma^2 n) + \sigma^4)}\sqrt{\frac{l}{4}}}$$

B   Multi-bit HODPA Against the AES S-Box Output (HW Model)

Given are d power signals $P(t_i)$ according to the noise-free Hamming weight model

$$P(t_1) = W(M_1)$$
$$\ldots \quad \ldots$$
$$P(t_{(d-1)}) = W(M_{(d-1)})$$
$$P(t_d) = W(S(X \oplus K) \oplus M)$$

with $M = M_1 \oplus \ldots \oplus M_{(d-1)}$. Let n be the bit length of all intermediate variables, i.e. in the case of AES $n = 8$. Then, the covariance of the product $\prod_{i=1}^{d} P(t_i)$ and the Hamming weight of the hypothesized S-box output $W(S(X \oplus K_h))$ is

$$COV[\prod_{i=1}^{d} P(t_i), W(S(X \oplus K_h))] = COV[\prod_{i=1}^{d} P(t_i), W(S(X \oplus K_h)) - \frac{n}{2}]$$

$$= E[\prod_{i=1}^{d} P(t_i) \cdot (W(S(X \oplus K_h)) - \frac{n}{2})] - E[\prod_{i=1}^{d} P(t_i)] \cdot \underbrace{E[(W(S(X \oplus K_h)) - \frac{n}{2})]}_{=0}$$

$$= E[\prod_{i=1}^{d} P(t_i) \cdot (W(S(X \oplus K_h))] - \frac{n}{2} \cdot \underbrace{E[\prod_{i=1}^{d} P(t_i)]}_{=(\frac{n}{2})^d}$$

$$= \underbrace{E[\prod_{i=1}^{d} P(t_i) \cdot (W(S(X \oplus K_h))]}_{=(\frac{n}{2})^{(d+1)}, if K_h \neq K \quad and \quad 2^{-(d+1)}(n^{(d+1)}-n)+n2^{-d}(d \bmod 2), if K_h = K} \qquad -(\frac{n}{2})^{(d+1)}$$

The variance of the Hamming weight of the hypothesized S-box output is

$$V[W(S(X \oplus K_h))] = V[W(S(X \oplus K_h)) - \frac{n}{2}] = \frac{n}{4}$$

The variance of the product $\prod_{i=1}^{d} P(t_i)$ is

$$V[\prod_{i=1}^{d} P(t_i)] = E[\prod_{i=1}^{d} P^2(t_i)] - E[\prod_{i=1}^{d} P(t_i)]^2$$

$$= \underbrace{E[W^2(M_1) \cdot \ldots \cdot W^2(M_{(d-1)}) \cdot W^2(S(X \oplus K_h))]}_{(\frac{n}{4}+\frac{n^2}{4})^d}$$

$$- \underbrace{E[W(M_1) \cdot \ldots \cdot W(M_{(d-1)}) \cdot W(S(X \oplus K_h))]^2}_{(\frac{n}{2})^{(2d)}}$$

$$= (\frac{n}{4} + \frac{n^2}{4})^d - (\frac{n}{2})^{(2d)}$$

This results in a correlation coefficient

$$\rho(\prod_{i=1}^{d} P(t_i), W(X \oplus K_h)) = \frac{COV[\prod_{i=1}^{d} P(t_i), W(X \oplus K_h)]}{\sqrt{V[\prod_{i=1}^{d} P(t_i)]} \cdot \sqrt{V[W(X \oplus K_h)]}}$$

$$= \frac{COV[\prod_{i=1}^{d} P(t_i), W(X \oplus K_h) - \frac{n}{2}]}{\sqrt{V[\prod_{i=1}^{d} P(t_i)]} \cdot \sqrt{V[W(X \oplus K_h) - \frac{n}{2}]}}$$

$$= \frac{2^{-(d+1)}(n^{(d+1)} - n) + n2^{-d}(d \bmod 2) - (\frac{n}{2})^{(d+1)}}{\sqrt{\left((\frac{n}{4} + \frac{n^2}{4})^d - (\frac{n}{2})^{(2d)}\right)\left(\frac{n}{4}\right)}}$$

C   Single-Bit HODPA Against the AES S-Box Output (General Model)

Given are d power signals $P(t_i)$ according to the general model

$$P(t_1) = \left(2\left(S(X \oplus K) \oplus M\right)[j] - 1\right)\epsilon + \sigma N_1$$

$$P(t_2) = \left(2M_1[j] - 1\right)\epsilon + \sigma N_2$$

$$\ldots \quad \ldots$$

$$P(t_d) = \left(2M_{(d-1)}[j] - 1\right)\epsilon + \sigma N_d \quad \text{with}$$

with $M = M_1 \oplus ... \oplus M_{(d-1)}$ and $N_1, ..., N_d \sim N(0, 1)$ and $0 \le j \le 7$. The correlation coefficient is defined as

$$\rho\left(\prod_{i=1}^{d} P(t_i), S(X \oplus K_h)[j]\right)$$

$$= \frac{E[\prod_{i=1}^{d} P(t_i) S(X \oplus K_h)[j]] - E[\prod_{i=1}^{d} P(t_i)] E[S(X \oplus K_h)[j]]}{\sqrt{V[\prod_{i=1}^{d} P(t_i)] V[S(X \oplus K_h)[j]]}}$$

where $S(X \oplus K_h)[j]$ denotes the state of bit j of a hypothesized S-box output $S(X \oplus K_h)$. The expectation values in the numerator are

$$E[\prod_{i=1}^{d} P(t_i) S(X \oplus K_h)[j]]$$

$$= E[\sum_{S(X \oplus K) \oplus M = 0} P(t_1) \sum_{M_1 = 0}^{1} P(t_2) \ ... \ \sum_{M_{d-1} = 0}^{1} P(t_d) \sum_{S(X \oplus K_h)[j] = 0}^{1} S(X \oplus K_h)[j]]$$

$$= 2^{-(d+1)}\Big(((-\epsilon) + \epsilon)((-\epsilon) + \epsilon)...((-\epsilon) + \epsilon)(0 + 1)\Big) = 0 \qquad \text{if} \quad K_h \ne K$$

$$= 2^{-(d)}\Big(\frac{2^d}{2}\epsilon^d(-1)^{d+1}\Big) = \frac{1}{2}\epsilon^d(-1)^{d+1} \qquad \text{if} \quad K_h = K$$

$$E[\prod_{i=1}^{d} P(t_i)] = \prod_{i=1}^{d} E[P(t_i)] = \prod_{i=1}^{d}(0.5(-\epsilon) + 0.5\epsilon) = 0$$
$$E[S(X \oplus K_h)[j]] = 0.5$$

The variances in the denominator are

$$V[S(X \oplus K_h)[j]] = 0.25$$

$$V[\prod_{i=1}^{d} P(t_i)] = E[\prod_{i=1}^{d} P^2(t_i)] - \underbrace{E[\prod_{i=1}^{d} P(t_i)]^2}_{=0}$$

$$= \prod_{i=1}^{d} E[P^2(t_i)] = \prod_{i=1}^{d} E[(\epsilon + \sigma N)^2] = \prod_{i=1}^{d} E[\epsilon^2 + 2\epsilon\sigma N + \sigma^2 N^2]$$

$$= (\epsilon^2 + \sigma^2)^d \quad \text{with} \quad N \sim N(0, 1) \quad \text{and} \quad \chi^2 = N^2 \sim \chi^2(1, 2)$$

This results in

$$\rho\left(\prod_{i=1}^{d} P(t_i), S(X \oplus K_h)[j]\right) = \frac{(-1)^{(d+1)}\epsilon^d}{\sqrt{(\epsilon^2 + \sigma^2)^d}} \qquad \text{if} \quad K_h = K$$

$$= 0 \qquad \text{if} \quad K_h \ne K$$

Chosen Ciphertext Secure Public Key Threshold Encryption Without Random Oracles

Dan Boneh[1,*], Xavier Boyen[2], and Shai Halevi[3]

[1] Stanford University, Stanford, CA
dabo@cs.stanford.edu
[2] Voltage Security, Palo Alto, CA
xb@boyen.org
[3] IBM, T.J. Watson, NY
shaih@alum.mit.edu

Abstract. We present a non-interactive chosen ciphertext secure threshold encryption system. The proof of security is set in the standard model and does not use random oracles. Our construction uses the recent identity based encryption system of Boneh and Boyen and the chosen ciphertext secure construction of Canetti, Halevi, and Katz.

1 Introduction

A threshold public key encryption system [14, 16, 13, 19] is a public key system where the private key is distributed among n decryption servers so that at least k servers are needed for decryption. In a threshold encryption system an entity, called the *combiner*, has a ciphertext C that it wishes to decrypt. The combiner sends C to the decryption servers, and receives partial decryption shares from at least k out of the n decryption servers. It then combines these k partial decryptions into a complete decryption of C. Ideally, there is no other interaction in the system, namely the servers need not talk to each other during decryption. Such threshold systems are called non-interactive. Often one requires that threshold decryption be *robust* [22, 18], namely if threshold decryption of a valid ciphertext fails, the combiner can identify the decryption servers that supplied invalid partial decryptions.

In this paper we study threshold encryption systems secure against chosen ciphertext attacks (CCA). The first such system, using random oracles, was given by Shoup and Gennaro [34]. Without random random oracles, this problem is much harder and was left as an open problem in [34]. Further work on this problem is discussed later in the introduction.

We present a very efficient non-interactive CCA threshold encryption system without random oracles. Our construction proceeds in two steps. First, we extend the CCA construction of Canetti et al. [10] to threshold systems. Second, we give a robust threshold version of a recent Identity Based Encryption (IBE) due to Boneh and Boyen [3]. We achieve robustness by adding a number of internal checks to the system. Our main construction is obtained by composing these

---

* Supported by NSF and the Packard Foundation.

D. Pointcheval (Ed.): CT-RSA 2006, LNCS 3860, pp. 226–243, 2006.

two results. This approach was outlined in the full versions of [3] and [10] and here we flesh out the full details. In Section 6 we briefly discuss several extensions such as proactive refresh [30, 24, 17] and distributed key generation [31, 21].

Related Work. Recall that the Cramer-Shoup system [11] and its variants [33, 28] provide efficient chosen ciphertext secure encryption without random oracles. All these systems require that the private key be used to test ciphertext validity during decryption. In a threshold environment none of the decryption servers possess the private key needed to perform this validity test. Consequently, constructing a threshold version of the Cramer-Shoup system is non-trivial. The first such construction is due to Canetti and Goldwasser [8]; other threshold versions of Cramer-Shoup are given in [1, 25].

These systems, however, are more complicated than the system in this paper: they require either a large degree of interaction between the decryption servers, or storage of a large number of pre-shared secrets. More recent constructions [29, 15, 12] are non-interactive, but are far less efficient than the construction in this paper. We refer to [34] for a comprehensive survey of the related work as well as the many applications of threshold encryption.

Our Contribution. This paper shows that CCA-secure threshold public key systems (in the standard model) are easier to derive from semantically secure Identity Based Encryption than from the Cramer-Shoup paradigm. In the non-threshold setting, the latest variant of either approach give public key systems that have similar encryption performance, whether IBE-based [6] or CS-based [28]. On the the other hand, in the threshold setting, the IBE approach appears to offer substantial benefits in terms of efficiency. The main reason is that in the IBE-to-CCA transformation from [10], the validity test performed during decryption requires only the public key. Consequently, each decryption server can test ciphertext validity on its own and only release a partial decryption of valid ciphertexts. (The more efficient transformation of Boneh and Katz [6] does not have this property, and is thus less suitable for threshold encryption.)

We extend [10] to give a generic transformation from threshold IBE to threshold public key encryption, and present a concrete construction based on a threshold version of the Boneh-Boyen IBE [3]. We add a number of internal checks to provide robustness against misbehaving decryption servers. The basic idea of this paper was originally suggested in the expanded versions of [3] and [10], but without any detail. This work gives an explicit account of the construction including all the additional checks that one has to perform.

We note that Boyen, Mei, and Waters [7] very recently gave a particularly simple and efficient CCA2-secure key encapsulation mechanism based on the Boneh-Boyen IBE framework. It is self-contained, by contrast to the generic CHK and BK transformations, which require additional ingredients. As with CHK, the BMW method supports public ciphertext verification, and is thus suitable for non-interactive threshold decryption. Likewise, they adapt our present construction to realize an efficient CCA2-secure non-interactive threshold KEM that eschews the need for signatures.

2   Definitions

As a preamble to our results, we recall the definitions of threshold PKE and IBE, and secure signatures.

As usual, we say that a function $f : \mathbb{Z} \to \mathbb{R}$ is negligible if for all $c > 0$ there exists $N \in \mathbb{Z}$ such that $|f(x)| < 1/x^c$ for all $x > N$.

2.1   Threshold Public Key Encryption

We define chosen ciphertext secure (CCA2) threshold public key encryption for a static adversary. We mostly follow the notation from Shoup and Genarro [34]. A Threshold Public Key Encryption (TPKE) system consists of five algorithms.

Setup(n, k, Λ)**:** Takes as input the number of decryption servers n, a threshold k where $1 \le k \le n$, and a security parameter $\Lambda \in \mathbb{Z}$. It outputs a triple $(\mathsf{PK}, \mathsf{VK}, \mathbf{SK})$ where PK is called the public key, VK is called the verification key, and $\mathbf{SK} = (\mathsf{SK}_1, \ldots, \mathsf{SK}_n)$ is a vector of n private key shares. Decryption server i is given the private key share (i, SK_i) and uses it to derive a decryption share for a given ciphertext. The verification key VK is used to check validity of responses from decryption servers.

Encrypt(PK, M)**:** Takes as input a public key PK and a message M. It outputs a ciphertext.

ShareDecrypt$(\mathsf{PK}, i, \mathsf{SK}_i, C)$**:** Takes as input the public key PK, a ciphertext C, and one of the n private key shares in $\mathbf{SK}$. It outputs a decryption share $\mu = (i, \hat{\mu})$ of the enciphered message, or a special symbol $(i, \perp)$.

ShareVerify$(\mathsf{PK}, \mathsf{VK}, C, \mu)$**:** Takes as input PK, the verification key VK, a ciphertext C, and a decryption share μ. It outputs valid or invalid. When the output is valid we say that μ is a valid decryption share of C.

Combine$(\mathsf{PK}, \mathsf{VK}, C, \{\mu_1, \ldots, \mu_k\})$**:** Takes as input PK, VK, a ciphertext C, and k decryption shares $\{\mu_1, \ldots, \mu_k\}$. It outputs a cleartext M or $\perp$.

Consistency Requirements. Let $(\mathsf{PK}, \mathsf{VK}, \mathbf{SK})$ be the output of *Setup*(n, k, Λ). We require the following two consistency properties:

1. For any ciphertext C, if $\mu = ShareDecrypt(\mathsf{PK}, i, \mathsf{SK}_i, C)$ where SK_i is the i-th private key share in $\mathbf{SK}$, then *ShareVerify*$(\mathsf{PK}, \mathsf{VK}, C, \mu) = $ valid.
2. If C is the output of *Encrypt*(PK, M) and $S = \{\mu_1, \ldots, \mu_k\}$ is a set of decryption shares $\mu_i = ShareDecrypt(\mathsf{PK}, i, \mathsf{SK}_i, C)$ for k distinct private keys in $\mathbf{SK}$, then we require that *Combine*$(\mathsf{PK}, \mathsf{VK}, C, S) = M$.

Security. Security of TPKE is defined using two properties: security against chosen ciphertext attacks, and consistency of decryptions.

Chosen Ciphertext Security. Security against chosen ciphertext attacks is defined using the following game between a challenger and a static adversary $\mathcal{A}$. Both are given n, k, and a security parameter $\Lambda \in \mathbb{Z}^+$ as input.

> **Init.** The adversary outputs a set $S \subset \{1, \ldots, n\}$ of $k - 1$ decryption servers to corrupt.

Setup. The challenger runs $Setup(n, k, \Lambda)$ to obtain a random instance $(\mathsf{PK}, \mathsf{VK}, \mathbf{SK})$ where $\mathbf{SK} = (\mathsf{SK}_1, \ldots, \mathsf{SK}_n)$. It gives the adversary PK, VK, and all (j, SK_j) for $j \in S$.

Query phase 1. The adversary adaptively issues decryption queries (C, i) where $C \in \{0, 1\}^*$ and $i \in \{1, \ldots, n\}$. The challenger responds with $ShareDecrypt(\mathsf{PK}, i, \mathsf{SK}_i, C)$.

Challenge. The adversary outputs two messages M_0, M_1 of equal length. The challenger picks a random $b \in \{0, 1\}$ and lets $C^* = Encrypt(\mathsf{PK}, M_b)$. It gives C^* to the adversary.

Query phase 2. The adversary issues further decryption queries (C, i), under the constraint that $C \neq C^*$. The challenger responds as in phase 1.

Guess. Algorithm $\mathcal{A}$ outputs its guess $b' \in \{0, 1\}$ for b and wins the game if $b = b'$.

We define the advantage of $\mathcal{A}$ as $\mathsf{AdvCCA}_{\mathcal{A}, n, k}(\Lambda) = |\Pr[b = b'] - \frac{1}{2}|$.

Decryption Consistency. Consistency of decryption is defined using the following game. The game starts with the *Init, Setup*, and *Query phase 1* steps as in the game above. The adversary then outputs a ciphertext C and two sets of decryption shares $S = \{\mu_1, \ldots, \mu_k\}$ and $S' = \{\mu'_1, \ldots, \mu'_k\}$ each of size k. Let VK be the verification key generated in the Setup step. The adversary wins if:

1. The shares in S and S' are valid decryption shares for C under VK;
2. S and S' each contain decryption shares from k distinct servers; and
3. $Combine(\mathsf{PK}, \mathsf{VK}, C, S) \neq Combine(\mathsf{PK}, \mathsf{VK}, C, S')$.

We let $\mathsf{AdvCD}_{\mathcal{A}, n, k}(\Lambda)$ denote the adversary's advantage in winning this game.

Definition 1. *We say that a TPKE system is secure if for any n and k where $0 < k \leq n$, and any polynomial time algorithm $\mathcal{A}$, the functions $\mathsf{AdvCCA}_{\mathcal{A}, n, k}(\Lambda)$ and $\mathsf{AdvCD}_{\mathcal{A}, n, k}(\Lambda)$ are negligible.*

2.2 IBE with Threshold Key Generation

Next, we define IBE with threshold key generation. Here we are only concerned with semantic security and ignore chosen ciphertext attacks. A Threshold Identity Based Encryption (TIBE) system consists of seven algorithms.

Setup(n, k, Λ): Takes as input the number of decryption servers n, a threshold k where $1 \leq k \leq n$, and a security parameter $\Lambda \in \mathbb{Z}$. It outputs a triple $(\mathsf{PK}, \mathsf{VK}, \mathbf{SK})$ where PK is called the system parameters, VK is called a verification key, and $\mathbf{SK} = (\mathsf{SK}_1, \ldots, \mathsf{SK}_n)$ is a vector of master key shares analogous to the private key shares in the definition of TPKE. Decryption server i is given the master key share (i, SK_i).

ShareKeyGen(PK, i, SK_i, ID): Takes as input the system parameters PK, an identity ID, and a master key share (i, SK_i). It outputs a private key share $\theta = (i, \hat{\theta})$ for ID.

ShareVerify(PK, VK, ID, θ): Takes as input the system parameters PK, the verification key VK, an identity ID, and a private key share θ. It outputs `valid` or `invalid`.

Combine(PK, VK, ID, $\{\theta_1, \ldots, \theta_k\}$): Takes as input PK, VK, an identity ID, and k private key shares $\{\theta_1, \ldots, \theta_k\}$. It outputs a private key d_{ID} or $\perp$.

Encrypt(PK, ID, M): Takes as input PK, an identity ID, and a message M, and outputs a ciphertext C.

ValidateCT(PK, ID, C): Takes as input PK, an identity ID, and a ciphertext C. It outputs `valid` or `invalid`. If `valid` we say that C is a valid encryption under ID.

Decrypt(PK, ID, d_{ID}, C): Takes as input PK, ID, a private key d_{ID}, and a ciphertext C. It outputs a message M or $\perp$.

Note that, unlike the previous section, decryption is not distributed. Only key generation is distributed.

Consistency Requirements. Let (PK, VK, **SK**) be the output of *Setup*(n, k, Λ). We require consistency properties as for TPKE systems:

1. For any identity ID, if $\theta = ShareKeyGen(\text{PK}, i, \text{SK}_i, C)$ where SK_i is one of the private key shares in **SK**, then *ShareVerify*(PK, VK, ID, θ) = `valid`.
2. For any ID, if $S = \{\theta_1, \ldots, \theta_k\}$ where $\theta_i = ShareKeyGen(\text{PK}, i, \text{SK}_i, \text{ID})$ for k distinct private keys in **SK**, and d_{ID} is the output of *Combine*(PK, VK, ID, S), then we require that for any M and $C = Encrypt(\text{PK}, \text{ID}, M)$ we have *ValidateCT*(PK, ID, C) = `valid` and *Decrypt*(PK, d_{ID}, C) = M.

Security. Security of a TIBE is defined using two properties: security against chosen identity attacks and consistency of key generation. There are two ways to define chosen identity attacks against IBE schemes, depending on whether the adversary chooses the target identity adaptively (an adaptive-ID attack [5]) or selects it in advance (a selective-ID attack [9]); we only need the latter for our purposes.

Selective-ID Security. Semantic security against a selective identity attack is defined using the following game:

Init. The adversary outputs an identity ID^* that it wishes to attack and a set of $k - 1$ decryption servers $S \subset \{1, \ldots, n\}$ that it wishes to corrupt.

Setup. The challenger runs *Setup*(n, k, Λ) to obtain a random instance (PK, VK, **SK**) where **SK** = $(\text{SK}_1, \ldots, \text{SK}_n)$. It gives the adversary PK, VK, and all (j, SK_j) for $j \in S$.

Query phase 1. The adversary adaptively issues chosen identity queries (ID, i) where $\text{ID} \in \{0, 1\}^*$ and $i \in \{1, \ldots, n\}$. The only constraint is that $\text{ID} \neq \text{ID}^*$. The challenger responds with $ShareKeyGen(\text{PK}, i, \text{SK}_i, \text{ID})$.

Challenge. The adversary outputs two messages M_0, M_1 of equal length. The challenger picks a random $b \in \{0, 1\}$ and sets the challenge ciphertext to $C^* = Encrypt(\text{PK}, \text{ID}^*, M)$. It gives C^* to the adversary.

Query phase 2. The adversary and the challenger interact as in phase 1.
Guess. Algorithm $\mathcal{A}$ outputs its guess $b' \in \{0,1\}$ for b and wins the game if $b = b'$.

We define the advantage of $\mathcal{A}$ as $\mathsf{AdvIND\text{-}ID}_{\mathcal{A},n,k}(\Lambda) = |\Pr[b = b'] - \frac{1}{2}|$.

Key Generation Consistency. Consistency of key generation (and decryption) is defined using the following game. The game starts with the *Init*, *Setup*, and *Query phase 1* steps as in the game above. The adversary then outputs an identity ID, a ciphertext C, and two sets of private key shares $S = \{\theta_1, \ldots, \theta_k\}$ and $S' = \{\theta'_1, \ldots, \theta'_k\}$ each of size k. Let PK and VK be the system parameters and the verification key generated in the Setup step. The adversary wins if:

1. the shares in S and S' are valid private key shares for ID under VK;
2. S and S' each contain private key shares from k distinct servers;
3. C is valid for the given ID, i.e., *ValidateCT*(PK, ID, C) = `valid`;
4. the keys $d_{\mathsf{ID}} = Combine(\mathsf{PK}, \mathsf{VK}, \mathsf{ID}, S)$ and $d'_{\mathsf{ID}} = Combine(\mathsf{PK}, \mathsf{VK}, \mathsf{ID}, S')$ are such that $\perp \neq d_{\mathsf{ID}} \neq d'_{\mathsf{ID}} \neq \perp$;
5. *Decrypt*(PK, ID, d_{ID}, C) $\neq$ *Decrypt*(PK, ID, d'_{ID}, C).

Let $\mathsf{AdvCD\text{-}ID}_{\mathcal{A},n,k}(\Lambda)$ be the adversary's advantage in winning the game.

Definition 2. *We say that a TIBE system is selective-ID secure if for any n, k (where $0 < k \leq n$), and any polynomial time $\mathcal{A}$, the functions $\mathsf{AdvIND\text{-}ID}_{\mathcal{A},n,k}(\Lambda)$ and $\mathsf{AdvCD\text{-}ID}_{\mathcal{A},n,k}(\Lambda)$ are negligible.*

2.3   Strong Existentially Unforgeable Signatures

A signature scheme is made up of three algorithms, *SigKeyGen*, *Sign*, and *SigVerify*, for generating a key pair, signing a message, and verifying a signature, respectively.

The standard notion of security for a signature scheme is called existential unforgeability under a chosen message attack [23]. We need a slightly stronger notion of security, called strong existential unforgeability [2]. We define strong existential unforgeability under a "one chosen message" attack using the following game between a challenger and an adversary $\mathcal{A}$:

Setup. The challenger runs algorithm *SigKeyGen*(Λ) to obtain a public key *VerK* and a private key *SigK*. The adversary $\mathcal{A}$ is given *VerK*.
Query. The adversary $\mathcal{A}$ requests a signature on a single messages of its choice, $M \in \{0,1\}^*$, under *VerK*. The challenger responds with a signature $\sigma = Sign(SigK, M)$.
Output. The adversary $\mathcal{A}$ outputs a pair (M', σ') and wins the game if $(M', \sigma') \neq (M, \sigma)$ and *SigVerify*(*VerK*, M', σ') = `valid`.

We define $\mathsf{AdvSig}_{\mathcal{A}}(\Lambda)$ to be the probability that $\mathcal{A}$ wins in the above game.

Definition 3. *A signature scheme is existentially unforgeable under a one chosen message attack if for any probabilistic polynomial time algorithm $\mathcal{A}$ the function $\mathsf{AdvSig}_{\mathcal{A}}(\Lambda)$ is negligible.*

Efficient constructions for such signatures schemes, without random oracles, are known using the Strong-RSA assumption [20] and the 2-Strong-Diffie-Hellman (2-SDH) assumption [4].

2.4   Bilinear Maps

We briefly review the necessary facts about bilinear groups and bilinear maps, also called pairings, using the following notation:

1. $\mathbb{G}$ and $\mathbb{G}_1$ are two (multiplicative) cyclic groups of prime order p;
2. g is a generator of $\mathbb{G}$;
3. e is a bilinear map $e : \mathbb{G} \times \mathbb{G} \to \mathbb{G}_1$;
4. $\mathcal{GG}(\Lambda)$ is a bilinear Group Generator as described below.

A pairing is a map $e : \mathbb{G} \times \mathbb{G} \to \mathbb{G}_1$ with the following properties [26, 27, 5]:

1. Bilinearity: for all $u, v \in \mathbb{G}$ and $a, b \in \mathbb{Z}$, we have $e(u^a, v^b) = e(u, v)^{ab}$.
2. Non-degeneracy: $e(g, g) \neq 1$.

Algorithm $\mathcal{GG}(\Lambda)$ is a bilinear Group Generator that takes a security parameter $\Lambda \in \mathbb{Z}$ as input and outputs the description of groups $\mathbb{G}$ and $\mathbb{G}_1$ and a bilinear map $e : \mathbb{G} \times \mathbb{G} \to \mathbb{G}_1$ where the group operation in $\mathbb{G}$ and $\mathbb{G}_1$ as well as the map e can be computed in polynomial time in Λ. To simplify the notation we use $\mathbb{G} \xleftarrow{\text{R}} \mathcal{GG}(\Lambda)$ to denote the output of a random execution of $\mathcal{GG}$ on input Λ, and posit that the output $\mathcal{GG}(\Lambda)$ contains a description of p, $\mathbb{G}$, $\mathbb{G}_1$, and e.

2.5   Bilinear Diffie-Hellman Assumption

We say that an algorithm $\mathcal{B}$ that outputs $b \in \{0, 1\}$ has advantage $\epsilon(\Lambda)$ in solving the *decision* BDH problem [26, 32, 5] for the bilinear group generator $\mathcal{GG}$ if

$$\left| \Pr\left[\mathcal{B}(\mathbb{G}, g, g^a, g^b, g^c, e(g, g)^{abc}) = 0 \right] - \Pr\left[\mathcal{B}(\mathbb{G}, g, g^a, g^b, g^c, T) = 0 \right] \right| \geq \epsilon(\Lambda)$$

where the probability is over the random choice of group $\mathbb{G} \xleftarrow{\text{R}} \mathcal{GG}(\Lambda)$, the random choice of generator g in $\mathbb{G}$, the random choice of a, b, c in $\mathbb{Z}_p$, the random choice of $T \in \mathbb{G}_1$, and the random bits consumed by $\mathcal{B}$. We refer to the distribution on the left as $\mathcal{P}_{BDH}$, and on the right as $\mathcal{R}_{BDH}$.

Definition 4. *We say that the Decision-BDH assumption holds for $\mathcal{GG}$ if any polynomial time algorithm has negligible advantage in solving the Decision BDH problem for $\mathcal{GG}$.*

3   A Threshold Identity Based Encryption System

We start with a description of a concrete Threshold IBE (TIBE) system and prove its semantic security against selective identity attacks without random oracles. For robustness against misbehaving servers we need to add several internal checks to the scheme from [3]. In a later section we show how this construction leads to a non-interactive threshold PKE with chosen ciphertext security. The TIBE system works as follows:

Setup(n, k, Λ). Run the group generator $\mathcal{GG}(\Lambda)$ to obtain a bilinear group $\mathbb{G}$ of prime order $p > n$. Select random generators g, g_2, h_1 in $\mathbb{G}$, and a random degree $k - 1$ polynomial $f \in \mathbb{Z}_p[X]$. Set $\alpha = f(0) \in \mathbb{Z}_p$ and $g_1 = g^\alpha$.

The system parameters PK consist of PK $= (\mathbb{G}, g, g_1, g_2, h_1)$. For $i = 1, \ldots, n$ the master key share (i, SK_i) of server i is defined as $\mathsf{SK}_i = g_2^{f(i)}$. The public verification key VK consists of the n-tuple $(g^{f(1)}, \ldots, g^{f(n)})$.

ShareKeyGen(PK, i, SK_i, ID). Let PK $= (\mathbb{G}, g, g_1, g_2, h_1)$ and pick a random $r \in \mathbb{Z}_p$. Output the private key share $\theta_i = (i, (w_{i,0}, w_{i,1}))$ calculated as

$$w_{i,0} = \mathsf{SK}_i \cdot (g_1^{\mathsf{ID}} h_1)^r, \qquad w_{i,1} = g^r .$$

ShareVerify(PK, VK, ID, θ_i). To verify that θ_i is a valid private key share for identity ID, let VK $= (u_1, \ldots, u_n)$ where $u_i = g^{f(i)}$, and $\theta_i = (i, (w_{i,0}, w_{i,1}))$. Output valid or invalid according to the truth of the following condition:

$$e(u_i, g_2) \cdot e(g_1^{\mathsf{ID}} h_1, w_{i,1}) = e(g, w_{i,0})$$

Combine(PK, VK, ID, $(\theta_1, \ldots, \theta_k)$). If one of $\theta_1, \ldots, \theta_k$ is invalid, or if two shares θ_i and θ_j bear the same server index, then output $\perp$ and exit. Otherwise, let $\theta_i = (i, (w_{i,0}, w_{i,1})$. Without loss of generality we assume that decryption servers $i = 1, \ldots, k$ were used to generate $\theta_1, \ldots, \theta_k$. To derive the private key for ID let $\lambda_1, \ldots, \lambda_k \in \mathbb{Z}_p$ be the Lagrange coefficients so that $\alpha = f(0) = \sum_{i=1}^k \lambda_i f(i)$. Output the reconstituted private key $d_{\mathsf{ID}} = (w_0, w_1)$ given by

$$w_0 = \prod_{i=1}^k w_{i,0}^{\lambda_i} , \qquad w_1 = \prod_{i=1}^k w_{i,1}^{\lambda_i}$$

Encrypt(PK, ID, M). To encrypt $M \in \mathbb{G}_1$ for identity ID, pick a random $s \in \mathbb{Z}_p$ and output

$$C = \left(e(g_1, g_2)^s \cdot M, \quad g^s, \quad g_1^{s \cdot \mathsf{ID}} h_1^s \right)$$

ValidateCT(PK, ID, C). To validate a ciphertext $C = (A, B, C_1)$ with respect to an identity ID, output valid or invalid depending on whether

$$e(B, g_1^{\mathsf{ID}} h_1) = e(C_1, g)$$

Decrypt(PK, ID, d_{ID}, C). To decrypt $C = (A, B, C_1)$ using a private key $d_{\mathsf{ID}} = (w_0, w_1)$, first check that *ValidateCT*(PK, ID, C) = valid and that $e(g_1, g_2) \cdot e(g_1^{\mathsf{ID}} h_1, w_1) = e(g, w_0)$. If either check fails, output $\perp$ and exit. Otherwise, output the plaintext

$$A \cdot e(C_1, w_1)/e(B, w_0)$$

The two checks during decryption ensure that C is a valid ciphertext under ID and that d_{ID} is a valid private key for ID. These checks are needed to ensure consistency of key generation in case some servers misbehave. If these conditions are fullfilled, then the decryption is correct, because $(w_0, w_1) = (g_2^\alpha (g_1^{\mathsf{ID}} h_1)^{\bar{r}}, g^{\bar{r}})$ for some $\bar{r} \in \mathbb{Z}_p$, and

$$A \cdot \frac{e(C_1, w_1)}{e(B, w_0)} = M \cdot e(g_1, g_2)^s \cdot \frac{e(g_1^{\mathsf{ID}}, g)^{s\bar{r}} \cdot e(h_1, g)^{s\bar{r}}}{e(g, g_2^\alpha)^s \cdot e(g, g_1^{\mathsf{ID}})^{s\bar{r}} \cdot e(g, h_1)^{s\bar{r}}} = M.$$

3.1 Security

We now prove the semantic security of this threshold IBE against selective identity attacks. The proof is based on the proof in [3] and gives a tight reduction.

As in [3], the key to the simulation is the construction of a public key $(..., h_1, ...)$ that allows the simulator to calculate private key shares for any identity except ID^*. A difference from the proof in [3] is that the simulator must be able to extract private key shares (i.e., elements of the vector SK). In addition, the simulator must produce a valid verification key VK, which only exists in the threshold setting. In order to construct the components of VK that correspond to the corrupted servers, the simulator does interpolation in the exponent. The Lagrange interpolation coefficients are blinded and yet carry over unaffected through the bilinear map in the verification equation. The details follow.

Theorem 1. *Suppose the Decision BDH assumption holds for $\mathcal{GG}$. Then the TIBE system above is semantically secure against selective identity, chosen plaintext attacks.*

Proof. First, suppose $\mathcal{A}$ has advantage $\mathsf{AdvIND\text{-}ID}_{\mathcal{A},n,k} > \epsilon$ in attacking the threshold IBE system for a given value of the security parameter Λ. We build an algorithm $\mathcal{B}$ that solves the Decision BDH problem in a random instance $\mathbb{G} \xleftarrow{\text{R}} \mathcal{GG}(\Lambda)$ with advantage ϵ.

Let there thus be a random bilinear group $\mathbb{G} \xleftarrow{\text{R}} \mathcal{GG}(\Lambda)$ and a random generator $g \in \mathbb{G}^*$ of $\mathbb{G}$. Algorithm $\mathcal{B}$ is given as input a random tuple $(\mathbb{G}, g, g^a, g^b, g^c, T)$ that is either sampled from $\mathcal{P}_{BDH}$ (where $T = e(g,g)^{abc}$) or from $\mathcal{R}_{BDH}$ (where T is uniform and independent in $\mathbb{G}_1$). Algorithm $\mathcal{B}$'s goal is to output 1 if $T = e(g,g)^{abc}$ and 0 otherwise. Set $g_1 = g^a$, $g_2 = g^b$, $g_3 = g^c$. Algorithm $\mathcal{B}$ works by interacting with $\mathcal{A}$ in a threshold selective-ID game as follows:

Initialization. The adversary $\mathcal{A}$ chooses a set S of $k-1$ decryption servers that it wants to corrupt. Let $S = \{s_1, \ldots, s_{k-1}\} \subset \{1, \ldots, n\}$. The adversary $\mathcal{A}$ also announces the identity ID^* it wants to attack.

Setup. $\mathcal{B}$ does the following:

1. First, $\mathcal{B}$ picks a random integer $\gamma \in \mathbb{Z}_p$ and defines $h_1 = g_1^{-\mathsf{ID}^*} g^\gamma \in \mathbb{G}$. Algorithm $\mathcal{B}$ gives $\mathcal{A}$ the public key $\mathsf{PK} = (\mathbb{G}, g, g_1, g_2, h_1)$. Note that the corresponding master key, which is unknown to $\mathcal{B}$, is $g_2^a = g^{ab} \in \mathbb{G}$.

2. Next, $\mathcal{B}$ generates the master key shares for the $k-1$ corrupt servers in S. To do so, $\mathcal{B}$ first picks $k-1$ random integers $\alpha_1, \ldots, \alpha_{k-1} \in \mathbb{Z}_p$. Let $f \in \mathbb{Z}_p[X]$ be the degree $k-1$ polynomial implicitly defined to satisfy $f(0) = a$ and $f(s_i) = \alpha_i$ for $i = 1, \ldots, k-1$; note that $\mathcal{B}$ does not know f since it does not know a. Algorithm $\mathcal{B}$ gives $\mathcal{A}$ the $k-1$ master key shares $\mathsf{SK}_{s_i} = g_2^{\alpha_i}$. These keys are consistent with this polynomial f since $\mathsf{SK}_{s_i} = g_2^{f(s_i)}$ for $i = 1, \ldots, k-1$.

3. Finally, $\mathcal{B}$ constructs the verification key, which is a n-vector $(u_1, \ldots, u_n)$ such that $u_i = g^{f(i)}$ for the polynomial f defined above, as follows.
   - For $i \in S$, computing u_i is easy since $f(i)$ is equal to one of the $\alpha_1, \ldots, \alpha_{k-1}$, which are known to $\mathcal{B}$. Thus, $u_{s_1}, \ldots, u_{s_k} \in \mathbb{G}$ are easy for $\mathcal{B}$ to compute.

- For $i \notin S$, algorithm $\mathcal{B}$ needs to compute the Lagrange coefficients $\lambda_0, \lambda_1, \ldots, \lambda_{k-1} \in \mathbb{Z}_p$ such that $f(i) = \lambda_0 f(0) + \sum_{j=1}^{k-1} \lambda_j f(s_j)$; these Lagrange coefficients are easily calculated since they do not depend on f. Algorithm $\mathcal{B}$ then sets $u_i = g_1^{\lambda_0} u_{s_1}^{\lambda_1} \cdots u_{s_{k-1}}^{\lambda_{k-1}}$, which entails that $u_i = g^{f(i)}$ as required.

Once it has computed all the u_i's, $\mathcal{B}$ gives to $\mathcal{A}$ the verification key $\mathsf{VK} = (u_1, \ldots, u_n)$.

Phase 1. $\mathcal{A}$ issues up to q_s private key share generation queries to the uncorrupt servers. Consider a key generation query to server $i \notin S$ for the identity $\mathsf{ID} \neq \mathsf{ID}^*$.

Algorithm $\mathcal{B}$ needs to return $(i, (w_{i,0}, w_{i,1}))$ where $w_{i,0} = \mathsf{SK}_i (g_1^{\mathsf{ID}} h_1)^r$ and $w_{i,1} = g^r$ for some random $r \in \mathbb{Z}_p$. To do so, $\mathcal{B}$ first computes the Lagrange coefficients $\lambda_0, \lambda_1, \ldots, \lambda_{k-1} \in \mathbb{Z}_p$ such that $f(i) = \lambda_0 f(0) + \sum_{j=1}^{k-1} \lambda_j f(s_j)$. Next, $\mathcal{B}$ picks a random $r \in \mathbb{Z}_p$ and sets

$$w_{i,0} = g_2^{\frac{-\gamma \lambda_0}{\mathsf{ID}-\mathsf{ID}^*}} (g_1^{\mathsf{ID}} h_1)^r \cdot \prod_{j=1}^{k-1} g_2^{\lambda_j \alpha_j}, \quad w_{i,1} = g_2^{\frac{-\lambda_0}{\mathsf{ID}-\mathsf{ID}^*}} g^r$$

We claim that $(w_{i,0}, w_{i,1})$ are a valid response to this decryption query. To see this, let $\tilde{r} = r - \frac{b\lambda_0}{\mathsf{ID}-\mathsf{ID}^*}$. Then we have that

$$g_2^{\frac{-\gamma \lambda_0}{(\mathsf{ID}-\mathsf{ID}^*)}} (g_1^{\mathsf{ID}} h_1)^r = g_2^{\frac{-\gamma \lambda_0}{(\mathsf{ID}-\mathsf{ID}^*)}} (g_1^{\mathsf{ID}-\mathsf{ID}^*} g^\gamma)^r$$

$$= g_2^{\lambda_0 a} (g_1^{\mathsf{ID}-\mathsf{ID}^*} g^\gamma)^{r - \frac{b\lambda_0}{\mathsf{ID}-\mathsf{ID}^*}} = g_2^{\lambda_0 a} (g_1^{\mathsf{ID}} h_1)^{\tilde{r}}$$

It follows that the private key share $(i, (w_{i,0}, w_{i,1}))$ defined above satisfies

$$w_{i,0} = g_2^{f(i)} \cdot (g_1^{\mathsf{ID}} h_1)^{\tilde{r}}, \qquad w_{i,1} = g^{\tilde{r}}$$

and $\tilde{r}$ is uniform in $\mathbb{Z}_p$ as required. Hence, $(i, (w_{i,0}, w_{i,1}))$ is a valid response to $\mathcal{A}$.

Challenge. $\mathcal{A}$ outputs two same-length messages M_0 and M_1 on which it wishes to be challenged. $\mathcal{B}$ flips a fair coin $b \in \{0, 1\}$, and responds with the challenge ciphertext

$$C = (T \cdot M_b, \ g_3, \ g_3^\gamma)$$

Since $C = (T \cdot M_b, \ g^c, \ g_1^{c \cdot \mathsf{ID}^*} h_1^c)$, the challenge ciphertext is a valid encryption of M_b with the correct distribution whenever $T = e(g, g)^{abc} = e(g_1, g_2)^c$ (as is the case when the input 5-tuple is sampled from $\mathcal{P}_{BDH}$). On the other hand, when T is uniform and independent in $\mathbb{G}_1$ (which occurs when the input 5-tuple is sampled from $\mathcal{R}_{BDH}$) the challenge ciphertext C is independent of b in the adversary's view.

Phase 2. $\mathcal{A}$ issues additional queries as in Phase 1, to which algorithm $\mathcal{B}$ responds as before.

Guess. Finally, $\mathcal{A}$ outputs a guess $b' \in \{0, 1\}$. Algorithm $\mathcal{B}$ concludes its own game by outputting a guess as follows. If $b = b'$ then $\mathcal{B}$ outputs 1 meaning $T = e(g, g)^{abc}$. Otherwise, it outputs 0 meaning $T \neq e(g, g)^{abc}$.

When the input 5-tuple is sampled from $\mathcal{P}_{BDH}$ (where $T = e(g,g)^{abc}$) then $\mathcal{A}$'s view is identical to its view in a real attack game and therefore $\mathcal{A}$ must satisfy $|\Pr[b = b'] - 1/2| > \epsilon$. On the other hand, when the input 5-tuple is sampled from $\mathcal{R}_{BDH}$ (where T is uniform in $\mathbb{G}_1$) then $\Pr[b = b'] = 1/2$. Therefore, with uniformly chosen g in $\mathbb{G}^*$, uniformly chosen a, b, c in $\mathbb{Z}_p$, and uniformly chosen T in $\mathbb{G}_1$, we have, as required, that

$$\left| \begin{array}{l} \Pr\left[\mathcal{B}(\mathbb{G}, g, g^a, g^b, g^c, e(g,g)^{abc}) = 0\right] \\ - \Pr\left[\mathcal{B}(\mathbb{G}, g, g^a, g^b, g^c, T) = 0\right] \end{array} \right| \geq \left| \left(\frac{1}{2} \pm \epsilon \right) - \frac{1}{2} \right| = \epsilon$$

To complete the proof of Theorem 1 it remains to prove consistency of key generation. We argue that for any algorithm $\mathcal{A}$ we have $\mathsf{AdvCD\text{-}ID}_{\mathcal{A},n,k}(\Lambda) = 0$. To see this, observe that the two tests performed during decryption ensure that $Decrypt(\mathsf{PK}, \mathsf{ID}, d_{\mathsf{ID}}, C)$ outputs the same value for all reconstituted keys d_{ID} that pass the tests. Furthermore, conditions (1)–(4) needed for the adversary to win the consistency of key generation game ensure that both tests succeed. Hence, $Decrypt$ will output the same value no matter which key d_{ID} is given as input, and thus $\mathsf{AdvCD\text{-}ID}_{\mathcal{A},n,k}(\Lambda) = 0$.

4   Threshold Public Key Encryption from Threshold IBE

In this section, we use the techniques of Canetti et al. [10] to show that any semantically secure TIBE gives a chosen ciphertext secure TPKE. Later, we apply this transformation on our TIBE to obtain an efficient chosen ciphertext secure TPKE in the standard model.

Let $\mathcal{E}_{TIBE} = (Setup_{TIBE}, ShareKeyGen_{TIBE}, ShareVerify_{TIBE}, Combine_{TIBE}, \ldots)$ be a TIBE system. Let $\mathcal{S} = (SigKeyGen, Sign, SigVerify)$ be a signature system. We construct a TPKE as follows:

$Setup_{TPKE}(n, k, \Lambda)$. To generate a TPKE key set, execute $Setup_{TIBE}(n, k, \Lambda)$ from the TIBE system and output the resulting tuple (PK, VK, **SK**).

$Encrypt_{TPKE}(\mathsf{PK}, M)$. To encrypt a message M under the public key PK, first run $SigKeyGen(\Lambda)$ to obtain a signing/verification key pair $(SigK, VerK)$. Next, run $Encrypt_{TIBE}(\mathsf{PK}, VerK, M)$ to obtain a ciphertext C_0, i.e., using $VerK$ as the identity to encrypt to. Then, run $Sign(SigK, C_0)$ to obtain a signature σ. Output the triple $C = (C_0, VerK, \sigma)$ as the complete ciphertext.

$ShareDecrypt_{TPKE}(\mathsf{PK}, i, \mathsf{SK}_i, C)$. To obtain a partial decryption of a ciphertext $C = (C_0, VerK, \sigma)$ under private key share SK_i, do the following:

1. Run $SigVerify(VerK, C_0, \sigma)$. If the verification fails, output $(i, \perp)$ and exit.
2. Run $ValidateCT_{TIBE}(\mathsf{PK}, VerK, C_0)$. If the validation fails, output $(i, \perp)$ and exit.
3. Run $ShareKeyGen_{TIBE}(\mathsf{PK}, i, \mathsf{SK}_i, VerK)$ to obtain a TIBE private key share μ for the identity $VerK$. Output μ as the decryption share.

$ShareVerify_{TPKE}(\mathsf{PK}, \mathsf{VK}, C, \mu)$. To verify a decryption share μ with respect to a ciphertext $C = (C_0, VerK, \sigma)$ under verification key VK, do the following:

1. Run $SigVerify(VerK, C_0, \sigma)$ and $ValidateCT_{TIBE}(\mathsf{PK}, VerK, C_0)$. If either test fails do: if $\mu = (i, \bot)$ then output `valid` and exit and if not then output `invalid` and exit.
2. Otherwise, both tests succeeded. Run $ShareVerify_{TIBE}(\mathsf{PK}, \mathsf{VK}, VerK, \mu)$ and output the result.

$Combine_{TPKE}(\mathsf{PK}, \mathsf{VK}, C, \{\mu_1, \ldots, \mu_k\})$. To obtain a full decryption of a cipher-text $C = (C_0, VerK, \sigma)$ given k partial decryption shares $\mu_1, \ldots, \mu_k$, first check that all shares are valid and none are of the form $(i, \bot)$. Output $\bot$ and exit if not. Next, run $Combine_{TIBE}(\mathsf{PK}, \mathsf{VK}, VerK, \{\mu_1, \ldots, \mu_k\})$ to obtain a private key d for identity $VerK$. If $d = \bot$, output $\bot$ and exit. Otherwise, run $Decrypt_{TIBE}(\mathsf{PK}, VerK, d, C_0)$ and output the result.

4.1   Security

The following theorem proves security of this system. The proof is based on the proof in [10].

Theorem 2. *Suppose $\mathcal{E}_{TIBE}$ is a selective-ID secure TIBE and $\mathcal{S}$ is existentially unforgeable under a one chosen message attack. Then the TPKE system above is chosen ciphertext secure.*

Proof. Suppose $\mathcal{A}$ has non-negligible advantage in attacking the TPKE above. First, suppose $\mathsf{AdvCCA}_{\mathcal{A},n,k}(\Lambda) > 1/\Lambda^c$ for some $c > 0$, and sufficiently large Λ. We build an algorithm that either breaks the TIBE or breaks the signature scheme. We start with an algorithm $\mathcal{B}$ that breaks the TIBE. Algorithm $\mathcal{B}$ uses $\mathcal{A}$ to interact with a TIBE challenger as follows:

Initialization. Algorithm $\mathcal{B}$ runs $\mathcal{A}$ to obtain a list $S \subset \{1, \ldots, n\}$ of the $k-1$ servers that $\mathcal{A}$ wishes to corrupt. Next, $\mathcal{B}$ runs $SigKeyGen(\Lambda)$ to obtain a signing key $SigK^*$ and a verification key $VerK^*$. It outputs the set S and the identity $\mathsf{ID}^* = VerK^*$ to the TIBE challenger.
Setup. The TIBE challenger runs $Setup(n, k, \Lambda)$ to obtain $(\mathsf{PK}, \mathsf{VK}, \mathsf{SK})$. It gives $\mathcal{B}$ the values PK, VK, and all (j, SK_j) for $j \in S$. Algorithm $\mathcal{B}$ forwards these values to $\mathcal{A}$.
Query phase 1. $\mathcal{A}$ adaptively issues decryption queries of the form (C, i) where $C = (C_0, VerK, \sigma)$ and $i \in \{1, \ldots, n\}$. For each such query:

1. $\mathcal{B}$ runs $SigVerify(VerK, C_0, \sigma)$ and $ValidateCT_{TIBE}(\mathsf{PK}, VerK, C_0)$. If either output is `invalid`, algorithm $\mathcal{B}$ responds to $\mathcal{A}$'s query with $\mu = (i, \bot)$.
2. Otherwise, in the unlikely event that $VerK = VerK^*$, algorithm $\mathcal{B}$ moves to the challenge phase, picks a random $b' \in \{0, 1\}$ as its guess for b, outputs b', and aborts the simulation.
3. Otherwise, $\mathcal{B}$ issues an identity query $(\mathsf{ID} = VerK, i)$ to the TIBE challenger and obtains a private key share θ in return. It gives the decryption share $\mu = \theta$ to $\mathcal{A}$.

Challenge. $\mathcal{A}$ outputs two equal length messages M_0 and M_1. Algorithm $\mathcal{B}$ forwards M_0 and M_1 to the TIBE challenger. Recall that the challenge identity ID^* was set during initialization to $\mathsf{ID}^* = VerK^*$. The TIBE challenger responds with the encryption C_0^* of M_b under ID^* for some $b \in \{0, 1\}$. Algorithm $\mathcal{B}$ then runs $Sign(SigK^*, C_0^*)$ to obtain a signature σ^*. It gives $\mathcal{A}$ the challenge ciphertext $C^* = (C_0^*, VerK^*, \sigma^*)$.

Query phase 2. Algorithm $\mathcal{A}$ continues to issue decryption queries (C, i) where $C = (C_0, VerK, \sigma)$ and $C \neq C^*$. Algorithm $\mathcal{B}$ responds as in the query phase 1; in particular if $VerK = VerK^*$ then $\mathcal{B}$ picks a random $b' \in \{0, 1\}$ as its guess for b, outputs b' and aborts the simulation.

Guess. Eventually, $\mathcal{A}$ outputs its guess $b' \in \{0, 1\}$ for b. Algorithm $\mathcal{B}$ forwards b' to the TIBE challenger and wins the game if $b = b'$.

This completes the description of algorithm $\mathcal{B}$. Let $\mathsf{AdvIND\text{-}ID}_{\mathcal{B},n,k}(\Lambda)$ be $\mathcal{B}$'s advantage in winning the TIBE game above. Let $\mathsf{AdvCCA}_{\mathcal{A},n,k}(\Lambda)$ be $\mathcal{A}$'s advantage in winning the TPKE game. Let $\mathbf{abort}$ be the event that $\mathcal{B}$ aborted during the simulation in query phase 1 or 2.

As long as event $\mathbf{abort}$ does not happen, $\mathcal{B}$'s simulation of a TPKE challenger is perfect. Therefore,

$$|\mathsf{AdvIND\text{-}ID}_{\mathcal{B},n,k}(\Lambda) - \mathsf{AdvCCA}_{\mathcal{A},n,k}(\Lambda)| < \Pr[\mathbf{abort}] \tag{1}$$

Now, observe that when event $\mathbf{abort}$ happens, then $\mathcal{B}$ obtains an existential forgery for the signature public key $VerK^*$. If $\mathbf{abort}$ happens in query phase 1 then the forgery is obtained with no chosen message queries. If $\mathbf{abort}$ happens in query phase 2 then the forgery is obtained after one chosen message query. Either way, we obtain an algorithm, $\mathcal{C}$, that produces an existential forgery on the signature scheme $\mathcal{S}$ with probability $\Pr[\mathbf{abort}]$ using at most one chosen message query. Hence, $\mathsf{AdvSig}_{\mathcal{C}} = \Pr[\mathbf{abort}]$. It now follows from (1) that

$$\mathsf{AdvIND\text{-}ID}_{\mathcal{B},n,k}(\Lambda) + \mathsf{AdvSig}_{\mathcal{C}}(\Lambda) > \mathsf{AdvCCA}_{\mathcal{A},n,k}(\Lambda)$$

Therefore, if $\mathsf{AdvCCA}_{\mathcal{A},n,k}(\Lambda)$ is a non-negligible function then at least one of $\mathsf{AdvIND\text{-}ID}_{\mathcal{B},n,k}(\Lambda)$ or $\mathsf{AdvSig}_{\mathcal{C}}(\Lambda)$ must also be non-negligible, as required.

To complete the proof of Theorem 2 we need to argue that $\mathsf{AdvCD}_{\mathcal{A},n,k}(\Lambda)$ is a negligible function. Suppose, $\mathsf{AdvCD}_{\mathcal{A},n,k}(\Lambda)$ is non-negligible. Then we immediately obtain an algorithm $\mathcal{B}$ for which $\mathsf{AdvCD\text{-}ID}_{\mathcal{B},n,k}(\Lambda)$ is non-negligible contradicting the fact that $\mathcal{E}_{TIBE}$ is a secure TIBE. To see this, suppose $\mathcal{A}$ outputs (C, S, S') that lets $\mathcal{A}$ win the TPKE decryption consistency game. Let $C = (C_0, VerK, \sigma)$. Then $ValidateCT_{TIBE}(\mathsf{PK}, VerK, C_0) = \mathtt{valid}$, since otherwise all shares in S and S' must be of the form $(j, \bot)$. Furthermore, $ShareVerify_{TIBE}(\mathsf{PK}, \mathsf{VK}, VerK, \mu) = \mathtt{valid}$ for all shares $\mu \in S, S'$. Therefore, the decryption shares in S and S' are valid private key shares for $\mathsf{ID} = VerK$. It now follows that $(VerK, C_0, S, S')$ is a tuple that wins the TIBE key generation consistency game as required. This completes the proof of Theorem 2.

5   A Concrete Threshold Public Key System

Our full non-interactive, CCA2-secure, threshold PKE system is immediately obtained by applying the generic transformation of Sections 4 to the threshold IBE system of Section 3. The construction in described in Appendix A. We outline the properties of the system.

Recall that the TIBE of Section 3 worked for identities in $\mathbb{Z}_p^*$ where p was the order of the bilinear groups $\mathbb{G}$. To apply the conversion method we need identities that are public keys of a signature system. Such identities may not be elements of $\mathbb{Z}_p^*$. Therefore, the threshold system described in the appendix uses a collision resistant hash H to hash arbitrary identities into $\mathbb{Z}_p^*$. Security, of course, depends on the BDH assumption, security of the signature system, and the collision resistance of H.

The security of the TPKE scheme follows immediately from that of the underlying TIBE of Section 3 and the generic conversion from TIBE to TPKE from Section 4. We thus have the following corollary.

Corollary 3. *The system in Appendix A is chosen ciphertext secure assuming the BDH assumption holds for $\mathcal{GG}$, the signature scheme is existentially unforgeable under a one chosen message attack, and the hash function H is collision resistant.*

Thus, we are able to construct a CCA2-secure threshold public key system, without random oracles, in which there is no interaction needed between the decryption parties. The reason we are able to avoid interaction is that using the method of [10] anyone can check that a ciphertext is valid. In the Cramer-Shoup framework only parties possessing the private key can check ciphertext validity, which makes threshold decryption non-trivial.

The system includes additional tests during *ShareDecrypt* and *Decrypt* to provide robustness against misbehaving servers. These tests are possible with no additional information due to the fact that the DDH problem is easy in bilinear groups. In particular, we are able to test that a given IBE private key is valid for a given identity and that a given IBE ciphertext is a valid encryption under a given identity.

We note that a more efficient transformation from IBE to CCA2-secure public-key encryption was presented by Boneh and Katz [6]. Because that transformation uses MACs and commitments instead of signatures, only parties possessing the private key can check ciphertext validity. As a result, the method of [6] does not lend itself to the construction of non-interactive CCA2-secure threshold systems. This is the primary reason why, in the above construction, we had to use the original transformation of [10] based on signatures (or one-time signatures). An elegant alternative was recently proposed in [7].

6   Extensions

Distributed key generation. In the TPKE system of Section 5 one need not rely on a trusted dealer to issue shares to the decryption servers. One can generate a public key and shares of a private key using standard distributed key generation techniques used for ElGamal encryption [31, 21].

Proactive refresh. Proactive refresh enables the decryption servers to refresh their shares of the secret decryption key, without changing the key. Periodic proactive refresh make it harder for an adversary to recover k shares of the secret key, since he must recover all k shares within one time period. The standard proactive refresh techniques of [30, 24, 17] used for ElGamal encryption also apply to our Threshold PKE.

7   Conclusions

We presented a simple non-interactive threshold encryption system that is chosen ciphertext secure without random oracles. The construction illustrates the benefits of building chosen ciphertext security from identity based encryption.

References

1. M. Abe. Robust distributed multiplication without interaction. In *Proceedings of Crypto 1999*, pages 130–47, 1999.
2. J. H. An, Y. Dodis, and T. Rabin. On the security of joint signature and encryption. In *Proceedings of Eurocrypt 2002*, volume 2332 of *LNCS*. Springer-Verlag, 2002.
3. D. Boneh and X. Boyen. Efficient selective-ID identity based encryption without random oracles. In *Proceedings of Eurocrypt 2004*, volume 3027 of *LNCS*, pages 223–38. Springer-Verlag, 2004.
4. D. Boneh and X. Boyen. Short signatures without random oracles. In *Proceedings of Eurocrypt 2004*, volume 3027 of *LNCS*, pages 56–73. Springer-Verlag, 2004.
5. D. Boneh and M. Franklin. Identity-based encryption from the Weil pairing. In J. Kilian, editor, *Proceedings of Crypto 2001*, volume 2139 of *LNCS*, pages 213–29. Springer-Verlag, 2001.
6. D. Boneh and J. Katz. Improved efficiency for CCA-secure cryptosystems built using identity based encryption. In *Proceedings of RSA 2005*, LNCS. Springer-Verlag, 2005.
7. X. Boyen, Q. Mei, and B. Waters. Direct chosen ciphertext security from identity-based techniques. In *ACM Conference on Computer and Communications Security—CCS 2005*. ACM Press, 2005. Full version available at http://eprint.iacr.org/2005/288.
8. R. Canetti and S. Goldwasser. An efficient threshold public key cryptosystem secure against adaptive chosen ciphertext attack. In *Proceedings of Eurocrypt 1999*, pages 90–106, 1999.
9. R. Canetti, S. Halevi, and J. Katz. A forward-secure public-key encryption scheme. In *Proceedings of Eurocrypt 2003*, volume 2656 of *LNCS*. Springer-Verlag, 2003.
10. R. Canetti, S. Halevi, and J. Katz. Chosen-ciphertext security from identity-based encryption. In *Proceedings of Eurocrypt 2004*, volume 3027 of *LNCS*, pages 207–22. Springer-Verlag, 2004.
11. R. Cramer and V. Shoup. Design and analysis of practical public-key encryption schemes secure against adaptive chosen ciphertext attack. *SIAM Journal of Computing*, 33:167–226, 2003. Extended abstract in Crypto 1998.

12. I. Damgard, N. Fazio, and A. Nicolosi. Secret-key zero-knowledge protocols for NP and applications to threshold cryptography. manuscript, 2004.
13. A. DeSantis, Y. Desmedt, Y. Frankel, and M. Yung. How to share a function securely. In *Proceedings of STOC 1994*, pages 522–33, 1994.
14. Y. Desmedt and Y. Frankel. Threshold cryptosystems. In *Proceedings of Crypto 1989*, pages 307–15, 1989.
15. Y. Dodis and J. Katz. Chosen-ciphertext security of multiple encryption. In *Proceedings of TCC 2005*, LNCS. Springer-Verlag, 2005.
16. Y. Frankel. A practical protocol for large group oriented networks. In *Proceedings of Eurocrypt 1989*, pages 56–61, 1989.
17. Y. Frankel, P. Gemmell, P. MacKenzie, and M. Yung. Optimal resilience proactive public key cryptosystems. In *Proceedings of FOCS 1997*, pages 384–93, 1997.
18. Y. Frankel, P. Gemmell, and M. Yung. Witness-based cryptographic program checking. In *Proceedings of STOC 1996*, pages 499–08, 1996.
19. P. Gemmel. An introduction to threshold cryptography. *RSA CryptoBytes*, 2(3):7–12, 1997.
20. R. Gennaro, S. Halevi, and T. Rabin. Secure hash-and-sign signatures without the random oracle. In *Proceedings of Eurocrypt 1999*, LNCS, pages 123–39. Springer-Verlag, 1999.
21. R. Gennaro, S. Jarecki, H. Krawczyk, and T. Rabin. Secure distributed key generation for discrete-log based cryptosystems. In J. Stern, editor, *Proceedings of Eurocrypt 1999*, volume 1592 of *LNCS*, pages 295–310. Springer-Verlag, 1999.
22. R. Gennaro, T. Rabin, S. Jarecki, and H. Krawczyk. Robust and efficient sharing of RSA functions. *J. Cryptology*, 13(2):273–300, 2000.
23. S. Goldwasser, S. Micali, and R. Rivest. A digital signature scheme secure against adaptive chosen-message attacks. *SIAM J. Computing*, 17(2):281–308, 1988.
24. A. Herzberg, S. Jarecki, H. Krawczyk, and M. Yung. Proactive secret sharing, or how to cope with perpetual leakage. In *Proceedings of Crypto 1995*, 1995.
25. S. Jarecki and A. Lysyanskaya. Adaptively secure threshold cryptography: introducing concurrency, removing erasures. In *Proceedings of Eurocrypt 2000*, pages 221–42, 2000.
26. A. Joux. A one round protocol for tripartite Diffie-Hellman. In W. Bosma, editor, *Proceedings of ANTS IV*, volume 1838 of *LNCS*, pages 385–94. Springer-Verlag, 2000.
27. A. Joux and K. Nguyen. Separating decision Diffie-Hellman from Diffie-Hellman in cryptographic groups. *Journal of Cryptology*, 16(4):239–47, 2003.
28. K. Kurosawa and Y. Desmedt. A new paradigm of hybrid encryption scheme. In *Proceedings of Crypto 2004*, volume 3152 of *LNCS*, pages 426–42. Springer-Verlag, 2004.
29. P. MacKenzie. An efficient two-party public key cryptosystem secure against adaptive chosen ciphertext attack. In *Proceedings of PKC 2003*, 2003.
30. R. Ostrovsky and M. Yung. How to withstand mobile virus attacks. In *Proceedings of PODC 1991*, pages 51–61, 1991.
31. T. Pederson. A threshold cryptosystem without a trusted party. In *Proceedings of Eurocrypt 1991*, volume 547 of *LNCS*, pages 522–26, 1991.
32. R. Sakai, K. Ohgishi, and M. Kasahara. Cryptosystems based on pairings. In *Proceedings of the Symposium on Cryptography and Information Security—SCIS 2000*, Japan, 2000.
33. V. Shoup and R. Cramer. Universal hash proofs and a paradigm for chosen ciphertext secure public key encryption. In *Proceedings of Eurocrypt 2002*, 2002.
34. V. Shoup and R. Gennaro. Securing threshold cryptosystems against chosen ciphertext attack. *Journal of Cryptology*, 15(2):75–96, 2002. extended abstract in Eurocrypt 1998.

A   Description of the Full TPKE System

We give an explicit description of the full non-interactive CCA2-secure threshold PKE system from Section 5. It is obtained by directly composing the constructions given in Sections 3 and 4. The system works as follows:

Setup(n, k, Λ). Run the group generator $\mathcal{GG}(\Lambda)$ to obtain a bilinear group $\mathbb{G}$ of prime order $p > n$. Select random generators g, g_2, h_1 in $\mathbb{G}$, and a random degree $k - 1$ polynomial $f \in \mathbb{Z}_p[X]$. Set $\alpha = f(0) \in \mathbb{Z}_p$ and $g_1 = g^\alpha$.

The public key PK consist of PK $= (\mathbb{G}, g, g_1, g_2, h_1)$. For $i = 1, \ldots, n$ the secret key SK_i of server i is defined as $\mathsf{SK}_i = g_2^{f(i)}$. The public verification key VK consists of PK along with the n-tuple $(g^{f(1)}, \ldots, g^{f(n)})$.

We will also need a collision resistant hash function H that outputs digests in $\mathbb{Z}_p$, and a signature scheme $(SigKeyGen, Sign, SigVerify)$ that is strongly existentially unforgeable against one chosen message attacks. Both H and the signature scheme are part of PK, but we leave them as implicit members to simplify the presentation.

Encrypt(PK, M). To encrypt a message $M \in \mathbb{G}_1$ under the public key PK $= (g, g_1, g_2, h_1)$, first run $SigKeyGen$ to obtain a signing key $SigK$ and a signature verification key $VerK$. Let $\mathsf{ID} = H(VerK)$. Next, pick a random $s \in \mathbb{Z}_p$ and compute

$$C_0 = \left(e(g_1, g_2)^s \cdot M, \quad g^s, \quad g_1^{s \cdot \mathsf{ID}} h_1^s \right)$$

Let $\sigma = Sign(SigK, C_0)$ be a signature on C_0 using the signing key $SigK$. Output the ciphertext $C = (C_0, VerK, \sigma)$.

ShareDecrypt$(\mathsf{PK}, \mathsf{SK}_i, C)$. Decryption server i uses its private key share SK_i to partially decrypt a ciphertext $C = (C_0, VerK, \sigma)$ as follows. First, run algorithm $SigVerify(VerK, C_0, \sigma)$ to check that σ is a valid signature of C under $VerK$. Also let $\mathsf{ID} = H(VerK)$ and test whether $e(B, g_1^{\mathsf{ID}} h_1) = e(C_1, g)$. If either condition fails, output $\mu = (i, \perp)$ and exit.

Otherwise, C is well-formed, and the decryption server i needs to output a share of the private key needed to decrypt C_0. To do so, it picks a random r in $\mathbb{Z}_p$, and outputs the decryption share $\mu_i = (i, (w_0, w_1))$, where

$$w_0 = \mathsf{SK}_i \cdot (g_1^{\mathsf{ID}} h_1)^r \qquad \text{and} \qquad w_1 = g^r$$

Notice that (w_0, w_1) is an IBE private key share corresponding to the identity $\mathsf{ID} = H(VerK)$.

ShareVerify$(\mathsf{PK}, \mathsf{VK}, C, \mu_i)$. To verify that μ_i is a correct partial decryption of the ciphertext $C = (C_0, VerK, \sigma) = ((A, B, C_1), VerK, \sigma)$, first run algorithm $SigVerify(VerK, C_0, \sigma)$ to check that σ is a valid signature of C_0 under $VerK$. Also let $\mathsf{ID} = H(VerK)$ and test whether $e(B, g_1^{\mathsf{ID}} h_1) = e(C_1, g)$. We say that C is well-formed if both tests succeed.

1. If C is not well-formed: if μ_i is of the form $(i, \perp)$ then output `valid` and exit, otherwise output `invalid` and exit.
2. If C is well-formed *and* μ_i is of the form $(i, \perp)$, then output `invalid` and exit.

3. Otherwise, C is well-formed and $\mu_i = (i, (w_0, w_1))$. In this case, let $\mathsf{VK} = (u_1, \ldots, u_n)$ where $u_i = g^{f(i)}$, and output $\mathtt{valid}$ or $\mathtt{invalid}$ according to whether the following equation holds or not:

$$e(u_i, g_2) \cdot e(g_1^{\mathsf{ID}} h_1, w_1) = e(g, w_0)$$

Combine$(\mathsf{PK}, \mathsf{VK}, C, \{\mu_1, \ldots, \mu_k\})$. To decrypt a ciphertext $C = (C_0, VerK, \sigma)$ using the partial decryptions $\mu_1, \ldots, \mu_k$, first check that all shares $\mu_i = (i, \hat{\mu}_i)$ bear distinct server indices i, and that they are all *valid*, i.e., that all *ShareVerify*$(\mathsf{PK}, \mathsf{VK}, C_0, \mu_i) = \mathtt{valid}$; otherwise output $\perp$ and exit.

Without loss of generality, assume that the shares $\mu_1, \ldots, \mu_k$ were generated by the decryption servers $i = 1, \ldots, k$, respectively. The combiner proceeds as follows:

1. If any partial decryption μ_i is of the form $(i, \perp)$, then output $\perp$ and exit.
2. Otherwise, all shares $\mu_1, \ldots, \mu_k$ are of the form $\mu_i = (i, (w_{i,0}, w_{i,1}))$ with distinct i, and *SigVerify*$(VerK, C_0, \sigma)$ and *ValidateCT*$(\mathsf{PK}, H(VerK), C_0)$ must both succeed. Determine the Lagrange coefficients $\lambda_1, \ldots, \lambda_k \in \mathbb{Z}_p$ so that $\alpha = f(0) = \sum_{i=1}^{k} \lambda_i f(i)$, and set

$$w_0 = \prod_{i=1}^{k} w_{i,0}^{\lambda_i} \qquad \text{and} \qquad w_1 = \prod_{i=1}^{k} w_{i,1}^{\lambda_i}$$

3. Use (w_0, w_1) to decrypt $C_0 = (A, B, C_1)$, as

$$M = A \cdot e(B, w_0)/e(C_1, w_1)$$

Observe that the above decryption goes through since as we observed earlier $w_0 = g_2^\alpha \cdot (g_1^{\mathsf{ID}} h_1)^{\bar{r}}$ and $w_1 = g^{\bar{r}}$ for some $\bar{r} \in \mathbb{Z}_p$, hence (w_0, w_1) is an IBE private key corresponding to $\mathsf{ID} = H(VerK)$.

How to Construct Multicast Cryptosystems Provably Secure Against Adaptive Chosen Ciphertext Attack*

Yitao Duan and John Canny

Computer Science Division, University of California, Berkeley,
Berkeley, CA 94720, USA
{duan, jfc}@cs.berkeley.edu

Abstract. In this paper we present a general framework for constructing efficient multicast cryptosystems with provable security and show that a line of previous work on multicast encryption are all special cases of this general approach. We provide new methods for building such cryptosystems with various levels of security (e.g., IND-CPA, IND-CCA2). The results we obtained enable the construction of a whole class of new multicast schemes with guaranteed security using a broader range of common primitives such as OAEP. Moreover, we show that multicast cryptosystems with high level of security (e.g. IND-CCA2) can be based upon public key cryptosystems with weaker (e.g. CPA) security as long as the decryption can be securely and efficiently "shared". Our constructions feature truly constant-size decryption keys whereas the lengths of both the encryption key and ciphertext are independent of group size.

1 Introduction

Multicast offers an efficient way to deliver the same message to a group of receivers and has become the basis of many applications. The Internet today supports a basic form of multicast service. On the Internet, a multicast group is identified by a Class D IP address and any receivers can join or leave a multicast group by sending IGMP (Internet Group Management Protocol) [1] messages to their local router. Any sender can send message to a multicast group by addressing the message to the group address.

The current IP Multicast service does not provide mechanisms to restrict message delivery to a specified set of receivers therefore other means have to be used to secure the communication. A multicast encryption system provides *confidentiality* for multicast data – ensuring that any parties other than the intended recipients should not be able to access the message. To this end, most of the existing work use one of two approaches. The first is represented by the work in

---

* This work was supported by National Science Foundation award #EIA-0122599 (Title: "ITR/SI: Societal Scale Information Systems: Technologies, Design, and Applications").

network research that is concerned with *multicast security*. In this approach symmetric key encryption is used and the data is encrypted with a traffic encryption key (TEK) that is known only to the multicast group members. The difficulty here is key management: The TEK may have to be changed when members join or leave the group. This is known as re-keying. Early schemes (e.g., Group Key Management Protocol (GKMP) [2]) let the group controller or the sender share a pairwise key with each group member and distribute keys to them on a one-to-one basis. For obvious reasons this cannot scale to large groups.

Some work has been done to improve the scalability of such schemes. Among the efficient solutions, the Logical Key Hierarchy (LKH) (or Key Graph) was independently discovered in [3] and [4] and has been an inspiration for many subsequent works [5, 6, 7, 8, 9, 10]. In these schemes, individual and auxiliary keys are organized into a hierarchy and each group member is assigned to a leaf and holds all the keys from its leaf to the root. The root key is shared by all group members and used as the TEK. New TEK is distributed by encrypting it with keys that deleted members do not have. So far $O(\log n)$ seems to be the best storage (for both center and members) and communication complexity the LKH-based schemes achieved, where n is the size of the multicast group.

The problem with this approach is that revoking a single user involves changing the keys for all others and the receivers must be stateful and always online in order to receive the latest TEK.

The second approach uses asymmetric key cryptosystem and allows the receivers to be stateless. This includes the work in cryptography such as *traitor tracing*, a concept introduced by Chor, Fiat and Naor [11], and *broadcast encryption*, initiated by Fiat and Naor [12]. Both are based on encryption schemes where a ciphertext can be decrypted by multiple parties with different keys. The scheme in [12] requires $O(t \log t \log n)$ keys per user and the transmission of $O(t^2 \log^2 t \log n)$ messages where t is the number of revoked users. Subsequent work proposed a number of other schemes including [13, 14, 15, 16, 17], and [18, 19] which achieved $O(t)$ message complexity and $O(\log^{1+\epsilon} n)$ keys per user. Boneh and Franklin's scheme proposed in [13] is based on Reed-Solomon codes and the representation problem for discrete logs. They also presented a modification, using techniques by Cramer and Shoup [20], that was provably secure against adaptive chosen ciphertext attack.

Recently Boneh et al. presented a broadcast encryption scheme based on bilinear map with constant-size ciphertexts and private keys (and $O(n)$-size public key) [21]. However, in this system, the decryption requires the public key and the knowledge of the set of legitimate recipients. Therefore the "effective" decryption key and/or ciphertext in a real application actually become linear in the total number of receivers.

There is a line of work in the second approach that we classify as Asymmetric Threshold Decryption-based (ATD-based) multicast encryption. This includes [14, 17, 22, 23, 24], although none of them explicitly formalized their schemes this way. In these schemes a private key is shared using a $(t + 1, n + t)$-threshold scheme and the shares are distributed *asymmetrically*. Namely the center is given t shares and each user is given 1 share. The center broadcasts a ciphertext together with t partial decryptions. Any member with a valid share of the private key can produce another decryption share and recover the message. With such schemes, user only needs to store a key of constant length. And both the message complexity and sender storage are $O(t)$, independent of the group size.

1.1 Our Results

We focus on the ATD-based multicast encryption cryptosystems and introduce a general framework for constructing such systems with guaranteed security. As we will show later, all existing ones are special cases of our constructions. In particular, they are all based on specific ElGamal encryption that relies on specific assumptions (e.g. DDH). The results we obtained in this paper, on the other hand, are more general. The main contributions are: (1) We show that *any* threshold encryption scheme can be used to construct a multicast cryptosystem that retains the same level of security (e.g. IND-CPA, IND-CCA2) as the underlying threshold encryption. (2) We obtain new results that improve over existing ATD-based schemes in both security and efficiency. Specifically, the resulting scheme from our construction can be made CCA-secure even if the underlying threshold scheme is not. (3) Furthermore, we show that an IND-CCA2 secure multicast scheme can be constructed from a public key cryptosystem that does *not* have a secure threshold implementation (such as OAEP) or has only weaker security (e.g. only IND-CPA), provided the decryption can be securely and efficiently shared (to be elaborated in Sect. 4.4). All of our security proofs are in the same (standard or random oracle) model as the underlying threshold scheme or public key cryptosystem.

These general security results can be used to analyze existing systems in a more unified framework and provide guidelines for constructing future schemes with guaranteed security. This frees the system designer from the burden of security consideration and allows them to focus on other aspects of their schemes.

2 Preliminaries

We consider the scenario where a single party, called *the center*, sends messages, over insecure channels, to a group U of n parties who are denoted *members* of the group. In such a setting, the center often has a special role. Since it is often distributing information of its own choice, it is assumed to have control over the group membership, i.e., the center is allowed to make decisions about who can join the group and whose membership should be revoked. This is in line with almost all multicast schemes such as [25, 4, 26, 27, 22, 23, 24].

We assume a computationally bounded adversary who is allowed to attack the system from both outside and *inside* the group. The insider's attack is modelled by allowing the adversary to corrupt and gain total control of up to t group members where t is a predefined threshold. We only consider non-adaptive adversary who chooses what members to corrupt before the key generation.

The multicast communication we are considering in this paper is assumed to be "closed", i.e., we only provide *the center* with the ability to encrypt messages (and of course only the intended recipients can decrypt them). This is different from the public key systems such as [17, 22, 23, 24] where the information to encrypt a message is public. The openness is unnecessary for some applications and unacceptable for some others (e.g. military communication). By "closing" the communication, we can provide more flexible constructions that can make use of a broader range of primitives. The price for this flexibility is the loss of the public key feature, which should not be a problem for many applications. However we observe that in many instantiations of our constructions, it is easy to "publicize" the encryption key, without affecting the security of the scheme, as demonstrated by works such as [22, 23, 24]. This effectively turns the scheme into a public key system and all the openness features are reinstalled.

3   Multicast Cryptosystem

Definition 1. *An n-way multicast encryption scheme* $\mathcal{ME} = (\mathsf{KeyGen}, \mathsf{Reg}, \mathsf{E}, \mathsf{D})$ *consists of the following set of algorithms:*

1. *Key Generation* KeyGen: *a probabilistic polynomial-time (in k) algorithm which takes as inputs a security parameter 1^k, a threshold t, the number of (initial) group members n, and generates global information I, the encryption key Σ and the master secret key Γ.*
2. *Registration algorithm* Reg: *a probabilistic algorithm to compute the secret initialization data for a new user subscribing to the system. Reg receives as input the master key Γ and a new index i associated with the user; it returns the user's secret key Γ_i.*
3. *Encryption* E: *a probabilistic polynomial-time algorithm that, on inputs Σ, the encryption key, and a string $m \in \{0,1\}^k$, and a set R of revoked users (with $|R| \leq t$) and their keys, produces as output $\psi \in \{0,1\}^*$ called the ciphertext[1].*
4. *Decryption* D: *a deterministic polynomial-time algorithm such that $\forall m \in \{0,1\}^k$, $\forall i \in U \setminus R$, $\mathsf{D}(\Gamma_i, \mathsf{E}(\Sigma, \{(j, \Gamma_j)|j \in R\}, m)) = m$. On all other inputs it outputs a special symbol $\perp$.*

KeyGen and Reg should be run by the center and the two can also be executed together with an initial set of n members as input. Admitting new members is relatively trivial, at least for all the construction we will be presenting, so in the following we simply omit Reg and use $(I, \Sigma, \Gamma, \Gamma) \leftarrow \mathsf{KeyGen}(1^k, t, n)$ to denote this process, where $\Gamma = (\Gamma_1, \ldots, \Gamma_n)$ is a vector of secret keys for the n members.

3.1   Notion of Security

The communication paradigm we are considering shares similarities with both symmetric key and public key cryptosystems. On one hand the communication

---

[1] Note that member revocation is implicitly embedded in the encryption algorithm.

is "closed" in that we only allow the center to send messages to the group. On the other hand the keys are "asymmetric" since now there are multiple recipients and our definition includes member revocation which means the encryption key and the decryption keys must be different.

Dodis and Fazio [23] first precisely formalized the notion of adaptive security for public key multicast encryption schemes, which allow anyone having access to the public key to send messages to the group, at both CPA and CCA2 levels. Since our setting is different from the "public key" paradigm, we adopt a slightly modified definition. The major difference is that, we do not *explicitly* allow the adversary to see the sender's keys since ours is not a public key cryptosystem. Instead the adversary can obtain encryptions of arbitrary messages by querying an *encryption oracle* who also encrypts the target message later. This is similar to the security definition based on indistinguishability for symmetric key cryptosystems. The ability to handle member revocation is modelled by allowing the adversary to corrupt members and obtain their secret keys. This formalization is general and captures the security notions of many multicast schemes such as those LKH schemes [3, 4] which are based on symmetric key cryptography. However we note that in all the construction we introduce later, the secret keys of the revoked members constitute the actual *encryption key*. In essence in our constructions the exposure of encryption key can be modelled as corrupting members. This effectively turns our scheme into a "public key" paradigm from the adversary's point of view and the security definitions from [23] are appropriate.

Formal Model. Given a multicast encryption scheme $\mathcal{ME} = (\mathsf{KeyGen}, \mathsf{E}, \mathsf{D})$, a polynomial time adversary $\mathcal{A}$'s attack is modelled by the following game:

Game ME:

M1. The adversary $\mathcal{A}$ chooses to corrupt a fixed set R of t members.

M2. $(I, \Sigma, \Gamma, \boldsymbol{\Gamma}) \leftarrow \mathsf{KeyGen}(1^k, t, n)$ is run and $\mathcal{A}$ is given the public information I and the secret keys of corrupted members. User i receives Γ_i. The center is given R and their keys.

M3. The adversary interacts with the center, who acts as the encryption oracle, in an arbitrary fashion. On any query m from $\mathcal{A}$, the center returns its encryption.

M4. $\mathcal{A}$ chooses two plaintexts m_0 and m_1 of the same length and gives them to the center who chooses $b \in \{0, 1\}$ at random, and gives the "target" ciphertext $\psi' = \mathsf{E}(\Sigma, \{(j, \Gamma_j) | j \in R\}, m_b)$ to $\mathcal{A}$.

M5. $\mathcal{A}$ continues to interact with the center.

M6. At the end of the game, $\mathcal{A}$ outputs $b' \in \{0, 1\}$.

The advantage of $\mathcal{A}$ is defined as

$$\mathsf{Adv}^{CPA}_{\mathcal{ME}, \mathcal{A}}(k) = |Pr(b' = b) - 1/2|$$

In addition, in the case of a adaptive chosen ciphertext attack (CCA2)[2], in both stages M3 and M5, $\mathcal{A}$ is also allowed to interact in an arbitrary manner with the group members who act as the decryption oracles. On a query ψ from $\mathcal{A}$, member i returns $\mathsf{D}(\Gamma_i, \psi)$. The only restriction on the interaction is that the target ciphertext ψ' cannot be one of the queries made to any of the decryption oracles. As before, $\mathcal{A}$'s advantage in the CCA2 case is defined as

$$\mathsf{Adv}_{\mathcal{ME},\mathcal{A}}^{CCA2}(k) = |Pr(b' = b) - 1/2|$$

Definition 2 (t-Resilient Multicast Encryption Scheme). *Let* $\mu \in \{CPA,$ *CCA2*$\}$. *A multicast encryption scheme* $\mathcal{ME}$ *is t-resilient against a* μ-*type attack if the advantage,* $\mathsf{Adv}_{\mathcal{ME},\mathcal{A}}^{\mu}(k)$, *of any probabilistic polynomial time adversary* $\mathcal{A}$ *is a negligible function of* k.

4 ATD-Based Multicast Encryption

In this section we define two constructions and show that a line of previous work on multicast or broadcast encryption can actually be characterized as special cases of these constructions.

4.1 Threshold Decryption Scheme

A $(t+1, n)$-threshold cryptosystem $\mathcal{TD} = (\mathsf{KeyGen_{TD}}, \mathsf{D_{TD}}, \mathsf{V_{TD}}, \eta, \mathsf{E_{TD}})$ consists of the following algorithms:

- Key generation algorithm $(PK, VK, \boldsymbol{SK}) \leftarrow \mathsf{KeyGen_T}(1^k, t, n)$: a probabilistic algorithm that, given a security parameter 1^k, a threshold t, and the number of players n, generates a public key, PK, a verification key VK, and n private keys $\boldsymbol{SK} = (SK_1, \ldots, SK_n)$. PK and VK are made public while SK_i is known only to player i, $i = 1, 2, \ldots, n$.
- Share computation $\mathsf{D_{TD}}$: a probabilistic algorithm that, given a private key SK_i and the ciphertext c, $\mathsf{D_{TD}}$ computes $\rho = \mathsf{D_{TD}}(SK_i, c)$, called a *decryption share*.
- Share verification V: a deterministic algorithm that takes as input the public verification key VK, the ciphertext c, and a share ρ, and outputs $\mathsf{V}(VK, c, \rho)$ $\in \{0, 1\}$.
- Share combination algorithm η: given the verification key VK, the ciphertext c, and a set Λ of $t + 1$ shares, η either outputs the corresponding result $r = \eta(VK, c, \Lambda)$ or a special symbol $\perp$ that is different from all possible correct results.
- Encryption algorithm $\mathsf{E_{TD}}$: the "opposite" of $\mathsf{D_{TD}}$. This function is carried out in the normal manner by a single party and should follow the same definition as the encryption algorithm in a standard public key cryptosystem.

---

[2] We do not explicitly consider non-adaptive chosen ciphertext attack (CCA1). It should be easy to see that all the discussions and proofs still hold in the case of CCA1, by simply restricting the adversary from interacting with the decryption oracles after the target ciphertext is generated in both Game ME and TD (Sect. 4.1).

The operation of a threshold decryption scheme can be modelled as follows. There is a trusted dealer (e.g. the center) and a set of n decryption servers indexed $1, \ldots, n$. In an initialization phase, the dealer runs the key generation algorithm and creates PK, VK and $\boldsymbol{SK}$. SK_i is given to server i. To decrypt a ciphertext ψ, a client gives ψ to the servers, requesting a decryption share from each of them. It can verify the validity of the shares using the given verification key. Once the client collects valid shares from at least $t+1$ servers, she can apply η to obtain the decryption.

Threshold cryptosystems are part of a general approach known as threshold cryptography, introduced by Boyd [28], Desmedt [29], and Desmedt and Frankel [30]. There are schemes based on both Diffie-Hellman problem [30] and RSA [31]. All these schemes can be shown to be secure against chosen plaintext attack, but they are not known to withstand chosen ciphertext attack. After Cramer and Shoup discovered the first truly practical public key cryptosystem that is provably secure against chosen ciphertext attack *without* using random oracles [20], several of its threshold implementations have been proposed and proved CCA2 secure (also without using the random oracle model) [32, 33, 34]. Shoup and Gennaro presented a more efficient threshold scheme in [35] that is proven CCA2 secure in the random oracle model.

We adopt Shoup and Gennaro's definition of security for threshold decryption schemes from [35], which is a natural extension of security for a public key cryptosystem, and define the security of a $(t+1, n)$-threshold decryption scheme $\mathcal{TD} = (\mathsf{KeyGen_{TD}}, \mathsf{D_{TD}}, \mathsf{V_{TD}}, \eta, \mathsf{E_{TD}})$ with respect to the following game:

Game TD:

TD1. The adversary $\mathcal{A}$ chooses to corrupt a fixed set of t servers.
TD2. The key generation algorithm is run. The public key, verification key and the private keys of the corrupted servers are given to $\mathcal{A}$. Other private keys are given to the uncorrupted servers.
TD3. $\mathcal{A}$ chooses two plaintexts m_0 and m_1 of the same length and gives them to an "encryption oracle" that chooses $b \in \{0, 1\}$ at random, and gives the "target" ciphertext $\psi' = \mathsf{E_{TD}}(PK, m_b)$ to $\mathcal{A}$.
TD4. At the end of the game, the adversary outputs $b' \in \{0, 1\}$.

This game defines the attack scenario for CPA security. The adversary's advantage is defined to be the absolute difference between $1/2$ and the probability that $b' = b$:

$$\mathsf{Adv}^{CPA}_{\mathcal{TD}, \mathcal{A}}(k) = |Pr(b' = b) - 1/2|$$

For CCA2 attacks, $\mathcal{A}$ is allowed to interact with uncorrupted decryption servers, who act as the decryption oracles, in an arbitrary fashion, feeding them ciphertexts $\psi \neq \psi'$, and obtaining decryption shares. The calls to the decryption oracles can happen at any point during the execution of the game, both before and after stage TD3, and be arbitrarily interleaved with other oracle calls. $\mathcal{A}$'s advantage is defined as

$$\mathsf{Adv}^{CCA2}_{\mathcal{TD}, \mathcal{A}}(k) = |Pr(b' = b) - 1/2|$$

Definition 3 (t-Resilient Threshold Decryption Scheme). *Let $\mu \in \{CPA,$ $CCA2\}$. A threshold decryption scheme $\mathcal{TD}$ is t-resilient against μ-type attacks if the advantage, $\mathsf{Adv}^{\mu}_{\mathcal{TD},\mathcal{A}}(k)$, of any probabilistic polynomial time adversary $\mathcal{A}$ is a negligible function of k.*

4.2   Basic Construction

Construction 1 (ME1). *Given a threshold decryption $\mathcal{TD} = (\mathsf{KeyGen_{TD}}, \mathsf{D_{TD}},$ $\mathsf{V_{TD}}, \eta, \mathsf{E_{TD}})$, a security parameter 1^k, a threshold t and the number of (initial) members n, a multicast encryption scheme $\mathcal{ME}^{\mathcal{TD}}_{C1} = (\mathsf{KeyGen}, \mathsf{E}, \mathsf{D})$ can be constructed as follows:*

1. *Key Generation* KeyGen: *Run* $(PK, VK, \boldsymbol{SK}) \leftarrow \mathsf{KeyGen_{TD}}(1^k, t, n+t)$. *Set $I = (PK, VK)$ and the encryption key $\Sigma = \{(j, SK_j) : j = n+1, \ldots, n+t\}$. Σ is given to the center. Member i receives secret key $\Gamma_i = (i, SK_i)$. The master secret key is $\boldsymbol{\Gamma} = (\Gamma_1, \ldots, \Gamma_{n+t})$.*

2. *Encryption* E: *Given a set R of revoked members, and their secret keys, with $|R| \leq t$, a message m, the encryption proceeds as follows. Let $T = \{n+1, \ldots, n+t\}$. The encryptor randomly selects a subset of T with $t - |R|$ elements, denoted T', and computes the ciphertext $\psi = (c, \{(j, c_j) : j \in T' \cup R\})$ where $c = \mathsf{E_{TD}}(PK, m)$ and $c_j = \mathsf{D_{TD}}(SK_j, c)$.*

3. *Decryption* D: *Given a secret key Γ_i and a ciphertext ψ, the ciphertext is first parsed into $\psi = (c, \Lambda')$ where $\Lambda' = \{(j, c_j) : j \in T' \cup R\}$ with $c_j = \mathsf{D_{TD}}(SK_j, c)$. For all $j \in T' \cup R$, the decryption first test $v_j = \mathsf{V_{TD}}(VK, c, c_j)$. If any $v_j = 0$, D returns $\perp$. Otherwise it returns*

$$m = \eta(VK, c, \Lambda' \cup \{(i, \mathsf{D_{TD}}(SK_i, c))\}) \tag{1}$$

With this construction, the multicast ciphertext essentially consists of the ciphertext of the underlying threshold scheme, together with t partial decryptions produced using the keys of revoked members. To decrypt, a legitimate member combines the partial decryptions embedded in the ciphertext with another one computed using her own share of the private key. As we will show, this construction preserves the security of the underlying threshold scheme.

Theorem 1 (Security Inheritance). *Let $\mu \in \{CPA, CCA2\}$. Given a threshold decryption scheme $\mathcal{TD} = (\mathsf{KeyGen_{TD}}, \mathsf{D_{TD}}, \mathsf{V_{TD}}, \eta, \mathsf{E_{TD}})$ that is t-resilient against μ-type attacks, the multicast encryption scheme $\mathcal{ME}^{\mathcal{TD}}_{C1}$ constructed using Construction 1 with threshold t and (initial) group size n is t-resilient against μ-type attacks.*

The proof of this theorem is similar to that of Theorem 2, which is more interesting and is presented later, and is omitted from this paper.

Many existing multicast schemes can be shown to be special cases of our Construction 1 and their security can be readily predicted by Theorem 1. The Revocation method 1 in [17] and the group key distribution scheme in [14] are just Construction 1 instantiated with a special use of threshold ElGamal[3]. The basic scheme in [22], the "public key (multicast) encryption" from [17] and the CPA secure scheme from [23] can all be shown to be Construction 1 with a standard threshold ElGamal cryptosystem. These schemes are shown to be secure against chosen plaintext attacks in their individual papers. The same conclusion can be reached immediately through Theorem 1.

Theorem 1 also provides guidelines for constructing *new* multicast encryption schemes with guaranteed security. For example, some threshold schemes are known to be CCA2 secure (e.g. [35, 32, 33, 34] and the IND-CCA2 threshold ElGamal in [36]) and a multicast encryption constructed via Construction 1 using one of these schemes is therefore guaranteed to be CCA2 secure too. In addition, all existing ATD-based multicast encryption schemes [17, 14, 22, 23, 24] are based on discrete logarithm. Theorem 1 provides security guarantee for constructing multicast encryption using any other assumptions. For example, [31] provides a threshold RSA scheme with CPA security. Such scheme can be used to construct a RSA-based CPA secure multicast cryptosystem. Another example of factorization-based scheme is the threshold version of Paillier cryptosystem [37] presented in [36]. [36] provides techniques to make this scheme IND-CCA2. A multicast cryptosystem with the same level of security based on Paillier cryptosystem can thus be constructed using Construction 1. All the above examples have never been proposed before. They are the natural products of Construction 1 and their security is guaranteed by Theorem 1.

4.3   Extension to Construction 1

Construction 1 provides a simple way to utilize a threshold scheme to construct multicast encryption and we have shown that the resulting scheme is as secure as the underlying threshold scheme. It is basically an "encrypt-then-decrypt-t-times" scheme. It can be improved both in efficiency and security with simple extension.

In Construction 1, the encryptor has access to what are equivalent to t decryption shares in $\mathcal{TD}$ which are not available to an encryptor in the underlying threshold scheme. This gives her a chance to "protect" these shares and, as a result, the resulting multicast encryption can be made more secure than $\mathcal{TD}$. This can be seen as an extension of Construction 1:

Construction 1e (ME1e). *Same as Construction 1 except for the following:*

- *The encryption* E *produces ciphertext as* $\psi = (c, \{(j, \mathsf{D}_{\mathsf{TD}}(SK_j, c)) : j \in T' \cup R\}, v)$ *where* $c = \mathsf{E}_{\mathsf{TD}}(PK, m)$ *and* $v = \mathsf{Tag}(c, \Sigma, I)$ *is a "tag" for the ciphertext.*

---

[3] Their scheme uses this construction not to encrypt any useful messages. Instead, it is basically a distributed Diffie-Hellman key exchange which is equivalent to producing an ElGamal encryption of an arbitrary message (which is ignored) and allowing any member with proper keys to derive from the ciphertext, and partial decryptions, a secret key that can be used to encrypt actual data.

– *The decryption* D *first computes* Valid(Γ_i, ψ, I) *where* Valid *is a checking function outputting 0, or 1. If* Valid *outputs 0,* D *returns* $\perp$. *Otherwise it proceeds the same as Construction 1.*

This construction can be used to build a multicast scheme with higher security than the underlying threshold scheme. This is essentially what was done in [23] and [24]. The protection mechanism (i.e. Tag and Valid) depends on the threshold scheme and the security goal. In [23], the standard techniques of [20] (which attaches tags to the ciphertext so that the recipients with proper keys can verify its validity) was applied to protect the decryption shares and the security achieved is what [23] called gCCA2 (Generalized CCA) which is a variant, and weaker version, of CCA2. To achieve real CCA2 security, [23] used secure message authentication code (MAC) to make the verification tags non-malleable. And [24] essentially used a threshold version of M-CS [32].

4.4 Sharable Trapdoor Permutation-Based Construction

A whole class of public key cryptosystems are based on trapdoor permutations. Let $f_{PK} : \{0,1\}^k \to \{0,1\}^k$ be a k-bit to k-bit trapdoor (one-way) permutation with inverse f_{SK}^{-1}, defined by the public-private key pair (PK, SK). A public key cryptosystem $\mathcal{E}^{f,g,h}$ encrypts a message m as $\mathsf{E}(m) = h(f_{PK}(g(m)))$ where g and h are probabilistic, invertible functions that specify pre- and post-encoding operations, respectively. Given a ciphertext c, the decryption algorithm D computes $u = h^{-1}(c), v = f_{SK}^{-1}(u)$ and $m = g^{-1}(c, u, v)^4$. Depending on the security, the decryption may involve computing Valid$(c, u, v) \in \{0, 1\}$ which is the verification of the encoding. The decryption returns $\perp$ if Valid$(c, u, v) = 0$. We denote such cryptosystem as $\mathcal{E}^{f,g,h} = (\mathsf{KeyGen}, \mathsf{E}, \mathsf{D}, \mathsf{Valid})$ where KeyGen generates (PK, SK) on given security parameter 1^k. In the following, the keys will be dropped from the notations when there is no need to make them explicit.

Such cryptosystems are prevalent in practice. One example is the RSA Public Key Cryptography Standard # 1 [38], where $g(m)$ is essentially m padded with a string of random non-zero bytes in the high-order bit positions and post-encoding is simply omitted. Other schemes make use of hash functions. Let $G : \{0,1\}^* \to \{0,1\}^\infty$ be a random number generator and $H : \{0,1\}^* \to \{0,1\}^{k_0}$ be a hash function where $l = k - k_0$ is the length of the message. In [39] Bellare and Rogaway proposed the scheme $\mathcal{E}_{BR}^G$ where $\mathsf{E}(m) = f(r) \parallel G(r) \oplus m$ with $r \leftarrow_R \{0,1\}^k$. [39] showed that it is semantically secure in the random oracle model. [39] also presented another scheme, denoted $\mathcal{E}_{BR}^{G,H}$, that is shown to be CCA2 secure, also in the random oracle model. In $\mathcal{E}_{BR}^{G,H}$, message m is encrypted as $\mathsf{E}(m) = (f(r), m \oplus G(r), H(r, m))$ where $r \leftarrow_R \{0,1\}^k$. Given a ciphertext (s, c, v), the decryption algorithm computes $r = f^{-1}(s), m = G(r) \oplus c$, and $v' = H(r, m)$. If $v' = v$, it outputs m, and $\perp$ otherwise.

---

[4] Note that g and h are easily invertible and do not require trapdoors. Also note that both g and h are probabilistic and $g(m)$ maybe independent of m. In this case simply inverting v does not reveal m. However, these decryptions all have the following property: once the pre-image of the trapdoor permutation is recovered, it is easy to compute m. We simply use $g^{-1}(\cdot)$ to denote this process.

Another popular scheme is the OAEP scheme introduced in [40]. In this scheme, to encrypt a message m of length l bits, one selects a random value $r \leftarrow_R \{0,1\}^{k_0}$ and computes $s = (m \parallel 0^{k_1}) \oplus G(r)$ and $t = r \oplus H(s)$ where $k_1 = k - l - k_0$. The ciphertext is $c = f(s,t)$. To decrypt a ciphertext c, the decryptor extracts (s,t) using the private key $(s,t) = f^{-1}(c)$ and computes $r = t \oplus H(s)$ and $M = s \oplus G(r)$. If $[M]_{k_1} = 0^{k_1}$, it returns $[M]^l$. Otherwise it returns $\perp$. In the above, $[M]_l$ (resp. $[M]^l$) denotes the l least (resp. most) significant bits of M.

In [40], Bellare and Rogaway proved that OAEP construction together with any trapdoor one-way permutation is IND-CCA1. OAEP was widely believed to achieve stronger security (i.e. IND-CCA2). But Shoup showed in [41] that it is unlikely such security proof exists, for any trapdoor permutation. However, he proved that, when instantiated with low-exponent RSA, OAEP was IND-CCA2. This result was extended to arbitrary exponent RSA in [42].

All these schemes provide practical public key cryptosystems with various security and efficiency. (The OAEP scheme provides optimal bit complexity in that the ciphertext size is only slightly greater than that of plaintext.) However, they do not have threshold implementations that retain the same security, especially at CCA2 level. As Shoup and Gennaro noted in [35], the difficulty in transforming a non-threshold CCA secure public key encryption scheme, $\mathcal{E}$, into a CCA secure threshold scheme is that $\mathcal{E}$'s security proof can rely in a critical way on the fact that the decryption algorithm makes the "validity test" before generating an output. In a distributed setting, this means the test can only be performed *after* the individual decryption shares are combined. A single decryption server is unable to carry out such test. Both $\mathcal{E}_{BR}^{G,H}$ [39] and OAEP can be easily shown to have this difficulty.

One way to address this difficulty is to introduce a validity test that is publicly checkable so that a decryptor can perform the check before carrying out the decryption. This was suggested in [43] and followed by systems such as [35] which used non-interactive zero-knowledge proofs of membership to construct such check which is costly.

An ATD-based multicast encryption scheme, on the other hand, does not suffer from this difficulty at all. This is because in such a scheme, the decryptor is presented with what are equivalent to t decryption shares in the underlying sharing scheme. She can proceed to combine these shares with the one produced using her private key and perform the simple validity test as in the original public key cryptosystem (not the expensive publicly checkable threshold version) *before* emitting any output. As we show in Theorem 2, this construction preserves the CCA security of the public key cryptosystem even though its threshold implementation does not.

Our new construction is based on sharable trapdoor functions.

Definition 4 ($((t+1,n)$-Secure Sharing Scheme). *Let f be a trapdoor function with inverse f^{-1} defined by the public-private key pair (PK, SK). A sharing scheme $\mathcal{SS}^f = (\mathsf{S}, \eta)$ for f consists of two polynomial time algorithms:*

- S: *Given* (PK, SK), *a threshold* t *and an integer* $n > t$, S *generates* $SK_1, \ldots,$ SK_n *(in the same space as* SK*), called* shares *of* SK.
- η: *Given the public key* PK, *a set* Λ *of* $t+1$ *evaluations* $f_{SK_i}^{-1}(u)$, *for any* u *in the domain of* f_{PK}, η *computes* $f_{SK}^{-1}(u)$.

And SS^f *is* $(t+1, n)$*-secure if for all* $\{i_1, \ldots, i_j\} \subset U$ *where* $0 \le j \le t < n$, *for all probabilistic polynomial time algorithm* A, *for all polynomial poly*$(\cdot)$, *for all* k *large enough*

$$Pr[f_{PK}(u) = w : (SK_1, \ldots, SK_n) \leftarrow S(PK, SK, t, n);$$
$$w \in_R \{0, 1\}^k; u \leftarrow A(1^k, w, H, SK_{i_1}, \ldots SK_{i_j})] < 1/poly(k)$$

where H *is the history tape of length polynomial in* k *containing all the partial evaluations the players generated so far.*

And f *is* $(t+1, n)$*-sharable if it has one* $(t+1, n)$*-secure sharing scheme.*

This is essentially the same definition as $(t+1, n)$-secure function sharing primitive in [31]. [31] also showed how to implement such sharing with trapdoor permutations such as RSA. We show that using this primitive we can construct efficient multicast encryption schemes with high security.

Construction 2 (ME2). *Let* $\mathcal{E}^{f,g,h} = (\mathsf{KeyGen_E}, \mathsf{E_E}, \mathsf{D_E}, \mathsf{Valid})$ *be a public key cryptosystem based on* $(t + 1, n)$*-sharable trapdoor permutation* f *with sharing scheme* $SS^f = (\mathsf{S}, \eta)$. *Given a security parameter* 1^k, *a threshold* t *and the number of (initial) members* n, *a multicast encryption scheme* $\mathcal{ME}_{C2}^{\mathcal{E}^f} = (\mathsf{KeyGen}, \mathsf{E}, \mathsf{D})$ *can be constructed as follows:*

1. *Key Generation* KeyGen: *The center runs* $\mathsf{KeyGen_E}$ *with parameter* 1^k, *and obtains* $(PK, SK) \leftarrow \mathsf{KeyGen_E}(1^k)$. *It sets* $I = PK$ *and shares* SK *using the sharing algorithm* S *with parameter* $(t+1, n+t)$ *to obtain* $\mathbf{SK} = \mathsf{S}(PK, SK, t, n+t)$. *The encryption key is* $\Sigma = \{(j, SK_j) : j \in T\}$ *where* $T = \{n+1, \ldots, n+t\}$. Σ *is given to the center. Member* i *receives secret key* $\Gamma_i = (i, SK_i)$. *The master secret key is* $\boldsymbol{\Gamma} = (\Gamma_1, \ldots, \Gamma_{n+t})$.
2. *Encryption* E: *Given a set* R *of revoked members, and their secret keys, with* $|R| \le t$, *a message* m, *the encryptor randomly selects a subset of* T *with* $t - |R|$ *elements, denoted* T', *and computes the ciphertext*

$$\psi = (c, \{(j, f_{SK_j}^{-1}(u)) : j \in T' \cup R\}) \tag{2}$$

   where $c = \mathsf{E_E}(PK, m)$ *and* $u = h^{-1}(c)$.
3. *Decryption* D: *Given a secret key* Γ_i *and a ciphertext* ψ, *the ciphertext is first parsed into* $\psi = (c, \Lambda')$ *where* $\Lambda' = \{(j, f_{SK_j}^{-1}(u)) : j \in T' \cup R\}$. *The decryptor computes* $u = h^{-1}(c)$ *and* $v = \eta(u, \Lambda' \cup \{(i, f_{SK_i}^{-1}(u))\})$. *If all these steps are successful, it computes* $w = \mathsf{Valid}(c, u, v)$. *If* $w = 0$, *it returns* $\perp$. *Otherwise it returns* $m = g^{-1}(c, u, v)$.

Theorem 2. *Let* $\mu \in \{CPA, CCA2\}$. *If a public key cryptosystem* $\mathcal{E}^{f,g,h} = (\mathsf{KeyGen_E}, \mathsf{E_E}, \mathsf{D_E}, \mathsf{Valid})$ *based on* $(t+1, n)$*-sharable trapdoor permutation* f *with* *sharing scheme $SS^f = (S, \eta)$ is secure against μ type attacks, then a multicast encryption scheme $M\mathcal{E}_{C2}^{\mathcal{E}} = (\mathsf{KeyGen}, \mathsf{E}, \mathsf{D})$ constructed using Construction 2 with threshold t and (initial) group size n is t-resilient against μ-type attacks.*

Proof. First note that it is trivial to verify that the scheme is correct – i.e., the decryption produces the correct plaintext given a valid ciphertext. We prove its security by showing that if $M\mathcal{E}_{C2}^{\mathcal{E}}$ is not t-resilient against μ-type attacks, neither is $\mathcal{E}^{f,g,h}$. Let $\mathcal{A}^{M\mathcal{E}}$ be a polynomial time adversary that wins the game ME with non-negligible advantage. We can construct another polynomial time adversary $\mathcal{A}^{\mathcal{E}}$ that breaks $\mathcal{E}^{f,g,h}$ with at least the same advantage. $\mathcal{A}^{\mathcal{E}}$ achieves this by simulating a game ME and running $\mathcal{A}^{M\mathcal{E}}$ to win.

$(PK, SK) \leftarrow \mathsf{KeyGen}_{\mathsf{E}}(1^k)$ is run and PK is given to $\mathcal{A}^{\mathcal{E}}$ while SK is kept secret from it. $\mathcal{A}^{\mathcal{E}}$ selects randomly t numbers $SK_1, \ldots, SK_t$ from the space of SK. $\mathcal{A}^{\mathcal{E}}$ starts Game ME and lets $\mathcal{A}^{M\mathcal{E}}$ select t members to corrupt. Without loss of generality, let $T = \{1, 2, \ldots, t\}$ be the indexes of the members $\mathcal{A}^{M\mathcal{E}}$ chooses to corrupt. $\mathcal{A}^{\mathcal{E}}$ simulates the key generation process in game ME and gives $\Sigma = ((1, SK_1), \ldots, (t, SK_t))$ as the corrupted keys and $I = PK$ as the public information to $\mathcal{A}^{M\mathcal{E}}$.

$\mathcal{A}^{\mathcal{E}}$ lets $\mathcal{A}^{M\mathcal{E}}$ run and simulates the rest of game ME as follows:

- Whenever $\mathcal{A}^{M\mathcal{E}}$ queries the encryption oracle with message m, $\mathcal{A}^{\mathcal{E}}$ returns ψ computed using Equation 2 with $T' \cup R$ replaced by T.
- $\mathcal{A}^{\mathcal{E}}$ chooses whatever $\mathcal{A}^{M\mathcal{E}}$ choose as the two test plaintexts m_0 and m_1. Whenever $\mathcal{A}^{M\mathcal{E}}$ makes a query to the encryption oracle with m_0 and m_1, $\mathcal{A}^{T\mathcal{D}}$ passes them to its own encryption oracle in its game attacking $\mathcal{E}^{f,g,h}$ (denoted game E). Let c' be the result returned by the encryption oracle in game E. $\mathcal{A}^{\mathcal{E}}$ computes and returns the following to $\mathcal{A}^{M\mathcal{E}}$:

$$\psi' = (c', \{(j, f_{SK_j}^{-1}(u)) : j \in T\}) \tag{3}$$

  where $c = \mathsf{E}_{\mathsf{E}}(PK, m)$ and $u = h^{-1}(c')$. This corresponds to the target ciphertext in game ME.
- In the case of $\mu = $ CCA2, whenever $\mathcal{A}^{M\mathcal{E}}$ makes a query to one of the decryption oracles with ciphertext ψ, $\mathcal{A}^{\mathcal{E}}$ first parses ψ into a form as specified by Equation 2. Let $\{(j, u_j) : j \in T\}$ be the shares embedded in ψ. $\mathcal{A}^{\mathcal{E}}$ then verifies these shares by checking whether $u_j = f_{SK_j}^{-1}(u)$, where $u = h^{-1}(c)$, holds. If any of the tests fails it returns $\perp$ to $\mathcal{A}^{M\mathcal{E}}$. Otherwise it forwards c to its own decryption oracle and passes whatever the decryption oracle returns to $\mathcal{A}^{M\mathcal{E}}$.

$\mathcal{A}^{\mathcal{E}}$ stops when $\mathcal{A}^{M\mathcal{E}}$ stops and outputs whatever the latter does.

We need to show that $\mathcal{A}^{M\mathcal{E}}$ simulated by $\mathcal{A}^{\mathcal{E}}$ has all the information it would have in a real game ME and that its interaction with the simulated oracles is indistinguishable from that in a real game. First note that here, although the encryption key for $\mathcal{A}^{M\mathcal{E}}$, $SK_1, \ldots, SK_t$, are not actually generated by running S ($\mathcal{A}^{\mathcal{E}}$ does not have access to SK), they are just as good: the encryption key given to $\mathcal{A}^{M\mathcal{E}}$ is not distinguishable from that in a real game ME and does not affect its ability to win the game. This follows Lemma 1 from [31].

Second, $\mathcal{A}^{\mathcal{ME}}$ will receive $\perp$ on ciphertext ψ in the simulated game ME in one of the following two cases: (1) $\mathcal{A}^{\mathcal{E}}$'s decryption oracle returns $\perp$ on c; and (2) one of the tests on $u_j = f_{SK_j}^{-1}(u)$ fails. In the first case, $\mathcal{A}^{\mathcal{ME}}$ will receive $\perp$ in a real game ME, as specified by the decryption in Construction 2. In the second case $f_{SK_j}^{-1}(u)$, together with any partial evaluation of one of the decryption oracles in a real game ME, will combine to a u' that is not consistent with c and will fail Valid (otherwise it can be shown that either f is not $(t+1, n)$-sharable or $\mathcal{E}^{f,g,h}$ is not IND-CCA2). Again $\mathcal{A}^{\mathcal{ME}}$ will receive $\perp$ in a real game ME.

And in all other cases $\mathcal{A}^{\mathcal{ME}}$ will receive the correct decryption in both real and the simulated game ME. So if $\mathcal{A}^{\mathcal{ME}}$ can win a real game, it can win the simulated one.

It is easy to verify that if $\mathcal{A}^{\mathcal{ME}}$ wins the simulated game ME, $\mathcal{A}^{\mathcal{E}}$ distinguishes the two target ciphertexts with at least the same advantage. This is because, by definition of Construction 2, if ψ' in Equation 3 is the encryption of $m_{b'}$ in $\mathcal{ME}$, c' must be the encryption of $m_{b'}$ in $\mathcal{E}$.

Finally $\mathcal{A}^{\mathcal{E}}$'s running time is polynomial in that of $\mathcal{A}^{\mathcal{ME}}$ which itself is a polynomial in k. So $\mathcal{A}^{\mathcal{E}}$'s running time is also polynomial in k.

This is very powerful result because securing threshold scheme is hard so it is not always possible to use Construction 1 to construct multicast cryptosystems with high security. Construction 2 and Theorem 2 offer a simple method to construct multicast schemes with guaranteed security using a whole class of existing primitives. For instance, both RSA-OAEP [42] and $\mathcal{E}_{BR}^{G,H}$ [39], which have been shown to be difficult to obtain threshold implementations with the same level of security, can be used to build multicast scheme with CCA2 security. This has never been achieved before.

Besides security, Construction 2 also enjoys higher efficiency than Construction 1, which directly uses a threshold scheme. Note that in a sharing scheme used by Construction 2, there is neither decryption share verification nor publicly checkable validity test on ciphertext, both of which are essential for a threshold scheme or a real function sharing application to achieve robustness (as in e.g. [44]) and CCA security. With Construction 2, both can be omitted and the encoding verification that is part of the public key cryptosystem used can achieve both goals.

4.5   From IND-CPA to IND-CCA: Generic Conversion

In Construction 2, the security of $\mathcal{ME}_{C2}^{\mathcal{E}}$ relies on that of $\mathcal{E}^{f,g,h}$. Combined with results from previous work, we show that $\mathcal{ME}_{C2}^{\mathcal{E}}$ can be IND-CCA even if $\mathcal{E}^{f,g,h}$ is only IND-CPA.

In [45] Naor and Yung presented a generic conversion from an IND-CPA public key cryptosystem to one secure against "lunch-time" attack (a.k.a. non-adaptive chosen ciphertext attack, CCA1). The conversion used a twin-encryption paradigm and non-interactive zero-knowledge proof (NIZKP) of language membership in the common random string setting to show the consistency of the ciphertext. Rackoff and Simon later [46] improved this construction to be secure against adaptive chosen ciphertext attack (CCA2). Their solution involves replacing one of the twin encryption keys with the *sender*'s public key and providing a NIZKP of knowledge of the plaintext. [36] also provided similar conversion, in the random oracle model, that also works directly with *threshold* cryptosystems. The NIZKPs used in [45, 46, 36] are all publicly verifiable thus can be readily used in a threshold setting.

Putting all these together, we have the following whose proof immediately follows the results of [36, 46, 45] and ours.

Corollary 1. *If a public key cryptosystem $\mathcal{E}^{f,g,h}$ based on $(t+1,n)$-sharable trapdoor permutation f with sharing scheme $\mathcal{SS}^f$ is secure against chosen plaintext attacks, then there exists a multicast encryption scheme $\mathcal{ME}$ by Construction 2 with threshold t and (initial) group size n that is t-resilient against chosen ciphertext attacks.*

SUMMARY. Figure 1 summarizes the possible conversions covered in this paper between various primitives, including public key cryptosystem (PKC), threshold decryption scheme (TD) and multicast encryption (ME), at different security levels such as IND-CPA, IND-CCA (1 and 2). A solid arrow from A to B indicates "generic conversion", meaning that, under some reasonable assumptions, *any* A can be transformed into B. A dashed arrow, on the other hand, denotes "existential conversion", meaning that *some* A can be transformed into B. The conditions under which such conversions can succeed were stated in the literature. Some of the relevant ones covered in this paper are labelled on the arrows.

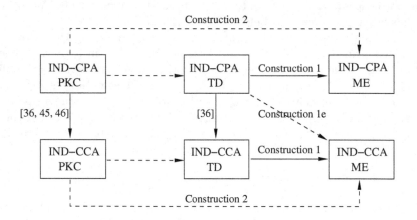

Fig. 1. Conversions

5 Conclusion

In this paper we have presented a general framework for constructing efficient multicast cryptosystems with provable security. Our constructions are based on asymmetric use of threshold schemes and we showed that a line of previous work on multicast encryption are all special cases of this general approach. We provided new methods for constructing multicast cryptosystems that achieve various levels of security (e.g., IND-CPA, IND-CCA2) from primitives with even weaker security. Using our scheme, each member only needs to store a key of constant length while both the encryption key size and the ciphertext length are $O(t)$ which is independent of the group size.

Acknowledgements. The first author would like to thank David Wagner for his encouragement and suggestions during the inception of this work. The authors thank the anonymous reviewers for their valuable comments.

References

1. Fenner, W.: Internet group management protocol, version 2. RFC-2236 (1997)
2. Harney, H., Muckenhirn, C.: Group key management protocol (gkmp) architecture. IETF Request for Comments, RFC 2094 (1997)
3. Wallner, D., Harder, E., Agee, R.: Key management for multicast: Issues and architectures. IETF Request For Comments, RFC 2627 (1999)
4. Wong, C.K., Gouda, M., Lam, S.S.: Secure group communications using key graphs. IEEE/ACM Trans. Netw. **8** (2000) 16–30
5. Canetti, R., Garay, J., Itkis, G., Micciancio, D., Naor, M., Pinkas, B.: Multicast security: A taxonomy and some efficient constructions. In: INFOCOMM'99. (1999)
6. Chang, I., Engel, R., Kandlur, D., Pendarakis, D., Saha, D.: Key management for secure internet multicast using boolean function minimization techniques. In: Proceedings IEEE Infocomm'99. Volume 2. (1999) 689–698
7. Wong, C.K., Lam, S.S.: Keystone: A group key management service. In: International Conference on Telecommunications, ICT 2000. (2000)
8. Li, X.S., Yang, Y.R., Gouda, M.G., Lam, S.S.: Batch rekeying for secure group communications. In: Proceedings of the tenth international World Wide Web conference on World Wide Web, Orlando, FL USA (2001) 525–534
9. Setia, S., Koussih, S., Jajodia, S., Harder, E.: Kronos: A scalable group re-keying approach for secure multicast. In: IEEE Symposium on Security and Privacy. (2000) 215–228
10. Yang, Y.R., Li, X.S., Zhang, X.B., Lam, S.S.: Reliable group rekeying: a performance analysis. In: Proceedings of the 2001 conference on Applications, technologies, architectures, and protocols for computer communications, ACM Press (2001) 27–38
11. Chor, B., Fiat, A., Naor, M.: Tracing traitors. In: CRYPTO 1994. Volume 839 of Lecture Notes in Computer Science., Springer-Verlag (1994) 257–270
12. Fiat, A., Naor, M.: Broadcast encryption. In: CRYPTO 1993. Volume 773 of Lecture Notes in Computer Science., Springer-Verlag (1994) 480–491
13. Boneh, D., Franklin, M.: An efficient public key traitor tracing scheme. In: CRYPTO 1999. Volume 1666 of Lecture Notes in Computer Science., Springer-Verlag (1999) 338–353
14. Anzai, J., Matsuzaki, N., Matsumoto, T.: A quick group key distribution scheme with "entity revocation". In: ASIACRYPT 1999. Volume 1716 of Lecture Notes in Computer Science., Singapore, Springer (1999) 333–347
15. Luby, M., Staddon, J.: Combinatorial bounds for broadcast encryption. In: EUROCRYPT 1998. Volume 1403 of Lecture Notes in Computer Science., Springer-Verlag (1998) 512–526

16. Garay, J.A., Staddon, J., Wool, A.: Long-lived broadcast encryption. In: CRYPTO 2000. Volume 1880 of Lecture Notes in Computer Science., Springer-Verlag (2000) 333–352
17. Naor, M., Pinkas, B.: Efficient trace and revoke schemes. In: Proceedings of Financial Crypto 2000. (2000)
18. Halevy, D., Shamir, A.: The LSD broadcast encryption scheme. In: CRYPTO 2002. Volume 2442 of Lecture Notes in Computer Science., Springer-Verlag (2002) 47–60
19. Naor, D., Naor, M., Lotspiech, J.B.: Revocation and tracing schemes for stateless receivers. In: CRYPTO 2001. Volume 2139 of Lecture Notes in Computer Science., Springer-Verlag (2001) 41–62
20. Cramer, R., Shoup, V.: A practical public key cryptosystem provably secure against adaptive chosen ciphertext attack. In: CRYPTO 1998. Volume 1462 of Lecture Notes in Computer Science., Springer-Verlag (1998) 13–25
21. Boneh, D., Gentry, C., Waters, B.: Collusion resistant broadcast encryption with short ciphertexts and private keys. In: CRYPTO 2005. Volume 3621 of Lecture Notes in Computer Science., Springer-Verlag (2005) 258–275
22. Tzeng, W.G., Tzeng, Z.J.: A public-key traitor tracing scheme with revocation using dynamic shares. In: Proceedings of the 4th International Workshop on Practice and Theory in Public Key Cryptography, Springer-Verlag (2001) 207–224
23. Dodis, Y., Fazio, N.: Public key trace and revoke scheme secure against adaptive chosen ciphertext attack. In: Workshop on Public Key Cryptography – PKC '03. Volume 2567 of Lecture Notes in Computer Science. (2003) 100–115
24. Kim, C.H., Hwang, Y.H., Lee, P.J.: An efficient public key trace and revoke scheme secure against adaptive chosen ciphertext attack. In: ASIACRYPT 2003. Volume 2894 of Lecture Notes in Computer Science., Springer-Verlag (2003) 359–373
25. Liu, D., Ning, P., Sun, K.: Efficient self-healing group key distribution with revocation capability. In: Proceedings of the 10th ACM conference on Computer and communication security, ACM Press (2003) 231–240
26. Staddon, J., Miner, S., Franklin, M., Balfanz, D., Malkin, M., Dean, D.: Self-healing key distribution with revocation. In: Proceedings of the 2002 IEEE Symposium on Security and Privacy, IEEE Computer Society (2002) 241
27. Wang, H.: Resilient lkh: Secure multicast key distribution schemes. In: Proceedings of the 2003 International Workshop on Advanced Developments in Software and Systems Security (WADIS). (2003)
28. Boyd, C.: Digital multisignatures. Cryptography and Coding (1986) 241–246
29. Desmedt, Y.: Society and group oriented cryptography: A new concept. In: CRYPTO 1987. Volume 293 of Lecture Notes in Computer Science., Springer-Verlag (1987) 120–127
30. Desmedt, Y.G., Frankel, Y.: Threshold cryptosystems. In: CRYPTO 1989. Volume 435 of Lecture Notes in Computer Science., Springer-Verlag (1989) 307–315
31. De Santis, A., Desmedt, Y., Frankel, Y., Yung, M.: How to share a function securely. In: Proceedings of the twenty-sixth annual ACM symposium on Theory of computing, ACM Press (1994) 522–533
32. Canetti, R., Goldwasser, S.: An efficient threshold public key cryptosystem secure against adaptive chosen ciphertext attack. In: EUROCRYPT 1999. Volume 1592 of Lecture Notes in Computer Science., Springer-Verlag (1999) 90–106
33. Abe, M.: Robust distributed multiplication without interaction. In: CRYPTO 1999. Volume 1666 of Lecture Notes in Computer Science., Springer-Verlag (1999) 130–147

34. Jarecki, S., Lysyanskaya, A.: Adaptively secure threshold cryptography: Introducing concurrency, removing erasures (extended abstract). In: Proceedings of Eurocrypt 2000. Volume 1807 of Lecture Notes in Computer Science., Springer-Verlag (2000) 221–242

35. Shoup, V., Gennaro, R.: Securing threshold cryptosystems against chosen ciphertext attack. J. Cryptology **15** (2002) 75–96

36. Fouque, P.A., Pointcheval, D.: Threshold cryptosystems secure against chosen-ciphertext attacks. In: ASIACRYPT 2001. Volume 2248 of Lecture Notes in Computer Science., Springer-Verlag (2001) 351–368

37. Paillier, P.: Public-key cryptosystems based on discrete logarithms residues. In: EUROCRYPT 1999. Volume 1592 of Lecture Notes in Computer Science., Springer-Verlag (1999) 223–238

38. RSA Labs: PKCS#1 v2.1: RSA cryptography standard (2002)

39. Bellare, M., Rogaway, P.: Random oracles are practical: a paradigm for designing efficient protocols. In: Proceedings of the 1st ACM conference on Computer and communications security, ACM Press (1993) 62–73

40. Bellare, M., Rogaway, P.: Optimal asymmetric encryption – how to encrypt with RSA. In: EUROCRYPT 1994. Volume 950 of Lecture Notes in Computer Science., Springer-Verlag (1994) 92–111

41. Shoup, V.: OAEP reconsidered. In: CRYPTO 2001. Volume 2139 of Lecture Notes in Computer Science., Springer-Verlag (2001) 239–259

42. Fujisaki, E., Okamoto, T., Pointcheval, D., Stern, J.: RSA-OAEP is secure under the rsa assumption. In: CRYPTO 2001. Volume 2139 of Lecture Notes in Computer Science., Springer-Verlag (2001) 260–274

43. Lim, C.H., Lee, P.J.: Another method for attaining security against adaptively chosen ciphertext attacks. In: CRYPTO 1993. Volume 773 of Lecture Notes in Computer Science., Springer-Verlag (1993) 420–434

44. Gennaro, R., Jarecki, S., Krawczyk, H., Rabin, T.: Robust and efficient sharing of RSA functions. In: CRYPTO 1996. Volume 1109 of Lecture Notes in Computer Science., Springer-Verlag (1996) 157–172

45. Naor, M., Yung, M.: Public-key cryptosystems provably secure against chosen ciphertext attacks. In: Proceedings of the twenty-second annual ACM symposium on Theory of computing, ACM Press (1990) 427–437

46. Rackoff, C., Simon, D.R.: Non-interactive zero-knowledge proof of knowledge and chosen ciphertext attack. In: CRYPTO 1991. Volume 576 of Lecture Notes in Computer Science., Springer-Verlag (1992) 433–444

On the (Im)possibility of Blind Message Authentication Codes

Michel Abdalla[1], Chanathip Namprempre[2], and Gregory Neven[1,3]

[1] Departement d'Informatique,
École normale supérieure,
45 Rue d'Ulm, 75230 Paris Cedex 05, France
Michel.Abdalla@ens.fr
http://www.di.ens.fr/~mabdalla
[2] Electrical Engineering Department,
Thammasat University,
Klong Luang, Patumtani 12121, Thailand
cnamprem@engr.tu.ac.th
http://www.engr.tu.ac.th/~cnamprem
[3] Department of Electrical Engineering,
Katholieke Universiteit Leuven,
Kasteelpark Arenberg 10, B-3001 Heverlee-Leuven, Belgium
Gregory.Neven@esat.kuleuven.ac.be
http://www.neven.org

Abstract. Blind signatures allow a signer to digitally sign a document without being able to glean any information about the document. In this paper, we investigate the symmetric analog of blind signatures, namely *blind message authentication codes* (blind MACs). One may hope to get the same efficiency gain from blind MAC constructions as is usually obtained when moving from asymmetric to symmetric cryptosystems. Our main result is a negative one however: we show that the natural symmetric analogs of the unforgeability and blindness requirements cannot be simultaneously satisfied. Faced with this impossibility, we show that blind MACs do exist (under the one-more RSA assumption in the random oracle model) in a more restrictive setting where users can share common state information. Our construction, however, is only meant to demonstrate the existence; it uses an underlying blind signature scheme, and hence does not achieve the desired performance benefits. The construction of an efficient blind MAC scheme in this restrictive setting is left as an open problem.

Keywords: provable security, blind signatures, blind MACs.

1 Introduction

THE CONCEPT. Blind signatures [7, 8] allow a signer to digitally sign a document while preventing the signer from seeing the content of the document, or even from recognizing the signature when faced with it later on. Blind signatures

D. Pointcheval (Ed.): CT-RSA 2006, LNCS 3860, pp. 262–279, 2006.

form a crucial anonymity-providing ingredient in digital cash protocols [7, 9], and have also been applied in a number of electronic voting schemes [7, 10, 13] to protect voters' privacy. Since their first introduction in 1982, blind signatures have become a well-studied primitive with formal security notions [12, 15], practical schemes realizing these notions under various assumptions [8, 15, 2, 5, 6], and a theoretical construction based on the existence of trapdoor one-way permutations [12].

In the same way that message authentication codes (MACs) can be seen as the symmetric-key equivalent of digital signatures, Pinkas [14] suggested *blind MACs* as the symmetric analog of blind signatures – leaving the construction of such schemes as an open problem however. In a blind MAC scheme, a user interacts with a tagger that knows a secret key K to obtain a valid tag τ for a message M, but without leaking any information about M to the tagger in the process. At a later point in time, the tagger can use K to check the validity of a given message-tag pair (M, τ), but cannot link it back to the session during which the tag was created.

MOTIVATION. The main motivation for blind MACs is efficiency. As is the case for standard MACs, one could hope to construct blind MACs from purely symmetric primitives, so that they can provide a more efficient alternative in applications where not all of the properties provided by digital signatures are needed. Good candidate applications are those where signatures are verified by the same entity that created them. In particular, we are interested in applications in which a "signer" does not need to convince others that it has generated (or has approved) the data in question, but only needs to convince itself at some later time that the data have not been modified. In other words, only the integrity of the data, not the non-repudiation of the data source, is of interest.

The first mention in public literature (to the best of our knowledge) of blind MACs was made by Pinkas [14] in the context of a fairness-providing transformation of Yao's secure two-party computation protocol [17]. The evaluator of the circuit commits to a number of 0 and 1-bits, and has these commitments blindly signed by the circuit constructor. The constructor then puts the blinded signatures in the output tables of the garbled circuit. At the end of the protocol, the constructor and evaluator gradually open their commitments. The constructor can verify that he indeed signed the commitments being opened by the evaluator, which prevents the latter from opening a commitment to some random value instead of the real output. Pinkas noted that, since the signatures are generated and verified by the same party, blind MACs could be used instead of blind signatures. He did not provide any formal definitions of the concept however, and left an actual construction as an open problem.

Blind MACs could also be used in Chaum's original online digital cash protocol [7]. A coin in this protocol is essentially a unique identifying string that is blindly signed by the bank. When the coin is spent, the merchant verifies the bank's signature and forwards the coin to the bank. The bank checks the validity of the signature again, and looks up in a database whether the coin is being double-spent. If not, it transfers the correct amount to the merchant's account, and adds the coin's identifying string to the database. Since the bank has to be online at the time the coin is spent anyway, the merchant may just as well leave the verification entirely up to the bank, so that the latter can use blind MACs instead of blind signatures (assuming that the clients' bank is the same as the merchants' bank). The gain in efficiency will reduce the infrastructural requirements brought about by online payment processing, and may actually make the protocol feasible in practice. In fact, we recently learned that blind MACs were already considered in this particular context by the Digicash research team [16]. They did not further pursue this idea because they suspected blind MACs to be impossible, without proving this fact however.

A third instance where blind MACs could take the place of blind signatures is in certain electronic voting schemes. The protocol of Fujioka et al. [10] for example works as follows. Voters commit to their votes, and have the commitment blindly signed by an administrator who checks their right to vote. All voters then send the signed commitment through an anonymous channel to a second authority called the counter. The counter verifies the administrator's signature and publishes all commitments on a bulletin board. At the end of the voting stage, each voter checks that his/her vote is posted on the bulletin board, and publicly complains if it is not. Finally, voters anonymously send the opening information for their commitments to the counter, who publishes everything on the bulletin board and announces the result of the election.

Note that in this protocol, the signer and verifier are not the same entity. Nevertheless, the administrator could use a blind MAC scheme to tag the voters' commitments, and reveal the tagging key after the end of the voting stage. MAC values are more efficiently verified than signatures, thus lowering the computational threshold for citizens to perform an independent audit of the election. A disadvantage is that the counter cannot verify the validity of commitments before posting them on the bulletin board, possibly resulting in more "junk" votes being published there. This problem however is also present in the original scheme, since voters can publish false complaints, of which the validity has to be checked as well. Moreover, if the counter can be trusted not to create fake registrations, then the administrator could give him the secret key at the start of the election already, allowing him to "weed out" junk votes earlier on.

OUR RESULTS. We first give proper definitions for the syntax and security of blind MACs, modeled after those of blind signatures. Our main result is a negative one: in Theorem 4, we show that the natural symmetric analogs of the one-more unforgeability [15] and blindness [12] requirements are contradictory, meaning that blind MACs satisfying both properties simultaneously cannot exist. Intuitively, the problem is that, because of the absence of a public key, the user has no way to check whether the tagger is using the same key throughout different tagging sessions. We present a universal adversary that breaks the blindness of any blind MAC scheme by using different keys in different tagging sessions, and we show that this attacker always succeeds, unless the scheme is forgeable.

Faced with the impossibility of blind MACs in their most general definition, we investigate whether they can exist under a more restrictive, yet still somewhat useful definition. In Section 5, we give a provably secure blind MAC construction in a setting where users share common state information. Whether this setting is realistic depends on the application. For Pinkas' two-party computation protocol [14], this is a perfectly reasonable assumption since there is only one user, the circuit evaluator, who can easily maintain state throughout different signing sessions. For digital cash and voting schemes however, it may be less realistic to assume the availability of common state information.

The sole purpose of our construction is to demonstrate the existence of blind MACs in this restrictive setting. It is based on an underlying blind signature scheme, and therefore does *not* achieve the performance benefits one would hope to get from a blind MAC scheme. We argue however that, before trying to come up with efficient constructions, it is important to understand what it exactly is that we are trying to construct, and whether it can be constructed at all. The fact that blind MACs can be constructed from blind signatures may sound rather unsurprising at first, but is not trivial: firstly, our impossibility result shows that not even such a "trivial" construction exists in the most natural definition of blind MACs, and secondly, our construction needs a special form of blindness from the underlying blind signature scheme, which we had to prove to be satisfied by a slight variant of Chaum's scheme [8].

ORGANIZATION. Section 2 recalls the definition of blind signatures and the security notions. Section 3 presents the definition and security notions for blind MACs. Section 4 states and proves the impossibility result. Section 5 describes the weaker model with state-sharing users, and shows that a secure blind MAC scheme exists in this model. Section 6 considers extensions to concurrent attack scenarios. Section 7 lists a few open problems.

2 Blind Signatures

NOTATION. We let $\mathbb{N} = \{1, 2, 3, \ldots\}$ denote the set of natural numbers. If $k \in \mathbb{N}$, then 1^k is the string of k ones. The empty string is denoted ε. If x, y are strings, then $|x|$ is the length of x and $x\|y$ is the concatenation of x and y. If S is a set, then $|S|$ is its cardinality. If A is a randomized algorithm, then $\mathsf{A}(x_1, x_2, \ldots : O_1, O_2, \ldots)$ means that A has inputs $x_1, x_2, \ldots$ and access to oracles $O_1, O_2, \ldots$. Also $y \xleftarrow{\$} \mathsf{A}(x_1, x_2, \ldots : O_1, O_2, \ldots)$ means that we run the randomized algorithm A on inputs $x_1, x_2, \ldots$ and with access to oracles $O_1, O_2, \ldots$, and let y denote the output obtained.

An *interactive algorithm* is a stateful algorithm that on input an incoming message M_{in} (this is ε if the party is initiating the protocol) and state information St outputs an outgoing message M_{out} and updated state St'. For an interactive algorithm A having access to oracles $O_1, O_2, \ldots$, this is written as $(M_{\mathrm{out}}, St') \xleftarrow{\$} \mathsf{A}(M_{\mathrm{in}}, St : O_1, O_2, \ldots)$. Two interactive algorithms A and B are said to *interact* when the outgoing messages of A are passed as incoming messages to B, and vice versa, until both algorithms enter either the `halt` or the `fail` state. We write $(M_A, St_A, M_B, St_B) \overset{\$}{\leftarrow} [A(St_A) \leftrightarrow B(St_B)]$ to denote the final outgoing messages and states after an interaction between A and B when run on initial states St_A and St_B, respectively. More formally, it is the outcome of the following experiment:

$M_B \leftarrow \varepsilon$
Repeat
 $(M_A, St_A) \overset{\$}{\leftarrow} A(M_B, St_A)$; $(M_B, St_B) \overset{\$}{\leftarrow} B(M_A, St_B)$
Until $\{St_A, St_B\} \subseteq \{\texttt{halt}, \texttt{fail}\}$
Return (M_A, St_A, M_B, St_B)

SYNTAX OF BLIND SIGNATURES. We repeat the definition of blind signatures as proposed by Juels et al. [12]. A blind signature scheme $\mathcal{BS}$ is a tuple of four polynomial-time algorithms (Kg, User, Sign, Vf) where

- the randomized *key generation* algorithm Kg, on input a security parameter 1^k with $k \in \mathbb{N}$, outputs a public key pk and a corresponding secret key sk.
- User and Sign are possibly randomized interactive algorithms called the *user* and *signer* algorithm, respectively. The user runs the User algorithm on an initial state consisting of a public key pk and a message $M \in \{0,1\}^*$, and lets it interact with the Sign algorithm that is run by the signer on initial state a secret key sk. At the end of the protocol, the User algorithm either enters the halt state and outputs a signature σ as its last outgoing message, or enters the fail state to indicate failure. The Sign algorithm simply enters the halt state at the end of the protocol, without generating any output.
- the deterministic *verification* algorithm Vf takes a public key pk, a message $M \in \{0,1\}^*$ and a signature σ as input, and outputs acc or rej to indicate acceptance or rejection of the signature, respectively.

Correctness of a blind signature scheme requires that for all $k \in \mathbb{N}$ and for all $M \in \{0,1\}^*$, it holds that $St_{\mathsf{User}} = \texttt{halt}$ and $\mathsf{Vf}(pk, M, \sigma) = \texttt{acc}$ when $(pk, sk) \overset{\$}{\leftarrow} \mathsf{Kg}(1^k)$ and $(M_{\mathsf{Sign}}, St_{\mathsf{Sign}}, \sigma, St_{\mathsf{User}}) \overset{\$}{\leftarrow} [\mathsf{Sign}(sk) \leftrightarrow \mathsf{User}((pk, M))]$ with probability 1.

UNFORGEABILITY OF BLIND SIGNATURES. The security of a blind signature scheme is twofold: on the one hand, a user should not be able to forge signatures (*unforgeability*), and on the other hand, the signer should not be able to see the message that is being signed, or even be able to relate signed messages to previous protocol sessions (*blindness*).

The standard definition of existential unforgeability under chosen-message attack [11] does not apply to blind signatures: the signer doesn't see the messages he signs, and hence the experiment has no way of telling whether the forgery is on a new message or on a message that was signed before. Therefore, we use the notion of *one-more unforgeability* as introduced by Pointcheval and Stern [15]. Let $\mathcal{BS} = (\mathsf{Kg}, \mathsf{User}, \mathsf{Sign}, \mathsf{Vf})$ be a blind signature scheme, let $k \in \mathbb{N}$ be the security parameter, and let A be a forging algorithm. The experiment first generates a fresh key pair $(pk, sk) \overset{\$}{\leftarrow} \mathsf{Kg}(1^k)$, and runs A on input $(1^k, pk)$. The adversary has access to a signing oracle that runs the $\mathsf{Sign}(sk, \cdot)$ algorithm and maintains state across invocations. (In a *sequential attack*, only one signing session can be active at the same time, while a *parallel attack* allows arbitrarily interleaved sessions. For simplicity, we concentrate on sequential attacks first, and postpone the discussion of parallel attacks to Section 6.) At the end of its execution, the adversary outputs a set of message-signature pairs $\{(M_1, \sigma_1), \ldots, (M_m, \sigma_m)\}$. Let n be the number of completed signing sessions during A's attack. Then A is said to win the game if $\mathsf{Vf}(pk, M_i, \sigma_i) = \mathsf{acc}$ for all $1 \leq i \leq m$, all M_i are different and $m > n$.

The advantage function $\mathbf{Adv}^{\mathrm{omu\text{-}sa}}_{\mathcal{BS}, \mathsf{A}}(k)$ is defined as A's probability of winning the above game, and $\mathcal{BS}$ is said to be *one-more unforgeable under sequential attacks* (omu-sa-secure) if this is a negligible function for all polynomial-time adversaries A. We note here that, in the definition above and in the rest of the paper, the "time complexity" is the worst case total execution time of the experiment plus the code size of the adversary in some fixed RAM model of computation.

BLINDNESS OF BLIND SIGNATURES. We present a sequential variant of the blindness notion as introduced by Juels et al. [12]. The adversary now plays the role of a cheating signer, who is trying to distinguish between two signatures created in different signing sessions. The experiment chooses a random bit b, generates a fresh key pair (pk, sk) and runs the adversary A on input $(1^k, pk, sk)$. The adversary outputs two challenge messages M_0 and M_1. Then, the adversary plays the role of the signer in two sequential interactions with a User algorithm. If $b = 0$, then the first interaction is with $\mathsf{User}(pk, M_0)$ and the second is with $\mathsf{User}(pk, M_1)$; if $b = 1$, then A first interacts with $\mathsf{User}(pk, M_1)$ and then with $\mathsf{User}(pk, M_0)$. If in both sessions the User algorithms accept, then A is additionally given the resulting signatures σ_0, σ_1 for messages M_0, M_1. The adversary outputs its guess d and wins the game if $b = d$. The advantage $\mathbf{Adv}^{\mathrm{blind\text{-}sa}}_{\mathcal{BS}, \mathsf{A}}(k)$ is defined as $2p - 1$, where p is the probability that A wins this game. The scheme $\mathcal{BS}$ is said to be *blind under sequential attacks* (blind-sa-secure) if this is a negligible function for all polynomial-time adversaries A. We refer to the full version [1] for formal descriptions of the experiments defining security for blind signatures.

3   Blind MACs

SYNTAX OF BLIND MACS. We define the syntax and security of blind MAC schemes in analogy to those of blind signatures.

Definition 1 [Syntax of a blind MAC scheme]. A blind MAC scheme $\mathcal{BMAC}$ is a tuple of four polynomial-time algorithms $(\mathsf{Kg}, \mathsf{User}, \mathsf{Tag}, \mathsf{Vf})$ where

- the randomized *key generation* algorithm Kg, on input a security parameter 1^k with $k \in \mathbb{N}$, outputs a key K.
- User and Tag are possibly randomized interactive algorithms called the *user* and *tagging* algorithm, respectively. The user runs the User algorithm on an

initial state containing the security parameter 1^k and a message $M \in \{0,1\}^*$, and lets it interact with the Tag algorithm that is run by the tagger on initial state the key K. [1] At the end of the protocol, the User algorithm either enters the halt state and outputs a MAC value τ as its outgoing message, or enters the fail state to indicate failure. The Tag algorithm simply enters the halt state at the end of the protocol, without generating any output.

– the deterministic *verification* algorithm Vf takes a key K, a message $M \in \{0,1\}^*$ and a MAC value τ as input, and outputs acc or rej to indicate acceptance or rejection of the MAC value, respectively.

Correctness of a blind MAC scheme requires that for all $k \in \mathbb{N}$ and for all $M \in \{0,1\}^*$, with probability 1 it holds that $St_{\mathsf{User}} = \mathtt{halt}$ and $\mathsf{Vf}(K, M, \tau) = \mathtt{acc}$ whenever $K \xleftarrow{\$} \mathsf{Kg}(1^k)$ and $(M_{\mathsf{Tag}}, St_{\mathsf{Tag}}, \tau, St_{\mathsf{User}}) \xleftarrow{\$} [\mathsf{Tag}(K) \leftrightarrow \mathsf{User}((1^k, M)]$.

SECURITY OF BLIND MACs. Analogously to the security of blind signatures, the security of a blind MAC scheme consists of an unforgeability and a blindness requirement. The game defining unforgeability works as follows. The experiment generates a fresh key $K \xleftarrow{\$} \mathsf{Kg}(1^k)$, and runs the adversary A on input 1^k. The adversary can interact in sequential sessions with a tagging oracle that runs the Tag algorithm initialized with key K. At the end of its execution, A outputs m message-tag pairs. The adversary wins the game if all messages are different, all tags are valid under key K, and $m > n$, where n is the number of completed tagging sessions during the attack. We give a more formal description of the definition below.

Definition 2 [Unforgeability of a blind MAC scheme]. Let $\mathcal{BMAC} = (\mathsf{Kg}, \mathsf{User}, \mathsf{Tag}, \mathsf{Vf})$ be a blind message authentication scheme. Let $k \in \mathbb{N}$, and let A be a forger with access to the tagging oracle. Consider the following experiment.

Experiment $\mathbf{Exp}^{\mathrm{omu\text{-}sa}}_{\mathcal{BMAC}, \mathsf{A}}(k)$:
> $K \xleftarrow{\$} \mathsf{Kg}(1^k)$; $n \leftarrow 0$
> $\{(M_1, \tau_1), \dots, (M_m, \tau_m)\} \xleftarrow{\$} \mathsf{A}(1^k : \mathrm{TAG}(\cdot))$
> If $\mathsf{Vf}(K, M_i, \tau_i) = \mathtt{acc}$ for all $1 \le i \le m$
> and $m > n$ and $M_i \ne M_j$ for all $1 \le i < j \le m$
> then return 1 else return 0,

where A's queries to the tagging oracle are answered as follows:

Oracle $\mathrm{TAG}(M_{\mathrm{in}})$:
> If $M_{\mathrm{in}} = \perp$ then $St_{\mathsf{Tag}} \leftarrow K$; $M_{\mathrm{out}} \leftarrow \perp$
> else $(M_{\mathrm{out}}, St_{\mathsf{Tag}}) \xleftarrow{\$} \mathsf{Tag}(M_{\mathrm{in}}, St_{\mathsf{Tag}}[s])$
> If $St_{\mathsf{Tag}} = \mathtt{halt}$ then $n \leftarrow n + 1$
> Return M_{out}

---

[1] We need to pass 1^k as a parameter to the User algorithm, because otherwise it would no longer be a polynomial-time algorithm if the message is of logarithmic length. Moreover, since the user does not know the key itself, it is reasonable to give it 1^k so that at least it can check whether the tagger is using a key of the correct size.

The omu-sa advantage of A in breaking $\mathcal{BMAC}$ is defined as the probability that the above experiment returns 1:

$$\mathbf{Adv}^{\text{omu-sa}}_{\mathcal{BMAC},\mathsf{A}}(k) \;=\; \Pr\left[\mathbf{Exp}^{\text{omu-sa}}_{\mathcal{BMAC},\mathsf{A}}(k) = 1\right],$$

and $\mathcal{BMAC}$ is said to be *one-more unforgeable under sequential attacks* (omu-sa-secure) if the advantage $\mathbf{Adv}^{\text{omu-sa}}_{\mathcal{BMAC},\mathsf{A}}(k)$ is a negligible function in the security parameter k for all adversaries A with time complexity polynomial in k.   ∎

In the blindness game, the experiment chooses a random bit b and generates a fresh key $K \xleftarrow{\$} \mathsf{Kg}(1^k)$. On input $(1^k, K)$, the adversary A first outputs two messages M_0, M_1. The adversary then sequentially interacts with two User sessions, playing the role of the tagger. If $b = 0$, then the first user session is initialized with message M_0, and the second with M_1; if $b = 1$, then the first session is initialized with message M_1, and the second with M_0. If both User algorithms accept, the adversary gets to see both resulting tags τ_0, τ_1 for messages M_0, M_1. The adversary has to guess the value of b.

We stress that the experiment does not enforce the resulting tags to be valid under key K. While we could include such restriction in the formal security notion, it would be out of touch with reality: the secret key K is not known to the users, so there is nobody to enforce this restriction in the real world. In fact, as we will see in the next section, it is exactly this lack of verifiability of tags that plays a central role in the proof of impossibility of blind MACs. We give a formal blindness definition below.

Definition 3 [Blindness of a blind MAC scheme]. Let $\mathcal{BMAC} = (\mathsf{Kg}, \mathsf{User}, \mathsf{Tag}, \mathsf{Vf})$ be a blind message authentication scheme. Let $k \in \mathbb{N}$, and let A be an adversary. Consider the following experiment.

Experiment $\mathbf{Exp}^{\text{blind-sa}}_{\mathcal{BMAC},\mathsf{A}}(k)$:
$\quad b \xleftarrow{\$} \{0,1\}\,;\; K \xleftarrow{\$} \mathsf{Kg}(1^k)$
$\quad ((M_0, M_1), St_\mathsf{A}) \xleftarrow{\$} \mathsf{A}(\varepsilon, (1^k, K))$
$\quad (M_\mathsf{A}, St_\mathsf{A}, \tau_b, St_b) \xleftarrow{\$} [\mathsf{A}(St_\mathsf{A}) \leftrightarrow \mathsf{User}((1^k, M_b))]$
$\quad (M_\mathsf{A}, St_\mathsf{A}, \tau_{1-b}, St_{1-b}) \xleftarrow{\$} [\mathsf{A}(St_\mathsf{A}) \leftrightarrow \mathsf{User}((1^k, M_{1-b}))]$
$\quad$ If $St_0 = \mathtt{fail}$ or $St_1 = \mathtt{fail}$ then $\tau \leftarrow \mathtt{fail}$ else $\tau \leftarrow (\tau_0, \tau_1)$
$\quad d \xleftarrow{\$} \mathsf{A}(\tau, St_\mathsf{A})$
$\quad$ If $b = d$ then return 1 else return 0

The blind-sa advantage of A in breaking $\mathcal{BMAC}$ is defined as

$$\mathbf{Adv}^{\text{blind-sa}}_{\mathcal{BMAC},\mathsf{A}}(k) \;=\; 2 \cdot \Pr\left[\mathbf{Exp}^{\text{blind-sa}}_{\mathcal{BMAC},\mathsf{A}}(k) = 1\right] - 1$$

and $\mathcal{BMAC}$ is said to be *blind under sequential attacks* (blind-sa-secure) if the advantage $\mathbf{Adv}^{\text{blind-sa}}_{\mathcal{BMAC},\mathsf{A}}(k)$ is a negligible function in the security parameter k for all adversaries A with time complexity polynomial in k.   ∎

4   Impossibility of Blind MACs

In this section, we show that blind MAC schemes simultaneously satisfying the one-more unforgeability and blindness requirements cannot exist. We do so by demonstrating a universal blindness adversary A and a universal forger F so that for any candidate scheme, one of them always has a non-negligible chance of success.

Theorem 4. [Secure blind MAC schemes do not exist]. Let $\mathcal{BMAC}$ be a blind MAC scheme. Either $\mathcal{BMAC}$ is one-more forgeable under sequential attacks, or it is not blind under sequential attacks.

Proof (Theorem 4). We define an adversary A breaking the blindness of $\mathcal{BMAC}$ and an adversary F breaking the one-more unforgeability of $\mathcal{BMAC}$, both under sequential attacks, so that

$$\mathbf{Adv}^{\text{blind-sa}}_{\mathcal{BMAC}, A}(k) + \mathbf{Adv}^{\text{omu-sa}}_{\mathcal{BMAC}, F}(k) = 1 \ ,$$

from which the theorem follows.

The key idea in constructing A is from the observation that, in a blind MAC scheme, the user has no way of telling under which key a tag is computed. Our adversary exploits this fact by using two different keys to generate the tags for the two user sessions. Then, it only needs validate one of the final message-tag pairs to determine during which user session the tag was computed. The possibility that a tag computed with the second key is also valid under the first key, or that both keys happen to be identical, is ruled out by the existence of a forger F that is successful in exactly these cases.

We now present both adversaries in more detail. Algorithm A, on initial state $(1^k, K)$, generates a second key $K' \xleftarrow{\$} \mathsf{Kg}(1^k)$ and outputs challenge messages $M_0 = 0$ and $M_1 = 1$. (In fact, any two distinct challenge messages would do.) It interacts with the first User algorithm by honestly running $\mathsf{Tag}(K)$, and with the second by running $\mathsf{Tag}(K')$. Since both K and K' are keys generated by the Kg algorithm, the correctness requirement for $\mathcal{BMAC}$ implies that neither of the user sessions fails, and hence that A gets back tags (τ_0, τ_1). If $\mathsf{Vf}(K, M_0, \tau_0) = \mathsf{acc}$, the adversary returns $d = 0$, else it returns $d = 1$.

The forger F works as follows: on input 1^k, it generates a fresh random key $K' \xleftarrow{\$} \mathsf{Kg}(1^k)$. It simulates an interaction $(M, St_{\mathsf{Tag}}, \tau, St_{\mathsf{User}}) \xleftarrow{\$} [\mathsf{Tag}(K') \leftrightarrow \mathsf{User}((1^k, M_0))]$ in which a tagger uses key K' to tag message $M_0 = 0$ (or whichever message M_0 algorithm A used above). It then outputs $\{(M_0, \tau)\}$ as its single forgery without making any tagging oracle call.

Now, we analyze the success probability of A and F. From Definition 3,

$$\mathbf{Adv}^{\text{blind-sa}}_{\mathcal{BMAC}, A}(k) = 2 \cdot \Pr\left[\mathbf{Exp}^{\text{blind-sa}}_{\mathcal{BMAC}, A}(k) = 1 \right] - 1$$

$$= \Pr\left[\mathbf{Exp}^{\text{blind-sa}}_{\mathcal{BMAC}, A}(k) = 1 \mid b = 1 \right] + \Pr\left[\mathbf{Exp}^{\text{blind-sa}}_{\mathcal{BMAC}, A}(k) = 1 \mid b = 0 \right] - 1$$

$$= \Pr\left[\mathbf{Exp}^{\text{blind-sa}}_{\mathcal{BMAC}, A}(k) = 1 \mid b = 1 \right]$$

$$= 1 - \mathbf{Adv}^{\text{omu-sa}}_{\mathcal{BMAC}, F}(k)$$

The second equality follows easily from simple algebra and the fact that b is a randomly chosen bit. The third equality follows from the correctness requirement of $\mathcal{BMAC}$: if $b = 0$, then during the first user session, A tags message M_0 with key K. Hence, verification of the resulting tag with key K must always succeed, making A output the correct guess $d = 0$ with probability 1. Looking closely at the probability on the third line, we see that it is one minus the probability that a tag τ_0 obtained from an interaction $[\mathsf{Tag}(K') \leftrightarrow \mathsf{User}((1^k, M_0))]$ also verifies correctly under an independently generated key K. This however is exactly the success probability of our forger F, leading to the last equation, which concludes the proof.

5 Blind MACs for State-Sharing Users

The attack in Section 4 is due to the fact that, unlike in the case of blind signatures, the user has no public key based on which it can check whether the tagger is behaving honestly, and in particular, whether he's using the correct key to tag the message. The attack only holds however for user sessions that are completely isolated from each other, and does not exclude the existence of blind MACs when user sessions can communicate with one another. Depending on the application, it may be unrealistic to assume that all users are connected through secure communication channels (or even know of each other's existence), but it may be more reasonable to assume that small groups of user sessions can share some common state information. We ask ourselves whether a weaker form of blindness is achievable, where anonymity is guaranteed among messages tagged in state-sharing user sessions. For applications like electronic cash and voting, this would provide a rather limited form of anonymity. In Pinkas' two-party computation protocol however [14], there is only one user (the circuit evaluator), so it is perfectly safe to assume that the different user sessions share common state information.

In the following, we describe a provably secure construction of a blind MAC scheme in the state-sharing users setting. The main purpose of the construction, however, is to prove the existence of blind MACs in this restrictive setting: it is based on an underlying blind signature scheme, and hence does not achieve the performance benefits that were the original motivation for blind MACs. The secret key of the blind MAC scheme contains both the public and the private key of the underlying blind signature scheme. In the first move of the tagging protocol, the tagger sends the public key to the user. If the common state information is empty, then the user stores this public key in the common state information; otherwise, the user compares the public key to the one that is stored in the common state, and rejects if the keys are different. The rest of the protocol is identical to that of the blind signature scheme. To prove the security of the construction, we introduce a new (and actually, more natural) blindness notion for blind signatures that we call *dishonest-key blindness*, where the public key can be maliciously constructed by the adversary, rather than being honestly generated through the key generation algorithm. Then, we show that Chaum's blind signature scheme with a prime encryption exponent whose value is larger than the RSA modulus is (unconditionally) dishonest-key blind. Together with the known fact that this scheme is one-more unforgeable in the random oracle model under the one-more RSA assumption [4], this implies that a blind MAC scheme with state-sharing users exists in the random oracle model if the one-more RSA-inversion problem [4] is hard.

SYNTAX AND SECURITY OF BLIND MACS WITH STATE-SHARING USERS. We model the common state information as a third input string CSt that is given to the User algorithm, and that the user can update through a third output string CSt'. We add this common state as an input to the user in the blindness experiment in Definition 1. The common state is initialized to ε and maintained between both user sessions. The rest of the experiment remains the same. The unforgeability notion as stated in Definition 2 remains unchanged.

A CONSTRUCTION BASED ON BLIND SIGNATURES. The main idea for our blind MAC construction is to store the public key for the base blind signature scheme in the users' common state information. Then, we use the algorithms of the blind signature scheme in a natural way.

Construction 5 [A blind MAC scheme for state-sharing users]. Let $\mathcal{BS} = (\mathsf{Kg}_s, \mathsf{User}_s, \mathsf{Sign}, \mathsf{Vf}_s)$ be a blind signature scheme. We associate to it a blind MAC scheme $\mathcal{BMAC} = (\mathsf{Kg}_m, \mathsf{User}_m, \mathsf{Tag}, \mathsf{Vf}_m)$ as follows:

- On input 1^k, the key generation algorithm Kg_m runs $\mathsf{Kg}_s(1^k)$ to obtain a key pair (pk, sk), sets $K \leftarrow (pk, sk)$ and returns K.
- On input K, the tagging algorithm Tag starts the interaction with User_m by parsing K as (pk, sk), sends pk to User_m, runs Sign on initial state sk interacting with User_m to completion. It sets its state to whatever Sign does.
- On inputs an initial state 1^k, a message M, and an initial shared-state CSt, the algorithm User_m first receives pk from Tag. If $CSt = \varepsilon$, then User_m sets $CSt \leftarrow pk$. Otherwise, it sets $pk \leftarrow CSt$ and runs User_s on the initial state (pk, M) interacting with Tag until the interaction completes. It sets its state and output to those of User_s.
- On input a key K, a message M, and a MAC value τ, the algorithm Vf_m parses K as (pk, sk), and returns $\mathsf{Vf}_s(pk, M, \tau)$. ∎

DISHONEST-KEY BLINDNESS FOR BLIND SIGNATURES. Before stating the security of our blind MAC construction, we briefly describe here the concept of dishonest-key blindness, which is needed to prove its security. Recall that the standard blindness notion for blind signatures assumes that the adversary is given a key pair generated properly through the key generation algorithm. This however does not cover attacks where the signer creates a public key in a special, malicious way that allows him to break the blindness of the scheme. The dishonest-key blindness notion that we propose gives the adversary more power by letting it dictate the public key to be used. This public key need not be generated by the Kg_s algorithm, nor does the adversary need to know the corresponding secret key. The adversary gets as only input 1^k, and outputs challenge messages

M_0, M_1 along with the public key pk. The rest of the experiment is unchanged: the adversary engages in two sequential User sessions that are initialized with $(1^k, pk, M_0)$ and $(1^k, pk, M_1)$, the order depending on the experiment's choice for bit b.

It is based on this stronger security requirement of the underlying blind signature scheme that we construct a secure blind MAC in the state-sharing model. Let $\mathbf{Adv}_{\mathcal{BS},A}^{\text{dk-blind-sa}}(k)$ be the advantage of an adversary A in winning the above game against $\mathcal{BS}$ in a sequential attack. We say that $\mathcal{BS}$ is dk-blind-sa-secure if this advantage is a negligible function in k for all polynomial-time algorithms A. We refer to Appendix A for a formal definition of dishonest-key blindness.

SECURITY. The following theorem states that, if the underlying blind signature scheme is one-more unforgeable and *dishonest-key blind*, then the resulting blind MAC scheme is secure.

Theorem 6. If a blind signature scheme $\mathcal{BS}$ is one-more unforgeable and dishonest-key blind under sequential attacks, then the blind MAC scheme with state-sharing users $\mathcal{BMAC}$ associated to $\mathcal{BS}$ as per Construction 5 is one-more unforgeable and blind under sequential attacks.

Theorem 6 follows directly from the following two lemmas.

Lemma 7. If a blind signature scheme $\mathcal{BS}$ is omu-sa secure, then the blind MAC scheme with state-sharing users $\mathcal{BMAC}$ associated to $\mathcal{BS}$ as per Construction 5 is also omu-sa secure.

Lemma 8. If a blind signature scheme $\mathcal{BS}$ is dk-blind-sa secure, then the blind MAC scheme with state-sharing users $\mathcal{BMAC}$ associated to $\mathcal{BS}$ as per Construction 5 is blind-sa-secure.

Proof (Lemma 7). We prove the lemma via a standard reduction, namely, we assume the existence of a forger F_m mounting an attack against $\mathcal{BMAC}$, and construct a forger F_s mounting an attack against $\mathcal{BS}$ so that, if the success probability of the former is non-negligible, then so is that of the latter. The idea is for F_s to run F_m using its signing oracle to simulate F_m's tagging oracle TAG($\cdot$). Since the only difference between a tagger-user interaction in $\mathcal{BMAC}$ and a signer-user interaction in $\mathcal{BS}$ is in the public key that the tagger sends to the user as the first message, this simulation can be done perfectly. Thus, if F_m is able to produce one more valid message-tag pair than the number of finished interactive sessions with its tagging oracle, then so can F_s with respect to its signing oracle.

Now we provide more details of how F_s works. Let $\mathcal{BS} = (\text{Kg}_s, \text{User}_s, \text{Sign}, \text{Vf}_s)$ and let $\mathcal{BMAC} = (\text{Kg}_m, \text{User}_m, \text{Tag}, \text{Vf}_m)$. On input $(1^k, pk)$, it runs $F_m(1^k)$. For each tagging session that F_m runs, F_s starts the interaction by sending pk to F_m as the first message, then simply relays messages between F_m and its own signing oracle. When F_m eventually halts, F_s outputs whatever F_m does.

Forger F_s perfectly simulates the environment for F_m. To see this, let pk be F_s's input public key, and let sk be the matching secret key used by its signing

oracle. Notice that from the definition of $\mathcal{BMAC}$ in Construction 5, each interaction in the transcript of messages between the tagger $\mathsf{Tag}(pk, sk)$ and a user
$\mathsf{User}_m(1^k, M)$ is composed of pk followed by other messages generated through
the interaction between the signer $\mathsf{Sign}(sk)$ and $\mathsf{User}_s(pk, M)$ for any message
M. Since all F_s does is to first send pk and then to relay messages between the
signing oracle and F_m (who is acting in the role of User_m), F_s simulates F_m in
the exact same environment as that of the experiment in Definition 2.

Furthermore, let $(M_1, \sigma_1), \ldots, (M_m, \sigma_m)$ be the outputs of F_s. By definition
of Vf_m, it is the case that, for all $1 \leq i \leq m$, $\mathsf{Vf}_m((pk, sk), M_i, \sigma_i) = \mathsf{acc}$ if and
only if $\mathsf{Vf}_s(pk, M_i, \sigma_i) = \mathsf{acc}$. Thus, if F_m's outputs are valid message-tag pairs
under $K = (pk, sk)$, then F_s's outputs are also valid message-signature pairs
under pk. Since F_s interacts with its oracle the same number of sessions as F_m
does, if F_m uses strictly fewer sessions than the number of output pairs, then so
does F_s. Thus, if F_m succeeds, then so does F_s, or

$$\mathbf{Adv}^{\mathrm{omu\text{-}sa}}_{\mathcal{BMAC}, \mathsf{F}_m}(k) \leq \mathbf{Adv}^{\mathrm{omu\text{-}sa}}_{\mathcal{BS}, \mathsf{F}_s}(k) \, ,$$

which proves the lemma.

Proof (Lemma 8). We prove the lemma via a standard reduction, namely, we
assume the existence of an adversary A_m attacking the blindness of $\mathcal{BMAC}$, and
then construct an adversary A_s attacking the dishonest-key blindness of $\mathcal{BS}$ so
that, if the success probability of the former is non-negligible, then so is that
of the latter. The idea is for A_s to first run A_m, and to output the public key
contained in the first message of A_m's first user interaction as the public key
with which both User_s sessions should be run. The rest of the messages are then
relayed faithfully between A_m and the User_s sessions. In A_m's second User_m
interaction, the first outgoing message from A_m is simply dropped.

Now we provide more details of how A_s works. We emphasize that A_s operates in the dishonest-key model. On input 1^k, the adversary A_s generates a key
pair (pk, sk) via $\mathsf{Kg}(1^k)$, runs $\mathsf{A}_m(1^k, (pk, sk))$, obtains A_m's challenge messages
M_0, M_1, and waits until A_m outputs its first outgoing message pk' as part of a
User_m session. Then, A_s outputs pk' as the public key for the users along with
the same challenge messages M_0, M_1. Adversary A_s relays messages faithfully
between User_s and A_m (who is acting in the role of the tagger) for the rest of the
interaction. The interaction with the second user is similar: A_s drops the first
message from A_m and simply relays following messages to and from its second
User_s session. Finally, when given $\sigma = (\sigma_0, \sigma_1)$ or $\mathtt{fail}$, A_s forwards σ to A_m
and outputs A_m's guess d as its own.

We first argue that A_s simulates A_m in the same environment as that in
Definition 3. Consider the three phases in A_m's attack: starting, interacting with
users, and guessing. In the first phase, A_s starts A_m with a legitimate key pair
which is indeed what A_m expects. Since, by definition of $\mathcal{BMAC}$, User_s outputs
whatever User_m outputs, the tags that A_s gives to A_m in the last phase are
also correctly distributed. For the second phase, recall that A_s drops the first
message received from A_m and relays messages between User_s and A_m. Thus, the

messages relayed to A_m are exactly what A_m would see in its role as a tagger. Therefore, this phase also follows the correct distribution.

Now suppose that A_m succeeds. We argue that A_s does too. Let pk' be the first outgoing message that A_m outputs to start the session with the "first" user. Let $b \in \{0,1\}$ such that interaction $[A_s \leftrightarrow User_s(pk', M_b)]$ starts first. Recall that A_s simulates $User_m(1^k, \cdot)$ using $User_s(pk', \cdot)$. This means that the interaction $[A_m \leftrightarrow User_m(pk', M_b)]$ also starts first. Since A_s outputs the same answer as A_m, A_s guess correctly whenever A_m does. So we have

$$\mathbf{Adv}_{\mathcal{BMAC}, A_m}^{\text{blind-sa}}(k) \leq \mathbf{Adv}_{\mathcal{BS}, A_s}^{\text{dk-blind-sa}}(k) \ ,$$

which concludes the proof.

EXISTENCE OF DISHONEST-KEY BLIND SIGNATURE SCHEMES. We describe a variant of Chaum's blind signature scheme here. Theorem 9 below states that this scheme is one-more unforgeable and dishonest-key blind. Recall that in Chaum's RSA-based blind signature scheme, the public key is (N, e) and the private key is (N, d) where N is an RSA modulus, e is an RSA encryption exponent, and d is the corresponding RSA decryption exponent. On inputs a public key (N, e) and a message M, the user computes $\overline{M} \leftarrow r^e \cdot H(M) \bmod N$, where r is a random value in $\mathbb{Z}_N^*$ and $H : \{0,1\}^* \rightarrow \mathbb{Z}_N^*$ is a public hash function, then submits $\overline{M}$ to the signer. The signer then responds with $\overline{\sigma} \leftarrow \overline{M}^d \bmod N$. Finally, the user computes and outputs $\sigma \leftarrow r^{-1} \cdot \overline{\sigma} \bmod N$. A message-signature pair (M, σ) is valid if and only if $\sigma^e \equiv H(M) \bmod N$. The variant that we are interested in is Chaum's scheme with the additional requirements that e is prime and that $e > N$. The user checks that these requirements hold before starting the protocol, and checks that $\sigma \in \mathbb{Z}_N^*$ and $\sigma^e \equiv H(M) \bmod N$ at the end of the protocol. If any of these checks fail, the User algorithm terminates in a `fail` state. We note that this check can be done in deterministic polynomial time [3].

Theorem 9. [Security of modified Chaum scheme]. Let $H : \{0,1\}^* \rightarrow \mathbb{Z}_N^*$ be a random oracle, and let $\mathcal{BS}$ be Chaum's blind signature scheme with prime encryption exponent $e > N$. Then, $\mathcal{BS}$ is one-more unforgeable under sequential attacks in the random oracle model assuming that the one-more-RSA-inversion problem is hard. Furthermore, $\mathcal{BS}$ is unconditionally dishonest-key blind under sequential attacks.

Proof (Theorem 9). Bellare et al. proved in [4] that Chaum's scheme is one-more unforgeable in the random oracle model assuming that the one-more-RSA-inversion problem is hard. Their proof does not make additional assumptions about the encryption exponent e. Thus, the same security result holds for our variant of Chaum's scheme.

Now we prove the blindness result. Let A be a dishonest-key blindness adversary. Over the course of the experiment, A's inputs are the incoming messages from the two users and the two resulting signatures. Consider the two worlds dictated by which message is signed first (i.e. $b = 0$ or $b = 1$) and regard each input of A as a random variable. We argue that each of these random variables has the same distribution in both worlds. We consider them one by one. First, we consider an incoming message $\overline{M}$, which is computed as $r^e \cdot H(M) \bmod N$ where r is a random value in $\mathbb{Z}_N^*$. Since e is prime and $e > N$, we have that $\gcd(e, \phi(N)) = 1$ where $\phi(N)$ is the Euler's totient function. Thus, the map $f : \mathbb{Z}_N^* \to \mathbb{Z}_N^*$ defined as $f(x) = x^e \bmod N$ is a permutation on $\mathbb{Z}_N^*$. Consequently, given that r is a random value in $\mathbb{Z}_N^*$, we have that r^e is also a random value in $\mathbb{Z}_N^*$. Thus, so is $\overline{M}$. This is true regardless of the value of b. Therefore, the random variable $\overline{M}$ follows the same distribution in both worlds, namely a uniform distribution over $\mathbb{Z}_N^*$.

Second, we consider a signature σ resulting from A's interaction with a user. At the end of the protocol, the user verified that σ is an element of $\mathbb{Z}_N^*$ such that $\sigma^e \equiv H(M) \bmod N$. Since $f(x)$ is a permutation over $\mathbb{Z}_N^*$, there is only one such element σ. Therefore, σ is uniquely determined by (N, e, M), and in particular does not contain any information about during which session it was created. Thus, $\mathcal{BS}$ is dishonest-key blind.

As a corollary, it follows that blind MAC schemes with state-sharing users that are at the same time one-more unforgeable (omu-sa secure) and blind (blind-sa secure) exist in the random oracle model if the one-more RSA-inversion problem is hard.

6   Parallel Attacks

We note that all our results can be extended to parallel attacks, i.e. attacks where the adversary can interact with signers, taggers or users in an arbitrarily interleaved way. We refer to the full version [1] for notation and security notions under parallel attacks, and simply summarize the results here.

Since any blind MAC scheme that is secure under parallel attacks is also secure under sequential attacks, our impossibility result of Theorem 4 directly implies that secure blind MACs under parallel attacks do not exist either.

In the state-sharing users setting, the result of Theorem 6 easily extends to parallel attacks: if the underlying blind signature scheme is one-more unforgeable and dishonest-key blind under parallel attacks, then the blind MAC scheme of Construction 5 is one-more unforgeable and blind under parallel attacks. Moreover, since the signing protocol in Chaum's scheme only has two moves, security under sequential and parallel attacks are equivalent, and the result of Theorem 9 holds for parallel attacks as well.

7   Future Work

In forthcoming work, we will further explore the notion of dishonest-key blindness for other schemes than the modified Chaum scheme presented in Section 5. The latter relies on random oracles and the one-more RSA-inversion assumption; we will investigate which other schemes satisfy the stronger notion, and whether a general transformation exists that converts any honest-key blind signature scheme into a dishonest-key blind signature scheme.

As previously stated, the sole purpose of the construction in Section 5 is to demonstrate the existence of blind MAC schemes in the setting in which the users share a common state information. Finding efficient constructions in this setting is left as an open problem. Also, one could investigate the existence of blind MACs in other models, such as a model in which users can collude with a cheating signer, or one in which all users have access to a verification oracle.

Acknowledgements

We would like to thank Mihir Bellare and the anonymous reviewers for their valuable suggestions. The first and third author were supported in part by the French RNRT Project Crypto++ and by the European Commission through the IST Program under Contract IST-2002-507932 ECRYPT. The third author is a Postdoctoral Fellow of the Research Foundation – Flanders (FWO-Vlaanderen), and was supported in part by the Flemish Government under GOA Mefisto 2006/06 and Ambiorix 2005/11, and by the European Commission through the IST Project PRIME.

References

1. Michel Abdalla, Chanathip Namprempre, and Gregory Neven.   On the (im)possibility of blind message authentication codes. Full version of current paper. Available from authors' web pages.
2. Masayuki Abe.   A secure three-move blind signature scheme for polynomially many signatures. In Birgit Pfitzmann, editor, *Advances in Cryptology – EURO-CRYPT 2001*, volume 2045 of *Lecture Notes in Computer Science*, pages 136–151, Innsbruck, Austria, May 6–10, 2001. Springer-Verlag, Berlin, Germany.
3. Manindra Agrawal, Neeraj Kayal, and Nitin Saxena.   PRIMES is in P. http://www.cse.iitk.ac.in/users/manindra/primality.ps, August 2002.
4. Mihir Bellare, Chanathip Namprempre, David Pointcheval, and Michael Semanko. The one-more-RSA-inversion problems and the security of Chaum's blind signature scheme. *Journal of Cryptology*, 16(3):185–215, 2003.
5. Alexandra Boldyreva. Threshold signatures, multisignatures and blind signatures based on the gap-Diffie-Hellman-group signature scheme. In Yvo Desmedt, editor, *PKC 2003: 6th International Workshop on Theory and Practice in Public Key Cryptography*, volume 2567 of *Lecture Notes in Computer Science*, pages 31–46, Miami, USA, January 6–8, 2003. Springer-Verlag, Berlin, Germany.
6. Jan Camenisch, Maciej Koprowski, and Bogdan Warinschi. Efficient blind signatures without random oracles. In Carlo Blundo and Stelvio Cimato, editors, *SCN 04: 4th International Conference on Security in Communication Networks*, Lecture Notes in Computer Science, pages 134–148, Amalfi, Italy, September 8–10, 2005. Springer-Verlag, Berlin, Germany.
7. David Chaum.   Blind signatures for untraceable payments.   In David Chaum, Ronald L. Rivest, and Alan T. Sherman, editors, *Advances in Cryptology – CRYPTO'82*, pages 199–203, Santa Barbara, CA, USA, 1983. Plenum Press, New York, USA.

8. David Chaum. Blind signature system. In David Chaum, editor, *Advances in Cryptology – CRYPTO'83*, page 153, Santa Barbara, CA, USA, 1984. Plenum Press, New York, USA.
9. David Chaum, Amos Fiat, and Moni Naor. Untraceable electronic cash. In Shafi Goldwasser, editor, *Advances in Cryptology – CRYPTO'88*, volume 403 of *Lecture Notes in Computer Science*, pages 319–327, Santa Barbara, CA, USA, August 21–25, 1990. Springer-Verlag, Berlin, Germany.
10. Atsushi Fujioka, Tatsuaki Okamoto, and Kazuo Ohta. A practical secret voting scheme for large scale elections. In Jennifer Seberry and Josef Pieprzyk, editors, *Advances in Cryptology – AUSCRYPT ' 92*, volume 718 of *Lecture Notes in Computer Science*, pages 244–251. Springer-Verlag, Berlin, Germany, 1993.
11. Shafi Goldwasser, Silvio Micali, and Ronald L. Rivest. A digital signature scheme secure against adaptive chosen-message attacks. *SIAM Journal on Computing*, 17(2):281–308, April 1988.
12. Ari Juels, Michael Luby, and Rafail Ostrovsky. Security of blind digital signatures (Extended abstract). In Burton S. Kaliski Jr., editor, *Advances in Cryptology – CRYPTO'97*, volume 1294 of *Lecture Notes in Computer Science*, pages 150–164, Santa Barbara, CA, USA, August 17–21, 1997. Springer-Verlag, Berlin, Germany.
13. Tatsuaki Okamoto. Receipt-free electronic voting schemes for large scale elections. In Bruce Christianson, Bruno Crispo, T. Mark A. Lomas, and Michael Roe, editors, *Security Protocols, 5th International Workshop, Paris, France, April 7-9, 1997, Proceedings*, volume 1361 of *Lecture Notes in Computer Science*, pages 25–35. Springer-Verlag, Berlin, Germany, 1998.
14. Benny Pinkas. Fair secure two-party computation. In Eli Biham, editor, *Advances in Cryptology – EUROCRYPT 2003*, volume 2656 of *Lecture Notes in Computer Science*, pages 87–105, Warsaw, Poland, May 4–8, 2003. Springer-Verlag, Berlin, Germany.
15. David Pointcheval and Jacques Stern. Security arguments for digital signatures and blind signatures. *Journal of Cryptology*, 13(3):361–396, 2000.
16. Berry Schoenmakers. Personal Communication, August 2005.
17. Andrew C. Yao. Protocols for secure computations. In *23rd Annual Symposium on Foundations of Computer Science*, pages 160–164, Chicago, Illinois, November 3–5, 1982. IEEE Computer Society Press.

A   Formal Definition of Dishonest-Key Blindness

The concept of dishonest-key blindness for blind signature schemes is an extension of the classical notion of blindness in which the adversary is allowed to choose the public key used by the user algorithm when trying to break the blindness of the scheme. In particular, in the experiment defining this new notion, no key generation is performed and no key pair is given to the adversary as input to its first phase. Instead, the adversary outputs the public key of its choice along with the challenge messages at the end of its first stage. It is this public key that is given as input to the users during the second phase of the experiment defining dishonest-key blindness.

Definition 10 [Dishonest-key blindness of a blind signature scheme]. Let $\mathcal{BS} = (\mathsf{Kg}, \mathsf{User}, \mathsf{Sign}, \mathsf{Vf})$ be a blind signature scheme. Let $k \in \mathbb{N}$, and let A be an adversary. Consider the following experiment.

Experiment $\mathbf{Exp}^{\text{dk-blind-sa}}_{\mathcal{BS}, \mathsf{A}}(k)$:

$b \xleftarrow{\$} \{0,1\}$

$((M_0, M_1, pk), St_\mathsf{A}) \xleftarrow{\$} \mathsf{A}(\varepsilon, 1^k)$ // A outputs pk of its choice
                                                          // both users use pk output by A during the attack

$(M_\mathsf{A}, St_\mathsf{A}, \tau_b, St_b) \xleftarrow{\$} [\mathsf{A}(St_\mathsf{A}) \leftrightarrow \mathsf{User}((pk, M_b))]$

$(M_\mathsf{A}, St_\mathsf{A}, \tau_{1-b}, St_{1-b}) \xleftarrow{\$} [\mathsf{A}(St_\mathsf{A}) \leftrightarrow \mathsf{User}((pk, M_{1-b}))]$

If $St_0 = \mathtt{fail}$ or $St_1 = \mathtt{fail}$ then $\tau \leftarrow \mathtt{fail}$ else $\tau \leftarrow (\tau_0, \tau_1)$

$d \xleftarrow{\$} \mathsf{A}(\tau, St_\mathsf{A})$

If $b = d$ then return 1 else return 0

The dk-blind-sa-advantage of A in breaking $\mathcal{BS}$ is defined as

$$\mathbf{Adv}^{\text{dk-blind-sa}}_{\mathcal{BS}, \mathsf{A}}(k) = 2 \cdot \Pr\left[\mathbf{Exp}^{\text{dk-blind-sa}}_{\mathcal{BS}, \mathsf{A}}(k) = 1\right] - 1,$$

and $\mathcal{BS}$ is said to be *dishonest-key blind under sequential attacks or* dk-blind-sa-*secure* if $\mathbf{Adv}^{\text{dk-blind-sa}}_{\mathcal{BS}, \mathsf{A}}(k)$ is a negligible function in the security parameter k for all adversaries A with time complexity polynomial in k.   ∎

An Optimal Non-interactive Message Authentication Protocol

Sylvain Pasini and Serge Vaudenay

EPFL, CH-1015 Lausanne, Switzerland
http://lasecwww.epfl.ch

Abstract. Vaudenay recently proposed a message authentication protocol which is interactive and based on short authenticated strings (SAS). We study here SAS-based non-interactive message authentication protocols (NIMAP). We start by the analysis of two popular non-interactive message authentication protocols. The first one is based on a collision-resistant hash function and was presented by Balfanz et al. The second protocol is based on a universal hash function family and was proposed by Gehrmann, Mitchell, and Nyberg. It uses much less authenticated bits but requires a stronger authenticated channel.

We propose a protocol which can achieve the same security as the first protocol but using less authenticated bits, without any stronger communication model, and without requiring a hash function to be collision-resistant. Finally, we demonstrate the optimality of our protocol.

1 Introduction

Message authentication protocols are typically used to exchange public keys so that *secure* communications can be set up. For a better usability, a non-interactive protocol is preferred. It should be noted that the protocol uses two separate channels. The first one is a broadband insecure channel (e.g. an email or a wireless channel) and the second one is a narrowband authenticated channel (e.g. authentication by a human voice or a manual authentication by a human operator).

In SSH and in GPG, the simple folklore protocol used to exchange public keys is presented in Balfanz et al. [BSSW02]. It is non-interactive and based on a collision-resistant hash function. The authenticated string is the k-bit hashed value of the input message m. We recall that this protocol is typically weak against offline attacks, such as birthday attacks, which have a complexity of $2^{k/2}$ and that hash functions which resist to collision attacks are threatened species these days [BCJ$^+$05, WLF$^+$05, WYY05b, WYY05a, WY05]. For instance, it is possible to forge two different RSA keys with the same MD5 hash as shown in [LWdW05, LdW05].

Another protocol is MANA I which was proposed by Gehrmann-Mitchell-Nyberg [GMN04]. It is based on an universal hash function family. This protocol is more resistant against offline attacks since it uses an authenticated value which has a random part K. The second part is the hashed value (using K as key) of the input message m. The protocol requires to send the hashed value "at once". Hence, even if an adversary has an infinite complexity, his probability of success is at most 2^{-k} where k is the size

D. Pointcheval (Ed.): CT-RSA 2006, LNCS 3860, pp. 280–294, 2006.
© Springer-Verlag Berlin Heidelberg 2006

of K and the size of the hash. However, the requirement renders the protocol "less non-interactive" by imposing a strong assumption on the communication model.

We propose a protocol which has the same security than the one presented by Balfanz et al. [BSSW02] but using less authenticated bits and without requiring the hash function to be collision-resistant. Our protocol is based on a trapdoor commitment scheme in the Common Reference String (CRS) model or in the Random Oracle model.

Finally, we propose a definition of the optimality of a message authentication protocol and we analyze the three above protocols.

2   Preliminaries

The considered model is a communication network made up of devices which use insecure broadband communication channels between them. In addition, they can use a narrowband channel which can be used to authenticate short messages, i.e. short authenticated strings (SAS).

Fig. 1. NIMAP Channels

Communication devices are located on nodes n of given identity ID_n and can run several instances which are formally denoted by a unique instance tag π_n^i. We concentrate on non-interactive message authentication protocols (NIMAP).

2.1   Adversarial Model Against NIMAP

A *message authentication protocol* has an input m on the side of the claimant Alice of identity ID and an output $\widehat{\text{ID}}\|\hat{m}$ on the side of the verifier Bob. Authentication is successful if the output is $\widehat{\text{ID}} = \text{ID}$ and $\hat{m} = m$. The protocol is non-interactive if it only uses messages send by Alice to Bob.

We assume that adversaries have full control on the broadband communication channel. Indeed, an attacker can read messages from the channel, he can prevent a message from being delivered, he can delay it, replay it, modify it, and change its recipient address. Here, we adopt the security model from Vaudenay [Vau05] based on Bellare-Rogaway [BR93]. The adversary has full control on which node launches a new instance of a protocol, on the input of the protocol, and on which protocol instance runs a new step of the protocol. Namely, we assume that the adversary has access to a launch(n,r,x) oracle where n is a node, r is a character, Alice or Bob, and x is the input. This oracle returns a unique instance tag π_n^i. Since a node can a priori run concurrent protocols, there may be several instances related to the same node n. The adversary also has access to the oracle receive(π_A^i) which returns a message m which is meant to be sent to Bob and to the oracle send(π_B^i, m) which sends a message m to a given instance of Bob.

Typically, a NIMAP between nodes A and B with input m on the side of Alice and using two messages runs as follows.

1. $\pi_A \leftarrow \mathsf{launch}(A, \mathsf{Alice}, m)$
2. $p_1 \leftarrow \mathsf{receive}(\pi_A)$
3. $p_2 \leftarrow \mathsf{receive}(\pi_A)$
4. $\pi_B \leftarrow \mathsf{launch}(B, \mathsf{Bob}, \emptyset)$
5. $\mathsf{send}(\pi_B, p_1)$
6. $\widehat{\mathsf{ID}} || \hat{m} \leftarrow \mathsf{send}(\pi_B, p_2)$

By convention, we describe protocols by putting a *hat* on the notation for Bob's received messages (i.e. inputs of the send oracle) which are not authenticated since they can differ from Alice's sent messages (i.e. outputs of the receive oracle) in the case of an active attack.

On a global perspective, several $\mathsf{launch}(A_k, \mathsf{Alice}, m_k)$ and $\mathsf{launch}(B_\ell, \mathsf{Bob}, \emptyset)$ can be queried. These queries create several $\pi_{A_k}^{i_k}$ instances of Alice (authentication claims) and several $\pi_{B_\ell}^{j_\ell}$ instances of Bob (authentication verifications). We may have a perfect matching between the k's and ℓ's such that related instances have matching conversations which fully follow the protocol specifications, and the $\pi_{B_\ell}^{j_\ell}$ ends with output $\mathsf{ID}_{A_k} || m_k$ for the matching k. In any other case, we say that an attack occurred. We say that an *attack is successful* if there exists at least an instance $\pi_{B_\ell}^{j_\ell}$ which terminated and output $\widehat{\mathsf{ID}} || \hat{m}$ such that there is no k for which $\widehat{\mathsf{ID}} = \mathsf{ID}_{A_k}$ and $\hat{m} = m_k$. Note that many protocol instances can endlessly stay in an unterminated state or turn in an abort state. We call *one-shot attacks* the attacks which launch a single instance of Alice and Bob. The *attack cost* is measured by

- the number Q of launched instances of Alice, i.e. the *online complexity*.
- the additional complexity C, i.e. the *offline complexity*.
- the probability of success p.

Here is a useful lemma taken from [Vau05].

Lemma 1. *We consider a message authentication protocol with claimant Alice and verifier Bob in which a single SAS is sent. We denote by μ_A (resp. μ_B) the complexity of Alice's (resp. Bob's) part. We consider adversaries such that the number of instances of Alice (resp. Bob) is at most Q_A (resp. Q_B). We further denote T_0 and p_0 their time complexity and probability of success, respectively. There is generic transformation which, for any Q_A, Q_B, and any adversary, transforms it into a one-shot adversary with complexity $T \le T_0 + \mu_A Q_A + \mu_B Q_B$ and probability of success $p \ge p_0 / Q_A Q_B$.*

Assuming that no adversary running a one-shot attack has a probability of success larger than p, using Lemma 1, we can upper bound the probability of success of an attack which uses Q_A, resp. Q_B, instances of Alice, resp. Bob, by $Q_A Q_B p$.

2.2 Authenticated Channels

When referring to "channel", we refer by default to an insecure broadband channel without any assumption. As mentioned before, the devices can use an authenticated channel. An *authenticated channel* is related to a node identity ID. Formally, an authenticated channel from a node n has an identifier ID_n. It allows the recipient of a message to know the identity of the node from which the message has been sent as is. Note that an adversary cannot modify it (i.e. integrity is implicitly protected), but she can delay it, remove it, or replay it, and of course, read it. Precisely, an authenticated channel does not provide confidentiality. By convention, we denote $\text{authenticate}_{\text{ID}_n}(x)$ a message x which has been sent from node n through the authenticated channel.

The receive oracle maintains unordered sets of authenticated messages in every channel ID_n from node n. Only receive oracles with a π_n^i instance can insert a new message in this set. When a send oracle is queried with any message $\text{authenticate}_{\text{ID}_n}(x)$, it is accepted by the oracle only if x is in the set related to channel ID_n. Note that concurrent or successive instances related to the same node write in the same channel, i.e. in the same set. Thus, when an instance of Alice sends a message, Bob can only authenticate the node from which it has been sent, i.e. n, but not the connection to the right instance.

Weak Authenticated Channels. By default, authenticated channels without any other assumption are called *weak*. This means that an adversary can delay a message, remove it, or replay it. In particular, the owner of the message has not the insurance that the message has been delivered to the recipient.

Stronger Authenticated Channels. In some cases we need special assumptions on the authenticated channel. We can consider *stronger authentication channels*, namely channels in which additional properties are achieved as proposed by Vaudenay [Vau05]. In the following, we use one possible property that can be assumed on a stronger authentication channel. A *stall-free transmission* assumes that when a message is released by a receive oracle either it is used as input in the immediately following send oracle query or it is never used. Namely, we cannot wait for a new message from Alice before delivering the authenticated message to Bob.

For instance, a *face to face conversation* and a *telephone call* are clearly authenticated channels. When one talks to the other one, the recipient further knows that the message has not been recorded since interactivity implies coherent conversations (stall-free). *Mail, e-mail,* and *voice mail* can be stalled and released in a different order. Note that an e-mail without any cryptographic appendix such as a GPG signature is in fact not an authenticated channel since it can easily be forged.

2.3 Hash Functions

Collision-Resistant Hash Functions (CRHF). A collision-resistant hash function is a hash function in which it should be hard to find two inputs x and y such that $H(x) = H(y)$ and $x \neq y$. Due to the birthday attacks, the hash length must be at least of 160 bits.

Weakly Collision-Resistant Hash Functions (WCRHF). Weak collision resistance means that the game of Fig. 2 is hard. Assume a (T, ε)-weakly collision-resistant hash function H defined on a finite set X. Any adversary $\mathcal{A}$ bounded by a complexity T wins the WCR game on Fig. 2 with probability at most ε.

Universal Hash Functions Families (UHFF). An ε-universal hash function family is a collection of functions H_K from a message space to a finite set $\{0, 1\}^k$ which depends on a random parameter K such that for any $x \neq y$ we have

$$\Pr[H_K(x) = H_K(y)] \leq \varepsilon$$

where the probability is over the random selection of K.

$$\mathcal{A} \qquad\qquad\qquad\qquad C$$

$$\xleftarrow{\qquad x \qquad} \text{pick } x \in_U X$$

$$\xrightarrow{\qquad y \qquad}$$

winning condition: $H(y) = H(x)$ and $y \neq x$

Fig. 2. WCR game

2.4   Commitment Schemes

We can formalize a *commitment scheme* by two algorithms commit and open. For any message m we have $(c,d) \leftarrow \text{commit}(m)$. The c value is called the *commit* value and the d value the *decommit* value. Knowing both c and d, the message can be recovered using the open oracle, i.e. $m \leftarrow \text{open}(c,d)$. Intuitively, a commitment scheme should be *hiding*, meaning that for any c, it is hard to deduce any information about the corresponding message m, and *binding*, meaning that one cannot find c,d,d' such that (c,d) and (c,d') open to two different messages. We also introduce *keyed commitment schemes* which have in addition a setup oracle to initialize a pair of keys, i.e. $(K_p, K_s) \leftarrow \text{setup}()$. The public key K_p is used in commit and open oracles. Keyed commitment schemes should be understood as working in the Common Reference String (CRS) model. Namely, K_p is a common reference string set up once for all and K_s is unknown to anyone.

Binding Property. The semantic binding (SB) game of Fig. 3 must be hard, i.e. for any message m and any commit value c one cannot find two decommit values d and d' such that $m \leftarrow \text{open}(K_p, c, d)$ and $m' \leftarrow \text{open}(K_p, c, d')$ with $m \neq m'$. The scheme is (T, ε)-semantically binding if any adversaries $\mathcal{A}$ bounded by a complexity T has a probability to find two decommit values d and d' which is at most ε.

$$\mathcal{A} \qquad\qquad\qquad\qquad C$$

$$\xleftarrow{\qquad K_p \qquad} (K_p, K_s) \leftarrow \text{setup}()$$

$$\xrightarrow{\quad m||c||d||d' \quad} m \leftarrow \text{open}(K_p, c, d)$$

$$m' \leftarrow \text{open}(K_p, c, d')$$

Winning condition: $m, m' \neq \perp$ and $m' \neq m$

Fig. 3. SB Game

Trapdoor Commitment Model. The notion of *trapdoor commitment* was introduced by Brassard, Chaum, and Crepeau [BCC88]. We define (T, ε)-trapdoor commitment schemes by four algorithms setup, commit, open, and equivocate. The first three work as before. The algorithm equivocate defeats the binding property by using the secret key K_s. More precisely, for any $(K_p, K_s) \leftarrow \text{setup}()$ we have

- for any m and any $(c,d) \leftarrow \text{commit}(K_p, m)$ we have $m \leftarrow \text{open}(K_p, c, d)$,
- for any m, by running $(c,d) \leftarrow \text{commit}(K_p, m)$, c is uniformly distributed,
- for any m, any $\hat{c}$, and any $\hat{d} \leftarrow \text{equivocate}(K_s, m, \hat{c})$, the $\text{open}(K_p, \hat{c}, \hat{d})$ algorithm yields m.
- for any adversary bounded by a complexity T in the SB game, the winning probability is smaller than ε.

Note that this primitive is a particular case of *strongly equivocable commitment* as described by Damgård-Groth [DG03].

Trapdoor commitment schemes are perfectly hiding and computationnaly binding commitment schemes. Note that for any (K_p, K_s) and any m, the distribution of (c, d), which has been yield using the commit algorithm, is equal to the distribution of $(\hat{c}, \hat{d})$, which have been yield choosing a $\hat{c}$ with uniform distribution and using the equivocate algorithm.

For instance, a trapdoor commitment based on the discrete logarithm problem was proposed by Boyar and Kurtz [BK90]. Another trapdoor commitment scheme was proposed by Catalano et al. [CGHGN01] based on the Paillier's trapdoor permutation [Pai99]. The proposed scheme uses an RSA modulus $N = pq$ and a value $h \in \mathbb{Z}_{N^2}$ such that its order is a multiple of N. The public key is $K_p = (N, h)$ and the private key is $K_s = (p, q)$. The commit algorithm of a message m picks uniformly two random values r, s and outputs $c \leftarrow (1 + mN)r^N h^s \bmod N^2$ and $d = (r, s)$. Note that the commit value c is uniformly distributed for any m since r and s are uniformly distributed and $(r, s) \mapsto r^N h^s \bmod N^2$ is the Paillier trapdoor permutation (see [Pai99]). We denote $\mathcal{F}_h(r, s)$ this permutation. The decommit algorithm simply checks that $c = \mathrm{commit}(K_p, m)$ with $d = (r, s)$. The trapdoor is the collision-finding function: given a commit $\hat{c}$ and a message m, one can find $\hat{d} = (\hat{r}, \hat{s})$ such that $\hat{c} = (1 + mN)\mathcal{F}_h(\hat{r}, \hat{s}) \bmod N^2$ by using the trapdoor on the Paillier permutation and knowing p, q, i.e. $(\hat{r}, \hat{s}) \leftarrow \mathcal{F}_h^{-1}(\hat{c}(1 + mN)^{-1})$. Thus, given a $\hat{c}$, an adversary can find $\hat{d}$ for any message m and thus defeats the binding property.

Oracle Trapdoor Commitment. Finally, we consider trapdoor commitment schemes in which commit, open, and equivocate are given as oracles (and not as algorithms). In such cases, access to equivocate with an input $\hat{c}$ equal to any c which was output by commit is prohibited.

There is a very simple oracle trapdoor commitment scheme in the random oracle model:

- The setup() algorithm is unused.
- The commit(m) oracle with input message m in $\{0, 1\}^k$ picks a random value e in $\{0, 1\}^\ell$, builds $d \leftarrow (m, e)$, and calls the random oracle $c \leftarrow H(m, e)$.
- The open(c, d) oracle simply extracts m from d and checks that $c = H(m, e)$.
- The equivocate(m, c) oracle yields a decommit value $d = (m, e)$ such that $c = H(m, e)$ by modifying the table of H. This is possible without modifying the final distribution of H, except with probability less than $(Q + C)(2^{-\ell} + 2^{-k})$ since c is independent from previous oracle calls.

3   Previous Non-interactive Authentication Protocols

3.1   A NIMAP Based on a Collision-Resistant Hash Functions

We first present a protocol taken from Balfanz et al. [BSSW02] based on a collision resistant hash function.

Note that the authenticated string is constant for all instances of the protocol which use the same input m, i.e. the authenticated string is $H(m)$. This characteristic allows

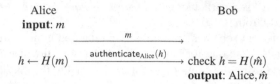

Alice   Bob
input: m

$$\xrightarrow{\quad\quad m \quad\quad}$$

$h \leftarrow H(m) \xrightarrow{\text{authenticate}_{\text{Alice}}(h)}$ check $h = H(\hat{m})$
                                                   output: Alice, $\hat{m}$

Fig. 4. Non-Interactive Message Authentication using a CRHF

adversaries to run completely offline attacks. An attacker has *simply* to find a collision on the hash function between two messages m_1 and m_2 and then succeeds with probability 1.

Theorem 2 ([Vau05]). *Let μ be the overall time complexity of the message authentication protocol in Fig. 4 using weak authentication. We denote by T, Q, and p the time complexity, number of oracle queries launch, and probability of success of adversaries, respectively. There is a generic transformation which transforms any adversary into a collision finder on H whose complexity is $T + \mu Q$ and probability of success is p.*

In short, the best known offline attack against this protocol is the collision attack. An adversary has a probability of success of $1 - e^{-\frac{1}{2}T^2 2^{-k}}$ by using T hashes computations. It clearly succeeds for $T = O(2^{k/2})$. Collision resistance requires the number of authenticated bits to be at least 160 and cannot be reduced considering offline attacks and using only weak authentication.

3.2 A NIMAP with Strong Authentication

The Gehrmann-Mitchell-Nyberg MANA I [GMN04] protocol is depicted in Fig. 5.[1]

MANA I uses a universal hash function family H. Proposed constructions lead to 16–20 bit long SAS values but require strong authentication. Indeed, using weak authentication, an adversary who gets authenticate$(K\|\mu)$ has enough time to find a message $\hat{m}$ such that $\mu = H_K(\hat{m})$ and to substitute m with $\hat{m}$. We can also achieve security with a stronger authenticated channel which achieves stall-free transmissions.

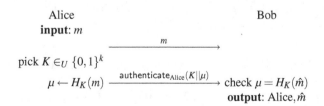

Alice   Bob
input: m

$$\xrightarrow{\quad\quad m \quad\quad}$$

pick $K \in_U \{0,1\}^k$
$\mu \leftarrow H_K(m) \xrightarrow{\text{authenticate}_{\text{Alice}}(K\|\mu)}$ check $\mu = H_K(\hat{m})$
                                                   output: Alice, $\hat{m}$

Fig. 5. The MANA I Protocol

---

[1] Note that the original MANA I protocol is followed by an authenticated acknowledgment from Bob to Alice in [GMN04].

Theorem 3. *Given an ε-universal hash function family H, any adversary which is bounded by a complexity T and by Q_A (resp. Q_B) instances of Alice (resp. Bob) against the protocol of Fig. 5 using stall-free authentication has a probability of success at most $Q_A Q_B ε$.*

Proof. A one-shot adversary has no advantage to send $\hat{m}$ before it has received m and he cannot send $\hat{m}$ after $K\|\mu$ is released. Indeed, he would not be able to send $\hat{m}$ after receiving $K\|\mu$ due to the stall-free assumption. Thus, the attacker must select m and $\hat{m}$ and hope that $H_K(\hat{m}) = H_K(m)$. Clearly, the assumption on H limits the probability of success to ε.

Now, consider powerful adversaries. Using Lemma 1, we can deduce that the probability of success of an adversary is at most $Q_A Q_B ε$.   □

4   A Proposed NIMAP with Weak Authentication

Consider the protocol depicted on Fig. 6 in which the message m is transmitted by sending $(c,d) \leftarrow \text{commit}(K_p,m)$. This message can be recovered by anyone using the open function. To authenticate this message, the hashed value of c is sent using an authenticated channel. We prove that this protocol is secure with authenticated strings which can be shorter than in the protocol of Fig. 4. Non-deterministic commitment scheme is the heart of the protocol since an attacker cannot predict the c value and thus cannot predict the $H(c)$ value which is the authenticated one.

Alice   Bob

input: m

$(c,d) \leftarrow \text{commit}(K_p,m)$ $\xrightarrow{\quad c\|d \quad}$ $\hat{m} \leftarrow \text{open}(K_p,\hat{c},\hat{d})$

$h \leftarrow H(c)$ $\xrightarrow{\text{authenticate}_{\text{Alice}}(h)}$ check $h = H(\hat{c})$

output: Alice, $\hat{m}$

Fig. 6. Non-Interactive Message Authentication Based on a WCRHF

Lemma 4. *Consider the message authentication protocol depicted in Fig. 6. We assume that the function H is a $(T + \mu, ε_h)$-weakly collision resistant hash function and the commitment scheme is a $(T + \mu, ε_c)$-trapdoor commitment scheme in the CRS model (resp. oracle commitment scheme). There exists a (small) constant μ such that for any T, any one-shot adversary against this message authentication protocol with complexity bounded by T has a probability of success p smaller than $ε_h + ε_c$.*

Recall that the c value is sent through the insecure broadband channel and thus has not to be minimized. Thus, we can use an $ε_c$ as small as desired since we can use any commitment scheme as secure as desired.

Assuming that H is optimally WCR, the best WCR attack using T hash computations has a probability of success $ε_h \approx 1 - e^{-T2^{-k}}$. So, we need $T = \Omega(2^k)$ to succeed with a one-shot attack. Thus, using the same amount of authenticated bits as the protocol of Fig. 4, our protocol has a better resistance against offline attacks. Equivalently, we can achieve the same security as the protocol of Fig. 4, but using only half amount of authenticated bits, e.g. 80 bits.

Proof. A one-shot adversary $\mathcal{A}$ against the protocol in Fig. 6 follows the game depicted on Fig. 7(a) in which it runs a man-in-the middle attack. Clearly, it can be reduced to an adversary $\mathcal{A}$ who plays the game described in Fig. 7(b).

$$
\begin{array}{ccc}
K_p & K_p & K_p \\
\downarrow & \downarrow & \downarrow \\
\text{Alice} & \mathcal{A} & \text{Bob}
\end{array}
$$

$$\xleftarrow{\quad m \quad}$$

$(c,d) \leftarrow \mathsf{commit}(K_p,m) \xrightarrow{\ c\|d\ } \qquad \xrightarrow{\ \hat{c}\|\hat{d}\ } \hat{m} \leftarrow \mathsf{open}(K_p,\hat{c},\hat{d})$

$h \leftarrow H(c) \xrightarrow{\qquad\qquad h \qquad\qquad}$

Winning condition: $H(\hat{c}) = h$ and $\hat{m} \neq m$

(a)

$$
\begin{array}{ccc}
\mathcal{A} & & C
\end{array}
$$

$\xleftarrow{\quad K_p \quad} (K_p, K_s) \leftarrow \mathsf{setup}()$

$\xrightarrow{\quad m \quad}$

$\xleftarrow{\quad c\|d \quad} (c,d) \leftarrow \mathsf{commit}(K_p,m)$

$\xrightarrow{\quad \hat{c}\|\hat{d} \quad} \hat{m} \leftarrow \mathsf{open}(K_p,\hat{c},\hat{d})$

Winning condition: $H(\hat{c}) = H(c)$ and $m \neq \hat{m}$

(b)

Fig. 7. Game Against the Proposed Protocol (a) and Reduced Game (b)

Assume a one-shot adversary $\mathcal{A}$ bounded by a complexity T. Given c, the adversary $\mathcal{A}$ has to find a $\hat{c}$ such that $H(\hat{c}) = H(c)$. In addition, it must find a $\hat{d}$ which opens to $\hat{m}$ (using $\hat{c}$) which is different from the input m. He can of course choose a $\hat{c}$ either equal or either different to c. We study the two cases.

Case 1. ($\hat{c} = c$) The adversary $\mathcal{A}$ chooses $\hat{c}$ equal to c and obviously fulfills the condition $H(\hat{c}) = H(c)$. As depicted on Fig. 8, we can reduce the adversary $\mathcal{A}$ to an adversary against the binding game of Fig. 6. We use an algorithm $\mathcal{B}$ bounded by complexity μ which plays the binding game with a challenger C on one side and simulates a challenger for $\mathcal{A}$ on the other side at the same time. Using adversary $\mathcal{A}$ and algorithm $\mathcal{B}$, we construct an adversary $\mathcal{AB}$ which plays the binding game. Note that adversary $\mathcal{AB}$ has a complexity bounded by $T + \mu$.

First, the challenger C generates the pair of keys (K_p, K_S) and sends K_p to $\mathcal{B}$. $\mathcal{B}$ sends it to $\mathcal{A}$ and receives a message m from $\mathcal{A}$. He computes (c,d) using the commit function with K_p and sends $c\|d$ to $\mathcal{A}$. As assumed, $\mathcal{A}$ chooses a $\hat{c}$ equal to c and also sends $\hat{c}\|\hat{d}$ to $\mathcal{B}$. $\mathcal{B}$ can now deduce $\hat{m}$ using the open function with inputs c and $\hat{d}$. Finally, $\mathcal{B}$ sends all required values to the challenger C.

$$\mathcal{A} \qquad\qquad\qquad\qquad \mathcal{B} \qquad\qquad\qquad\qquad C$$

$$\xleftarrow{\quad K_p \quad} \qquad\qquad \xleftarrow{\quad K_p \quad} \quad (K_p, K_s) \leftarrow \mathsf{setup}()$$

$$\xrightarrow{\quad m \quad}$$

$$\xleftarrow{\quad c \| d \quad} \quad (c,d) \leftarrow \mathsf{commit}(K_p, m)$$

$$(\hat{c} = c) \xrightarrow{\quad \hat{c} \| \hat{d} \quad} \quad \hat{m} \leftarrow \mathsf{open}(K_p, c, \hat{d}) \quad \xrightarrow{\; m \| \hat{m} \| c \| d \| \hat{d} \;} \; m = \mathsf{open}(K_p, c, d)$$
$$\hat{m} = \mathsf{open}(K_p, c, \hat{d})$$

Winning condition: $\hat{m}, m \neq \perp$ and $\hat{m} \neq m$

Fig. 8. Reduction to the SB game ($\hat{c} = c$)

$$\mathcal{A} \qquad\qquad\qquad\qquad \mathcal{B} \qquad\qquad\qquad\qquad C$$

$$\xleftarrow{\quad K_p \quad} \quad (K_p, K_s) \leftarrow \mathsf{setup}()$$

$$\xrightarrow{\quad m \quad}$$

$$\xleftarrow{\quad c \quad} \quad \mathsf{pick}\ c \in_U C$$

$$\xleftarrow{\quad c \| d \quad} \quad d \leftarrow \mathsf{equivocate}_{K_s}(m, c)$$

$$\xrightarrow{\quad \hat{c} \| \hat{d} \quad} \quad \hat{m} \leftarrow \mathsf{open}_{K_p}(\hat{c}, \hat{d})$$

$$\xrightarrow{\quad \hat{c} \quad}$$

Winning condition: $H(\hat{c}) = H(c)$ and $m \neq \hat{m}$

Fig. 9. Reduction to the WCR Game with Trapdoor Commitment ($\hat{c} \neq c$)

Note that $\mathcal{B}$ simulates perfectly a challenger for $\mathcal{A}$. Hence, $\mathcal{A}$ and $\mathcal{AB}$ win their respective game at the same time. Consequently, both win with the same probability of success. Recall that the probability of success of an adversary bounded by a complexity $T + \mu$ against the binding game of Fig. 6 is smaller than ε_c when the commitment scheme is a $(T + \mu, \varepsilon_c)$-trapdoor commitment. Hence, the probability that $\mathcal{A}$ succeeds and $c = \hat{c}$ is at most ε_c. Note that this case equally applies to trapdoor commitment schemes.

Case 2. ($\hat{c} \neq c$) The adversary $\mathcal{A}$ searches a $\hat{c}$ different from c. As depicted on Fig. 9, we can reduce the adversary $\mathcal{A}$ to an adversary against a second preimage search game. We use an algorithm $\mathcal{B}$ bounded by a complexity μ with the help of one query to the equivocate oracle. $\mathcal{B}$ plays the second preimage game with a challenger C on one side and simulate a challenger for $\mathcal{A}$ on the other side at the same time. Using adversary $\mathcal{A}$ and algorithm $\mathcal{B}$, we construct an adversary $\mathcal{AB}$ which plays the second preimage game with the challenger C. Note that adversary $\mathcal{AB}$ has a complexity bounded by $T + \mu$.

First, $\mathcal{B}$ generates the keys and sends K_p to $\mathcal{A}$. $\mathcal{B}$ receives a message m from $\mathcal{A}$ and receives a challenge c from C. $\mathcal{B}$ can deduce the decommit value d by calling the oracle $\mathsf{equivocate}(m, c)$. Note that c has been picked uniformly and consequently the distribution of (c, d) is the same as if they have been yield by the commit algorithm. Then, $\mathcal{B}$ can send $c \| d$ to $\mathcal{A}$. $\mathcal{A}$ sends a $\hat{c} \| \hat{d}$ to $\mathcal{B}$. Finally, $\mathcal{B}$ sends it to the challenger C.

Note that $\mathcal{B}$ simulates perfectly a challenger for $\mathcal{A}$. Hence, $\mathcal{A}$ and $\mathcal{AB}$ win their respective game at the same time and consequently with the same probability of success. Recall that the probability of success of an adversary against a second preimage game bounded by a complexity $T + \mu$ is smaller than ε_h when H is a $(T + \mu, \varepsilon_h)$-weakly collision-resistant hash function. Hence, the probability that $\mathcal{A}$ succeeds and $c \neq \hat{c}$ is at most ε_h. Note that the proof equally applies to oracle commitment schemes since it is unlikely that the challenge c was output by a commit oracle.

We conclude that any one-shot adversary bounded by a complexity T against the protocol of Fig. 6 has a probability of success smaller than $\varepsilon_c + \varepsilon_h$ when the protocol uses a $(T + \mu, \varepsilon_h)$-weakly collision resistant hash function H and a $(T + \mu, \varepsilon_c)$-trapdoor commitment scheme. □

We consider now powerful adversaries.

Theorem 5. *Consider the message authentication protocol of Fig. 6. We assume that the function H is a $(T + \mu, \varepsilon_h)$-weakly collision resistant hash function and the commitment scheme is a $(T + \mu, \varepsilon_c)$-trapdoor commitment scheme in the CRS model (resp. oracle commitment scheme). There exists a (small) constant μ such that for any T, any adversary against this message authentication protocol with complexity bounded by T and with number of Alice's (resp. Bob's) instances bounded by Q_A (resp. Q_B) has a probability of success p at most $Q_A(\varepsilon_h + \varepsilon_c)$.*

Assuming that WCR hash functions and trapdoor commitments such that $\varepsilon_c \ll \varepsilon_h = O(T2^{-k})$ exist, we have $p = O(T \cdot Q_A 2^{-k})$. As an example, assuming that an adversary is limited to $Q_A \leq 2^{10}$, $T \leq 2^{70}$, and that the security level requires $p \leq 2^{-20}$, the protocol of Fig. 4 requires $k \geq 160$ and our protocol requires $k \geq 100$. Using MD5 [Riv92], our protocol still achieves a quite luxurious security even though collisions have been found on MD5 [WY05].

Proof. Consider an adversary who launches Q_A instances of Alice and Q_B instances of Bob. Clearly, we can simulate all instances of Bob, pick one who will make the attack succeed, and launch only this one. Hence, we reduce to $Q_B = 1$. Recall from Lemma 4 that any one-shot adversary has a probability of success smaller than $\varepsilon_h + \varepsilon_c$. Using Lemma 1, we conclude that any adversary has a probability of success at most $Q_A(\varepsilon_h + \varepsilon_c)$. □

5   On the Required Entropy of Authenticated Communications

Using a weak authenticated channel, adversaries can delay or replay authenticated messages. With non-interactive protocols an adversary can run the catalog attack: i.e. he launches several instances of Alice and recover many authenticated SAS. He launches one Bob and use one SAS of the catalog.

We would like to upper bound the security of an arbitrary message authentication protocol given the amount of authenticated strings it uses. Assume that the protocol is used between Alice and Bob. We suppose that the protocol can use any sequence of authenticated messages in a given set S during the protocol. We call it a *transcript*. Note that authenticated strings are interleaved with regular messages which are not represented in the transcript. For any input message m, the used transcript during a protocol instance is picked in the set S of all possible transcripts with a distribution $\mathcal{D}_m$.

Theorem 6. *We consider an arbitrary message authentication protocol between Alice and Bob which uses an authenticated channel. Let S be the set of all possible protocol transcripts through the authentication channel for any input message. Let s be its cardinality. There exists a generic one-shot attack with probability of success at least $\frac{1}{s} - 2^{-t}$ which runs in polynomial time in terms of t.*

Proof. We consider a general man-in-the-middle attack in which the adversary first picks $m \in_U \{0,1\}^t$ and $\hat{m} \in_U \{0,1\}^t$ and launches Alice with input m. The attack runs synchronized protocols between Alice and a simulator for Bob, and a simulator for Alice with input $\hat{m}$ and Bob. Following the attack, every authenticated message which must be sent by the simulator is replaced by an authenticated message which has just been received by the simulator.

Let SAS_m be the (random) sequence of all authenticated strings (the transcript) which would be exchanged in the protocol between Alice and the simulator if the simulator where honest, and $SAS_{\hat{m}}$ be the similar sequence between the simulator and Bob. Clearly, if $SAS_{\hat{m}} = SAS_m$, the attack succeeds. Note that an attack makes sense only if $\hat{m}$ is different of m.

We have

$$\Pr[\text{success}] = \Pr[SAS_m = SAS_{\hat{m}} \text{ and } m \neq \hat{m}]$$
$$\geq \Pr[SAS_m = SAS_{\hat{m}}] - \Pr[m = \hat{m}].$$

Note that SAS_m and $SAS_{\hat{m}}$ are two identically distributed independent random variables whose support are included in S. Due to Lemma 8 (see Appendix) we can write $\Pr[SAS_m = SAS_{\hat{m}}] \geq \frac{1}{s}$. Since m and $\hat{m}$ are uniformly distributed in $\{0,1\}^t$, we have $\Pr[m = \hat{m}] = 2^{-t}$. Finally, we obtain

$$\Pr[\text{success}] \geq \frac{1}{s} - 2^{-t}$$

with equality if and only if the SAS distribution is uniform among the set S. $\square$

We finally provide a generic attack in the general case.

Theorem 7. *We consider an arbitrary NIMAP between Alice and Bob which uses a weak authenticated channel. Let S be the set of all possible protocol transcripts through the authentication channel for any input message. Let s be its cardinality. There exists a generic attack which uses Q_A instances of Alice and an offline complexity $O(T)$ with probability of success approximately $1 - e^{-\frac{T \cdot Q_A}{s}}$.*

Proof (Sketch). We consider the generic attack in which the adversary starts by simulating T Alice instances launched with random inputs $\hat{m}_i$ and obtains a list of possible SAS, i.e. $\widehat{SAS}_i$. Then, he launches Q_A real instances of Alice with random inputs m_j and consequently obtains Q_A authenticated SAS, i.e. SAS_j. The attack succeeds when at least one authenticated SAS released by Alice corresponds to a computed one, i.e.

there exists k, ℓ such that $SAS_k = \widehat{SAS}_\ell$. The adversary can launch a single Bob with input $\hat{m}_\ell$ by simulating Alice and can use SAS_k for the authentication when needed.

If the distribution of all SAS is uniform, we have a birthday effect and thus the probability of success is approximately $1 - e^{-\frac{T \cdot Q_A}{s}}$. When the distribution is not uniform, the probability is even larger (see Appendix B of [Pas05]). $\square$

Theorem 6 says that there exists a one-shot attack against *any* message authentication protocol which succeeds with probability essentially $\frac{1}{s}$ where s is the size of S. Theorem 7 says that there exists a generic attack against *any* NIMAP which uses a weak authenticated channel which succeeds with probability essentially $1 - e^{-\frac{T \cdot Q_A}{s}}$ where Q_A is the number of instances of Alice used. Hence, they cannot be secure unless $T \cdot Q_A$ is negligible against s. Thus, any NIMAP which is secure for $T \cdot Q_A \ll s$ is optimal.

Consequently, our proposed protocol is optimal due to Theorem 5 provided that WCR hash functions and trapdoor commitment schemes such that $\varepsilon_c \ll \varepsilon_h = O(T2^{-k})$ exist. By comparison with our protocol, we can note that the protocol of Fig. 4 is not optimal.

6 Applications

One key issue in cryptography is to setup secure communications over insecure channels, such as Internet. We know that using public key cryptography it is possible by exchanging public keys in an authenticated way. The proposed protocol is used in this case for public key authentications, e.g. GPG public keys. Typical applications where public key cryptography is used, and consequently public key authentication is required, are

- distant hosts authentication, e.g. SSH
- e-mail authentication, e.g. GPG signature
- secure e-mail, e.g. GPG encryption
- secure voice over IP, e.g. PGPfone

Another possible application can be authentication of legal documents. For instance, if two persons would exchange a document without complex appendix, such as GPG signature, they can simply send the corresponding commit and decommit values and then authenticate the hashed commit value. The recipient can check whether or not it is correct. Note that integrity is protected.

7 Conclusion

In this paper, we have proposed a new non-interactive message authentication protocol based on a commitment scheme. It has the same security as the currently used in SSH against one-shot attacks but using only half authenticated bits, e.g. 80 bits. 100 bits only are required against more general attacks. Indeed, due to the commitment scheme, the authenticated value is not foreseeable and the protocol is resistant to collision attacks. The latter theorem proposes that our protocol is optimal. We can in addition conclude on the non-optimality of the protocol used today, but the question about MANA I is still opened. Finally, we stress that the security of our protocol relies essentially on the hardness of the SB game of the commitment scheme and on the hardness on the WCR game of the hash function.

References

[BCC88]   Gilles Brassard, David Chaum, and Claude Crépeau. Minimum disclosure proofs of knowledge. *Journal of Computer and System Sciences*, 37(2):156–189, 1988.

[BCJ+05]   Eli Biham, Rafi Chen, Antoine Joux, Patrick Carribault, Christophe Lemuet, and William Jalby. Collisions of SHA-0 and reduced SHA-1. In *Advances in Cryptology – EUROCRYPT '05: 24th Annual International Conference on the Theory and Applications of Cryptographic Techniques*, Lecture Notes in Computer Science, pages 36–57, Aarhus, Denmark, 2005. Springer-Verlag.

[BK90]   Joan F. Boyar and Stuart A. Kurtz. A discrete logarithm implementation of perfect zero-knowledge blobs. *Journal of Cryptology*, 2(2):63–76, 1990.

[BR93]   Mihir Bellare and Phillip Rogaway. Entity authentication and key distribution. In Douglas R. Stinson, editor, *Advances in Cryptology – CRYPTO '93: 13th Annual International Cryptology Conference*, volume 773 of *Lecture Notes in Computer Science*, pages 232–249, Santa Barbara, California, U.S.A., 1993. Springer-Verlag.

[BSSW02]   Dirk Balfanz, Diana K. Smetters, Paul Stewart, and H. Chi Wong. Talking to strangers: Authentication in ad-hoc wireless networks. In *Proceedings of Network and Distributed System Security Symposium 2002 (NDSS'02)*, San Diego, California, U.S.A, February 2002.

[CGHGN01]   Dario Catalano, Rosario Gennaro, Nick Howgrave-Graham, and Phong Q. Nguyen. Paillier's cryptosystem revisited. In *CCS '01: Proceedings of the 8th ACM conference on Computer and Communications Security*, pages 206–214, Philadelphia, Pennsylvania, U.S.A., 2001. ACM Press.

[DG03]   Ivan Damgård and Jens Groth. Non-interactive and reusable non-malleable commitment schemes. In *STOC '03: Proceedings of the thirty-fifth annual ACM symposium on Theory of computing*, pages 426–437, San Diego, California, U.S.A., 2003. ACM Press.

[GMN04]   Christian Gehrmann, Chris J. Mitchell, and Kaisa Nyberg. Manual authentication for wireless devices. *RSA Cryptobytes*, 7(1):29–37, January 2004.

[LdW05]   Arjen K. Lenstra and Benne de Weger. On the possibility of constructing meaningful hash collisions for public keys. In Colin Boyd and Juan Manuel González Nieto, editors, *ACISP '05: The 10th Australasian Conference on Information Security and Privacy*, volume 3574 of *Lecture Notes in Computer Science*, pages 267–279, Brisbane, Australia, 2005. Springer-Verlag.

[LWdW05]   Arjen Lenstra, Xiaoyun Wang, and Benne de Weger. Colliding X.509 certificates. Cryptology ePrint Archive, Report 2005/067, 2005. http://eprint.iacr.org/.

[Pai99]   Pascal Paillier. Public-key cryptosystems based on composite degree residuosity classes. In Jacques Stern, editor, *Advances in Cryptology – EUROCRYPT '99: International Conference on the Theory and Application of Cryptographic Techniques*, volume 1592 of *Lecture Notes in Computer Science*, pages 223–238, Prague, Czech Republic, May 1999. Springer.

[Pas05]   Sylvain Pasini. Secure communications over insecure channels using an authenticated channel. Master's thesis, Swiss Federal Institute of Technology (EPFL), 2005. http://lasecwww.epfl.ch/php_code/publications/search.php?ref=Pas05.

[Riv92]   Ronald L. Rivest. The MD5 message digest algorithm. Technical Report Internet RFC-1321,IETF, 1992.

[Vau05]   Serge Vaudenay. Secure communications over insecure channels based on short authenticated strings. In Victor Shoup, editor, *Advances in Cryptology – CRYPTO '05: The 25th Annual International Cryptology Conference*, volume 3621 of *Lecture Notes in Computer Science*, pages 309–326, Santa Barbara, California, U.S.A., August 2005. Springer-Verlag.

[WLF+05]   Xiaoyun Wang, Xuejia Lai, Dengguo Feng, Hui Chen, and Xiuyuan Yu. Cryptanalysis of the hash functions MD4 and RIPEMD. In Ronald Cramer, editor, *Advances in Cryptology – EUROCRYPT '05: 24th Annual International Conference on the Theory and Applications of Cryptographic Techniques*, Lecture Notes in Computer Science, pages 1–18, Aarhus, Denmark, 2005. Springer-Verlag.

[WY05]   Xiaoyun Wang and Hongbo Yu. How to break MD5 and other hash functions. In Ronald Cramer, editor, *Advances in Cryptology – EUROCRYPT '05: 24th Annual International Conference on the Theory and Applications of Cryptographic Techniques*, Lecture Notes in Computer Science, pages 19–35, Aarhus, Denmark, 2005. Springer-Verlag.

[WYY05a]   Xiaoyun Wang, Yiqun Yin, and Hongbo Yu. Finding collisions in the full SHA1. In Victor Shoup, editor, *Advances in Cryptology – CRYPTO '05: The 25th Annual International Cryptology Conference*, volume 3621 of *Lecture Notes in Computer Science*, pages 17–36, Santa Barbara, California, U.S.A., 2005. Springer-Verlag.

[WYY05b]   Xiaoyun Wang, Xiuyuan Yu, and L. Y. Yin. Efficient collision search attacks on SHA-0. In Victor Shoup, editor, *Advances in Cryptology – CRYPTO '05: The 25th Annual International Cryptology Conference*, volume 3621 of *Lecture Notes in Computer Science*, pages 1–16, Santa Barbara, California, U.S.A., 2005. Springer-Verlag.

Appendix

Lemma 8. *Let X and Y be two identically distributed independent random variables with distribution D over a support set S. We have*

$$\Pr[X = Y] \geq \frac{1}{\#S} \tag{1}$$

with equality if and only if D is the uniform distribution.

Proof. Let s be the size of the set S. We have

$$\Pr[X = Y] = \sum_{S_i \in S} \Pr[X = S_i] \cdot \Pr[Y = S_i] = \sum_{S_i \in S} p_i^2$$

where p_i is $\Pr[X = S_i]$.

Let us write $p_i = \frac{1}{s} + \rho_i$. Thus, we obtains

$$\sum_{S_i \in S} p_i^2 = (\frac{1}{s})^2 \sum_{S_i \in S} 1 + 2 \frac{1}{s} \sum_{S_i \in S} \rho_i + \sum_{S_i \in S} \rho_i^2.$$

Knowing that the sum of p_i equals to 1, we can easily deduce that the sum of ρ_i equals 0. Thus, $\sum_{S_i \in S} p_i^2$ equals $\frac{1}{s} + \sum_{S_i \in S} \rho_i^2$. The sum of ρ_i^2 is greater or equal to 0. Note that it is equal to 0 if and only if all ρ_i are null, i.e. D is uniform.   □

A New Criterion for Nonlinearity of Block Ciphers

Orr Dunkelman[1,*] and Nathan Keller[2]

[1] Computer Science Department, Technion,
Haifa 32000, Israel
orrd@cs.technion.ac.il
[2] Einstein Institute of Mathematics, Hebrew University,
Jerusalem 91940, Israel
nkeller@math.huji.ac.il

Abstract. For years, the cryptographic community has searched for good nonlinear functions. Bent functions, almost perfect nonlinear functions, and similar constructions have been suggested as a good base for cryptographic applications due to their highly nonlinear nature. In the first part of this paper we study these functions as block ciphers, and present several distinguishers between almost perfect nonlinear permutations and random permutations. The data complexity of the best distinguisher is $O(2^{n/3})$ and its time complexity is $O(2^{2n/3})$ for an n-bit block size, independent of the key size.

In the second part of the paper we suggest a criterion to measure the effective linearity of a given block cipher. We devise a distinguisher for general block ciphers based on their effective linearity. Finally, we show that for several constructions, our distinguishing attack is better than previously known techniques.

Keywords: Almost perfect nonlinear permutations, highly nonlinear functions, effective linearity, differential cryptanalysis.

1 Introduction

For years, highly nonlinear functions were extensively used in various cryptographic applications. Highly nonlinear functions were promoted since algorithms that are close to linear are susceptible to various approximation attacks. *Differential cryptanalysis* [5] and *linear cryptanalysis* [11] show that even partial approximations of the encryption algorithm by a linear function are sufficient to mount powerful distinguishing and key-recovery attacks.

In [12] the authors suggested to use (almost) perfect nonlinear functions (functions with a maximal distance from all linear structures) in block ciphers. The nonlinear functions can be used either as a building block in the structure of the block cipher (such as an S-box [15, 16, 17]) or as the entire block cipher (as discussed in [1]). Such (almost) perfect nonlinear constructions can be used to prove security against differential and linear cryptanalysis [18].

---

* The research presented in this paper was partially supported by the Clore scholarship programme.

D. Pointcheval (Ed.): CT-RSA 2006, LNCS 3860, pp. 295–312, 2006.

In [15] the following construction, later named in [18], was suggested: An almost perfect nonlinear permutation (APNP) is a permutation $f : GF(2^n) \rightarrow GF(2^n)$ such that for any $a \neq 0$, the function $g(x, a) = f(x) \oplus f(x \oplus a)$ assumes exactly 2^{n-1} different values. It means that for any two pairs of distinct input values with the same difference, the corresponding output pairs do not have the same difference. Due to this property, APNPs are considered secure against differential cryptanalysis [5] as well as against linear cryptanalysis [11].[1] A permutation that is very close to be an APNP is the S-box SubBytes of AES [7]. In the SubBytes permutation, for a non-zero input difference α, there are two pairs of inputs with difference α and the same output difference, while the other 126 pairs with input difference α have different output differences.

In the first part of this paper we analyze the concept of using highly nonlinear functions (such as APNPs) as the entire block cipher. We devise several distinguishers between an APNP and a random permutation based on the "too good" differential properties of APNPs. The data and memory complexities of the best distinguisher are $O(2^{n/3})$ with time complexity of $O(2^{2n/3})$ where n is the block size, independent of the key size. The distinguishers are then extended to ciphers with close to uniform difference distribution tables (that is, ciphers that are close to APNPs). This result leads to the conclusion that despite their favorable properties, highly nonlinear ciphers might possess inherent weakness.

In a way, the concept of this part of the paper is similar to the concept behind impossible differential cryptanalysis [4]. Usually, differential cryptanalysis uses differentials with high probability. The idea of impossible differential attacks is to exploit differentials with zero probability. The attacker utilizes the fact that the differential properties are "too strong". In the same way, our attacks recognize APNPs due to their "too high" level of nonlinearity.

In the second part of this paper we analyze the following encryption scheme: Let f be a permutation, f^{-1} be its inverse, and K be a randomly chosen non-zero key. We define $g(x) = f^{-1}(f(x) \oplus K)$, and show that g has several predictable differential properties. Those properties are related to the level of nonlinearity of the function f. We use those properties to define the *effective linearity* of the function f, and show that the effective linearity of almost perfect nonlinear permutations is 1, for random permutations it is 2, and is 2^n for linear permutations (where n is the block size). We note that the effective linearity of a block cipher can be approximately computed with complexity $O(2^{n/2})$. We present various scenarios in which this value can be used in distinguishing and key recovery attacks. For example, we show that a 2-round Feistel structure surrounded by two key dependent decorrelation modules can be easily distinguished from random permutations, regardless of the Feistel round function.

This paper is organized as follows: In Section 2 we give the definition and theoretical background of highly nonlinear functions. In Section 3 we present two distinguishing attacks on ciphers with uniform difference distribution tables.

---

[1] Note that there are two colliding definitions to the term nonlinearity — nonlinearity as distance from linear functions, and nonlinearity in the sense of almost perfect nonlinear permutations. In this paper we use the latter definition.

Section 4 examines differential properties of the construction $g = f^{-1}(f(x) \oplus K)$, and defines the effective linearity of a function. In Section 5 we use the effective linearity to mount distinguishing and key recovery attacks. In the appendices we bring various computations and proofs. We summarize the paper in Section 6, and discuss the implications of our findings.

2 Definitions and Theoretical Background

There are several possible notions of high nonlinearity of a boolean function. In this paper we use the following definition, presented in [15]:

Definition 1. *A function $f : GF(2^n) \to GF(2^m)$ is perfectly nonlinear if for any non-zero $w \in GF(2^n)$ the difference $f(x + w) - f(x)$ obtains all the values $y \in GF(2^m)$ exactly 2^{n-m} times each.*

In [15] the properties of such functions were studied, and it was shown that perfectly nonlinear functions $f : GF(2^n) \to GF(2^m)$ exist if and only if $n \geq 2m$. However, in real life designs many designers prefer to use functions in which the size of the input is equal to the size of the output. Thus, perfectly nonlinear functions cannot be used. This led to the introduction of *almost perfect nonlinear functions*, defined as following:

Definition 2. *A function $f : GF(2^n) \to GF(2^n)$ is almost perfectly nonlinear if for any $a \neq 0$, the function $g(x, a) = f(x) \oplus f(x \oplus a)$ assumes exactly 2^{n-1} different values.*

If such f is also a permutation, f is called "almost perfectly nonlinear permutation" (in the sequel we abbreviate this notation to "APNP").

Note, that since the characteristic of the field $GF(2^n)$ is 2, it follows that for any function $f : GF(2^n) \to GF(2^n)$, every value of $g(x, a)$ is assumed an even number of times. Therefore, 2^{n-1} is the maximal possible output size of g for a non-zero α.

A permutation that is close to be an APNP is the S-box SubBytes of AES [7]. This S-box is a permutation $f : GF(2^8) \to GF(2^8)$ and for every $a \neq 0$, the function $g(x, a) = f(x) \oplus f(x \oplus a)$ assumes $2^{8-1} - 1 = 127$ values.

As stated before, this notion of nonlinearity is closely related to differential properties of the function f and the definition of an APNP can be restated in terms of the *difference distribution table* of the function used in differential cryptanalysis [5]. First we recall the definition of the difference distribution table of a function:

Definition 3. *Let $f : GF(2^n) \to GF(2^m)$ be a general function. The difference distribution table (DDT) of f is an $(2^n) \times (2^m)$ matrix whose (i, j) entry is defined as $\#\{x \in GF(2^n) | f(x) \oplus f(x \oplus i) = j\}$.*

A function f is considered optimally secure against differential cryptanalysis if the entries in the DDT of f are the lowest possible ones. This is the case when f is an almost perfect nonlinear function. We can now rephrase the definition of an almost perfect nonlinear function in terms of the DDT:

Definition 4. *A function $f : GF(2^n) \rightarrow GF(2^n)$ is almost perfectly nonlinear if the highest entry in the DDT of f (except for the entry $(0 \rightarrow 0)$ that equals 2^n) is 2.*

This definition is closely related to the following definition of δ-uniformity:

Definition 5. *For an $n \times s$ bits S-box $S(\cdot)$ (where $n \geq s$), we denote by δ the highest entry in the difference distribution table (except for $(0,0)$ entry which is always 2^n), namely*

$$\delta = \max_{\alpha \in \{0,1\}^n, \alpha \neq 0, \beta \in \{0,1\}^s} \#\{x | S(x) \oplus S(x \oplus \alpha) = \beta\}$$

S is called differentially δ-uniform.

Hence, almost perfect nonlinear permutations are differentially 2-uniform.

We recall that differential cryptanalysis is mostly interested in differentials with high probability (or zero probability). This led various papers [17, 18, 21] to suggest using functions that are as differentially uniform as possible. If such functions are used, the cipher is expected to have fewer differentials with high probability, as well as less zero probability differentials.

We stress that usually APNPs are not used in real life ciphers. However, this is mostly due to implementation issues, as the common belief is that these functions are better than other constructions. For example, many ciphers use APNPs as building blocks, like the S-box used in the AES.

3 Distinguishing Highly Nonlinear Functions from Random Permutations

In this section we present two distinguishing attacks on highly nonlinear functions. Each is based on a different assumption and performs in a different model (known plaintext or chosen plaintext). These attacks are capable of identifying whether a given black box is a random permutation or an highly nonlinear function. Hence, if an APNP is used as the cipher, it can be distinguished from a random permutation. We discuss the possible applications of such an attack in Section 5.

We note that the Even-Mansour construction $E_{K_1,K_2}(P) = F(P \oplus K_1) \oplus K_2$ assumes that the underlying F is a pseudo random permutation [8]. Our attacks can distinguish the case where F is an APNP from the case that F is a random permutation, despite the commonly believed good security properties of APNPs.

3.1 A Chosen Plaintext Distinguisher

The first attack is a chosen plaintext attack based on the birthday paradox. Let $f : GF(2^n) \rightarrow GF(2^n)$ be a black box permutation for which we have to determine whether it is an APNP or a random permutation.

We perform the following algorithm with a parameter m (to be determined later):

1. Encrypt m distinct pairs of plaintexts (P_1, P_2), such that $P_1 \oplus P_2 = \alpha$ and $P_1 < P_2$ for some fixed non-zero value α by f to get the ciphertext pairs of the form $(C_1, C_2) = (f(P_1), f(P_2))$.
2. Store the XOR values of the ciphertexts, i.e., $C_1 \oplus C_2$ in a hash table.
3. If we obtain a collision in the hash table (two pairs with the same ciphertext difference), halt and conclude that f is not an APNP.
4. If no collisions are encountered, conclude that f is an APNP.

A collision is formed of two distinct pairs $(P_1, P_2 = P_1 \oplus \alpha)$ and $(P_3, P_4 = P_3 \oplus \alpha)$, whose corresponding ciphertexts (C_1, C_2) and (C_3, C_4), respectively, satisfy $C_1 \oplus C_2 = C_3 \oplus C_4$. Such a collision means that the equation $f(x) \oplus f(x \oplus \alpha) = C_1 \oplus C_2 = C_3 \oplus C_4$ has (at least) four solutions.

Recall that an APNP is a permutation for which the equation $f(x) \oplus f(x \oplus \alpha) = \beta$ for non-zero α and β has at most two solutions (x_0 and $x_0 \oplus \alpha$ for some x_0). Thus, for any value of m (even for $m = 2^{n-1}$), no such collision is expected.

For a random permutation, however, the algorithm is expected to find such an instance. And thus, once such an instance is found, the algorithm concludes that f is not an APNP.

3.2 Analysis of the Chosen Plaintext Attack

Recall, that the attack is based on encrypting pairs of plaintexts (P_1, P_2) that satisfy $P_1 \oplus P_2 = \alpha$ for some fixed arbitrary non-zero α. If there are two distinct plaintext pairs (P_1, P_2) and (P_3, P_4) whose corresponding ciphertexts (C_1, C_2) and (C_3, C_4), respectively, satisfy $C_1 \oplus C_2 = C_3 \oplus C_4$, then the black box permutation f is not an APNP for sure.

Let us examine the number of expected quartets for f. Consider the row corresponding to α in DDT^f, the difference distribution table of f. For every fixed output difference β, the value that corresponds to β in this row represents the number of pairs with input difference α and output difference β (recall that x and $x \oplus \alpha$ appear as two pairs $(x, x \oplus \alpha)$ and $(x \oplus \alpha, x)$). In other words, $DDT^f(\alpha, \beta)$ (entry (α, β) of DDT^f) is

$$DDT^f(\alpha, \beta) = |\{x \in GF(2^n) : f(x) \oplus f(x \oplus \alpha) = \beta\}|.$$

Note that a permutation is considered APNP if and only if its difference distribution table does not contain values greater than 2.

For a random permutation f, we may assume that values in any single row of the difference distribution table behave almost as Poisson random variables. That is, the values in the difference distribution table are distributed according to $2 \cdot Poi(1/2)$.[2] Thus, the value $2k$ is expected to appear in a given row about $2^n \cdot e^{-1/2} \cdot 2^{-k}/k!$ times.

Collisions in Step 3 can occur only for values of β whose corresponding entry of the difference distribution table is more than 2. Let us examine only values of β in the difference distribution table with 4 or more. Out of the 2^n possible β

---

[2] Recall that in an XOR difference distribution table all values are even.

entries, $0.0902 \cdot 2^n$ such entries exist. Due to the birthday paradox, when we want to have success rate of p, we require $p > 1 - e^{-m \cdot (m-1)/(2 \cdot 0.0902 \cdot 2^n)}$. Therefore, to ensure success probability of 0.8 we need $m > 0.1618 \cdot 2^{n/2}$ pairs of this kind.

However, the algorithm encrypts also pairs whose output difference β has 2 in the difference distribution table. Hence, the real number of pairs we need to examine is about 4 times larger, as only one out of 4 pairs (more precisely, about 23%) has an output difference meeting our requirement. Therefore, the data complexity of the algorithm is $N = 2m = 1.4070 \cdot 2^{n/2}$ plaintexts (or queries to the black box).

The time complexity of the algorithm is $N = 1.4070 \cdot 2^{n/2}$ encryptions and $m = 0.7035 \cdot 2^{n/2}$ memory accesses in the worst case. The memory requirements are $m = 0.7035 \cdot 2^{n/2}$ memory cells in the worst case.

Changing the attack scenario into a known plaintext attack does not change the attack significantly. The data complexity is $m = 1.3459 \cdot 2^{n/2}$ queries, and the time complexity is $O(2^{2n/3})$ using Wagner's algorithm for the generalized birthday paradox problem [24].

3.3   An Improvement to the Chosen Plaintext Attack

An improvement to the algorithm uses the fact that the above is true for any non-zero α. The attack requires m distinct plaintexts, such that the XOR value of any two of them is among a list of m values (for example, setting some of the bits of all plaintexts to be zero). In this case for each pair of plaintexts (P_1, P_2) we compute $(P_1 \oplus P_2, f(P_1) \oplus f(P_2))$, and insert it into a hash table. A collision in the hash table suggests a quartet of values (P_1, P_2) and (P_3, P_4) such that $P_1 \oplus P_2 = P_3 \oplus P_4$ and $f(P_1) \oplus f(P_2) = f(P_3) \oplus f(P_4)$. This cannot be achieved for an almost perfect nonlinear permutation, and thus, can be also used for distinguishing.

For m plaintexts chosen in this way, we have $m^2/2$ pairs, each producing a string of $2n$ bits. Not all 2^{2n} possible $2n$-bit strings are produced in this process. More precisely, the number of possible values for this string is $m \cdot 2^n$. If we choose m such that $(m^2/2)^2 > 1.17m2^n$, we have a chance of 50% to find a collision (in case of a random permutation) according to the birthday paradox.

Setting $m = 1.794 \cdot 2^{n/3}$ we expect to find such a collision with probability 0.8. Thus, the data complexity of this attack is $m = 1.794 \cdot 2^{n/3}$ chosen plaintexts, and the time complexity of the attack is $m^2/2 = 1.609 \cdot 2^{2n/3}$ memory accesses.

3.4   Other Kinds of Permutations

Permutations that are very close to APNPs are widely used in block ciphers. For this kind of permutations, the above attacks still succeed with almost the same success rate.

Another question which arises, is what happens when the permutation we wish to distinguish is not so close to be differentially 2-uniform. That is, what if there are many entries with value of 4 in the difference distribution table of the permutation. Formally, out of the 2^{n-1} pairs, assume that at most a ratio p of

the pairs are in entries with value of 4 in the difference distribution table, while
the other non-zero entries are 2 (up to the $0 \to 0$ entry).

For these functions, the above algorithms fail, as the probability to have two
pairs whose output difference is the same, is no longer negligible. This can be
solved when p is far from 0.23,[3] e.g., $p < 0.2$ or $p > 0.3$ (in case $p > 0.3$, we
require that at least p of the pairs are in entries with value of 4).

The transformation of the above algorithms to deal with such permutations
is changing the identification from "find such an instance, halt and output ..."
to "count how many instances there are, and compare this number to how many
should be". The analysis of the exact number of plaintexts m needed is quite
straightforward given p and the requested success rate.

4   Differential Properties of $f^{-1}(f(x) \oplus K)$ and Their Applications

In this section we consider the differential properties of some special structure
derived from a permutation f and show how to utilize these properties in order
to study the structure of f itself. Let $g(x) = f^{-1}(f(x) \oplus K)$ be a permutation
where K is some fixed key. First, we show that using the properties of g we can
determine whether f is an APNP or a random permutation. Then we show how
to generalize this result in order to classify functions according to their level of
nonlinearity. We formalize this classification by defining the *effective linearity*
coefficient (EL) of a permutation which corresponds to the level of linearity
determined by our method.

4.1   Theoretical Background

Let $f : GF(2^n) \to GF(2^n)$ be a black box permutation for which we have to
determine whether it is an APNP or a random permutation. Choose an arbitrary
non-zero $K \in GF(2^n)$, and define the permutation $g(x) = f^{-1}(f(x) \oplus K)$.

Let A, B be a pair of plaintexts with a non-zero input difference α (e.g.,
$A \oplus B = \alpha$) and consider $\gamma = g(A) \oplus g(B)$. We shall compute the probability
of the event $\gamma = \alpha$, and show that this probability can be used to distinguish
APNPs from random permutations.

Let $\beta = f(A) \oplus f(B)$. If f is an APNP, (A, B) is the only pair of plaintexts
with input difference α and output difference β. Now, consider the pair $(f(A) \oplus K, f(B) \oplus K)$. There are two cases:

1. $f(A) \oplus K = f(B)$. In this case, we have $g(A) = B$ and $g(B) = A$ and thus
   $\gamma = g(A) \oplus g(B) = B \oplus A = \alpha$. This case occurs when $\beta = f(A) \oplus f(B) = K$
   which happens with probability of 2^{-n}.

---

[3] The expected ratio of 4 or more in the difference distribution table of a random
   permutation is about 0.09 of the entries. However, entries with 6,8, or even more,
   contribute more quartets. The total number of quartets counted by the above algorithms for a random permutation is equal to the case where 0.23 of the entries of
   the table that are 4 (while the remaining are 2's and 0's).

2. $f(A) \oplus K \neq f(B)$. In this case, the pairs $(f(A), f(B))$ and $(f(A) \oplus K, f(B) \oplus K)$ differ, but still have the same XOR difference β. Thus, if f is an APNP, the difference $\gamma = g(A) \oplus g(B) = (f^{-1}(f(A) \oplus K)) \oplus (f^{-1}(f(B) \oplus K))$ cannot be equal to α. Therefore, in this case $\gamma \neq \alpha$ always.

Combining the two cases together we obtain

$$\Pr_{A,B,K \in GF(2^n), A \neq B, K \neq 0}[\gamma = \alpha] = 2^{-n}. \qquad (1)$$

The analysis presented in Appendix A shows that for a random permutation this probability ($\Pr[\gamma = \alpha]$) equals to $2 \cdot 2^{-n}$. The difference between the probabilities can be used in order to distinguish between an APNP and a random permutation.

We have experimentally verified that the value $2 \cdot 2^{-n}$ is the correct value for a random permutation and that 2^{-n} is the correct value for an APNP. This was done by generating sets of random permutations of 8,10,12,14 and 16 bits, and counting all possible quartets $(A, B, g(A), g(B))$ (for a large set of K values, for all α values).

4.2   An Adaptive Chosen Plaintext and Ciphertext Distinguisher

The algorithm of the distinguisher is as follows: Let $f : GF(2^n) \to GF(2^n)$ be a black box permutation for which we have to determine whether it is an APNP or a random permutation and $m, threshold$ be integers specified later.

1. Encrypt m distinct plaintexts P_i, for $i = 1, \ldots, m$.
2. Choose an arbitrary $K \in GF(2^n)$.
3. Decrypt the values $f(P_i) \oplus K$ to get $g(P_i) = f^{-1}(f(P_i) \oplus K)$.
4. Store the m values of the form $P_i \oplus g(P_i)$ into a hash table.
5. Count the number of collisions in the hash table. If the number of collisions is greater than $threshold$ output "random permutation". Otherwise, output "APNP".

We note that if $P_1 \oplus g(P_1) = P_2 \oplus g(P_2)$ then we have a right quartet $((P_1, P_2), (g(P_1), g(P_2)))$. Starting with m plaintexts, we get m values of $P \oplus g(P)$. Once there is a collision in the hash table, the colliding values suggest a quartet. If there are three values in the same entry of the hash table, then we get three quartets, or generally, if there are k values in the same entry, we get $k(k-1)/2$ quartets.

For an APNP, about 2^{-n} of the all possible quartets satisfy our conditions, while for a random permutation, about $2 \cdot 2^{-n}$ satisfy our conditions. For $m = 4 \cdot 2^{n/2}$ and $threshold = 10$ the success rate is 0.816. For $m = 2^{n/2}$ the success rate of this attack is 0.594.

This attack may seem less desirable, given the attacks of the previous section, as it has a similar data complexity but a more stern attack model. However, this attack can be easily extended to other cases, as we present in Section 4.3.

We remark that one can make a slight change in the attack such that only quartets of the form $(A, B, g(A), g(B))$ where $A \neq B \neq g(A) \neq g(B)$ are counted. Using this variant of the attack the number of expected collisions is 0 for APNP, and if we get even one collision the permutation is certainly not an APNP. The number of expected collisions for a random permutation is $\frac{1}{4} \cdot 2^{-n}$ of the total number of possible quartets.

This can be used to increase the success probability of the attack by setting $threshold = 1$. More accurately, for the same data complexity as before, the success rate of the attack is about 0.98. We can also reduce the data complexity by a factor of $\sqrt{2}$ and still have a success rate of 0.86.

4.3 The Effective Linearity of a Permutation

Following the previous attack, we define the effective linearity of a permutation.

Definition 6. *Let $f : GF(2^n) \to GF(2^n)$ be a permutation. The effective linearity of f is:*

$$EL(f) = 2^n \left(\frac{1}{2^n - 1} \right)^2 \cdot \sum_{K \in GF(2^n) \backslash \{0\}} \sum_{\alpha \in GF(2^n) \backslash \{0\}} \Pr \left[\alpha \xrightarrow{g(x) = f^{-1}(f(x) \oplus K)} \alpha \right]$$

Actually, $EL(f)$ is the average of the probabilities $\Pr[\alpha = \gamma]$ over all non-zero K's and α's multiplied by 2^n.

For a random permutation this value is expected to be close to 2 (as shown in Appendix B). If this value is not close to 2, then our attacks can be applied to the permutation and distinguish it from a random permutation.

We can either calculate $EL(f)$ analytically when the difference distribution table of f is known (like in the analysis for a random permutation), or by experimentally measuring it. Taking several sets of $O(2^{n/2})$ messages and using several K values, we can use statistical methods to evaluate $EL(f)$. Note that the $O(2^{n/2})$ complexity is achieved by using many (if not all) values of α simultaneously.

The effective linearity of f is not lower than 1 (as when $K = f(A) \oplus f(B)$ we get that $g(A) = B$ and $g(B) = A$) and cannot be higher than 2^n (which is the value for linear permutations). As the value for a random permutation is 2, we suggest designing ciphers with effective linearity close to 2.

It is possible to show (see Appendix B) that for a two round Feistel construction whose round functions are both APNP, the effective linearity is 3, while if the used functions are random permutations it is expected to be at least 8. An interesting observation regarding the effective linearity of Feistel constructions, is that after three Feistel rounds using random permutations as round functions, the effective linearity is 2. This might be viewed as another realization of the Luby-Rackoff result about Feistel constructions [10].

Another interesting remark about Feistel constructions, is that if the permutation of the first round p_1 has effective linearity $EL(p_1)$, and the second round's permutation p_2 has effective linearity $EL(p_2)$, then the effective linearity $EL(f)$

of the two round Feistel satisfies $EL(f) \geq EL(p_1) \cdot EL(p_2)$. The exact proof is given in Appendix B.

When the round functions are not bijective, the difference distribution table of the 2-round Feistel construction is expected to have more zero entries than usual. As the sum of every line in the difference distribution table is constant, it follows that the remaining entries are expected to be higher, leading to an higher effective linearity. Thus, the more entries having a zero value in the difference distribution table, the higher the effective linearity is expected to be. For example, we show in Appendix B that the effective linearity of 2-round DES is at least 220 (independent of the key).

5   Various Attacks Based on the Effective Linearity of Permutations

In this section we present several possible scenarios in which measuring the effective linearity of various permutations can be used in order to mount distinguishing and key recovery attacks.

5.1   Treating Decorrelation Modules

Let us consider a cipher of the form $E = DM_2 \circ F_2 \circ F_1 \circ DM_1$, where DM_i is a decorrelation module (with some key) [22], and F_i is a Feistel round with a random permutation as the round function (along with some key).

We recall that once the key is set the decorrelation module is linear, but when the key is random, the probability of any non-trivial differential going through the decorrelation module equals $1/(2^n - 1)$ on average. A similar condition can be proved with respect to linear cryptanalysis as well.

Due to the nature of the decorrelation module, any differential (even a truncated one) cannot have probability higher than the trivial one through the first decorrelation module. The same is true for linear approximations as well. Moreover, it is impossible to devise a SQUARE-like property for this cipher as the decorrelation module prevents the attacker from setting a good input set.

While all these methods fail, we can efficiently distinguish the above E from a random permutation. As the decorrelation modules are linear, and as we have two rounds of a Feistel structure, we can easily determine that the effective linearity of E is 8, while the effective linearity of random permutations is only 2.

We note that the minimal value of the effective linearity of 2-round Feistel construction is 3 (achieved by applying APNPs as the round function in the two Feistel rounds). Thus, even if the Feistel round functions are replaced, our attack still works.

Our technique is able to pass the decorrelation module as if it does not exist. This is due to the fact that we count on many possible differentials, and we do not restrict ourselves to differentials of some structure, or even to sets of plaintexts of a given structure.

5.2   Distinguishing Known Ciphers and Identifying Black Box Permutations

It is possible to precompute the effective linearity of ciphers in advance (also for reduced round variants). Then, given a black box the attacker computes its EL and if the black box is one of the previously known encryption schemes, he can detect it.

However, we still do not know whether this attack is applicable against encryption schemes that are actually used today. It may occur that the measured effective linearity values are too close to 2 and the distinguishing will become infeasible. Moreover, as we noted in Section 4.1, the number of detected quartets in the distinguisher slightly depends on K and thus in some cases the difference between two encryption schemes can be less than the difference between applications of the same scheme with different values of K. In this case the attack might fail.

The effective linearity of a permutation depends on its difference distribution table. When computing the difference distribution table of a permutation one usually computes the average probabilities over all the possible keys and assumes that the probability for any single key is close to the average (see [5]). However, in some ciphers there are classes of keys for that some differential properties of the cipher differ from the average case, like for IDEA [6]. Such classes are called "differentially weak key classes". Usually such classes can be detected only if some explicit differential characteristic is known for the whole cipher.

We can use our distinguisher in order to detect such classes when the entire differential structure differs for different keys, even if any concrete characteristic is unknown. For example, there is a differential weak key class of IDEA. In that weak key class there exists some differential with probability 1 of the form $\alpha \to \beta$. For this weak key class, the effective linearity is higher by 1 than the effective linearity of IDEA with key not in the weak key class. We note that using the differential is easier for the purpose of distinguishing whether the key is in the weak key class. However, if decorrelation modules are added before IDEA and after it, then the differential distinguisher is not applicable anymore, while our distinguisher still succeeds.

Our technique can be used also for key recovery attacks, by measuring the effective linearity of reduced round versions of the cipher. Then, for a given n-round construction, we can try all possible subkeys of the last round, and try to peel it off. If the peeling succeeds, the effective linearity of the obtained cipher equals to the one of $(n-1)$ rounds of the cipher (instead of the expected $(n+1)$ rounds in case of a wrong guess).

5.3   Attacks Against Encryption Schemes of the Form $f^{-1}(f(x) \oplus K)$

The distinguisher can be applied directly against encryption schemes of the form $h(x) = f^{-1}(f(x) \oplus K)$ for an arbitrary permutation f, when K is a secret key. In this scenario, the data requirements are even less than in the original distinguisher: the attacker can use known plaintexts instead of adaptively chosen plaintexts.

The attack is performed essentially in the same way as the original distinguisher. The difference is that we already have the values of $h(x)$ which correspond to $g(x)$ in the original distinguisher. The number of found quartets supplies the attacker with the probability $\Pr[h(x) \oplus h(y) = \alpha | x \oplus y = \alpha]$. Thus, we know the average of the probability of the differential $\alpha \to \alpha$ through $h(\cdot)$, even without constructing the difference distribution table of h itself.

If f is a random permutation, the expected result for h is $2 \cdot 2^{-n}$. This, in contrast to a random permutation $h'(\cdot)$ with the respective probability of 2^{-n}. Therefore, even if f is a perfectly random permutation, the attacker can distinguish between h and a random permutation.

We remark that constructions of the form $h(x)$ are not so rare. For example, two rounds of any involution cipher are of the form $h(x)$. KHAZAD [2] is one example of such a cipher.

Note that there is an adaptive chosen plaintext attack that requires only 2 plaintexts that distinguishes $h(\cdot)$ from a random permutation. Its transformation into a known plaintext attack requires $O(2^{n/2})$ known plaintexts which is equivalent to the data complexity of our attack. However, in the chosen plaintext model our attack has a lower data complexity of $O(2^{n/3})$ (instead of the $O(2^{n/2})$ required for the transformation of the basic attack into a chosen plaintext attack). This last statement is true whenever f is not an APNP.

We also note that this construction covers all block ciphers with cycle 2. This follows from the fact that all ciphers with cycle 2 can be described as $h(x) = f^{-1}(f(x) \oplus K)$ for some permutation f and a non-zero constant K.

6 Summary

In the first part of this paper we presented several distinguishers for highly nonlinear permutations and random permutations. We conclude that while using APNPs as part of the encryption scheme seems desirable, using APNPs as the entire cipher can possess inherent weakness.

In the second part of the paper we have examined the structure $f^{-1}(f(x) \oplus K)$ for various permutations f. We have shown how to use the differential properties of this construction in order to study the differential structure of f. We also proved that this construction can be used to effectively determine the average

Table 1. Various Constructions and their Effective Linearity

Construction	Round Function	Effective Linearty
APNP	APNP	1
Random Permutation	Random Permutation	2
Affine Permutation	Affine Permutation	2^n
2-round Feistel	APNP	3
	Random Permutation	≥ 8
	DES	≥ 220
3-round Feistel	Random Permutation	2

probability of a differential of f. Finally, we have defined the *effective linearity* of a permutation that measures this probability. Table 1 contains various constructions and their effective linearity.

The effective linearity can be used to distinguish between an f that is an almost perfect nonlinear permutation and an f that is a random permutation. On the other hand, it can be used to distinguish ciphers with a relatively close to linear structure from random permutations, even if no concrete differential is known. Our attacks have better performance compared to previously known attacks for several structures. For example, we can distinguish between a random permutation and a permutation formed by two Feistel rounds surrounded by two key-dependent decorrelation modules, regardless of the round functions of the 2-round Feistel construction.

Acknowledgments

The authors would like to thank Osnat Ordan and Dana Cohen for their help in conducting the experiments, which verified our claims. It is also a pleasure to acknowledge the references and ideas expressed by Serge Vaudenay, Jennifer Seberry, and Eli Biham. We would also like to thank the anonymous referees for their valuable comments and insightful suggestions.

References

1. Kazumaro Aoki, Serge Vaudenay, *On the Use of GF-Inversion as a Cryptographic Primitive*, proceedings of Selected Areas in Cryptography 2003, Lecture Notes in Computer Science 3006, pp. 234–247, Springer-Verlag, 2004.
2. Paulo S.L.M. Baretto, Vincent Rijmen, *The KHAZAD Block Cipher*, Submitted to NESSIE, available online at *http://www.nessie.eu.org*.
3. Thomas Beth, Cunsheng Ding, *On Almost Perfect Nonlinear Permutations*, Advances in Cryptography, proceedings of EUROCRYPT '93, Lecture Notes in Computer Science 765, pp. 65–76, Springer-Verlag, 1994.
4. Eli Biham, Alex Biryukov, Adi Shamir, *Cryptanalysis of Skipjack reduced to 31 rounds*, Advances in Cryptology, proceedings of EUROCRYPT '99, Lecture Notes in Computer Science 1592, pp. 12–23, Springer-Verlag, 1999.
5. Eli Biham, Adi Shamir, *Differential Cryptanalysis of the Data Encryption Standard*, Springer-Verlag, 1993.
6. Joan Daemen, Rene Govaerts, Joos Vandewalle, *Weak Keys for IDEA*, Advances in Cryptology, proceedings of CRYPTO '93, Lecture Notes in Computer Science 773, pp. 224–231, Springer-Verlag, 1994.
7. Joan Daemen, Vincent Rijmen *The design of Rijndael: AES — the Advanced Encryption Standard*, Springer-Verlag, 2002.
8. Shimon Even, Yishay Mansour, *A Construction of a Cipher from a Single Pseudorandom Permutation*, Journal of Cryptology, Vol. 10, Number 4, pp. 151–162, Springer-Verlag, 1997.
9. Philip Hawkes, Gregory G. Rose, *Primitive Specification for SOBER-t16 Submission to NESSIE* and *Primitive Specification for SOBER-t32 Submission to NESSIE*, Submitted to NESSIE, available online at *http://www.nessie.eu.org*.

10. Michael Luby, Charles Rackoff, *How to Construct Pseudorandom Permutations from Pseudorandom Functions*, SIAM journal of Computing, Volume 17. No. 2, pp. 373–386, 1988.
11. Mitsuru Matsui, *Linear Cryptanalysis Method for DES Cipher*, Advances in Cryptology, proceedings of EUROCRYPT '93, Lecture Notes in Computer Science 765, pp. 386–397, Springer-Verlag, 1994.
12. Willi Meier, Othmar Staffelbach, *Nonlinearity Criteria for Cryptographic Functions*, Advances in Cryptology, proceedings of EUROCRYPT '89, Lecture Notes in Computer Science 434, pp. 549–562, Springer-Verlag, 1990.
13. Willi Meier, Othmar Staffelbach, *Fast Correlation Attacks on Stream Ciphers (Extended Abstract)*, Advances in Cryptology, proceedings of EUROCRYPT '88, Lecture Notes in Computer Science 330, pp. 300–315, Springer-Verlag, 1988.
14. US National Bureau of Standards, *Data Encryption Standard*, Federal Information Processing Standards Publications No. 46, 1977.
15. Kaisa Nyberg, *Perfect nonlinear S-boxes*, Advances in Cryptology, proceedings of EUROCRYPT '91, Lecture Notes in Computer Science 547, pp. 378–386, Springer-Verlag, 1991.
16. Kaisa Nyberg, *On the construction of highly nonlinear permutations*, Advances in Cryptology, proceedings of EUROCRYPT '92, Lecture Notes in Computer Science 658, pp. 92–98, Springer-Verlag, 1993.
17. Kaisa Nyberg, *Differentially uniform mappings for cryptography*, Advances in Cryptology, proceedings of EUROCRYPT '93, Lecture Notes in Computer Science 765, pp. 55–64, Springer-Verlag, 1994.
18. Kaisa Nyberg, Lars R. Knudsen, *Provable Security Against Differential Cryptanalysis*, Advances in Cryptology, proceedings of CRYPTO '92, Lecture Notes in Computer Science 740, pp. 566–578, Springer-Verlag, 1993.
19. Oscar S. Rothaus, *On Bent Functions*, Journal of Combinatorial Theory, Series A, Vol. 20 (1976), pp. 305–310, 1976.
20. Jennifer Seberry, Xian-Mo Zhang, Yuliang Zheng, *Relationships Among Nonlinearity Criteria (Extended Abstract)*, Advances in Cryptology, proceedings of EUROCRYPT '94, Lecture Notes in Computer Science 950, pp. 376–388, Springer-Verlag, 1995.
21. Jennifer Seberry, Xian-Mo Zhang, Yuliang Zheng, *Pitfalls in Designing Substitution Boxes (Extended Abstract)*, Advances in Cryptology, proceedings of CRYPTO '94, Lecture Notes in Computer Science 839, pp. 383–396, Springer-Verlag, 1995.
22. Serge Vaudenay, *Provable Security for Block Ciphers by Decorrelation*, Journal of Cryptology, Vol. 16, Number 4, pp. 249–286, Springer-Verlag, 2003.
23. David Wagner, *The Boomerang Attack*, proceedings of Fast Software Encryption 6, Lecture Notes in Computer Science 1636, pp. 156–170, Springer-Verlag, 1999.
24. David Wagner, *A Generalized Birthday Problem (Extended Abstract)*, Advances in Cryptology, proceedings of CRYPTO '02, Lecture Notes in Computer Science 2442, pp. 288–304, Springer-Verlag, 2002.

A   Theoretical Analysis of the Effective Linearity of a Random Permutation

In this section we analyze the probability of a pair A, B and their respective $g(A), g(B)$ to satisfy $A \oplus B = g(A) \oplus g(B)$ for a permutation f where $g(x) = f^{-1}(f(x) \oplus K)$. Examine the amount of quartets of the form $(A, B, g(A), g(B))$

such that $\gamma = g(A) \oplus g(B) = A \oplus B = \alpha$. As in the analysis for APNP, we also have two cases:

When $g(A) = B$ the expression $\gamma = \alpha$ holds if and only if $g(B) = A$. As in the analysis of APNPs, this event occurs with probability 2^{-n}. Note, that this case also contains the case where $g(B) = A$.

When $g(A) \neq B$ all the elements of the quartet $(A, B, g(A), g(B))$ are distinct. We observe that any quartet of this kind (which we denote by a right quartet) can be uniquely represented by the two ciphertext pairs $(f(A), f(B))$, $(f(A) \oplus K, f(B) \oplus K)$. These two pairs satisfy the following system of equations:

$$\begin{cases} C \oplus D = \beta \\ f^{-1}(C) \oplus f^{-1}(D) = \alpha \end{cases} \tag{2}$$

for some fixed β.

On the other hand, for any fixed value β, consider pairs of inputs to the function f^{-1} that solve System 2. Every pair of such pairs (C, D), (C_1, D_1) can be used if and only if one of the following holds:

$$\begin{cases} C \oplus C_1 = D \oplus D_1 = K \\ C \oplus D_1 = D \oplus C_1 = K \end{cases} \tag{3}$$

The probability of this event is $2 \cdot 2^{-n}$.

For sake of simplicity we assume that this probability is independent of K. Actually there is a measure of dependence of this value on K. Moreover, the same computation can be rewritten in another way such that the dependence on α is neglected and the dependence on K is computed explicitly. The resulting formulae is similar to Equation 5 below when K is substituted instead of α. Actually, this is the case in our distinguisher since we look for the average of the results for different α values, that allows us to reduce the dependence on α. Since the expected result considers the average for all possible keys, changing the formulae to be dependent on K does not affect the expected result.

Denote by t the number of pairs of solutions of System 2 for a specific value of β, summed over all possible β. Then the expected number of right quartets is $2 \cdot 2^{-n} \cdot t$.

In order to compute the value of t, we have to consider the function f^{-1}. We consider some fixed value β_0 and the element of the difference distribution table of f^{-1}, $DDT^{f^{-1}}$, corresponding to the pair (β_0, α). If $DDT^{f^{-1}}(\beta_0, \alpha) = 2k$, there are k solutions of the system (2) and thus there are $k(k-1)/2$ pairs of solutions. Summing over all the possible values of β, we get the equation

$$t = \sum_{\beta \in GF(2)^n} \frac{DDT^{f^{-1}}(\beta, \alpha)/2 \cdot (DDT^{f^{-1}}(\beta, \alpha)/2 - 1)}{2}, \tag{4}$$

where $DDT^{f^{-1}}$ is the difference distribution table of f^{-1}.

Recall that the difference distribution table of the function f^{-1} is actually the transpose of the difference distribution table of f. Denoting the difference distribution table of f by DDT^f, we have $(DDT^f)^T = DDT^{f^{-1}}$. Thus, we are able to rewrite Equation 4 in terms of the difference distribution table of f as:

$$t = \sum_{\beta \in GF(2)^n} \frac{DDT^f(\alpha, \beta)/2 \cdot (DDT^f(\alpha, \beta)/2 - 1)}{2} \tag{5}$$

Note that the analysis which was performed for the case where f is an APNP is a partial case of the analysis of the general case presented here. Indeed, if f is an APNP then the elements of DDT^f are all equal to $0, 2$ and then we get $t = 0$ since all the elements in the sum equal to zero.

Now, assume that f is a random permutation. As was stated earlier, the elements of DDT^f are distributed according to $2 \cdot Poi(1/2)$. Thus, the value $2k$ is expected to appear in a given row (and in particular, in the row corresponding to α) about $2^n \cdot e^{-1/2} \cdot 2^{-k}/k!$ times. Substituting these figures to Equation 5, we get

$$t = \sum_{k=1}^{2^{n-1}} (2^n \cdot e^{-1/2} \cdot 2^{-k}/k!) \cdot (k \cdot (k-1)/2)$$

$$= 1/2 \cdot 2^n \left[\underbrace{\sum_{k=1}^{2^{n-1}} k^2 e^{-1/2} 2^{-k}/k!}_{A} - \underbrace{\sum_{k=1}^{2^{n-1}} k e^{-1/2} 2^{-k}/k!}_{B} \right] \tag{6}$$

Let X be a random variable distributed according to $Poisson(\frac{1}{2})$, then we have $A = E[X^2]$ and $B = E[X]$. For such X it is known that $E[X] = \frac{1}{2}$ and $Var[X] = E[X^2] - E[X]^2 = \frac{1}{2}$. Thus, $E[X^2] = \frac{1}{2} + \frac{1}{4} = \frac{3}{4}$. Hence we get

$$t = 1/2 \cdot 2^n \left[\sum_{k=1}^{2^{n-1}} k^2 e^{-1/2} 2^{-k}/k! - \sum_{k=1}^{2^{n-1}} k e^{-1/2} 2^{-k}/k! \right] = 1/2 \cdot 2^n \left[\frac{3}{4} - \frac{1}{2} \right] = 1/8 \cdot 2^n \tag{7}$$

Therefore, the expected value of t is $\frac{1}{8} \cdot 2^n$, and the expected number of quartets for a fixed value of α is $\frac{1}{8} \cdot 2^n \cdot 2 \cdot 2^{-n} = \frac{1}{4}$. To compute the probability of a quartet to be a right one, we have to compute the total amount of quartets. Each quartet is constructed by a pair (A, B) such that $A \oplus B = \alpha$. The total number of such pairs is 2^{n-1}. However, each quartet is suggested by the two pairs (A, B) and $(g(A), g(B))$ and thus the ratio should be doubled.

Taking this into consideration, we get that the probability of a quartet to be right is $(\frac{1}{4}/2^{n-1}) \cdot 2 = 2^{-n}$. Summing this result with the result of the first case, we get

$$\Pr[\gamma = \alpha] = 2 \cdot 2^{-n} \tag{8}$$

Summarizing this result, the probability $\Pr[g(A) \oplus g(B) = A \oplus B]$ equals 2^{-n} for APNPs and $2 \cdot 2^{-n}$ for random permutations. This fact can be used in order to distinguish between an APNP and a random permutation.

B   Effective Linearity of Feistel Constructions

Let us examine a 2-round Feistel construction $f : \{0,1\}^{2n} \rightarrow \{0,1\}^{2n}$ with a permutation p_1 as the first round function, and a permutation p_2 as the second round function. Both permutations are defined over the space $\{0,1\}^n$.

An input difference (α_L, α_R) that enters this encryption scheme becomes after the first round (without the swap, which has no effect on our results) into $(\alpha_L \oplus \beta_1, \alpha_R)$, where $\alpha_R \xrightarrow{p_1} \beta_1$. After the second round the output difference is $(\alpha_L \oplus \beta_1, \alpha_R \oplus \beta_2)$ where $\alpha_L \oplus \beta_1 \xrightarrow{p_2} \beta_2$.

If the two permutations p_1 and p_2 are independent then the probability of the event $(\alpha_L, \alpha_R) \rightarrow (\alpha_L \oplus \beta_1, \alpha_R \oplus \beta_2)$ is $\Pr[\alpha_R \xrightarrow{p_1} \beta_1] \cdot \Pr[\alpha_L \oplus \beta_1 \xrightarrow{p_2} \beta_2]$. Thus, the difference distribution table of the 2-round Feistel construction contains in any entry the multiplication of the two related entries from p_1's and p_2's difference distribution tables.

The first observation, is that if p_1 and p_2 are both APNPs, we get that all entries in the difference distribution table of the construction are either zero, 4, or 2^{n+1} (in 2^{n+1} out of the 2^{2n} entries in each row/column), up to the $0 \rightarrow 0$ entry. Thus, we can compute the effective linearity of two round Feistel construction with independent APNP round functions is 3.

Our second observation is a more general one in nature. When we inspect the difference distribution table of the 2-round construction (from $2n$ bits to $2n$ bits) in order to compute the effective linearity of the construction, we find that:

$$t_f = \sum_{\beta \in \{0,1\}^{2n}} \binom{DDT^f(\alpha, \beta)/2}{2} =$$

$$= \sum_{\beta_1, \beta_2 \in \{0,1\}^n} \binom{DDT^{p_1}(\alpha_R, \beta_1) DDT^{p_2}(\alpha_L \oplus \beta_1, \beta_2)}{2} \geq$$

$$\geq 2 \cdot \sum_{\beta_1 \in \{0,1\}^n} \binom{DDT^{p_1}(\alpha_R, \beta_1)}{2} \cdot \sum_{\beta_2 \in \{0,1\}^n} \binom{DDT^{p_2}(\alpha_L, \beta_2)}{2} =$$

$$= 2 \cdot t_{p_1} \cdot t_{p_2}$$

We note that this is done under the assumption that the probability $\Pr[\gamma = \alpha]$ is quite independent with α (this is required in order to omit the β_1 difference from the sum).

Thus, the value of t_f of the construction is at least twice the multiplied values t_{p_1}, t_{p_2} of the permutations p_1 and p_2. As the effective linearity is related to twice the value of t, then the effective linearity of the construction is the multiplication of the two effective linearities. Or formally:

$$EL(f) \geq EL(p_1) \cdot EL(p_2)$$

Our third observation is that not only $EL(f) \geq EL(p_1) \cdot EL(p_2)$, in many cases $EL(f) \geq EL(p_1) \cdot EL(p_2) + EL(p_2)$. The proof will be given in the final version of this paper. The definition of the effective linearity requires that the entire value K is non-zero. In the Feistel construction, it is possible that the last round will be canceled even if the constant is non-zero (if the left half is 0 and the right half is non-zero). In that case, the left half remains constant during the computation of g (i.e., the left half of $g(x)$ is the same as the left half of x). This case occurs with probability 2^{-n}.

The last observation is valid only if the second permutation is not a linear permutation. If it is a linear permutation, then the second round has no effect on the linearity of the construction.

The same reasoning can be used when the round functions are non-bijective. In that case, as the round functions are not bijective, the number of zero entries in the difference distribution table is greater, and there is a possibility that $p_1(x) = p_1(y)$ even if $x \neq y$. Obviously, this implies that the average probability of $\alpha = \gamma$ is higher.

For example, when considering 2-round DES, we get from [5] that the difference distribution table of a DES round contains about 80% zero entries. This means, that given that an entry is non-zero, its expected value is 6. As the difference distribution table of 2-round DES is related to the multiplication of two 1-round difference distribution tables, the expected entry in non-zero entries is 36. After considering the number of non-zero entries, we get that the effective linearity of such a permutation is about 220. This is done under the assumption that the difference distribution table is uniform (all non-zero entries but the $0 \rightarrow 0$ one are 36). In case it is not uniform (which is more likely) the effective linearity is higher (as the effective linearity is proportional to the sum of squares of the entries, and by Jensen's inequalities is expected to be higher).

Our last result regarding Feistel constructions refers to a 3-round Feistel construction. If the round functions are random permutations, then the left half of the output difference is expected to behave randomly and uniformly. Thus, the difference distribution table, which is a multiplication of the difference distribution table of the left half and the right half, should have the same behavior as of the right side — of a random permutation. Thus, its effective linearity is predicted to be 2.

Block Ciphers Sensitive to Gröbner Basis Attacks

Johannes Buchmann, Andrei Pyshkin*, and Ralf-Philipp Weinmann

Technische Universität Darmstadt, Fachbereich Informatik,
Hochschulstr. 10, D-64289 Darmstadt, Germany
{buchmann, pyshkin, weinmann}@cdc.informatik.tu-darmstadt.de

Abstract. We construct and analyze Feistel and SPN ciphers that have
a sound design strategy against linear and differential attacks but for
which the encryption process can be described by very simple polynomial equations. For a block and key size of 128 bits, we present ciphers
for which practical Gröbner basis attacks can recover the full cipher key
requiring only a minimal number of plaintext/ciphertext pairs. We show
how Gröbner bases for a subset of these ciphers can be constructed with
neglegible computational effort. This reduces the key–recovery problem
to a Gröbner basis conversion problem. By bounding the running time
of a Gröbner basis conversion algorithm, FGLM, we demonstrate the
existence of block ciphers resistant against differential and linear cryptanalysis but vulnerable against Gröbner basis attacks.

1   Introduction

Since the publication of Courtois' and Pieprzyk's XSL method [9] and Murphy
and Robshaw's embedding of the AES [21], a considerable interest in algebraic
attacks on block ciphers has been provoked. While linearization based attacks on
stream ciphers have been shown to be very successful, the claimed attacks on the
AES and Serpent have thus far been highly controversial, if not outright refuted
[5]. Gröbner bases however are a proven tool for solving polynomial systems. Cid,
Murphy and Robshaw [6] recently did a first step of investigating the viability
of an algebraic attack using Gröbner bases on scaled-down versions of the AES.

The goal of this paper is to show that non-trivial iterated block ciphers with
a reasonable block and key length – in our case 128 bits – can be constructed
that are resistant against linear and differential cryptanalysis but which can be
broken by computing an appropriate Gröbner basis.

The paper is organized as follows. In Section 2 we present two families of
ciphers, FLURRY, a Feistel network and CURRY, a SPN construction, together
with suitable parameters. We explain how to obtain polynomial equations describing the key recovery problem. Section 3 then introduces the methodology
to obtain estimates on the complexity of attacks using linear and differential
cryptanalysis. Section 4 details how Gröbner bases can be used break the ciphers and gives experimental results for selected examples. Finally we show how

---

* Supported by a stipend of the Marga und Kurt Möllgaard-Stiftung.

D. Pointcheval (Ed.): CT-RSA 2006, LNCS 3860, pp. 313–331, 2006.
© Springer-Verlag Berlin Heidelberg 2006

the key recovery problem for a subset of these ciphers is related to the problem of Gröbner basis conversion.

1.1   Notation

We define the notation that we will be used throughout the rest of this paper.

All operations of the block ciphers described in this paper are carried out over a finite field $F := GF(2^n)$ with $n \in \{8, 16, 32, 64\}$. We fix θ to be a generating element of F over $GF(2)$, i.e. $F := GF(2)(\theta)$.

The internal state of our cipher consists of multiple elements of F. To refer to individual elements of the state after the execution of a complete round transformation we use the following conventions:

- For Feistel ciphers, the internal state is represented by a vector. We use variables $x_i^{(e)}$ to denote elements of the internal state of the cipher after the eth application of the round function and variables $k_i^{(e)}$ to denote elements of the expanded key used in round e.
- For SPN ciphers, the internal state is represented by a square matrix. We denote the *internal state variables* after the eth application of the round function by $x_{i,j}^{(e)}$ and the *expanded key variables* by $k_{i,j}^{(e)}$.

We define the state of round 0 to be the initial state and call the variables of the initial state *plaintext variables*. Correspondingly the variables referring to the state after the execution of the last round are called *ciphertext variables*. The set of state variables of a cipher is denoted by $\mathcal{X}$, the set of expanded key variables by $\mathcal{K}$. All polynomials considered are then elements of the polynomial ring $R = F[\mathcal{X} \cup \mathcal{K}]$.

A power product of variables of $(\mathcal{X} \cup \mathcal{K})$ shall be called a *term*, whilst the product of a *term* and a *coefficient* $c \in F$ shall be called a *monomial*.

2   Description of the Cipher Families

In this section we give blueprints for Feistel and SPN ciphers that allow for simple algebraic representations. For these we select parameters sets offering a high resistance against differential and linear cryptanalysis and describe how to construct systems of polynomial equations for them.

2.1   The Feistel Case: FLURRY

We construct the family FLURRY(n, m, r, f, D) of Feistel ciphers. The parameters used are:

- $m \in \mathbb{N}$: the plaintext space, the ciphertext space and the cipher key space are F^{2m}.
- $r \in \mathbb{N}$: the number of rounds
- $f : F \to F$: a non-linear mapping giving the S-Box of the round function
- $D = (d_{i,j}) \in F^{m \times m}$: a matrix describing the linear diffusion mapping of the round function.

We set $R = (r_1, \ldots, r_m) \in F^m$, $L = (l_1, \ldots, l_m) \in F^m$ and $K = (k_1, \ldots, k_m) \in F^m$. The round function $\rho : F^m \times F^m \times F^m \to F^m \times F^m$ of a FLURRY cipher is then defined as:

$$\rho(L, R, K) = (R, G(R, K) + L)$$

with $G : F^m \times F^m \to F^m$ being the parallel application of m S-Boxes followed by a linear transform:

$$G(r_1, \ldots, r_m, k_1, \ldots, k_m) = D \times \begin{pmatrix} f(r_1 + k_1) \\ f(r_2 + k_2) \\ \vdots \\ f(r_m + k_m) \end{pmatrix}.$$

A plaintext (L_0, R_0) is encrypted into a ciphertext (L_r, R_r) by iterating the round function ρ over r rounds:

$$(L_i, R_i) = \rho(L_{i-1}, R_{i-1}, K_{i-1}) \qquad i = 1, 2, \ldots, r - 1$$
$$(L_r, R_r) = \rho(L_{r-1}, R_{r-1}, K_{r-1}) + (K_r, K_{r+1})$$

After the last round transformation, an additional key addition is performed on both halves of the state. Analogously, using the inverse round function ρ^{-1}

$$\rho^{-1}(L, R, K) = (G(L, K) + R, L)$$

we can decrypt a ciphertext with the following sequence of steps:

$$(L_{r-1}, R_{r-1}) = \rho^{-1}(L_r + K_r, R_r + K_{r+1}, K_{r-1})$$
$$(L_{i-1}, R_{i-1}) = \rho^{-1}(L_i, R_i, K_{i-1}) \qquad i = r - 1, r - 2, \ldots, 1$$

The number of F-components of a cipher key, plaintext or ciphertext is denoted by $t = 2m$.

The key schedule. The key schedule is affine over F. We write the cipher key as a tuple of vectors $(K_0, K_1) \in F^m \times F^m$. Let the round keys for the first two rounds be K_0, K_1 and recursively compute subsequent round keys for $2 \le i \le r + 1$ as follows:

$$K_i = D \cdot K_{i-1}^T + K_{i-2} + v_i$$

where D is the same matrix used in the round function of the cipher and the v_i are round constants:

$$v_i = ((\theta + 1)^i, (\theta + 1)^{i+1}, \ldots, (\theta + 1)^{i+m-1})$$

2.2 The SPN Case: CURRY

In this section we construct a family $\text{CURRY}(n, m, r, f, D)$ of ciphers similar to SQUARE [11]. We explain the parameters used:

- $m \in \mathbb{N}$: the plaintext space, the ciphertext space and the cipher key space are $F^{m \times m}$.
- $r \in \mathbb{N}$: the number of rounds
- $f : F \to F$: a bijective non-linear mapping giving the S-Box of the round function
- $D = (d_{i,j}) \in F^{m \times m}$: an invertible matrix used for diffusion

The round function $\rho : F^{m \times m} \times F^{m \times m} \to F^{m \times m}$ of a CURRY cipher is defined as:

$$\rho(S, K) = D \cdot G(S + K)^T$$

with $G : F^{m \times m} \to F^{m \times m}$ being the parallel application of m^2 S-Boxes:

$$G((s_{i,j})) = (f(s_{i,j}))$$

A plaintext S_0 is encrypted into a ciphertext S_r by iterating the round function ρ exactly r times followed by an additional key addition after the last round:

$$S_i = \rho(S_{i-1}, K_{i-1}) \qquad i = 1, 2, \ldots, r - 1$$
$$S_r = \rho(S_{r-1}, K_{r-1}) + K_r$$

Analogously, using the inverse round function ρ^{-1}

$$\rho^{-1}(S, K) = G^{-1}((D^{-1} \cdot S)^T) + K$$

we can decrypt a ciphertext with the following sequence of steps:

$$S_{r-1} = \rho^{-1}(S_r + K_r, K_{r-1})$$
$$S_{i-1} = \rho^{-1}(S_i, K_i) \qquad i = r - 1, r - 2, \ldots, 1$$

Just as for FLURRY, let the number of F-components of a key, plaintext or ciphertext be denoted by t, this time $t = m^2$.

The key schedule. For CURRY the first round key is equivalent to the cipher key $K_0 \in F^{m \times m}$. Just as for FLURRY the key schedule is affine over F. Subsequent round keys $K_i, i \geq 1$ are recursively computed as follows:

$$K_i = D \cdot K_{i-1} + M_i$$

where D is the same matrix used in the round function and $M_i = ((a_{j,l}))$ with $a_{j,l} = \theta^{i+(j-1)m+l}$. The matrices M_i are round constants.

2.3   Selected Parameters

We will now specify suitable parameters for the S-Box function and the linear transformation. These will be used to more thoroughly investigate instances of our cipher constructions throughout this paper. The number of rounds shall be left unspecified for now.

The S-Box functions. The only non-linear components of FLURRY and CURRY are the S-Boxes. In order to achieve good resistance against differential and linear cryptanalysis even for low number of rounds these must be chosen very carefully. Two important characteristics of a S-Box are its differential uniformity and its nonlinearity. These are defined as follows:

Definition 1. *Let $f : F \to F$ be a mapping and*

$$\delta = \max_{\substack{a,b \in F \\ a \neq 0}} \#\{x \in F \ : \ f(x+a) = f(x) + b\}.$$

Then f is called differentially δ-uniform.

In the following definition we use the bijective map

$$F \to GF(2)^n, a = \sum_{i=0}^{n-1} \left(a_i \theta^i \right) \mapsto (a_0, \dots, a_{n-1})$$

to identify F with $GF(2)^n$. For $a = (a_0, \dots, a_{n-1})$, $b = (b_0, \dots, b_{n-1})$ we set

$$\langle a, b \rangle = \sum_{i=0}^{n-1} a_i b_i$$

Definition 2. *The nonlinearity of a function $f : F \to F$ is defined as*

$$\mathcal{N}(f) = \min_{\substack{a,b \in F \\ b \neq 0}} \#\{x \in F \mid \langle x, a \rangle \neq \langle f(x), b \rangle\}$$

For monomial functions as well as the multiplicative inverse over finite fields of characteristic two the δ-uniformity and the nonlinearity have been well studied in the literature [22, 2, 13]. We want to keep the degree of our S-Box functions low in order to make Gröbner basis attacks feasible. Table 1 shows the S-Box functions that we have picked.

Table 1. S-Box mappings over $GF(2^n)$ with $n \in \{8, 16, 32, 64\}$

function	mapping	bijective over $GF(2^n)$	δ-uniformity	$\mathcal{N}(f)$
f_{-1}	$x \mapsto \begin{cases} x^{-1} & \text{iff } x \neq 0 \\ 0 & \text{iff } x = 0 \end{cases}$	yes	4	$2^{n-1} - 2^{\frac{n}{2}}$
f_3	$x \mapsto x^3$	no	2	$\geq 2^{n-1} - 2^{\frac{n}{2}}$
f_5	$x \mapsto x^5$	no	4	$\geq 2^{n-1} - 2^{\frac{n}{2}+1}$
f_7	$x \mapsto x^7$	yes	≤ 6	$\geq 2^{n-1} - 3 \cdot 2^{\frac{n}{2}}$

We call f_3, f_5 and f_7 *monomial S-Boxes* and f_{-1} the *inversion S-box*.

Lemma 1. *1. f_3 is a 2-uniform mapping*
2. f_{-1} and f_5 are 4-uniform mappings.
3. f_7 has δ-uniformity of 6 or less.

Proof. Obviously for all $a, b \in F$ with $a \neq 0$ the equation $x^7 + (x + a)^7 = b$ has at most 6 roots. For claims 1 and 2, see [22].

Lemma 2.   *1. The nonlinearity of f_{-1} is $2^{n-2} - 2^{\frac{n}{2}}$.*
  2. For a polynomial function $f : F \rightarrow F$ of degree d the following holds true:
    $\mathcal{N}(f) \geq 2^{n-1} - \lfloor \frac{d-1}{2} \rfloor 2^{\frac{n}{2}}$

Proof. For claim 1, see [13], for claim 2 see [4].

The linear transformations. We use matrices of Maximum Distance Separable codes – *MDS matrices* for short – for the matrix D in the linear layer and the key schedule. We chose these types of linear transformations since they have optimal diffusion properties. This strategy is widely used in modern block cipher design; all ciphers following the wide-trail design use diffusion optimal matrices. The matrix D_4 below actually is the matrix used in the `MixColumns` step of Rijndael, D_2 is equivalent to a Pseudo-Hadamard Transform over F.

$$D_2 = \begin{pmatrix} \theta & 1 \\ 1 & 1 \end{pmatrix} \qquad D_4 = \begin{pmatrix} \theta & \theta+1 & 1 & 1 \\ 1 & \theta & \theta+1 & 1 \\ 1 & 1 & \theta & \theta+1 \\ \theta+1 & 1 & 1 & \theta \end{pmatrix}$$

Rijmen and Daemen introduced the notion of the *branch number* of a linear transformation to measure the quality of the diffusion provided. For a F-vector $X := (x_1, \ldots, x_m)$ we define $w(X)$ to be the hamming weight of X, i.e. the count of all non-zero coordinates of this vector. The following definition is according to [12]:

Definition 3. *Let $M \in F^{m \times m}$ be a matrix describing a be a linear map. The differential branch number $\mathcal{B}_d(M)$ of M is then defined as*

$$\mathcal{B}_d(M) = \min_{\substack{X \in F^m \\ X \neq 0}} (w(X) + w(MX))$$

while the linear branch number $\mathcal{B}_l(M)$ is defined as $\mathcal{B}_l(M) = \mathcal{B}_d(M^T)$.

For a symmetric matrix such as D_2, the linear and the differential branch number clearly coincide. For the circulant matrix D_4 the linear and differential branch number coincide as well [12]. Thus in our case it suffices to speak of *the branch number* $\mathcal{B}(M)$ of a matrix M. For MDS matrices the branch number is maximal [12], i.e. $\mathcal{B}(M) = m+1$ with m being the size of the matrix M. For block ciphers with $m = 1$ we use the identity matrix of size one, I_1, trivially resulting in $\mathcal{B}(I_1) = 2$.

2.4   Polynomial Representation of the Ciphers

In the following we will detail how to obtain a system of polynomial equations that describes the transformation of a plaintext into a ciphertext block round by round using intermediate state variables. Please note that our description is slightly simplified. For the sake of legibility we have omitted the round key addition after the final round; for our experiments the final key addition has of course been retained.

- FLURRY

  For Feistel ciphers the left half of the state in round e is identical to the right half of the state in round $e - 1$, giving rise to the following mr trivial linear equations:

$$x_j^{(e)} + x_{j+m}^{(e-1)} = 0$$

  Each monomial S-Box of the cipher induces a polynomial equation of degree $\deg(f)$. Thus we get a total of mr non-linear equations of form:

$$x_{m+j}^{(e)} + x_j^{(e-1)} + \sum_{l=1}^{m} d_{j,l} \cdot f\left(x_{m+l}^{(e-1)} + k_l^{(e-1)}\right) = 0$$

  with $1 \le e \le r$, $1 \le j \le m$. When using the inversion S-Box the polynomial system is correct only with probability $\left(\frac{2^n - 1}{2^n}\right)^{mr}$. The equations in this case are of a different form:

$$\left(x_j^{(e-1)} + x_{m+j}^{(e)}\right) \prod_{i=1}^{m} \left(x_{m+i}^{(e-1)} + k_i^{(e-1)}\right) + \sum_{l=1}^{m} d_{j,l} \prod_{\substack{i=1 \\ i \ne l}}^{m} \left(x_{m+i}^{(e-1)} + k_i^{(e-1)}\right) = 0$$

  The linear equations for the key schedule of FLURRY can be written as:

$$k_j^{(e)} + k_j^{(e-2)} + (\theta + 1)^{et+j} + \sum_{l=1}^{m} d_{j,l} k_l^{(e-1)} = 0$$

  with $2 \le e \le r$, $1 \le j \le m$.
- CURRY

  No trivial linear equations hold between intermediate state variables. Denote by $x_{(i,j)}^{(e)}$ the variable in row i, column j of the state in round e, analogously for $k_{(i,j)}^{(e)}$. Then for all rounds $e > 0$ the following equations hold with $1 \le i, j \le m$:

$$x_{i,j}^{(e)} + \sum_{l=1}^{m} d_{i,l} \cdot f\left(x_{j,l}^{(e-1)} + k_{j,l}^{(e-1)}\right) = 0$$

  Again for f_{-1} the non-linear equations look different:

$$x_{i,j}^{(e)} \prod_{u=1}^{m} \left(x_{j,u}^{(e-1)} + k_{j,u}^{(e-1)}\right) + \sum_{l=1}^{m} d_{i,l} \prod_{\substack{u=1 \\ u \ne l}}^{m} \left(x_{j,u}^{(e-1)} + k_{j,u}^{(e-1)}\right) = 0$$

Using the above equations, the polynomial system also does not hold with probability one but with probability $\left(\frac{2^n-1}{2^n}\right)^{m^2 r}$.

The linear equations for the key schedule can be expressed as follows:

$$k_{i,j}^{(e)} + (\theta)^{e+(i-1)m+j} + \sum_{l=1}^{m} d_{i,l} k_{l,j}^{(e-1)} = 0$$

with $1 \le e \le r$, $1 \le i, j \le m$.

Additionally, for each variable $\mathfrak{v} \in (\mathcal{X} \cup \mathcal{K})$ the relation $\mathfrak{v}^{2^n} + \mathfrak{v} = 0$ holds. These relations are called *field equations*; they will not be included in our polynomial system however.

3   Resistance Against Classical Attacks

In this section we determine the strength of our cipher constructions against differential and linear cryptanalysis. Differential cryptanalysis is a chosen-ciphertext attack due to Biham and Shamir and was the first successful attack on the DES [3]. This type of attack exploits biases in the first order derivative of the cipher. For carefully chosen plaintexts with specific differences a cryptanalyst makes assumption about their propagation through the cipher and predicts output differences in ciphertext pairs. If these predictions are correct with sufficiently high probability they allow an attacker to determine round key bits.

Linear cryptanalysis is a known plaintext attack that was devised by Matsui [20] to attack the DES. For this attack to succeed, the cryptanalyst has to construct a probable key-independent linear approximation for individual output bits of the cipher. By counting the number of time this linear approximation agrees with the actual output of the cipher she can establish which value for the key bit is more likely.

The notion of *practical security* of block ciphers against differential and linear cryptanalysis was introduced by Knudsen [19]. The exact definition of this notion is postponed to the end of Section 3.2. We will derive the number of rounds that will make our cipher practically secure against differential and linear cryptanalysis.

Note that our objective was not to evaluate the strength of our ciphers against all known attacks. Our ciphers may very well be vulnerable against one or several advanced attacks even if they resist standard linear and differential cryptanalysis. Indeed, as an example we argue that the choices we have made for the S-Boxes are very weak against interpolation attacks.

3.1   Estimating the Resistance Against Differential and Linear Cryptanalysis

A fundamental parameter that influences the complexity of differential and linear attacks is the minimum number of active S-Boxes N over consecutive rounds of the cipher. Kanda [18] gives useful results on both SPN ciphers and Feistel ciphers with a SP round function; from these we derive the following lemma:

Lemma 3. *The minimum number of active S-boxes in 4, 6, 8 consecutive rounds of a Feistel cipher with SP round function is lower bounded by $\mathcal{B}(D)$, $\mathcal{B}(D) + 2$ and $2\mathcal{B}(D) + 1$ respectively. For an SPN cipher the minimum number of active S-Boxes for 2r consecutive rounds is lower bounded by $r\mathcal{B}(D)$.*

In the following X denotes a uniformly distributed random variable in $GF(2)^n$ and $\rho : GF(2)^n \rightarrow GF(2)^n$ a function for which we wish to compute the linear and differential probability.

Definition 4. *The linear probability for a pair $(a, b) \in GF(2)^n \times GF(2)^n$ with $a \neq 0$ is defined as*

$$LP(a, b) = (2 \cdot \Pr_X \{\langle a, X \rangle = \langle b, \rho(X) \rangle\} - 1)^2$$

In the above definition, a is called *input mask* and b is called *output mask* of a round. A vector of masks $A = (a_1, \ldots, a_{r+1})$ with $a_i \neq 0$ for all $1 \leq i \leq r$ is called *linear characteristic* of a cipher.

Definition 5. *The differential probability for a pair $(\Delta x, \Delta y) \in GF(2)^n \times GF(2)^n$ with $\Delta x \neq 0$ is defined as*

$$DP(\Delta x, \Delta y) = \Pr_X \{\rho(X) + \rho(X + \Delta x) = \Delta y\}$$

The value Δx is called *input difference* of a round, while Δy is called *output difference*. A vector of differences $A = (a_1, \ldots, a_{r+1})$ with $a_i \neq 0$ for all $1 \leq i \leq r$ is called *differential characteristic* of a cipher.

Definition 6. *Let Ω_L be the set of all linear characteristics and Ω_D the set of all differential characteristics of a cipher C. The maximum linear characteristic probability (MLCP) of C is*

$$MLCP(C) = \max_{A \in \Omega_L} \prod_{i=1}^{r} LP(a_i, a_{i+1})$$

Analogously the maximum differential characteristic probability (MDCP) of C is

$$MDCP(C) = \max_{A \in \Omega_D} \prod_{i=1}^{r} DP(a_i, a_{i+1}).$$

3.2   Differential and Linear Cryptanalysis of FLURRY and CURRY

In this section we show how to compute upper bounds of MLCP and the MDCP of ciphers of the FLURRY and CURRY family. From these bounds we can deduce the number of rounds required to make an instance practically secure against differential and linear cryptanalysis.

The maximum differential probability of a function $f : F \rightarrow F$ can be calculated from δ as $p(f) = \frac{\delta}{\#F}$ where δ is according to Definition 1. The maximum linear probability of a mapping $f : F \rightarrow F$ can be computed as

$$q(f) = \left(1 - \frac{2\mathcal{N}(f)}{\#F}\right)^2$$

where $\mathcal{N}(f)$ is defined as in Section 2.3. For SPN ciphers and Feistel ciphers with a SP round function the MDCP is bounded by $p(f)^N$ while the MLCP is bounded by $q(f)^N$ [18], where N is the miminum number of active S-Boxes.

According to Knudsen [19], a block cipher with dependent round keys is practically secure against differential and linear cryptanalysis if the MLCP and the MDCP is too low for an attack to work under the assumption of independent round keys. Note however that for both r-round Feistel and r-round SPN ciphers, we need to consider the MLCP and MDCP of $r - 2$ rounds because of attacks that guess bits of the first and the last round key, so-called 2R attacks.

3.3 Interpolation Attacks

Jakobsen and Knudsen presented interpolation attacks in [16] as a counterpoint to the growing trend of using algebraic S-Boxes such as those proposed by Nyberg [22]. In fact, interpolation attacks can be seen as the first algebraic attacks on block ciphers. The underlying intuition of this attack is that the relationship between plaintext and ciphertext can be expressed as a tuple of polynomial expressions. If the degree of these polynomials is low enough, the coefficients of the polynomials can be interpolated from a number of plaintext/ciphertext pairs. A key–dependent equivalent of the encryption or the decryption algorithm has then been determined. In [16] upper bounds on the number of required pairs for known-plaintext interpolation attacks for selected examples are given. In general this number increases exponentially with the degree of the polynomial function describing the S-Box, the number of rounds and the number of elements in the internal state, while for the attacks we present in the next section it remains a constant quantity.

Courtois later improved on the work of Jakobsen and Knudsen and introduced an attack called General Linear Cryptanalysis [8]. In the same paper he also gives several examples of insecure ciphers based on inversion based S-Boxes that resist differential and linear cryptanalysis. His approach and his goals are quite different from ours however.

FLURRY and CURRY quite naturally are susceptible to interpolation attacks – their clean structure and the monomial S-boxes make them textbook examples. As a matter of fact, the cipher $\mathcal{PURE}$ presented in the original article is identical to the 64-bit cipher FLURRY$(32, 1, r, f_3, I_1)$ sans key scheduling.

4   Attacks Using Gröbner Bases

Gröbner bases are standard bases of polynomial ideals that can be used for solving systems of polynomial equations. What Gaussian elimination does for systems of linear equations, Gröbner basis algorithms try to emulate for polynomial systems. Unfortunately the computational complexity of Gröbner basis algorithms for nonlinear systems is no longer polynomial. In this paper we restrict ourselves to known-plaintext *Gröbner Basis attacks* that recover a secret key of a block cipher from a minimum number of plaintext/ciphertext pairs faster than a sequential exhaustive search of the key space – by computing Gröbner Bases.

We will briefly introduce the concepts necessary to explain our results. For a more thorough introduction to Gröbner bases we refer the reader to [1] and [10]. In the following we adopt the conventions of [1].

Definition 7 (Term order). *A term order $\leq$ is a linear order on the set of terms $\mathcal{T}(R)$ such that*

1. *$1 \leq t$ for all terms $t \in \mathcal{T}(R)$*
2. *for all terms $s, t_1, t_2 \in \mathcal{T}(R)$ whenever $t_1 \leq t_2$ then $st_1 \leq st_2$*

If a term order has been fixed, we define $\mathrm{HT}(f)$ to be the greatest term occuring in the polynomial $f \in R$ according to this order; this term is called the head term. Correspondingly $\mathrm{HM}(f)$ is the head monomial, i.e. the head term of f multiplied with the matching coefficient.

We will now introduce two useful and widely used term orders. To accomplish this we first need to define some technicalities: For a term $t = \mathfrak{v}_1^{e_1} \mathfrak{v}_2^{e_2} \cdots \mathfrak{v}_k^{e_k} \in \mathcal{T}(R)$ we define the *exponent vector* of t to be $\epsilon(t) = (e_1, e_2, \ldots, e_k) \in \mathbb{N}_0^k$. The total degree of the term t then is $\deg(t) = \sum_{i=1}^k e_i$.

Example 1 (Lexicographic term order). For terms s, t we define $s <_{lex} t$ iff there exists an i with $1 \leq i \leq k$ such that the first $i - 1$ components of $\epsilon(s)$ and $\epsilon(t)$ are equal but the ith component of $\epsilon(s)$ is smaller than the ith component of $\epsilon(t)$.

Example 2 (Degree reverse lexicographic term order). For terms s, t we define $s <_{DRL} t$ iff either $\deg(s) < \deg(t)$ or if $\deg(s) = \deg(t)$ and $s <_{lex} t$.

Definition 8 (Syzygy polynomial). *The syzygy polynomial of two polynomials f, g is defined as*

$$spol(f, g) = \frac{lcm(HM(f), HM(g))}{HM(f)} f - \frac{lcm(HM(f), HM(g))}{HM(g)} g$$

For a set of polynomials $G \subset R$ we can define the reduction of a polynomial $f \in R$ to a remainder r which we will denote by $f \to_G r$. The result of this operation may not be uniquely defined unless G is a Gröbner basis. In the following we will only be interested in polynomial divisions that leave no remainder.

Definition 9 (Reduction to zero). *A polynomial $f \in R$ reduces to zero modulo a set $G = \{g_1, \ldots, g_k\} \subset R$, if there exists a vector of polynomials $(m_1, \ldots, m_k)$ such that $f - \sum_{i=1}^k m_i g_i = 0$ with $HT(m_i g_i) \leq HT(f)$ for all $1 \leq i \leq k$.*

Definition 10 (Gröbner basis). *Let $\mathfrak{I}$ be an ideal of R. A finite set of polynomials $G \subset \mathfrak{I}$ is a Gröbner basis of $\mathfrak{I}$ if $f \to_G 0$ holds for every $f \in \mathfrak{I}$.*

Let $\mathcal{P}$ be a set of multivariate polynomial equations $p_i = 0$. For the ideal $\mathfrak{I}$ generated by the set $P = \{p_i\}$ computing the Gröbner basis relative to an appropriate term order, e.g the lexicographical term order enables us to solve the system $\mathcal{P}$.

Computing a Gröbner basis relative to a total-degree order however usually is faster than computing a lexicographical Gröbner basis of the same ideal. This was the reason for the development of algorithms that change the term order of a Gröbner basis. The two most prominent are the FGLM algorithm [15] and the Gröbner Walk [7]. While the FGLM algorithm as originally described only works for zero-dimensional ideals, i.e. when the number of solutions of $\mathcal{P}$ in the closure of F is finite, the Gröbner Walk does not have this restriction.

4.1   Key Recovery Using Gröbner Bases

Estimating the time and space complexity of Gröbner basis algorithms is no easy feat. For polynomial systems induced by block ciphers, no theoretical works estimating the performance of Gröbner basis algorithms are currently known. We therefore carried out experiments to study the resistance of our ciphers against Gröbner Basis attacks. Results of these experiments are presented and analysed in section 4.2.

The Gröbner basis attack we have successfully used on instances of FLURRY and CURRY to determine the secret key from a small number of plaintext/ciphertext pairs entailed the following steps:

1. Set up a polynomial system $\mathcal{P} = \{p_i = 0\}$ for the cipher in question with $p_i \in R$ as described in Section 2.4. The system $\mathcal{P}$ consists of both cipher and key schedule equations.
2. Request a plaintext/ciphertext pair $((P_1, \ldots P_t), (C_1, \ldots, C_t))$. This gives rise to the following additional system of linear equations $\mathcal{G} = \{g_i = 0\}$:

$$x_1^{(0)} + P_1 = 0 \qquad\qquad\qquad x_1^{(r)} + C_1 = 0$$
$$\vdots \qquad\qquad\qquad\qquad\qquad \vdots$$
$$x_t^{(0)} + P_t = 0 \qquad\qquad\qquad x_t^{(r)} + C_t = 0$$

  Let $\mathfrak{I}$ be the ideal generated by the set of polynomials $\mathcal{L} = (\bigcup_i \{p_i\}) \cup (\bigcup_i \{g_i\})$. We call this ideal the *key recovery ideal*.
3. Compute a degree-reverse lexicographic Gröbner basis G_{DRL} of $\mathfrak{I}$. For ciphers using a multiplicative inverse as S-Box function, the system may be inconsistent, resulting in $G_{DRL} = 1$.
4. If $G_{DRL} = 1$ go to Step 2, otherwise proceed.
5. Use a Gröbner basis conversion algorithm to obtain a lexicographical Gröbner basis G_{lex} from G_{DRL}. The variable ordering should be such that the key variables of the first round are the least elements.
6. Compute the variety Z of $\mathfrak{I}$ using the Gröbner basis G_{lex}.
7. Request another plaintext/ciphertext pair (P', C').

8. Try all elements $k \in Z$ as key candidates to encrypt P'. If k does not encrypt P' to C', remove k from Z, otherwise retain.
9. If Z contains more than one element, go to step 7.
10. Terminate

Considerable complexity is hidden in step 6. To compute the variety of an ideal using a lexicographical Gröbner basis, we need to successively eliminate variables by computing zeroes of univariate polynomials and back-substituting results. The complexity of this depends on the number of solutions of the polynomial system (zeroes of the ideal) and the complexity of the algorithm for finding roots of univariate polynomials. The best algorithm for factoring polynomials is due to Kaltofen and Shoup [17] and has a complexity of $O(d^{1.815}n)$ field operations, where d is the degree of the polynomial. This degree if bounded by $2^n - 1$. The number zeroes is equivalent to the number of distinct keys encrypting the plaintext to a ciphertext. In general we can expect this number to be small.

4.2 Experimental Results

We have performed experiments to analyze the resistance of FLURRY and CURRY using the computer algebra system MAGMA [23], version 2.11-8, on an AMD Athlon 64 3200+ equipped with 1024 Megabytes of RAM running Linux. MAGMA implements Faugére's F4 algorithm [14] and is widely considered the best publicly available tool for computing Gröbner bases. We have chosen n and m such that the ciphers evaluated are 128-bit block ciphers.

Table 2 lists a number of instantiations of FLURRY and CURRY ciphers for which we were able to successfully recover the secret key; the 6, 8 and 10 round FLURRY ciphers are resistant to linear and differential cryptanalysis. We see that ciphers with inversion-based S-boxes are easier to break than ciphers which use a monomial S-box, even if the monomial is of very low degree. Furthermore we were unable to determine an a priori indicator for selecting the most efficient Gröbner basis conversion algorithm – in some cases FGLM was faster, in other cases the Gröbner walk; the same holds for the memory consumption. As mentioned in Section 2.4 we did not add the field equations to our polynomial systems.

4.3 Gröbner Bases Without Polynomial Reductions

Sometimes one can determine whether a set of polynomials forms a Gröbner basis without computing normal forms. In the following let be $G \subset R$ be a finite set of polynomials with $0 \neq G$.

Proposition 1 (First Buchberger criterion). *Suppose that we have $f, g \in G$ such that*

$$lcm(HT(f), HT(g)) = HT(f) \cdot HT(g)$$

i.e the head terms of f and g are pairwise prime. Then $spol(f, g) \rightarrow_G 0$.

Proposition 1 is the first Buchberger criterion. Together with the following theorem given in [10], we can decide whether a sequence of polynomials is a Gröbner basis from looking at the head terms alone.

Table 2. Experimental results obtained with MAGMA

cipher	conversion	CPU time	memory used
FLURRY$(64, 1, 4, f_{-1}, I_1)$	Walk	0.011 s	3.48 MBytes
FLURRY$(64, 1, 4, f_{-1}, I_1)$	FGLM	0.011 s	3.48 MBytes
FLURRY$(64, 1, 4, f_3, I_1)$	Walk	0.04 s	3.48 MBytes
FLURRY$(64, 1, 4, f_3, I_1)$	FGLM	0.029 s	3.58 MBytes
FLURRY$(64, 1, 4, f_5, I_1)$	Walk	1.28 s	3.97 MBytes
FLURRY$(64, 1, 4, f_5, I_1)$	FGLM	2.3 s	6.36 MBytes
FLURRY$(64, 1, 4, f_7, I_1)$	Walk	13.61 s	6.22 MBytes
FLURRY$(64, 1, 4, f_7, I_1)$	FGLM	82.62 s	33.4 MBytes
FLURRY$(64, 1, 6, f_{-1}, I_1)$	Walk	0.15 s	3.58 MBytes
FLURRY$(64, 1, 6, f_{-1}, I_1)$	FGLM	0.059 s	3.58 MBytes
FLURRY$(64, 1, 6, f_3, I_1)$	Walk	59.91 s	10.63 MBytes
FLURRY$(64, 1, 6, f_3, I_1)$	FGLM	145.08 s	193.24 MBytes
FLURRY$(64, 1, 8, f_{-1}, I_1)$	Walk	3.43 s	4.51 MBytes
FLURRY$(64, 1, 8, f_{-1}, I_1)$	FGLM	1.46 s	4.46 MBytes
FLURRY$(64, 1, 10, f_{-1}, I_1)$	Walk	115.44 s	14.74 MBytes
FLURRY$(64, 1, 10, f_{-1}, I_1)$	FGLM	60.61 s	12.39 MBytes
FLURRY$(64, 1, 12, f_{-1}, I_1)$	Walk	4194.28 s	99.97 MBytes
FLURRY$(64, 1, 12, f_{-1}, I_1)$	FGLM	2064 s	142.90 MBytes
FLURRY$(32, 2, 4, f_{-1}, D_2)$	Walk	216.53 s	25.58 MBytes
FLURRY$(32, 2, 4, f_{-1}, D_2)$	FGLM	65.78 s	41.62 MBytes
FLURRY$(16, 4, 2, f_{-1}, D_4)$	Walk	264 s	37.13 MBytes
FLURRY$(16, 4, 2, f_{-1}, D_4)$	FGLM	26.119 s	18.56 MBytes
CURRY$(32, 2, 3, f_{-1}, D_2)$	Walk	1750.87 sec	138.77 MBytes
CURRY$(32, 2, 3, f_{-1}, D_2)$	FGLM	3676.26 sec	107.54 MBytes

Theorem 1. *The set G is a Gröbner basis iff $spol(f, g) \rightarrow_G 0$ for all $f, g \in G$ with $f \neq g$.*

When using polynomial S-boxes, this enables us to compute a degree-reverse lexicographic Gröbner bases of the key-recovery ideals of FLURRY and CURRY without performing polynomial reductions; the head terms of all polynomials of $\mathfrak{I}$ are univariate. For each polynomial of round e, either a power of a state variable of the preceeding round or a power of a key variable of the current round occur as head term. Some head terms however occur more than once.

By using an appropriate variable order we can force the set of head terms of each round to be disjunct from the set of head terms of all other rounds:

- CURRY
  For better legibility, we identify $x_{i,j}^{(e)}$ with $x_{et+im+j}$ and $k_{i,j}^{(e)}$ with $k_{et+im+j}$. We then fix the following variable order:

$$\underbrace{x_0 < \ldots < x_{t-1}}_{\text{plaintext variables}} < \underbrace{x_{tr} < \ldots < x_{t(r+1)-1}}_{\text{ciphertext variables}} < \underbrace{k_0 < \ldots < k_{t(r+1)-1}}_{\text{key variables}} < \underbrace{x_t < \ldots < x_{tr-1}}_{\substack{\text{internal state} \\ \text{variables}}}$$

– FLURRY

Again we decrease the number of indexes: we identify $x_i^{(e)}$ with x_{et+i} and $k_i^{(e)}$ with k_{et+i}. We then fix the following variable order:

$$\underbrace{x_0 < \ldots < x_{t-1}}_{\text{plaintext variables}} < \underbrace{x_{tr} < \ldots < x_{(t+1)r-1}}_{\text{ciphertext variables}} < \underbrace{x_{t(r-1)+m} < \ldots < x_{tr-1}}_{\substack{\text{state variables of the right} \\ \text{half of the second last round}}} <$$

$$\underbrace{k_0 < \ldots < k_{m-1}}_{\substack{\text{key variables of} \\ \text{the first round}}} < \underbrace{k_{m(r-1)} < \ldots < k_{mr-1}}_{\text{key variables of round } r} <$$

$$\underbrace{k_m < \ldots < k_{m(r-1)-1} < k_{mr} < \ldots < k_{m(r+2)-1}}_{\text{remaining key variables}} < \underbrace{x_t < \ldots < x_{t(r-1)+m-1}}_{\text{remaining state variables}}$$

To make the following linear transformation easier to describe we use a vectorial representation for FLURRY and a matrix representation for CURRY. The entries in the vector and matrix of each round are the left-hand side polynomials of the nonlinear cipher equations.

We can multiply the vectors respectively matrices of all rounds by D^{-1} to obtain pairwise prime head terms within each and across rounds. For CURRY this is sufficient. For FLURRY we also need to adjust the key schedule equations. The nonlinear polynomials of the first and the last round have powers of key variables as head terms. These key variables are of the first and the last round respectively. For the first round this poses no problem. However for the last round the key schedule polynomials that produce the last round key have the same head terms. Thus we rewrite the key schedule equations. We express all round keys except for the last round key as a linear combination of the first two round keys. Then we write the second round key as a linear combination of the first and the last round key. This results in all head terms being pairwise prime. In order for this to work for FLURRY, the order of the matrix used in the key schedule needs to be greater than the number of rounds.

We have shown how to make the head terms of all polynomials pairwise prime. Hence by Theorem 1, we have obtained a Gröbner basis. This strategy however does not work FLURRY and CURRY instances with inversion S-Boxes, as the head terms in these cases are never univariate.

4.4 Complexity of Gröbner Basis Conversions Using FGLM

The complexity of the FGLM algorithm hinges on two parameters of the input G: the number of variables of the polynomial ring R and the vector space dimension of the residue class ring $R/\mathfrak{I}$, where $\mathfrak{I}$ is the ideal generated by the Gröbner basis $G \subset R$. The following theorem [1] shows how this invariant of an ideal can be computed.

Theorem 2. *Let G be a Gröbner basis of the ideal $\mathfrak{I}$. Then*

$$\dim(R/\mathfrak{I}) = \#\{t \in \mathcal{T}(R) : HT(f) \nmid t \text{ for all } f \in G\}$$

Corollary 1. *Let* $G = \{g_1, \ldots, g_k\}$ *be a Gröbner basis for the ideal* $\Im \subset F[x_1, \ldots, x_k]$ *with head terms* $x_1^{d_1}, \ldots, x_k^{d_k}$. *Then* $\dim(R/\Im) = \prod_{i=1}^{k} d_i$.

Corollary 2. *Let* $\Im$ *be an ideal of an instantiation of either a* FLURRY *or a* CURRY *cipher as described in Section 2.4 and* f *a polynomial function. Then the following holds:*

1. $dim(R/\Im) = \deg(f)^{mr}$ *for* FLURRY(n, m, r, f, D).
2. $dim(R/\Im) = \deg(f)^{m^2r}$ *for* CURRY(n, m, r, f, D).

We restate Theorem 5.1 of [15].

Theorem 3. *Let* K *be a finite field and* $R = K[x_1, \ldots, x_k]$. *Furthermore* $G_1 \subset R$ *is the Gröebner basis relative to a term order* $<_1$ *of an ideal* $\Im$, *and* $d = dim(R/\Im)$. *We can then convert* G_1 *into a Gröbner basis* G_2 *relative to a term order* $<_2$ *in* $O(kd^3)$ *field operations.*

We conjecture the constant factor in the above estimate to be approximately one cipher operation. For the space complexity of the algorithm, no bound is given in the original paper. We note that the dominant memory requirement of the FGLM algorithm is a $d \times kd$ matrix over F. Thus the memory usage of the algorithm is upper bounded by $\lceil (kd^2n)/8 \rceil + o(1)$ bytes.

This allows us to estimate the maximum resistance of FLURRY and CURRY ciphers with polynomial S-Boxes against Gröbner basis attacks (see Table 3). Note that for the CURRY cipher we need to use a bijective S-Box in the round function; the lowest degree S-Box function that is bijective is f_7.

Table 3. Upper bounds on the complexity of breaking 128-bit FLURRY and CURRY ciphers with FGLM

cipher	n	$\dim(R/I)$	# of operations	memory required (bytes)
FLURRY$(32, 2, 4, f_3, D_2)$	8	$3^8 \approx 2^{12.68}$	$O(2^{41.0})$	$2^{30.4}$
FLURRY$(32, 2, 4, f_5, D_2)$	8	$5^8 \approx 2^{18.58}$	$O(2^{58.7})$	$2^{42.2}$
FLURRY$(32, 2, 4, f_7, D_2)$	8	$7^8 \approx 2^{22.46}$	$O(2^{70.4})$	$2^{49.9}$
FLURRY$(32, 2, 6, f_3, D_2)$	12	$3^{12} \approx 2^{19.02}$	$O(2^{60.6})$	$2^{43.2}$
FLURRY$(32, 2, 6, f_5, D_2)$	12	$5^{12} \approx 2^{27.86}$	$O(2^{87.2})$	$2^{61.3}$
FLURRY$(32, 2, 6, f_7, D_2)$	12	$7^{12} \approx 2^{33.69}$	$O(2^{104.7})$	$2^{73.0}$
FLURRY$(32, 2, 8, f_3, D_2)$	16	$3^{16} \approx 2^{25.36}$	$O(2^{80.0})$	$2^{56.7}$
FLURRY$(32, 2, 8, f_5, D_2)$	16	$5^{16} \approx 2^{37.15}$	$O(2^{115.5})$	$2^{80.3}$
FLURRY$(32, 2, 8, f_7, D_2)$	16	$7^{16} \approx 2^{44.92}$	$O(2^{138.8})$	$2^{95.8}$
FLURRY$(16, 4, 4, f_3, D_2)$	16	$3^{16} \approx 2^{25.36}$	$O(2^{80.0})$	$2^{55.7}$
FLURRY$(16, 4, 4, f_5, D_2)$	16	$5^{16} \approx 2^{37.15}$	$O(2^{115.5})$	$2^{79.3}$
FLURRY$(16, 4, 4, f_7, D_2)$	16	$7^{16} \approx 2^{44.92}$	$O(2^{138.8})$	$2^{94.8}$
CURRY$(32, 2, 3, f_7, D_2)$	12	$7^{12} \approx 2^{33.69}$	$O(2^{104.6})$	$2^{73.0}$

5   Conclusions

We have demonstrated that Gröbner basis algorithms can be used to successfully mount key-recovery attacks on algebraically simple block ciphers with a large block and key size; even when these ciphers are practically secure against differential and linear cryptanalysis. Key recovery can be accomplished with a minimal number of known plaintext/ciphertext pairs. Degree-reverse lexicographical Gröbner bases for our ciphers can be calculated by hand. These however do not give the solution to the polynomial system directly. Our contribution shows that the problem of recovering a key for these block ciphers can be reduced to a Gröbner basis conversion. By giving a formula for the vector space dimension of the polynomial ring modulo the key recovery ideal for all inversion-free ciphers considered we were able to estimate the complexity of a Gröbner basis conversion using the FGLM algorithm.

Acknowledgments. The authors would like to thank the anonymous referees for their comments. The third author acknowledges several fruitful discussions with Frederik Armknecht and Stefan Lucks.

References

1. Thomas Becker and Volker Weispfenning. *Gröbner Bases – A Computational Approach to Commutative Algebra*. Springer–Verlag, 1991.
2. Thomas Beth and Cunsheng Ding. On Almost Perfect Nonlinear Permutations. In Tor Helleseth, editor, *Advances in Cryptology – EUROCRYPT '93*, volume 765 of *Lecture Notes in Computer Science*, pages 65–76. Springer–Verlag, 1994.
3. Eli Biham and Adi Shamir. Differential Cryptanalysis of DES-like Cryptosystems. In Alfred Menezes and Scott A. Vanstone, editors, *Advances in Cryptology – CRYPTO '90*, volume 537 of *Lecture Notes in Computer Science*, pages 2–21. Springer–Verlag, 1991.
4. Jung Hee Cheon, Seongtaek Chee, and Choonsik Park. S-boxes with Controllable Nonlinearity. In Jacques Stern, editor, *Advances in Cryptology – EUROCRYPT '99*, volume 1592 of *Lecture Notes in Computer Science*, pages 286–294. Springer–Verlag, 1999.
5. Carlos Cid and Gaëtan Laurent. An Analysis of the XSL Algorithm. In C. Pandu Rangan, editor, *Advances in Cryptology – ASIACRYPT 2005*, volume 3788 of *Lecture Notes in Computer Science*, pages 333–353. Springer–Verlag, 2005.
6. Carlos Cid, Sean Murphy, and Matt Robshaw. Small Scale Variants of the AES. In Helena Handschuh and Henri Gilbert, editors, *Fast Software Encryption – FSE 2005*, Lecture Notes in Computer Science, pages 145–162. Springer–Verlag, 2005.
7. Stéphane Collart, Michael Kalkbrener, and Daniel Mall. Converting Bases with the Gröbner Walk. *Journal of Symbolic Computation*, 24(3/4):465–469, 1997.
8. Nicolas Courtois. The Inverse S-box, Non-linear Polynomial Relations and Cryptanalysis of Block Ciphers. In Hans Dobbertin, Vincent Rijmen, and Aleksandra Sowa, editors, *AES 4 Conference*, volume 3373 of *Lecture Notes in Computer Science*, pages 170–188. Springer–Verlag, 2005.

9. Nicolas Courtois and Josef Pieprzyk. Cryptanalysis of Block Ciphers with Overdefined Systems of Equations. In Yuliang Zheng, editor, *Advances in Cryptology – ASIACRYPT 2002*, volume 2501 of *Lecture Notes in Computer Science*, pages 267–287. Springer–Verlag, 2002.
10. David A. Cox, John B. Little, and Don O'Shea. *Ideals, Varieties, and Algorithms*. Springer–Verlag, NY, 2nd edition, 1996. 536 pages.
11. Joan Daemen, Lars Knudsen, and Vincent Rijmen. The block cipher Square. In Eli Biham, editor, *Fast Software Encryption – FSE 1997*, volume 1267 of *Lecture Notes in Computer Science*, pages 149–165. Springer–Verlag, 1997.
12. Joan Daemen and Vincent Rijmen. *The Design of Rijndael: The Wide Trail Strategy*. Springer–Verlag, 2001.
13. Hans Dobbertin. One-to-One Highly Nonlinear Power Functions on $GF(2^n)$. *Applicable Algebra in Engineering, Communication and Computing*, 9(2):139–152, 1998.
14. Jean-Charles Faugère. A New Efficient Algorithm for Computing Gröbner bases (F4). *Journal of Pure and Applied Algebra*, 139(1-3):61–88, June 1999.
15. Jean-Charles Faugère, P. Gianni, Daniel Lazard, and Teo Mora. Efficient Computation of Zero-Dimensional Gröbner Bases by Change of Ordering. *Journal of Symbolic Computation*, 16(4):329–344, 1993.
16. Thomas Jakobsen and Lars Knudsen. The Interpolation Attack on Block Ciphers. In Eli Biham, editor, *Fast Software Encryption – FSE 1997*, volume 1267 of *Lecture Notes in Computer Science*, pages 28–40. Springer–Verlag, 1997.
17. Erich Kaltofen and Victor Shoup. Subquadratic-time Factoring of Polynomials over Finite FIelds. *Mathematics of Computation*, 67(223):1179–1197, 1998.
18. Masayuki Kanda. Practical Security Evaluation against Differential and Linear Cryptanalyses for Feistel Ciphers with SPN Round Function. In Douglas R. Stinson and Stafford E. Tavares, editors, *Selected Areas in Cryptography – SAC 2000*, volume 2012 of *Lecture Notes in Computer Science*, pages 324–338. Springer–Verlag, 2001.
19. Lars R. Knudsen. Practically Secure Feistel Ciphers. In Ross J. Anderson, editor, *Fast Software Encryption – FSE 1993*, volume 809 of *Lecture Notes in Computer Science*, pages 211–221. Springer–Verlag, 1994.
20. M. Matsui. Linear Cryptanalysis Method for DES Cipher. In Douglas R. Stinson, editor, *Advances in Cryptology – CRYPTO '93*, volume 773 of *Lecture Notes in Computer Science*, pages 386–387. Springer–Verlag, 1994.
21. Sean Murphy and Matthew J.B. Robshaw. Essential Algebraic Structure within the AES. In Moti Yung, editor, *Advances in Cryptology – CRYPTO 2002*, volume 2442 of *Lecture Notes in Computer Science*, pages 1–16. Springer–Verlag, 2002.
22. Kaisa Nyberg. Differentially Uniform Mappings for Cryptography. In Tor Helleseth, editor, *Advances in Cryptology – EUROCRYPT '93*, volume 765 of *Lecture Notes in Computer Science*, pages 55–64. Springer–Verlag, 1994.
23. University of Sydney Computational Algebra Group. The Magma Computational Algebra System, 2004. http://magma.maths.usyd.edu.au/magma/.

A   A DRL Gröbner Basis for FLURRY$(32, 2, 4, f_3, D_2)$

The following sequence of polynomials G is a degree-reverse lexicographic Gröbner basis for a FLURRY$(32, 2, 4, f_3, D_2)$ for the following variable ordering:
$x_0 < x_1 < x_2 < x_3 < x_{16} < x_{17} < x_{18} < x_{19} < x_{14} < x_{15} < k_0 < k_1 < k_6 < k_7 < k_2 <$
$k_3 < k_4 < k_5 < k_8 < k_9 < k_{10} < k_{11} < x_4 < x_5 < x_6 < x_7 < x_8 < x_9 < x_{10} < x_{11} <$
$x_{12} < x_{13}$

$G = \{$

plaintext:
$x_0 + \theta^{31} + \theta^{29} + \theta^{27} + \theta^{24} + \theta^{22} + \theta^{21} + \theta^{19} + \theta^{13} + \theta^{11} + \theta^8 + \theta^7 + \theta^6 + \theta^4 + 1$
$x_1 + \theta^{31} + \theta^{30} + \theta^{29} + \theta^{22} + \theta^{21} + \theta^{15} + \theta^{14} + \theta^{11} + \theta^{10} + \theta^7 + \theta^6 + \theta^5 + \theta^3 + \theta$
$x_2 + \theta^{26} + \theta^{25} + \theta^{24} + \theta^{21} + \theta^{19} + \theta^{18} + \theta^{16} + \theta^{14} + \theta^8 + \theta^7 + \theta^6 + \theta^4 + \theta + 1$
$x_3 + \theta^{27} + \theta^{26} + \theta^{24} + \theta^{21} + \theta^{17} + \theta^{15} + \theta^{13} + \theta^{11} + \theta^9 + \theta^6 + \theta^4 + \theta$

ciphertext:
$x_{16} + \theta^{31} + \theta^{29} + \theta^{21} + \theta^{19} + \theta^{18} + \theta^{16} + \theta^{15} + \theta^{14} + \theta^{12} + \theta^4 + 1$
$x_{17} + \theta^{24} + \theta^{21} + \theta^{20} + \theta^{18} + \theta^{16} + \theta^{13} + \theta^{10} + \theta^9 + \theta^8 + \theta^6 + \theta^5 + \theta^3 + \theta + 1$
$x_{18} + \theta^{29} + \theta^{25} + \theta^{21} + \theta^{20} + \theta^{19} + \theta^{13} + \theta^{10} + \theta^9 + \theta^8 + \theta^7 + \theta^6 + \theta^5 + \theta^3$
$x_{19} + \theta^{29} + \theta^{27} + \theta^{26} + \theta^{20} + \theta^{13} + \theta^{10} + \theta^8 + \theta^5 + \theta^2$

round 1:
$x_4 + x_2$
$x_5 + x_3$
$k_0^3 + k_0^2 x_2 + k_0 x_2^2 + x_2^3 + C_1 x_7 + C_1 x_6 + C_1 x_1 + C_1 x_0$
$k_1^3 + k_1^2 x_3 + k_1 x_3^2 + x_3^3 + C_2 x_7 + C_1 x_6 + C_2 x_1 + C_1 x_0$

round 2:
$x_8 + x_6$
$x_9 + x_7$
$x_6^3 + x_6^2 k_2 + x_6 k_2^2 + k_2^3 + C_1 x_{11} + C_1 x_{10} + C_1 x_5 + C_1 x_4$
$x_7^3 + x_7^2 k_3 + x_7 k_3^2 + k_3^3 + C_2 x_{11} + C_1 x_{10} + C_2 x_5 + C_1 x_4$

round 3:
$x_{12} + x_{10}$
$x_{13} + x_{11}$
$x_{10}^3 + x_{10}^2 k_4 + x_{10} k_4^2 + k_4^3 + C_1 x_9 + C_1 x_8 + C_1 k_9 + C_1 k_8 + C_1 x_{15} + C_1 x_{14}$
$x_{11}^3 + x_{11}^2 k_5 + x_{11} k_5^2 + k_5^3 + C_2 x_9 + C_1 x_8 + C_2 k_9 + C_1 k_8 + C_2 x_{15} + C_1 x_{14}$

round 4:
$x_{14} + x_{16}$
$x_{15} + x_{17}$
$k_6^3 + k_6^2 x_{14} + k_6 x_{14}^2 + x_{14}^3 + C_1 x_{13} + C_1 x_{12} + C_1 k_{11} + C_1 k_{10} + C_1 x_{19} + C_1 x_{18}$
$k_7^3 + k_7^2 x_{15} + k_7 x_{15}^2 + x_{15}^3 + C_2 x_{13} + C_1 x_{12} + C_2 k_{11} + C_1 k_{10} + C_2 x_{19} + C_1 x_{18}$

key expansion:
$k_{11} + \theta^2 k_7 + (\theta^2 + \theta + 1) k_1 + \theta k_0 + \theta^4 + \theta^2$
$k_{10} + \theta^2 k_6 + \theta k_1 + k_0 + \theta^3 + \theta$
$k_9 + (\theta^2 + \theta) k_7 + (\theta + 1) k_6 + \theta^2 k_1 + (\theta + 1) k_0 + \theta^6 + \theta^5 + \theta^3 + 1$
$k_8 + (\theta + 1) k_7 + (\theta + 1) k_6 + (\theta + 1) k_1 + k_0 + \theta^5 + \theta^3$
$k_5 + (\theta^2 + \theta + 1) k_7 + \theta k_6 + \theta^2 k_1 + (\theta + 1) k_0 + \theta^6 + \theta^4 + \theta^3 + \theta$
$k_4 + \theta k_7 + k_6 + (\theta + 1) k_1 + k_0 + \theta^5 + \theta^4 + \theta^3 + 1$
$k_3 + \theta^2 k_7 + (\theta + 1) k_6 + (\theta^2 + \theta + 1) k_1 + \theta k_0 + \theta^6 + \theta^5 + \theta^4 + \theta$
$k_2 + (\theta + 1) k_7 + k_6 + \theta k_1 + k_0 + \theta^5 + \theta^2 + \theta + 1$

$\}$

with $C_1 = (\theta + 1)^{-1}$ and $C_2 = 1 + (\theta + 1)^{-1}$.

Universally Composable Oblivious Transfer in the Multi-party Setting

Marc Fischlin*

Institute for Theoretical Computer Science, ETH Zurich, Switzerland
marc.fischlin@inf.ethz.ch
http://www.fischlin.de/

Abstract. We construct efficient universally composable oblivious transfer protocols in the multi-party setting for honest majorities. Unlike previous proposals our protocols are designed in the plain model (i.e., without a common reference string), are secure against malicious adversaries from scratch (i.e., without requiring an expensive compiler), and are based on weaker cryptographic assumptions than comparable two-party protocols. Hence, the active participation of auxiliary parties pays off in terms of complexity. This is particularly true for the construction of one of our building blocks, an efficient universally composable *homomorphic* commitment scheme. Efficient solutions for this problem in the two-party setting are not known, not even in the common reference string model.

1 Introduction

Oblivious transfer (OT), originally defined by Rabin [23], is a two-party protocol between a sender and a receiver. In this protocol the receiver either obtains a message initially held by the sender, or gets the undefined symbol $\perp$ instead. Each event occurs with probability $1/2$, yet the sender remains unaware of the success of the transfer.

Oblivious transfer is a very important cryptographic primitive, for secure multi-party computations [18, 16] as well as a tool for more practical applications like anonymous buying over the Internet [2, 21, 1]. Often, these cases rely on a variant called chosen one-out-of-two oblivious transfer [13]. There, the sender holds two messages and the receiver gets to choose one (and only one), but the choice is hidden from the sender. Both versions of OT have been shown to be equally powerful [11].

Previous proposals for secure OT [13, 2, 4, 20] have often been investigated in an isolated setting where issues like concurrency of executions or side effects caused by other cryptographic protocols are not considered. As a noteworthy exception, Garay and MacKenzie [15] gave the first oblivious transfer protocol which is provably secure if run concurrently. Yet, even this protocol is not known to remain secure under more advanced attack models. For example, the adversary

---

* This work was supported by the Emmy Noether Programme Fi 940/1-1 of the German Research Foundation (DFG).

D. Pointcheval (Ed.): CT-RSA 2006, LNCS 3860, pp. 332–349, 2006.

could have some auxiliary information about the sender's messages (e.g., if the messages are used in other subprotocols), or the protocol may be executed in parallel with other cryptographic protocols.

Ideally, one would like to have an OT protocol which can be safely used as a building block within larger protocols, independently how the execution is interleaved with other steps. Such a security guarantee is provided by Canetti's universal composition (UC) framework [5]. In this framework one defines an idealized version of the primitive in question, capturing the desired security properties in an abstract way and ensuring that the functionality is secure in interdependent settings. For example, an idealized functionality for oblivious transfer is a trustworthy interface which waits to receive the two messages from the sender as well as the receiver's choice and then delivers the corresponding message to the receiver; no further information about one party's input is given to the other party.

Given an appropriate formalization of some functionality in the UC framework, one next shows that this functionality can be securely realized by an interactive protocol between the parties (without the trusted interface). Here, securely realizing means that, in any environment in which the protocol may be run, for this environment executions of the interactive protocol are indistinguishable from executions in the ideal model with the trustworthy functionality. The UC framework, notably the composition theorem, then guarantees that the protocol can indeed be securely deployed as a subroutine in more complex protocols and environments.

1.1 Previous Results

While some important stand-alone primitives like encryption and signatures basically preserve security in the UC framework [5, 6], other functionalities for commitment and oblivious transfer cannot be implemented by *any* protocol between two parties of which one can be dishonest [5, 8]. In particular, previously proposed OT protocols in a stand-alone setting (even if geared to be secure for concurrent executions like [15]) demonstrably fail to realize the aforementioned ideal OT functionality. In fact, Lindell [19], using a weaker security notion than universal composition called concurrent self-composition, shows that oblivious transfer (and other functionalities) cannot even be accomplished in this setting.

We stress that the impossibility results of [5, 8] refer to protocols between two parties in the plain model, i.e., without any auxiliary parties or setup assumptions. Indeed, Canetti [5] shows that any functionality can be realized in the UC framework by two or more parties if a majority of the players is honest (which for two parties implies that both parties cannot be corrupted). Although based on general cryptographic assumptions, this feasibility construction is computationally expensive. It requires to evaluate the circuit computing the functionality in a gate-wise manner, and also involves a general compiler lifting security in the presence of honest-but-curios adversaries to the case of malicious adversaries. This compiler usually relies on complex zero-knowledge proofs for general NP statements.

Another workaround for the impossibility results is to let the two parties have access to a common reference string (CRS) drawn according to some fixed distribution before the execution starts. This has been successfully applied to the case of commitments [8], as well to oblivious transfer [10]. The OT protocol in [10] is used as a building block to extend the aforementioned feasibility result of [5] to dishonest majorities. It consists of a two-level design which can be implemented by any trapdoor permutation. The first part is basically the OT protocol of Goldreich et al. [16] in the plain model but which is only secure against honest-but-curious adversaries. In the second step one patches the protocol to thwart malicious adversaries, using once more compiler techniques and zero-knowledge proofs. These zero-knowledge proofs then also use the common reference string model.

Recently, Garay et al. [17] utilized the CRS model, too, and proposed an extended committed oblivious transfer (ECOT) protocol which is universally composable. In such an ECOT protocol the parties run an oblivious transfer but they are also committed to their data; additionally, the sender can prove in zero-knowledge some relation about his committed values. The core of the ECOT protocol is of course a regular oblivious transfer and therefore the protocol in [17] realizes the OT functionality securely in the CRS model.

The solution in [17] does not rely on compiler techniques but is secure against malicious adversaries from scratch, under the decisional Diffie-Hellman assumption and the decisional composite residuosity assumption, and the strong RSA assumption or presuming chosen-message security of DSA. Yet, if implemented with the suggested efficient primitives the protocol is only known to be secure against adaptive corruptions if parties can erase internal data. Also, the protocol is geared towards evaluation of bit gates and therefore allows to transfer bit messages only; we are not aware if it can be extended easily to handle longer messages.

Another solution to bypass the two-party impossibility result, suggested in [22], is to lend super-polynomial power to the adversary in the real-life execution as well as in the idealized world, and to the environment trying to distinguish the two settings. This somewhat non-standard assumption about the computational power is done in a controlled way via so-called imaginary angels and it allows to overcome the need for a CRS in the general construction in [10]. The underlying oblivious transfer protocol in [22] is essentially identical to the one by Canetti et al. [10] and again needs a compiler and zero-knowledge proofs to handle malicious adversaries. Only the compiler is implemented differently by virtue of the imaginary angels, and can forgo the CRS.

1.2   Our Results

To overcome the two-party results of [5, 8] we work in the setting of honest majorities. As explained before, for two parties this trivially boils down to a protocol between honest users. We are therefore interested in the case of three or more parties.

Assuming an honest majority one could try to reduce the design of a multi-party OT protocol in the plain model to known two-party solutions in the CRS model. For example, in the three-party case (in which the adversary can corrupt only one party) the third party could pick the CRS and the sender and receiver then run the two-party protocol on this CRS *over a secure channel.*[1] We do not pursue this approach, though, because it would not improve over existing solutions. Instead, we try to make better use of the additional parties.

Moreover, for more than three parties the straightforward approach for letting the other parties generate the CRS would require a more advanced protocol than in the three-party case. Otherwise the adversary could corrupt the sender or the receiver in addition to some of the other players, possibly allowing the malicious sender or receiver to cheat for an adversarial chosen CRS. But easy protocols for jointly generating unbiased common reference strings are not known, especially since the CRS often contains non-trivial values like an RSA modulus with unknown factorization.

We present several protocols implementing UC oblivious transfer in the multi-party scenario, depending on the number of helper parties and, especially, on the maximum of dishonest players among them. For the case of $n = 3$ parties, among which there is an honest majority and thus at most $t \leq 1$ corrupt users, we present a basic protocol to realize universally composable OT very efficiently, requiring the parties to essentially perform only one or two encryptions/decryptions. In case of static adversaries or if we alternatively presume reliable data erasure, our protocol can be implemented with any CCA-secure public-key encryption (assuming authenticated channels between the parties).[2] As usual in multi-party computations, adaptive corruptions are dealt with using the more expensive but presumably inevitable non-committing encryption [9, 12]. Advantageously, in our protocol the number of bits which have to be encrypted in a non-committing way is limited by the length of the sender's messages and the receiver's choice, minimizing the usage of this encryption method.

We can extend our basic three-party protocol to more general n's and a limited number t of corrupt players. Although preserving the underlying cryptographic assumptions, the workload of the extended protocol increases exponentially with the number of corrupt players as we run many copies of our three-party protocol (yet, it remains within reasonable bounds for small t's). Hence, our protocol can tolerate up to $t = \mathcal{O}(\log k)$ dishonest players for security parameter k, where the exact bound on t depends on the relationship of the number n of honest users and t. For example, if we simply have an honest majority $n \geq 2t + 1$, then our protocol tolerates up to $t \approx \log \log k$ bad parties; if $n \geq t^2 + 2$ then we achieve the bound $t = \mathcal{O}(\log k)$. The description of this second protocol is omitted from this version and will appear in the full version.

In our third protocol, which is based on ideas developed by Bellare and Micali [4], we overcome the limitation $t = \mathcal{O}(\log k)$ by moving from secure encryp-

---

[1] The confidential transmission guarantees that no information is revealed to the adversary, even if the third party is corrupt and chooses the CRS in a malicious way.

[2] Observe that stand-alone oblivious transfer in the *two-party* setting cannot be constructed from black-box public-key encryption, even for static, honest-but-curious adversaries [14].

tion in general to the decisional Diffie-Hellman assumption. Our protocol utilizes UC *homomorphic* commitments which we show how to realize efficiently with Shamir's secret sharing and secure signatures in case of static corruptions and honest majorities. These homomorphic properties of commitments enable us to transfer well-known discrete-log based proof systems easily to the UC setting, resulting in an efficient OT protocol in connection with the DDH assumption. Since many practical RSA- or discrete-log-based proof systems in the stand-alone setting rely on similar homomorphic properties, our commitment protocol may be useful for the design of other efficient UC protocols in the multi-party setting.

The limitations of our commitment scheme, static corruptions and honest majorities, of course carry over to our OT protocol if implemented with this commitment protocol. However, given a UC homomorphic commitment scheme with stronger security properties, our OT protocol could tolerate any number of corrupt parties in principle. Yet, even then, our solution would only be secure against static corruptions —unless we allow reliable erasure in which case it would withstand adaptive adversaries as well.

2   Preliminaries

We work in the universal composition framework of [5]. In this section we give an overview over this framework, and refer the reader to [5] for a comprehensive introduction. Then we recall some useful basic functionalities.

2.1   UC Framework

As explained in the introduction, executions of a protocol which securely realizes some ideal functionality should be indistinguishable from executions with that functionality. This is formalized by considering two experiments as described below.

In the first experiment, the real-life execution, a probabilistic polynomial-time adversary $\mathcal{A}$ participates in a run of the interactive protocol π with a set of parties $P_1, \ldots, P_n$. All parties are connected through point-to-point communication channels. The channels are public, i.e., the adversary can read all data transmitted between parties. The adversary is also responsible for delivery of messages.

Each party is initially honest and follows the predetermined program of π. The adversary may corrupt parties, either at the outset only (*non-adaptive* or *static* adversaries) or at any point during the execution (*adaptive* adversaries). An adversary corrupting at most t parties during any possible execution is called *t-limited*. Once a party is corrupted by $\mathcal{A}$ the party hands over all internal data including its input, previous incoming and outgoing communication and the content of the random tape to the adversary. If we allow *reliable erasure* then the party may delete some of these data during the execution which then remains hidden from the adversary in case of a corruption. If a party gets corrupted by the adversary then the party follows the adversary's instructions from then on.

In particular, for so-called *malicious* adversaries the party may now deviate from its program.

In the second experiment, the ideal-model execution, a probabilistic polynomial-time adversary $\mathcal{S}$ (also called simulator) participates in an execution of (dummy) parties $P_1, \ldots, P_n$ with some ideal and trustworthy functionality $\mathcal{F}$. All parties are only connected to the functionality by secure channels and the simulator cannot read the content of transmissions. Once an honest dummy party gets some input it immediately forwards this input to the functionality which, at some point, may reply with output for some parties (including the simulator $\mathcal{S}$). If a party $P_1, \ldots, P_n$ receives such a message from the functionality it copies it to its output tape. We note that the simulator is responsible for delivery of these replies. Corruptions are dealt with as in the real-life setting.

In both settings an interactive distinguisher, the probabilistic polynomial-time environment $\mathcal{Z}$, is present. This environment can interact with honest parties by determining the inputs and by reading the output of these parties. Additionally, $\mathcal{Z}$ can communicate with the adversary $\mathcal{A}$ or $\mathcal{S}$, respectively. This interaction with the adversary may range from passing orders about corruptions to having the adversary report communications between parties. For both worlds the way the environment interacts with the adversary and the parties are identical. That is, $\mathcal{Z}$ only sees the input/output behavior of honest parties and the interactions with the adversary. In particular, if adversary $\mathcal{A}$ is t-limited and (non-)adaptive then so is the simulator $\mathcal{S}$.

At the end of an execution the environment should output a bit b indicating whether it thinks it observes an execution in the real-life world with protocol π and adversary $\mathcal{A}$ ($b = 0$), or in the ideal model with functionality $\mathcal{F}$ and adversary $\mathcal{S}$ ($b = 1$). The random variables describing the output distributions are denoted by $\mathrm{REAL}_{\pi,\mathcal{A},\mathcal{Z}}$ and $\mathrm{IDEAL}_{\mathcal{F},\mathcal{S},\mathcal{Z}}$, respectively. Informally, $\mathcal{A}$ should not have much more power attacking the interactive protocol π than $\mathcal{S}$ has in attacking the ideal functionality. Consequently,

Definition 1. *A protocol π securely realizes a functionality $\mathcal{F}$ if for every adversary $\mathcal{A}$ there exists a simulator $\mathcal{S}$ such that for every environment $\mathcal{Z}$ the random variables $\mathrm{REAL}_{\pi,\mathcal{A},\mathcal{Z}}$ and $\mathrm{IDEAL}_{\mathcal{F},\mathcal{S},\mathcal{Z}}$ are computationally indistinguishable. If the random variables are identically distributed then π securely realizes $\mathcal{F}$ in a perfect way.*

An important setting, which captures the intuition that a universally composable protocol can be used securely as a subprotocol, is the so-called hybrid model. There, an interactive protocol π is executed in presence of some ideal functionality $\mathcal{G}$, meaning that parties $P_1, \ldots, P_n$ and the adversary also have access to ideal functionality $\mathcal{G}$. Definition 1 straightforwardly carries over to this setting saying that $\mathrm{REAL}^{\mathcal{G}}_{\pi,\mathcal{A},\mathcal{Z}}$ and $\mathrm{IDEAL}_{\mathcal{F},\mathcal{S},\mathcal{Z}}$ should be indistinguishable.

The importance of the hybrid model becomes clear in light of the composition theorem [5]. If a protocol π securely realizes a functionality $\mathcal{F}$ in the $\mathcal{G}$-hybrid setting, and some protocol ρ securely realizes $\mathcal{G}$, then the protocol π^ρ (in which each call to $\mathcal{G}$ is replaced by running ρ) securely realizes $\mathcal{F}$. This can be extended to several functionalities $\mathcal{G}_1, \mathcal{G}_2, \ldots$ and protocols $\rho_1, \rho_2, \ldots$ realizing these functionalities. Additionally, nesting of functionalities (e.g., realizing $\mathcal{G}$ through ρ in some $\mathcal{H}$-hybrid setting and further realizing $\mathcal{H}$ by a protocol σ etc.) can be done up to constant depth.

On a technical note, protocol executions in the real-life and the ideal setting are always accompanied by session IDs. These IDs are provided and maintained by the system and enable the parties to distinguish between messages from different executions. Specifically, each invocation of a copy of some protocol or some functionality, respectively, is assigned a unique ID sid. For sake of readability we often omit mentioning these IDs for interactive protocols and note that any transmission in an interactive protocol is implicitly tagged by such a value sid as well as the identities of the sender and the receiver of the message.

2.2   Useful Functionalities

We usually show our OT protocols to be secure in the hybrid setting assuming some important functionalities as building blocks.

One important functionality for message transmissions is $\mathcal{F}_{\mathrm{auth}}$. This functionality provides integrity for transmissions in the sense that the adversary cannot tamper messages undetected, nor can the adversary inject additional messages. Yet, the adversary still gets to read the content of transmission between parties. This functionality is often assumed implicitly by presuming authenticated channels between parties.

Another important functionality adding confidentiality to authenticated transmissions is $\mathcal{F}_{\mathrm{smt}}$. This functionality can be securely realized (in the $\mathcal{F}_{\mathrm{auth}}$-hybrid model) by CPA- or CCA-secure public-key encryption for static adversaries, and by non-committing encryption for adaptive adversaries [5] (or by assuming reliable erasure for semantically secure encryption). In both cases the functionality merely reveals the length of the transmission to the adversary.

Our protocols use two other basic functionalities, $\mathcal{F}_{\mathrm{pke}}$ and $\mathcal{F}_{\mathrm{sig}}$, for secure public-key encryption and secure signatures. Namely, $\mathcal{F}_{\mathrm{pke}}$ allows to generate a public key enabling everyone to create ciphertexts which only the key generating party can decrypt. With $\mathcal{F}_{\mathrm{sig}}$ a signer party can generate a verification key allowing to publicly verify signatures only the signer can create. In [5, 6] it has been shown that $\mathcal{F}_{\mathrm{pke}}$ can be realized with (non-committing) CCA-secure encryption schemes, and $\mathcal{F}_{\mathrm{sig}}$ can be implemented through chosen-message secure signature schemes.

3   Universally Composable OT for Three Parties

We first discuss the case of three parties. Since we presume honest majorities the adversary can corrupt at most one of these three parties. The ideal functionality for oblivious transfer which our protocol should realize securely is given in Figure 1. We note that this protocol will then also provide the building block for larger n's and limited $t = \mathcal{O}(\log k)$.

Functionality $\mathcal{F}_{OT}$

$\mathcal{F}_{OT}$ proceeds as follows, running with two parties P_i, P_j and an adversary $\mathcal{S}$, and parameterized by a value κ:

- If receiving a message (ot-transfer, sid, P_i, P_j, m_0, m_1) with $m_0, m_1 \in \{0,1\}^\kappa$ from some party P_i store this message and ignore all further ot-transfer messages.
- If receiving a message (ot-choose, sid, P_i, P_j, b) from some party P_j check if a message (ot-transfer, sid, P_i, P_j, m_0, m_1) has been stored. If not, ignore this message. If so, send (ot-received, sid, P_i, P_j, m_b) to P_j and ignore all future ot-choose messages.

Fig. 1. $\binom{2}{1}$-Oblivious Transfer Functionality $\mathcal{F}_{OT}$ (adapted from [5])

3.1   The Protocol

The oblivious transfer protocol takes place between three parties: the sender S, holding two messages m_0, m_1, the receiver R with selection bit b, and a helper H with no input. If there are more than three parties then the helper position is filled in by the first party different from S and R (where we assume some order of the parties).

From a bird's eye view, the receiver and the helper each pick a random key and transfer the pair of keys to the sender. This is done such that, on the one hand, the receiver only knows one of the keys (where the order is determined by the receiver's choice) and, on the other hand, the sender remains oblivious about the owner of each key. The sender then encrypts each message with one of the keys and returns the ciphertext pair to the receiver. The receiver can decrypt message m_b with his secret key while the other message is protected by the third party's secret key.

We describe our protocol formally in Figure 2. We assume that the sender has already published a public key of an encryption scheme. Initially, the receiver and the helper both locally pick secret random string k_0 and k_1, respectively. Then they encrypt their string with the public key of S, confidentially exchange the ciphertexts and R also determines a random order of the ciphertexts which is only revealed to H. Both parties then transmit the re-ordered ciphertexts to the sender S.

In addition, R secretly sends a bit to S indicating another re-ordering of the ciphertexts. This bit, together with the first ciphertext re-arrangement between R and H, ensures that the receiver's string k_0 is encapsulated in the right ciphertext and that R later gets the message m_b. Viewed differently, the receiver's choice b is randomly split between the helper and the sender such that neither of the two parties alone can deduce b.

The sender waits to receive a ciphertext pair from each other party (and stops if it receives pairs that do not match). Then it sorts the ciphertexts according

---

Protocol OT$_3$ in the ($\mathcal{F}_{\text{pke}}$, $\mathcal{F}_{\text{smt}}$, $\mathcal{F}_{\text{auth}}$)-hybrid model

- **Setup:**
  - Upon receiving (ot-transfer, sid, S, R, m_0, m_1), the sender S generates a key pk by calling $\mathcal{F}_{\text{pke}}$. S sends pk to the receiver R and the helper party H, each time via $\mathcal{F}_{\text{auth}}$.
  - H echos the key received by S via $\mathcal{F}_{\text{auth}}$ to R, and R aborts if the keys do not match.

- **Key Exchange Step:**
  - Receiver R gets as input (ot-choose, sid, S, R, b). It picks a string $k_0 \leftarrow \{0,1\}^\kappa$ and a bit $\alpha \leftarrow \{0,1\}$, computes a ciphertext c_0 of k_0 under public key pk via $\mathcal{F}_{\text{pke}}$, and sends (c_0, α) to the helper H over the $\mathcal{F}_{\text{smt}}$ channel.
  - Having obtained (c_0, α) helper H chooses a string $k_1 \leftarrow \{0,1\}^\kappa$ and generates an $\mathcal{F}_{\text{pke}}$-ciphertext c_1 of k_1 under pk. It returns c_1 to R by $\mathcal{F}_{\text{smt}}$.
  - R and H then locally sort (c_0, c_1) according to α to $(c_\alpha, c_{\alpha \oplus 1})$ and send this ciphertext pair $(c_\alpha, c_{\alpha \oplus 1})$ to S via $\mathcal{F}_{\text{auth}}$.
  - The receiver also computes $\beta \leftarrow \alpha \oplus b$ and transmits β to S over $\mathcal{F}_{\text{smt}}$.

- **Transfer Step:**
  - The sender waits to receive the same pair $(c_\alpha, c_{\alpha \oplus 1})$ from R and H; if it receives distinct pairs, then the sender aborts. The sender also expects to get a bit β from R.
  - The sender S re-orders $(c_\alpha, c_{\alpha \oplus 1})$ to $(c_{\alpha \oplus \beta}, c_{\alpha \oplus \beta \oplus 1})$ and decrypts the pair to strings $(K_0, K_1) = (k_{\alpha \oplus \beta}, k_{\alpha \oplus \beta \oplus 1})$ by $\mathcal{F}_{\text{pke}}$. If any of the decryptions fails or does not yield a κ-bit string, then the sender aborts.
  - The sender masks the messages by $C_0 \leftarrow K_0 \oplus m_0$ and $C_1 \leftarrow K_1 \oplus m_1$, and sends (C_0, C_1) to receiver R over $\mathcal{F}_{\text{smt}}$.
  - Receiver R, upon getting (C_0, C_1), unmasks C_b with $K_b = k_{\alpha \oplus \beta \oplus b} = k_0$ to obtain message m_b. The receiver outputs (ot-received, sid, S, R, m_b).

Fig. 2. $\binom{2}{1}$-Oblivious Transfer Protocol for Three Parties

to R's request and decrypts them to obtain the strings $(K_0, K_1) = (k_b, k_{b \oplus 1})$ where b remains hidden from S. It masks the messages by $m_0 \oplus K_0$ and $m_1 \oplus K_1$ and confidentially returns them to R. The receiver unmasks the message m_b via $k_0 = K_b$, yet $m_{b \oplus 1}$ remains scrambled by H's secret string k_1.

3.2   Security

We prove security of our scheme in the ($\mathcal{F}_{\text{pke}}$, $\mathcal{F}_{\text{smt}}$, $\mathcal{F}_{\text{auth}}$)-hybrid model. We note that no further cryptographic assumption is required given these ideal interfaces, and executions of our protocol in this setting are even perfectly indistinguishable from ideal-model executions with functionality $\mathcal{F}_{\text{OT}}$. Of course, in order to realize functionality $\mathcal{F}_{\text{pke}}$ for example, cryptographic primitives are usually needed and the realization "only" guarantees computational indistinguishability.

Theorem 1. *Protocol* OT_3 *securely realizes functionality* $\mathcal{F}_{OT}$ *in the* $(\mathcal{F}_{pke},$ $\mathcal{F}_{smt}, \mathcal{F}_{auth})$-*hybrid model in a perfect way. This holds for* $n \geq 3$ *parties and* t-*limited malicious, adaptive adversaries,* $t \leq 1$.

Proof. (Sketch) We construct an ideal-model simulator $\mathcal{S}$ as follows. $\mathcal{S}$ runs a black-box simulation of the hybrid adversary $\mathcal{A}$ which is supposed to interact with ideal functionalities $\mathcal{F}_{pke}, \mathcal{F}_{smt}, \mathcal{F}_{auth}$ and the parties running the protocol. At the same time, $\mathcal{S}$ runs an execution with functionality $\mathcal{F}_{OT}$ in the ideal model. Recall that in this ideal setting none of the other functionalities is present.

In the hybrid setting the adversary and the honest parties communicate with the functionalities $\mathcal{F}_{pke}, \mathcal{F}_{smt}, \mathcal{F}_{auth}$ over which the simulator has full control in the black-box simulation. In particular, the only information available to the adversary $\mathcal{A}$ about the execution are data sent over $\mathcal{F}_{auth}$ between honest parties, the (length of) encryption requests forwarded by $\mathcal{F}_{pke}$ to $\mathcal{A}$, and internal data of corrupted parties. The latter may include information previously transmitted securely over $\mathcal{F}_{smt}$.

We next describe how the simulator emulates the adversary in the black-box simulation. For this we define the simulator's steps for each honest party and for corruptions of these parties:

Simulation of Sender. $\mathcal{S}$ simulates an honest S by simply following the prescribed program, i.e., if the dummy sender in the ideal model passes a message to the ideal functionality then $\mathcal{S}$ generates a key pair via $\mathcal{F}_{pke}$, waits to receive ciphertext pairs, decrypts them and masks the messages. The only exception lies in the final step if S is supposed to transmit (C_0, C_1) to the already *corrupted* receiver. Recall that this is the only case where $\mathcal{A}$ immediately learns these information sent over $\mathcal{F}_{smt}$ at this point. If both parties are honest then the simulator does not have to pass any information to $\mathcal{A}$ about this communication (as it is virtually invisible to $\mathcal{A}$ in the hybrid model).

To simulate the transmission of (C_0, C_1) to a corrupted receiver first note that $\mathcal{S}$ knows both α and β, from the communication between R and the simulated H, and the communication between R and S. Furthermore, it knows the strings K_0, K_1 from decrypting the ciphertexts. The simulator sets $b \leftarrow \alpha \oplus \beta$ and sends $(\mathtt{ot\text{-}choose}, sid, \mathsf{S}, \mathsf{R}, b)$ to the functionality $\mathcal{F}_{OT}$ in the ideal model to receive message m_b. The simulator sets $C_b \leftarrow K_b \oplus m_b$ and $C_{b \oplus 1} \leftarrow \{0, 1\}^\kappa$ and sends (C_0, C_1) in the name of S to R over $\mathcal{F}_{smt}$.

If the adversary requests to corrupt the sender, then $\mathcal{S}$ first corrupts S in the ideal model and learns the sender's input $(\mathtt{ot\text{-}transfer}, sid, \mathsf{S}, \mathsf{R}, m_0, m_1)$. The simulator then reveals the input and all the internal random coins to $\mathcal{A}$; no adaption is necessary for these genuine values. In addition, if (C_0, C_1) has already been sent then R must still be honest and the simulator can simply claim that the transmission over the adaptively secure channel was $(K_0 \oplus m_0, K_1 \oplus m_1)$.

Simulation of Receiver. In order to simulate an honest receiver the simulator runs a copy of R's program, with the only difference that $\mathcal{S}$ initially substitutes R's unknown input b by $\tilde{b} = 0$. This possibly causes the simulated receiver to later send $\tilde{\beta} = \alpha$ instead of $\beta = \alpha \oplus b$ (unless R is corrupted before).

If the adversary asks to corrupt R then $\mathcal{S}$ corrupts the party in the ideal model and learns the input (ot-choose, sid, S, R, b). If this corruption happens before the receiver is supposed to send β then handing over the internal coin tosses and b to the adversary complies with the transmitted data. Assume that the receiver has already sent the substituted $\tilde{\beta}$ to S. Since the sender is still honest the simulator is able to claim that $\beta = \alpha \oplus b$ has been transmitted over $\mathcal{F}_{smt}$ instead of $\tilde{\beta}$.

Finally, suppose the corruption of R takes place after S has supposedly sent the masked messages, in which case the adversary has not seen these values transmitted over $\mathcal{F}_{smt}$ (yet). Due to the corruption of the receiver in the ideal model the simulator learns the functionality's message (ot-received, sid, S, R, m_b). The other values including β are faked as before and this time the simulator also sets $C_b \leftarrow K_b \oplus m_b$ and $C_{b\oplus 1} \leftarrow \{0,1\}^\kappa$ and claims that R has received (C_0, C_1).

Simulation of Helper. The simulator simply runs a copy of H and reveals all internal data to $\mathcal{A}$ if H is corrupted; no external input is provided to H.

It is not hard to see that the black-box simulation above is perfectly indistinguishable from an actual attack in the hybrid setting, as long as the adversary can corrupt at most one of the parties. This proves the claim.   □

Several variations for the protocol apply. For instance, it is straightforward to extend the basic protocol into a chosen 1-out-of-N oblivious transfer. Analogously, to derive Rabin's OT functionality where the receiver either gets the sender's message m or receives $\bot$, let the sender use $m_0 = m$ and $m_1 = \bot$ in the protocol and let the sender choose the bit β at random instead.

4   DDH-Based UC Oblivious Transfer

As mentioned in the introduction the description of our extended three-party protocol to tolerate up to $t = \mathcal{O}(\log k)$ corrupt players is omitted. Instead, we present a UC oblivious transfer protocol which tolerates more than logarithmically many dishonest parties. While our protocol in principle withstands any number of corruptions given appropriate building blocks, our solution merely tolerates non-adaptive adversaries. Adaptive security can be achieved if we allow reliable data erasure.

Like the scheme in [17] the resulting protocol here is a derivation of the protocol by Bellare and Micali [4]. As in the Bellare-Micali protocol we have the sender first generate and send an element $X = g^x$ of a group $\langle g \rangle$, such that the discrete logarithm x is only know to the sender. The receiver, holding bit b, next generates a pair $W_0 = g^v X^{-b}$ and $W_1 = W_0 X$ such that it knows only one of the disrete logarithms, namely, $\log W_b = v$. The receiver returns the pair W_0, W_1 to the sender who thus remains oblivious about the bit b in an information-theoretical sense. The sender encrypts the messages m_0, m_1 with the ElGamal scheme, using W_0 and W_1, respectively, as the public keys. The receiver can then decrypt m_b from the encryption under key W_b.

In order to ensure universal composition of the basic protocol the transmissions of the value X and of the pair W_0, W_1 are each accompanied by proofs of knowledge. In the first case this is a Schnorr-type proof of knowledge of a discrete logarithm [24]; in the second case this corresponds to a proof of knowledge of one out of two discrete logarithms [7]. Both protocols follow the well-known commitment-challenge-response structure, and to implement them in the UC framework efficiently, we present a homomorphic UC commitment scheme that is used for the initial commitment.

Below, we start by presenting our homomorphic UC commitment protocol. Given such a functionality we explain how to efficiently prove statements about discrete logarithms in the UC framework, and then use these proofs to construct our oblivious transfer protocol. We note that helper parties are only required in the implementation of the homomorphic commitment funtionality, while the discrete-log based protocol can be realized between the sender and the receiver only (in presence of an ideal homomorphic commitment functionality).

Since the suggested implementation of the efficient UC homomorphic commitment scheme merely tolerates non-adaptive adversaries and dishonest minorities. Hence, if implemented with this commitment protocol, we require $n \geq 2t + 1$ players and the assembled protocol achieves security only against static corruptions, even if we presume reliable deletion of data.

4.1   UC Homomorphic Commitments in the Multi-party Case

In the multi-party setting with honest majority, a universally composable commitment scheme can be constructed along the lines of [3]. Namely, start with Shamir's $(t, 2t)$-threshold secret sharing scheme [25] where one can reconstruct a shared secret x from $t + 1$ shares, yet any t or less shares are independent of x. In this scheme the dealer chooses a random polynomial f of degree t over a sufficiently large field such that $f(0) = x$. The dealer then distributes $x_i \leftarrow f(P_i)$ to party P_i (for some unique identity $P_i \neq 0$). To reconstruct the secret one interpolates the polynomial from $t + 1$ shares and then evaluates it at 0.

Shamir's scheme has an additional feature useful in our context. Namely, the reconstruction algorithm is able to detect efficiently if, given more than $t + 1$ shares, any two subsets would reconstruct to distinct secrets (in which case the algorithm returns $\perp$). This can be checked by reconstructing the polynomial from $t + 1$ shares and then verifying that the polynomial evaluates to the right values for the other shares. We remark that this requires that identities of the parties are unchangeably associated to the shares.

For our homomorphic commitment scheme we also need a universally composable signature scheme $\mathcal{F}_{\text{sig}}$. According to [6] such a signature scheme can be derived from chosen-message secure signature schemes which, in turn, exist if one-way functions exist. We presume that the public verification key of the committer has already been reliably transmitted to each party, i.e., either by broadcast, or by sending it to each auxiliary party and to the receiver and letting the helpers forward a copy to the receiver for verification.

Basic UC Commitment Protocol. The basic version of a universally composable commitment protocol for $n \geq 2t + 1$ goes as follows. To commit to a value x to some party P_j, the committer P_i first computes shares $x_1, x_2, \ldots, x_{2t}$ of x and, for each share x_i, also derives a signature σ_i by $\mathcal{F}_{\text{sig}}$. The committer sends, over secure channels $\mathcal{F}_{\text{smt}}$, share x_i and signature σ_i to the i-th of the first $2t$ parties other than the sender (but possibly including the receiver P_j). Each share holder informs P_j when it has received a share with a valid signature but keeps the actual value and the signature secret. The receiver outputs a receipt about a commitment taking place if it has obtained $2t$ of such confirmations.

To open the commitment, party P_i requests the other parties to reveal their shares. All auxiliary parties then forward their previously obtained share together with the signature to P_j via $\mathcal{F}_{\text{smt}}$. If all $2t$ of these shares carry valid signatures then party P_j runs the reconstruction to derive some value x; it accepts x if and only if $x \neq \bot$.

We do not prove formally that the protocol above is a universally composable commitment scheme against *non-adaptive* adversaries. The basic properties for such a protocol [8], extraction and equivocability, can be easily seen. We note that the proof relies on static corruptions as we need to be able to derive the value x from shares sent by a malicious sender to honest parties in the commitment phase. Adaptive adversaries, however, may be able to send out inconsistent shares at first and adapt those values later after corrupting some parties.

Homomorphic UC Protocol. Shamir's polynomial-based scheme also allows to perform additions on shares, i.e., having shared $x_1, \ldots, x_n$ via $x_{\ell,1}, \ldots, x_{\ell,2t}$ among the same parties computing, say, $x_{\ell,1} + \cdots + x_{\ell,n}$ locally generates a share of $x_1 + \cdots + x_n$. This homomorphic property carries over to the universally composable commitment, allowing the committer to open any linear combination $\sum a_\ell x_\ell$ of committed values $x_1, \ldots, x_n$ for known $a_1, \ldots, a_n$.

When applying the homomorphic properties of the sharing scheme some care with regard to the signatures is necessary, though. For example, in the opening step the commitment scheme reveals only sums of the secrets but possibly not the individual values. We solve this by having the committer in the opening phase sign the sums $\sum a_\ell x_{\ell,k}$ of P_k's shares and send this signature to P_k. Then P_k can open the sum of the shares and prove correctness to the receiver by the additional signature. Party P_k will, however, disclose the individual shares with the initial signature if the sender's signature for the sum is invalid. The full protocol HCom is given in Figure 7 in Appendix A. The proof of the following proposition is omitted.

Proposition 1. *Protocol HCom securely realizes functionality $\mathcal{F}_{hcom}$ in the $(\mathcal{F}_{sig}, \mathcal{F}_{smt}, \mathcal{F}_{auth})$-hybrid model in a perfect way. This holds for $n \geq 2t + 1$ parties and t-limited malicious, non-adaptive adversaries.*

4.2   Efficient Proofs for Discrete Logarithms

Given a universally composable homomorphic commitment scheme we show how to build efficient proof systems for discrete-logarithm statements. In the sequel

---

Functionality $\mathcal{F}_{\mathrm{hcom}}$

$\mathcal{F}_{\mathrm{hcom}}$ proceeds as follows, running with parties $P_1, \ldots, P_n$ and adversary $\mathcal{S}$, and parameterized by an Abelian group $(\mathcal{A}, +)$.

- Upon receiving a message $(\mathbf{hcom\text{-}commit}, sid, P_i, P_j, x_1, \ldots, x_n)$ for $x_1, \ldots, x_n \in \mathcal{A}$ from some party P_i, send $(\mathbf{hcom\text{-}receipt}, sid, P_i, P_j, n)$ to P_j and $\mathcal{S}$. Ignore all further $\mathbf{hcom\text{-}commit}$ messages.
- If receiving a message $(\mathbf{hcom\text{-}open}, sid, P_i, P_j, a_1, \ldots, a_n)$ for $a_1, \ldots, a_n \in \mathbb{N}_0$ from party P_i, check that some message $(\mathbf{hcom\text{-}commit}, sid, P_i, P_j, x_1, \ldots, x_n)$ has been received from P_i before. If not, then ignore. Else compute $y \leftarrow \sum_{i=1}^n a_i x_i$ in $\mathcal{A}$, and send $(\mathbf{hcom\text{-}open}, sid, P_i, P_j, a_1, \ldots, a_n, y)$ to P_j and $\mathcal{S}$.

Fig. 3. Homomorphic Commitment Functionality $\mathcal{F}_{\mathrm{hcom}}$

we presume that q is a prime and that g is a generator of a group of order q in which the decisional Diffie-Hellman problem (given g^x, g^y, g^z decide if $z = xy \bmod q$) is intractable. Moreover, let the additive group of the homomorphic commitment scheme be $(\mathcal{A}, +) = (\mathbb{Z}_q, +)$.

Functionality $\mathcal{F}_{\mathrm{dlzk}}$

$\mathcal{F}_{\mathrm{dlzk}}$ proceeds as follows, running with two parties P_i, P_j and an adversary $\mathcal{S}$, and parameterized by a group $\langle g \rangle$ of order q generated by g:

- If receiving a message $(\mathbf{dlzk\text{-}verify}, sid, P_i, P_j, X)$ from a party P_j store this message and forward it to $\mathcal{S}$. Ignore all subsequent $\mathbf{dlzk\text{-}verify}$ messages.
- If receiving a message $(\mathbf{dlzk\text{-}prove}, sid, P_i, P_j, X, x)$ from some party P_i check that a message $(\mathbf{dlzk\text{-}verify}, sid, P_i, P_j, X)$ has been recorded. If not ignore, else verify that $g^x = X$. If so, deliver $(\mathbf{dlzk\text{-}verified}, sid, P_i, P_j, X)$ to $\mathcal{S}$ and P_j and halt. Else ignore the message.

Fig. 4. Proving Knowledge of Discrete Logarithms Through Functionality $\mathcal{F}_{\mathrm{dlzk}}$

The two functionalities we are trying to realize are given in Figures 4 and 5. The protocols to implement the functionalities are basically the Schnorr protocol [24] and the the Or-protocol of [7] which deploys the Schnorr protocol to prove knowledge of one out of two discrete logarithms. In this version of the paper we merely sketch the solutions; the formal protocols with the security proofs can be found in the full version.

To realize functionality $\mathcal{F}_{\mathrm{dlzk}}$ we run the well-known Schnorr protocol consisting of an initial commitment of the prover, a random challenge of the verifier and

---

Functionality $\mathcal{F}_{\text{dlor}}$

$\mathcal{F}_{\text{dlor}}$ proceeds as follows, running with two parties P_i, P_j and an adversary $\mathcal{S}$, and parameterized by a group $\langle g \rangle$ of order q generated by g:

- If receiving a message $(\texttt{dlor-verify}, sid, P_i, P_j, X, W_0, W_1)$ from some party P_j check that $W_1 = W_0 X$. If not ignore, else store this message and ignore all further $\texttt{dlor-verify}$ messages. Also, send $(\texttt{dlor-verify}, sid, P_i, P_j, X, W_0, W_1)$ to $\mathcal{S}$.
- If receiving a message $(\texttt{dlor-prove}, sid, P_i, P_j, X, W_0, W_1, w, b)$ from some party P_i check that a message $(\texttt{dlor-verify}, sid, P_i, P_j, X, W_0, W_1)$ has been stored. If so and $W_0 = g^w X^{-b}$, then deliver $(\texttt{dlor-verified}, sid, P_i, P_j, X, W_0, W_1)$ to P_j and $\mathcal{S}$ and halt. Else ignore the message.

Fig. 5. Proving Knowledge of One of Two Logarithms Through Functionality $\mathcal{F}_{\text{dlor}}$

a final response of the prover. Only this time we use our homomorphic commitment functionality (in the $\mathcal{F}_{\text{hcom}}$-hybrid model) to let the prover commit to the initial value. Furthermore, all transmissions are authenticated via $\mathcal{F}_{\text{auth}}$. Then the protocol realizes functionality $\mathcal{F}_{\text{dlzk}}$ in the UC framework. The same trick works for the Or-case, too, and we thus obtain:

Proposition 2. *There are protocols realizing functionalities $\mathcal{F}_{dlzk}$ and $\mathcal{F}_{dlor}$, respectively, in the $(\mathcal{F}_{hcom}, \mathcal{F}_{auth})$-hybrid model for any $t \leq n$ and t-limited malicious, adaptive adversary.*

Note that, in the proposition, we presume that we are given a commitment functionality $\mathcal{F}_{\text{hcom}}$ secure against adaptive corruptions. While such protocols can be designed in principle [5] our efficient solution in the previous section merely withstand non-adaptive adversaries. Hence, if implemented with this scheme our protocol here is also bound to static corruptions.

4.3   Oblivious Transfer Under the DDH Assumption

We show how to realize functionality $\mathcal{F}_{\text{OT}}$ in the $(\mathcal{F}_{\text{dlor}}, \mathcal{F}_{\text{dlzk}}, \mathcal{F}_{\text{auth}})$-hybrid model. Our protocol —as is— only withstands non-adaptive adversary; extending this to adaptive adversaries is possible if we allow reliable erasure, as discussed afterwards.

The full protocol is given in Figure 6. We have the sender create X with secret logarithm $x = \log_g X$ and the receiver chooses $W_0, W_1 = W_0 X$ such that it knows $\log_g W_b$ (but not $\log_g W_{b \oplus 1}$) for the selection bit b. We also let both parties prove knowledge via our ideal discrete-log functionalities. Then, the sender encrypts both messages such that one can decrypt m_a if and only if one knows $\log_g W_a$ for $a = 0, 1$. By this, it follows that the receiver can only retrieve one of the messages. In the protocol we assume for simplicity that the κ-bit messages have already been encoded in the group generated by g.

Protocol OT$_{\text{DDH}}$ in the $(\mathcal{F}_{\text{dlor}}, \mathcal{F}_{\text{dlzk}}, \mathcal{F}_{\text{auth}})$-hybrid model

- Upon receiving (ot-transfer, sid, S, R, m_0, m_1) the sender picks $x, y \leftarrow \mathbb{Z}_q$ at random and computes $X = g^x, Y = g^y$. It sends X to R over $\mathcal{F}_{\text{auth}}$ and both parties engage in a proof for X via $\mathcal{F}_{\text{dlzk}}$.
- The receiver gets (ot-choose, sid, S, R, b) as input and chooses $w \leftarrow \mathbb{Z}_q$ and computes $W_0 = g^w X^{-b}$ and $W_1 = W_0 X$. It sends (W_0, W_1) to S via $\mathcal{F}_{\text{auth}}$.
- The sender and receiver call functionality $\mathcal{F}_{\text{dlor}}$ with inputs (dlor-verify, sid, R, S, X, W_0, W_1) and (dlor-prove, sid, R, S, X, W_0, W_1, w, b), respectively.
- The sender computes $C_0 \leftarrow m_0 W_0^y$ and $C_1 \leftarrow m_1 W_1^y$ and transmits (Y, C_0, C_1) over $\mathcal{F}_{\text{auth}}$ to R.
- The receiver computes $m_b \leftarrow C_b Y^{-w}$ and outputs (ot-received, sid, S, R, m_b).

Fig. 6. Oblivious Transfer based on DDH

Theorem 2. *Protocol* OT$_{DDH}$ *securely realizes functionality* $\mathcal{F}_{OT}$ *in the* $(\mathcal{F}_{dlor}, \mathcal{F}_{dlzk}, \mathcal{F}_{auth})$-*hybrid model under the decisional Diffie-Hellman assumption. This holds for n parties and t-limited malicious, non-adaptive adversaries, $t \leq n$.*

The proof can be found in the full version. The protocol above remains secure against adaptive adversaries if we add another step where the sender erases $x, y \in \mathbb{Z}_q$ immediately after (Y, C_0, C_1) has been computed. We call this protocol OT$_{DDH}^{\text{erase}}$. The additional step guarantees that the simulated sender can deny to know the secrets to unmask $m_{b \oplus 1}$. Otherwise, the values X, Y pin down x, y and therefore $C_{b \oplus 1}$ and $m_{b \oplus 1}$.

Proposition 3. *Protocol* OT$_{DDH}^{erase}$ *securely realizes functionality* $\mathcal{F}_{OT}$ *in the* $(\mathcal{F}_{dlor}, \mathcal{F}_{dlzk}, \mathcal{F}_{auth})$-*hybrid model under the decisional Diffie-Hellman assumption. This holds for n parties and t-limited malicious, adaptive adversaries, $t \leq n$, assuming reliable erasure.*

Acknowledgments

We thank the anonymous reviewers for valuable comments.

References

1. William Aiello, Yuval Ishai, and Omer Reingold. *Priced Oblivious Transfer: How to Sell Digital Goods.* Eurocrypt 2001, Volume 2045 of LNCS, pages 119–135. Springer-Verlag, 2001.
2. Gilles Brassard, Claude Crpeau, and Jean-Marc Robert. *All-or-Nothing Disclosure of Secrets.* Crypto'86, Volume 263 of LNCS, pages 234–238. Springer-Verlag, 1987.
3. Michael Ben-Or, Shafi Goldwasser, and Avi Wigderson. *Completeness Theorems for Non-Cryptographic Fault-Tolerant Distributed Computation.* STOC 1988, pages 1–10. ACM Press, 1988.

4. Mihir Bellare and Silvio Micali. *Non-Interactive Oblivious Transfer and Applications.* Crypto'89, Volume 435 of LNCS, pages 547–557. Springer-Verlag, 1990.
5. Ran Canetti. *Universally Composable Security: A new Paradigm for Cryptographic Protocols.* FOCS 2001. IEEE Computer Society Press, 2001.
6. Ran Canetti. *On Universally Composable Notions of Security for Signature, Certification and Authentication.* CSFW 2004. IEEE Computer Society Press, 2004.
7. Ronald Cramer, Ivan Damgård, and Berry Schoenmakers. *Proofs of Partial Knowledge and Simplified Desing of Witness Hiding Protocols.* Crypto'94, Volume 839 of LNCS, pages 174–187. Springer-Verlag, 1995.
8. Ran Canetti and Marc Fischlin. *Universally Composable Commitments.* Crypto 2001, Volume 2139 of LNCS, pages 19–40. Springer-Verlag, 2001.
9. Ran Canetti, Uri Feige, Oded Goldreich, and Moni Naor. *Adaptively Secure Multi-Party Computation.* STOC 1996, pages 639–648. ACM Press, 1996.
10. Ran Canetti, Yehuda Lindell, Rafail Ostrovsky, and Amit Sahai. *Universally Composable Two-Party and Multi-Party Secure Computation.* STOC 2002, pages 494–503. ACM Press, 2002.
11. Claude Crepeau. *Equivalence Between Two Flavors of Oblivious Transfer.* Crypto'87, LNCS, pages 350–354. Springer-Verlag, 1987.
12. Ivan Damgård and Jesper Nielsen. *Improved Non-Committing Encryption Schemes Based on a General Complexity Assumption.* Crypto 2000, Volume 1880 of LNCS, pages 432–450. Springer-Verlag, 2000.
13. Shimon Even, Oded Goldreich, and Abraham Lempel. *A Randomized Protocol for Signing Contracts.* Comm. ACM, 28(6):637–647, 1985.
14. Yael Gertner, Sampath Kannan, Tal Malkin, Omer Reingold, and Mahesh Viswanathan. *The Relationship between Public Key Encryption and Oblivious Transfer.* FOCS 2000. IEEE Computer Society Press, 2000.
15. Juan Garay and Philip MacKenzie. *Concurrent Oblivious Transfer.* FOCS 2000, pages 314–324. IEEE Computer Society Press, 2000.
16. Oded Goldreich, Silvio Micali, and Avi Wigderson. *How to Play any Mental Game.* STOC 1987, pages 218–229. ACM Press, 1987.
17. Juan Garay, Philip MacKenzie, and Ke Yang. *Efficient and Universally Composable Committed Oblivious Transfer and Applications.* TCC 2004, Volume 2951 of LNCS, pages 297–316. Springer-Verlag, 2004.
18. Joe Kilian. *Founding Crytpography on Oblivious Transfer.* STOC 1988, pages 20–31. ACM Press, 1988.
19. Yehuda Lindell. *Lower Bounds for Concurrent Self Composition.* TCC 2004, Volume 2951 of LNCS, pages 203–222. Springer-Verlag, 2004.
20. Moni Naor and Benny Pinkas. *Efficient Oblivious Transfer Protocols.* SODA 2001, pages 448–457. ACM Press, 2001.
21. Moni Naor, Benny Pinkas, and Reuban Sumner. *Privacy Preserving Auctions and Mechanism Design.* Proceedings of the 1st Conference on Electronic Commerce, pages 129–139. ACM Press, 1999.
22. Manoj Prabhakaran and Amit Sahai. *New Notions of Security: Achieving Universal Composability without Trusted Setup.* STOC 2004, pages 242–251. ACM Press, 2004.
23. Michael Rabin. *How to Exchange Secrets by Oblivious Transfer.* Technical Report TR-81, Aiken Computation Laboratory, 1981.
24. C.P. Schnorr. *Efficient Signature Generation by Smart Cards. Journal of Cryptology,* 4:161–174, 1991.
25. Adi Shamir. *How to Share a Secret. Comm. ACM,* 22:612–613, 1979.

A   Homomorphic Commitment Protocol

In this section we present our UC homomorphic commitment protocol.

Protocol **HCom** for Abelian group $(\mathcal{A}, +)$ in the $(\mathcal{F}_{\text{sig}}, \mathcal{F}_{\text{smt}}, \mathcal{F}_{\text{auth}})$-hybrid model

– **Commitment:**
  - Upon receiving $(\text{hcom-commit}, sid, P_i, P_j, x_1, \ldots, x_n)$ the committer first generates a verification key vk of $\mathcal{F}_{\text{sig}}$ and sends it to the receiver P_i and to all $2t$ helper parties via $\mathcal{F}_{\text{auth}}$. The helper parties echo the key to the receiver over $\mathcal{F}_{\text{auth}}$ and the receiver stops if any of the keys do not match.
  - The committer then computes shares $x_{\ell,1}, x_{\ell,2}, \ldots, x_{\ell,2t}$ of all values $\ell = 1, 2, \ldots, n$ as well as signatures σ_k of $(x_{1,k}, \ldots, x_{n,k})$ via $\mathcal{F}_{\text{sig}}$ for $k = 1, 2, \ldots, 2t$. It sends $(commit, x_{1,k}, \ldots, x_{n,k}, \sigma_k)$ to party P_k over $\mathcal{F}_{\text{smt}}$ (where P_k is the k-th party different from the committer).
  - Receiving $(commit, x_{1,k}, \ldots, x_{n,k}, \sigma_k)$ from P_i party P_k first checks the signature σ_k by $\mathcal{F}_{\text{sig}}$. If the signature is invalid then P_k sends out $(commit\text{-}error, n)$ to the receiver over $\mathcal{F}_{\text{auth}}$. Otherwise, if the signature is valid, then party P_k sends $(commit\text{-}ok, n)$ to P_j over $\mathcal{F}_{\text{auth}}$.
  - Only if the receiver gets $2t$ messages $(commit\text{-}ok, n)$ from $P_1, \ldots, P_{2t}$ then it outputs $(\text{hcom-receipt}, sid, P_i, P_j, n)$.

– **Opening:**
  - When getting $(\text{hcom-open}, sid, P_i, P_j, a_1, \ldots, a_n)$ as input, the committer P_i computes $y \leftarrow \sum_{\ell=1}^n a_\ell x_\ell$ and $y_k \leftarrow \sum a_\ell x_{\ell,k}$ for $k = 1, 2, \ldots, 2t$. Additionally, it computes signatures τ_k of $(a_1, \ldots, a_n, y_k)$. Send $(open, a_1, \ldots, a_n, y_k, \tau_k)$ over $\mathcal{F}_{\text{smt}}$ to each P_k.
  - Party P_k, when receiving $(open, a_1, \ldots, a_n, y_k, \tau_k)$, verifies that $y_k = \sum a_\ell x_{\ell,k}$ for the previously received values and also checks the signature τ_k for $(a_1, \ldots, a_n, y_k)$. If all tests succeed then it sends $(open\text{-}ok, a_1, \ldots, a_n, y_k, \tau_k)$ to P_j over $\mathcal{F}_{\text{smt}}$; else it sends $(open\text{-}error, a_1, \ldots, a_n, x_{1,k}, \ldots, x_{n,k}, \sigma_k)$, i.e., reveals all shares including the signature.
  - The receiver waits to receive $2t$ messages including values $(a_1, \ldots, a_n)$, either of type *open-ok* or of type *open-error*. Having received such values the receiver checks the signature in each message with the help of $\mathcal{F}_{\text{sig}}$ and vk, i.e., the receiver verifies that τ_k is valid or that σ_k is valid. If so, for each *open-error* message the receiver reconstructs $y_k \leftarrow \sum a_\ell x_{\ell,k}$ and checks with these parts and the ones in *open-ok* that the reconstruction algorithms yields $y \neq \perp$. If all tests succeed then the receiver outputs $(\text{hcom-open}, sid, P_i, P_j, a_1, \ldots, a_n, y)$.

Fig. 7. Universally Composable Homomorphic Commitment Scheme

A Round and Communication Efficient Secure Ranking Protocol

Shaoquan Jiang[1,2] and Guang Gong[2]

[1] Department of Computer Science,
University of Electronic Science and Technology of China, Chengdu, China
[2] Department of Electrical and Computer Engineering,
University of Waterloo, Waterloo, ON, N2L 3G1, Canada
{jiangshq, ggong}@calliope.uwaterloo.ca

Abstract. In this work, we initiate the study of realizing a ranking functionality $(m_1, \cdots, m_n) \mapsto (r_1, \cdots, r_n)$ in the non-adaptive malicious model, where $r_i = 1 + \sharp\{m_j : m_j < m_i\}$. Generically, it has been solved by a general multi-party computation technique (via a circuit formulation). However, such a solution is inefficient in either round complexity or communication complexity. In this work, we propose an efficient construction without a circuit. Our protocol is constant round and efficient in communication complexity as well. Furthermore, we show it is directly secure in the non-adaptive malicious model (i.e., without a compiler, as is used in many general constructions).

1   Introduction

A general multi-party computation paradigm was initially studied by Yao [15]. In this paradigm, any cryptographic functionality can be solved as a special case. Such a functionality can be first formulated into some circuit. Then, a semi-honestly secure protocol for this circuit is constructed. In the semi-honest model, all the parties (including corrupted parties) strictly follow the protocol specification. This, of course, does not suffice for real applications. To obtain a realistically secure protocol, a compiler is proposed which, given a semi-honestly secure protocol, outputs a maliciously secure protocol. This approach is generally very powerful, as one need not take care of the problem's specific features (except the circuit itself). However, since a generic solution does not exploit the problem's special properties, it is not efficient in general. Therefore, it is very interesting and valuable to solve the specific problem without the generic approach. With this motivation in mind, we study a multi-party millionaire problem below. There are n parties, $P_1, \cdots, P_n$. Each P_i has m_i dollars for $1 \leq m_i \leq N$. The problem is how to admit P_i to get the rank of m_i but nothing beyond this. Formally, we are interested in realizing the *ranking* functionality $(m_1, \cdots, m_n) \mapsto (r_1, \cdots, r_n)$, where $r_i = 1 + \sharp\{m_j : m_j < m_i\}, i = 1, \cdots, n$.

1.1   Related Work

Since Yao [15], many authors [6, 9, 8, 5, 12] worked in the circuit paradigm. However, prior to the work by Cramer *et al.* [3] and by Jakobsson *et al.* [13], none

D. Pointcheval (Ed.): CT-RSA 2006, LNCS 3860, pp. 350–364, 2006.

of them have achieved both the non-adaptive malicious security and the communication complexity $O(kn|C|)$, where communication complexity is the total number of bits sent by all parties, $|C|$ is the circuit size, n is the number of parties, and k is the security parameter. Applying to a ranking functionality, one can achieve communication complexity $\tilde{O}(kn^3)$. However, [3] has a round complexity $O(d)$ and [13] has a round complexity $O(d+n)$, where d is the depth of circuit size and in case of ranking functionality $d = \Omega(\log n)^3$. All these results are evaluated in the broadcast channel. The currently most communication efficient and maliciously secure multi-party computation protocol is due to Hirt and Maurer [11], whose communication complexity is $O((mn^2 + n_I n^4 + n_o n + n^4)k)$ (in the pairwise channel), where m is the number of multiplication gates in the circuit, n_I and n_o are the number of inputs and outputs respectively. Applying to the ranking functionality, it implies a solution of communication complexity $O(kn^5)$. However, their protocol has a very high round complexity $O(n^2)$. Note a broadcast channel with communication complexity $O(n^2)$ and round complexity $O(n)$ can be constructed [10]. Therefore, no maliciously secure protocol for the ranking problem has previously achieved a result equivalent to a constant round with communication complexity $O(kn^3)$ in the broadcast channel.

1.2   Contribution

In this work, we construct an efficient n-party ranking protocol. Our protocol has a constant round complexity. Assume k is the security parameter and each party P_i has an input $m_i \in \{1, \cdots, N\}$. Then our protocol has a communication complexity $O(kn^2(n + N))$. Therefore, if $N = O(n)$ which is the typical case, the communication complexity is $O(kn^3)$. Thus, our protocol is efficient in both communication complexity and round complexity. In addition, we prove that our construction is secure in the non-adaptive malicious model.

This work is organized as follows. Section 2 introduces the security model. Section 3 introduces a building block: a zero sharing protocol. Section 4 introduces our ranking protocol containing some sub ideal functionalities and its security analysis. Section 5 considers how to realize the sub ideal functionalities and obtain a full functional ranking protocol.

2   Security Model

In this section, we introduce the security model for non-adaptive malicious adversary. In this model, we first have an ideal process in the non-adaptive malicious model, where the computation is achieved via a trusted third party. Then, we turn to the maliciously real process, which is a model for the execution of the real protocol. Finally, a protocol is said to be secure if the real process and ideal process essentially have an identical performance.

---

[3] This easily follows from two facts: (1) the circuit is acyclic and each node has at most two inputs; (2) the circuit size for ranking is trivially lower bounded by $\Omega(n)$.

Ideal Process. We consider the ideal process w.r.t. a non-adaptive malicious adversary $\mathcal{S}$. Let f be an ideal functionality and $\mathcal{F}$ be a trusted third party. Let $P_1, \cdots, P_n$ be n parties involved in the execution. $\mathcal{S}$ has an arbitrary auxiliary input z and P_i has an input $x_i \in D$. Before the protocol starts, $\mathcal{S}$ can select a set of parties $\Phi \subset \{P_1, \cdots, P_n\}$ for corruption. As a result, all the inputs of these parties are provided to $\mathcal{S}$. In addition, their future actions are fully taken by $\mathcal{S}$. After the protocol starts, the execution is described as follows.

- Upon input x_i, an uncorrupted P_i forwards it to $\mathcal{F}$. Upon receiving an output from $\mathcal{F}$, P_i outputs it directly.
- $\mathcal{S}$ can change the input x_i of a corrupted party P_i to $x_i' \in D \cup \{\bot\}$, and sends x_i' to $\mathcal{F}$.
- Upon receiving $x_1', x_2', \cdots, x_n'$ from all parties ($x_i' = x_i$ for an uncorrupted P_i), $\mathcal{F}$ may ask $\mathcal{S}$ for a message. Then, he follows f to compute output $(o_1, o_2, \cdots, o_n)$ from $(x_1', \cdots, x_n')$ and the response from $\mathcal{S}$ (if any). By default, if some $x_i' = \bot$, then $o_1 = \cdots = o_n = \bot$. Finally, $\mathcal{F}$ asks $\mathcal{S}$ to deliver o_i to P_i.
- $\mathcal{S}$ can deliver o_i or $\bot$ for an uncorrupted P_i. In any case, the secret part of o_i will be kept invisible from $\mathcal{S}$. Finally, $\mathcal{S}$ can output whatever he wishes.

Let r_0 and r_S be the random input of $\mathcal{F}$ and $\mathcal{S}$, respectively. The joint execution in the ideal process, denoted by $\text{IDEAL}_{\mathcal{F}, \mathcal{S}(z)}(\boldsymbol{x}; r_0, r_S)$, is a concatenation of the outputs for uncorrupted parties as well as the adversary $\mathcal{S}$. Later we will use random variable $\text{IDEAL}_{\mathcal{F}, \mathcal{S}(z)}(\boldsymbol{x})$ to denote $\text{IDEAL}_{\mathcal{F}, \mathcal{S}(z)}(\boldsymbol{x}; r_0, r_S)$ with random input r_0 and r_S.

Real Process. Let Γ be a protocol to realize a functionality f. Let $P_1, \cdots, P_n$ be n parties involved in the execution. Let $\mathcal{A}$ is a PPT adversary with an auxiliary input z. Before the protocol starts, $\mathcal{A}$ specifies a set of parties $\Phi \subset \{P_1, \cdots, P_n\}$ for corruption. As in the ideal process, once a party is corrupted, his secret input is provided to $\mathcal{A}$. In addition, his future action is fully taken by $\mathcal{A}$. After the protocol starts, the real process is described as follows.

- Upon input x_i, an uncorrupted P_i exactly follows the protocol Γ to answer the incoming message and generates its output o_i.
- The action for each corrupted P_i is fully taken by $\mathcal{A}$.

In addition, we assume the channel is authenticated with a guaranteed delivery. That is, for any message M from an uncorrupted P_i, $\mathcal{A}$ must deliver M to the specified receiver without any change. Let r_A, r_i, $(P_i \notin \Phi)$ are the random input for $\mathcal{A}$ and P_i, respectively. $o_A, o_i, (P_i \notin \Phi)$ are the outputs for $\mathcal{A}$, P_i, respectively. Similar to the ideal process, we can define the joint execution, denoted by $\text{REAL}_{\Gamma, \mathcal{A}(z)}(\boldsymbol{x}; r_A, \{r_i : P_i \notin \Phi\})$, to be a concatenation of outputs for uncorrupted parties and the output of $\mathcal{A}$. We use $\text{REAL}_{\Gamma, \mathcal{A}(z)}(\boldsymbol{x})$ to denote the variable of the joint execution with uniform random input.

Definition 1. *Let Γ be an n-party protocol to implement a functionality f. Γ is said to be secure in a malicious but non-adaptive model if for any probabilistic* *polynomial-time adversary $\mathcal{A}$ in this model there exists an expected polynomial-time ideal adversary $\mathcal{S}$ such that*

$$\left\{ \text{IDEAL}_{f,\mathcal{S}(z)}(\boldsymbol{x}) \right\}_{z,\boldsymbol{x}} \stackrel{c}{\equiv} \left\{ \text{REAL}_{\Gamma,\mathcal{A}(z)}(\boldsymbol{x}) \right\}_{z,\boldsymbol{x}},$$

where $\stackrel{c}{\equiv}$ means computational indistinguishability, z is auxiliary input and $\boldsymbol{x} = (x_1, \cdots, x_n)$ is the input vector for parties.

3   A Zero-Sharing Protocol

In our ranking protocol (Section 4), rank r_j is computed by pair-wise comparing m_j with other m_i and then summarizing the results. In order not to leak the comparison bit ($m_i < m_j$), we employ a blinding technique such that all blinding values (for a fixed i) collectively constitute a random sharing of zero. The zero-sharing protocol is constructed in this section, which will be introduced now. Formally, we propose a protocol for functionality $\mathcal{F}_0$: $(1^\kappa, \cdots, 1^\kappa) \mapsto \langle f_1(x), \cdots, f_n(x) \rangle$, where $f_i(x)$ is uniformly at random from $Z_q[x]$ of degree at most $n-1$ such that $\sum_{i=1}^n f_i(x) = 0$. In the ideal process, the adversary is allowed to choose $f_i(x)$ for corrupted parties. Our zero-sharing protocol is presented in Table 1, where p, q are primes with $p = 2q+1$, and $g, h \in Z_p^*$ have an order of q. Let $R_{2dl} = \{(g^{y_1}h^{y_2}, (y_1, y_2)) : y_1, y_2 \in Z_q\}$. The protocol employs a multi-party functionality $\mathcal{F}_{m2dl}$ for each party to prove the knowledge of his commitment. $\mathcal{F}_{m2dl}$ is defined as follows.

Definition 2. (Multi-party Functionality: $\mathcal{F}_{m2dl}$) *$\mathcal{F}_{m2dl}$ does the following with parties $P_1, \cdots, P_n$ and adversary $\mathcal{S}$, and is parameterized by relation R_{2dl}.*

- *Upon $(F_{jt}, (f_{jt}, f'_{jt}))_{t=1}^n$ from each P_j, $\mathcal{F}_{m2dl}$ checks if $(F_{jt}, (f_{jt}, f'_{jt})) \in R_{2dl}$. If it holds for all j and t, $\mathcal{F}_{m2dl}$ sends message **ok** to $P_1, \cdots, P_n$ and $\mathcal{S}$; otherwise, it does nothing.*

After the protocol execution, P_i can compute $u_{ij} = f_i(j)$, for each $j = 1, \cdots, n$. All parties can compute $U_{ij} = g^{u_{ij}} h^{u'_{ij}}$, where $u'_{ij} = f'_i(j)$. Note the sharing scheme has a property that for each j, $\sum_i u_{ij} = 0$. That is why we call it a zero-sharing protocol. In our ranking protocol (Section 4), u_{ij} will be used to blind the bit ($m_i < m_j$). The completeness of ranking will be guaranteed by the fact $\sum_i u_{ij} = 0$.

Lemma 1. *Our zero-sharing protocol securely realizes $\mathcal{F}_0$. Specifically, if Φ is the set of corrupted parties with $|\Phi| \leq n - 1$, then $f_i(x)$ for uncorrupted P_i is uniformly at random in $F[x]$ with degree at most $n - 1$ such that the only constraint is $\sum_{P_i \notin \Phi} f_i(x) = -\sum_{P_i \in \Phi} f_i(x)$. In addition, $\{u_{ij}\}_{i \notin \Phi, 1 \leq j \leq n}$ is uniform at random in $Z_q^{n \times (n-|\Phi|)}$ with only constraint $\sum_{i \in \Phi} u_{ij} = -\sum_{i \notin \Phi} u_{ij}$, for each $j = 1, \cdots, n$.*

Table 1. An Efficient Zero Sharing Protocol **Zero-Sharing**(n)

Params:	$p = 2q + 1, g, h;$
Output:	$f_i(x) \in Z_q[x]$ for each $P_i, i = 1, \cdots, n.$

1. For each $j = 1, \cdots, n$, P_i takes $f_{ij}(x) = \sum_{t=0}^{n-1} f_{ijt} x^t$, $f'_{ij}(x) = \sum_{t=0}^{n-1} f'_{ijt} x^t \in Z_q[x]$ randomly such that $\sum_j f_{ij}(x) = 0$ and $\sum_j f'_{ij}(x) = 0$. Then he computes $F_{ijt} = g^{f_{ijt}} h^{f'_{ijt}}$. He *privately* sends $(f_{ij}(x), f'_{ij}(x))$ to P_j, and makes F_{ijt} public.

2. P_j verifies if $\prod_{l=1}^{n} F_{ilt} = 1$ and if $F_{ijt} = g^{f_{ijt}} h^{f'_{ijt}}$ for all i, t. If the verification fails, P_j broadcasts a complaint toward unsuccessful P_i. P_i tries to resolve it by making $(f_{ij}(x), f'_{ij}(x))$ public. If the complaint still remains unresolved, by default set $(f_{ij}(x), f'_{ij}(x)) = (0, 0)$ for all j. P_j computes $f_j(x) = \sum_{i=1}^{n} f_{ij}(x)$, $f'_j(x) = \sum_{i=1}^{n} f'_{ij}(x)$.

3. $\mathcal{F}_{m2dl}$ is invoked such that P_j proves $(F_{jt}, (f_{jt}, f'_{jt})) \in R_{2dl}$ for all t, where $F_{jt} = \prod_{i=1}^{n} F_{ijt}$, $f_j(x) := \sum_{t=0}^{n-1} f_{it} x^t$, $f'_j(x) := \sum_{t=0}^{n-1} f'_{it} x^t$. If **ok** is received, P_i accepts and outputs $(f_i(x), f'_i(x))$; otherwise, he outputs $\perp$.

Proof. We now show that for any PPT real process adversary $\mathcal{A}$, there exists an expected polynomial-time ideal process adversary $\mathcal{S}$ such that the two executions are indistinguishable. $\mathcal{S}$ first prepares $p = 2q + 1, g$, and computes $h = g^a$, for $a \leftarrow Z_q$. Then he runs $\mathcal{A}$ with p, g, h. In turn, he will receive a set Φ for corruption. Then he corrupts Φ in the ideal process. For simplicity, we use P_i to represent a party in the internal simulation and $\tilde{P}_i$ the party in the external ideal process. $\mathcal{S}$ follows the real protocol execution to interact with $\mathcal{A}$ for Step one and Step two, at the end of which he will obtain $f_j(x)$ for each uncorrupted P_j. In Step three, upon $(F_{jt}, (f_{jt}, f'_{jt}))$ from each corrupted P_j (controlled by $\mathcal{A}$), $\mathcal{S}$ verifies if $(F_{jt}, (f_{jt}, f'_{jt})) \in R_{2dl}$. If yes, $\mathcal{S}$ sends message **ok** back to $\mathcal{A}^4$ and in the ideal process he sends f_{jt} for corrupted $\tilde{P}_j$ to $\mathcal{F}_0$. When $\mathcal{F}_0$ asks him to deliver the messages, he does it faithfully. On the other hand, if the verification fails, $\mathcal{S}$ sends $\perp$ for corrupted party $\tilde{P}_j$ to $\mathcal{F}_0$ (note if no party is corrupted, **ok** must occur). Finally, $\mathcal{S}$ outputs whatever $\mathcal{A}$ does. The difference between this simulated execution and real execution is that the outputs for uncorrupted parties might be different. In order to be consistent with the ideal process, the output for an uncorrupted P_i in the internal simulation is supposed to output $\tilde{f}_j(x)$, where $\tilde{f}_j(x)$ is the polynomial sent to $\tilde{P}_j$ from the ideal functionality (but invisible to $\mathcal{S}$). However, in the simulated execution, the output $f_j(x)$ of P_j might be different from $\tilde{f}_j(x)$. However, this is not a problem as given the adversary view in the simulated system, we can *consistently* reformulate the output of P_j to $\tilde{f}_j(x)$. Indeed, take a fixed party $P_r \notin \Phi$, reformulate $f_{rj}(x)$ to $\tilde{f}_{rj}(x) = f_{rj}(x) + \tilde{f}_j(x) - f_j(x)$ and $f'_{rj}(x)$ to $\tilde{f}'_{rj}(x) = f'_{rj}(x) - a^{-1}(\tilde{f}_j(x) - f_j(x))$, for each other $P_j \notin \Phi$. After the reformulation, adversary view does not change.

---

[4] Note in this case we must have $\sum_j f_{jt} = \sum_j f'_{jt} = 0$; otherwise, the adversary can be transformed to break Discrete Log problem as $1 = \prod_j F_{jt} = g^{\sum_j f_{jt}} h^{\sum_j f'_{jt}}$.

However, the distribution of the reformulated simulation is exactly according to the real protocol since for each $j = 1, \cdots, n$

$$\tilde{f}_{rj}(x) + \sum_{i \neq r} f_{ij}(x) = \tilde{f}_j(x) - f_j(x) + \sum_{i=1}^n f_{ij}(x) = \tilde{f}_j(x),$$
$$\tilde{f}'_{rj}(x) + \sum_{i \neq r} f'_{ij}(x) = \tilde{f}'_j(x) - f'_j(x) + \sum_{i=1}^n f'_{ij}(x) = \tilde{f}'_j(x).$$

Thus, our protocol realizes the functionality $\mathcal{F}_0$. The second statement is immediate from the first statement. ∎

4   Our Hybrid Ranking Protocol

In this section, we introduce our cryptographic ranking protocol in the hybrid model, see Table 2. Let p, q be large primes with $p = 2q + 1$. Let g, h, σ, z_1 be random elements in z_p^* of order q. These parameters are setup by a trusted third party. Before the ranking protocol, we assume that all parties have jointly run a zero sharing protocol. Consequently, each P_i has the secret output $f_i(x), f'_i(x)$ and public output F_{jt} for all j and t. Let $U_{ij} = \prod_{t=0}^{n-1} F_{it}^{j^t}$ (publicly computable), $u_{ij} = f_i(j), u'_{ij} = f'_i(j)$ (known to P_i). Then, $U_{ij} = g^{u_{ij}} h^{u'_{ij}}$. As mentioned before, u_{ij} will be used in the ranking protocol to blind the bit $(m_j < m_i)$. Our ranking protocol is divided into two phases: input commitment and rank computation.

In the input commitment phase, we first encode $m_i \in \{1, \cdots, N\}$ as a N-tuple $(0, \cdots, 0, 1, 1, \cdots, 1)$, where $'1'$ starts at the $(m_i + 1)$th component. Next, each P_i commits to the encoded N-tuple of m_i: $B_{i1} = h^{x_{i1}}, \cdots, B_{i(m_i)} = h^{x_{i(m_i)}}$, $B_{i(m_i+1)} = \sigma h^{x_{i(m_i+1)}}, \cdots, B_{iN} = \sigma h^{x_{iN}}$. Then, we ask each P_j to prove the knowledge of the commitment. Let $B'_{it} = B_{i(t+1)} B_{it}^{-1}, 1 \leq t \leq N$, where $B_{i(N+1)} = \sigma$ by default. Then, it suffices to prove: B_{i1} is commitment of bit 0, B'_{it} is a commitment of either 0 or 1, for each t. Indeed, once this is done, the digital committed in B_{it} is no more than that in $B_{i(t+1)}$. Since B_{i1} is a commitment of 0 and $B_{i(N+1)}$ is a commitment of 1, the tuple committed in $\{B_{it}\}_{t=1}^{N+1}$ must start with a sequence of $'0'$, followed by a sequence of $'1'$. Thus, the proof task can be formalized as an ideal functionality $\mathcal{F}_{mor+}$ below, where relations $R_{dl} = \{(h^x, x) : x \in Z_q\}$ and $R_{or} = R_{dl} \cup \{(\sigma h^x, x) : x \in Z_q\}$. In the protocol, $\mathcal{F}_{mor+}$ is invoked in which each P_i proves $(B'_{it}, x'_{it}) \in R_{or}$ for all t and $(B_{i1}, x_{i1}) \in R_{dl}$.

Definition 3. (Multi-party Functionality: $\mathcal{F}_{mor+}$) $\mathcal{F}_{mor+}$ does the following, running $P_1, \cdots, P_n$ and $\mathcal{S}$, and parameterized by relations R_{or} and R_{dl}.

- Upon receiving (B_{i1}, x_{i1}) and $\{(B'_{it}, x'_{it})\}_{t=1}^N$ from each P_i, $\mathcal{F}_{mor+}$ verifies if $(B'_{it}, x'_{it}) \in R_{or}$ and $(B_{i1}, x_{i1}) \in R_{dl}$. It holds for all i and t, $\mathcal{F}_{mor+}$ sends message **ok** to $P_1, \cdots, P_n$ and $\mathcal{S}$; otherwise, it does nothing.

If $\mathcal{F}_{mor+}$ does not send **ok**, each P_i aborts; otherwise, he is ready for phase two.

In the rank computation stage, each P_i essentially uses an oblivious transfer to send a blinded comparison bit $(m_i < m_j)$ to P_j. To do this, he does a fresh commitment to m_i using $\{B_{ijt}\}_{t=1}^N$, and then computes $D_{ijt} = (B_j\sigma^{-t})^{x_{ijt}}z_1^{u_{ij}} = h^{y_j x_{ijt}}z_1^{u_{ij}} \times \sigma^{(m_j-t)x_{ijt}}$. Then P_i sends $\{B_{ijt}|D_{ijt}\}_{t=1}^N$ to P_j. If P_j computed this message honestly, then $D_{ij(m_j)} = h^{x_{ijt}y_j}z_1^{u_{ij}}$. Thus, P_j can compute $\xi_i = B_{ij(m_j)} \cdot D_{ij(m_j)}^{-1/y_j} = \sigma^{\delta_{ij}}z_1^{-u_{ij}/y_j}$, where δ_{ij} is the bit $(m_i < m_j)$. Note $\sum_i \delta_{ij} = r_j - 1$ and $\sum_i u_{ij} = 0$. We have $\prod_{i=1}^n \xi_i = \sigma^{r_j-1}$. Finally, P_j can obtain r_j by trial. So what is left is for P_i to prove that $\{B_{ijt}|D_{ijt}\}_{t=1}^N$ is computed *appropriately*. To do this, define

$$R_{rc} = \left\{ \left(\{A_t||\tilde{A}_t||D_t\}_1^N||U||\Delta, \{x_t||\tilde{x}_t\}_1^N||u||u'\right) : \quad \tilde{A}_t A_t^{-1} = h^{\tilde{x}_t - x_t}, \right.$$
$$(A_t, x_t) \in R_{or}, D_t = (\Delta \cdot \sigma^{-t})^{\tilde{x}_t}z_1^u, U = g^u h^{u'},$$
$$\left. \Delta \in G_q, x_t, \tilde{x}_t, u, u' \in Z_q \right\}.$$

This relation is motivated by

$$\left(\{B_{it}||B_{ijt}||D_{ijt}\}_1^N||U_{ij}||B_j, \{x_t||x_{ijt}\}_1^N||u_{ij}||u'_{ij}\right) \in R_{rc},$$

which implies $\{B_{ijt}|D_{ijt}\}_{t=1}^N$ is computed appropriately. In the protocol, an ideal functionality $\mathcal{F}_{mrc}$ below is invoked for this purpose. This completes the description of our protocol.

Definition 4. (Multi-party Functionality: $\mathcal{F}_{mrc}$) $\mathcal{F}_{mrc}$ *does the following, running* $P_1, \cdots, P_n$ *and* $\mathcal{S}$, *and parameterized by* R_{rc}, σ *and* z_1.

- *Upon input*

$$\left(\{B_{it}||B_{ijt}||D_{ijt}\}_1^N||U_{ij}||B_j, \quad \{x_{it}||x_{ijt}\}_1^N||u_{ij}|u'_{ij}\right)$$

  from P_i, $\mathcal{F}_{mrc}$ *checks it is consistent w.r.t.* R_{rc}. *If it holds, then* $\mathcal{F}_{mrc}$ *sends message* $\mathbf{ok}(P_i, P_j)$ *to* P_j *and* $\mathcal{S}$; *otherwise, it ignores the verification for* (P_i, P_j).

Now we formally prove the security of our ranking protocol.

Theorem 1. *Under the Decisional Diffie-Hellman (DDH) assumption, our hybrid ranking protocol is secure in the non-adaptive malicious model.*

Proof. We show that for any PPT real process adversary $\mathcal{A}$, there exists an expected polynomial-time ideal process adversary $\mathcal{S}$ such that the joint executions in the two processes are indistinguishable. $\mathcal{S}$ internally runs the simulated real process with $\mathcal{A}$, playing the uncorrupted parties. At the same time, he is externally involved in the ideal process execution, on behalf of corrupted party $\tilde{P}_i \in \Phi$. The code of $\mathcal{S}$ is described as below.

- $\mathcal{S}$ first gets the input m_i for each $\tilde{P}_i \in \Phi$, and provides to $\mathcal{A}$ as the input for P_i. He prepares $p, g, h, \sigma = h^a$ for $a \leftarrow Z_q$ and runs $\mathcal{A}$ with it.
- $\mathcal{S}$ plays the role of uncorrupted P_i to compute $B_{it} = h^{x_{it}^*}$, where $x_{it}^* \leftarrow Z_q$, $1 \leq t \leq N$. Note that $\mathcal{S}$ does not know the real input m_i. Thus, he is unable to do a real commitment to m_i. Then, $\mathcal{S}$ simulates the ideal functionality

Table 2. Our Hybrid Ranking Protocol

Params: p, g, σ, h, z_1;	**Input:** m_i for $P_i, i = 1, \cdots, n$.

PHASE ONE: Input Commitment.

1. Each P_i takes $x_{it} \leftarrow Z_q$, computes $B_{it} = \begin{cases} h^{x_{it}} & \text{if } 1 \leq t \leq m_i, \\ \sigma h^{x_{it}} & \text{if } m_i < t \leq N \end{cases}$. Then P_i broadcasts $\langle B_{i1}, \cdots, B_{iN} \rangle$.

2. Let $B'_{it} = B_{i(t+1)} B_{it}^{-1}$ for $1 \leq t \leq N$, where by default $B_{i(N+1)} = \sigma$. Let $x'_{it} = x_{i(t+1)} - x_{it}$ and $x_{i(N+1)} = 0$. $\mathcal{F}_{mor+}$ is invoked in which each P_i proves $(B_{i1}, x_{i1}) \in R_{dl}$ and $(B'_{it}, x'_{it}) \in R_{or}$ for all t. If a message **ok** is not received from $\mathcal{F}_{mor+}$, P_i aborts; otherwise, he is ready for Phase Two. Let $B_i = \sigma^N \cdot \prod_{t=1}^N B_{it}^{-1} = \sigma^{m_i} h^{y_i}$, for $y_i = -\sum_{t=1}^N x_{it}$. P_i keeps y_i secret.

PHASE TWO: Rank Computation.

1. P_i prepares the outgoing message to P_j as follows.
   - P_i takes $x_{ijt} \leftarrow Z_q$, set $B_{ijt} = \begin{cases} h^{x_{ijt}} & \text{if } 1 \leq t \leq m_i, \\ \sigma h^{x_{ijt}} & \text{if } m_i < t \leq N \end{cases}$. In other words, P_i commits to m_i again.
   - P_i computes $D_{ijt} = (B_j \cdot \sigma^{-t})^{x_{ijt}} \cdot z_1^{u_{ij}}$, sends $\{D_{ijt}\}_{t=1}^N$ to P_j.

2. $\mathcal{F}_{mrc}$ is   invoked   in   which   each P_i proves $\left(\{B_{it} \| B_{ijt} \| D_{ijt}\}_{t=1}^N \| U_{ij} \| B_j, \{x_{it} \| x_{ijt}\}_{t=1}^N \| u_{ij} \| u'_{ij} \right) \in R_{rc}$ to P_j.

3. Receiving **ok**(P_i, P_j) for all i, P_j computes $\xi_i = B_{ij(m_j)} \cdot D_{ij(m_j)}^{-1/y_j}$. Finally, he derives $\xi = \prod_{i=1}^N \xi_i = \sigma^{r_j - 1}$ and obtains r_j by trial.

$\mathcal{F}_{mor+}$. If all the inputs from corrupted parties (controlled by $\mathcal{A}$) have been successfully verified, $\mathcal{S}$ (simulating $\mathcal{F}_{mor+}$) sends **ok** to $\mathcal{A}$; otherwise, he does nothing.

- $\mathcal{S}$ calculates $\tilde{m}_i$ from the input to $\mathcal{F}_{mor+}$ for corrupted P_i. He then externally sends $\tilde{m}_i$ for $\tilde{P}_i$ to $\mathcal{F}_0$. And in turn, he obtains the corresponding r_i. He then chooses arbitrary $\tilde{m}_j$ for uncorrupted P_j such that each $\tilde{m}_i$ for corrupted P_i has a rank r_i among $(\tilde{m}_1, \cdots, \tilde{m}_n)$. Then for uncorrupted P_j, he defines $x_{jt} = x_{jt}^*$ if $t \leq \tilde{m}_j$; $x_{jt} = x_{jt}^* - a$ if $\tilde{m}_j < t \leq N$. Under this formulation, $\{B_{it}\}_1^N$ is an encoding of $\tilde{m}_i$ as computed in Step 1 of Phase One.

- In Phase Two, $\mathcal{S}$ faithfully interacts with $\mathcal{A}$ by simulating $\mathcal{F}_{mrc}$ and all uncorrupted P_i with $\{x_{it}\}_1^N$. In the external execution (ideal process), $\mathcal{S}$ delivers r_j to uncorrupted $\tilde{P}_j$ if and only if **ok**(P_i, P_j) for all i has been computed in the internal execution for all corrupted P_i. Finally, $\mathcal{S}$ outputs whatever $\mathcal{A}$ does.

The view of $\mathcal{A}$ is different from the real process: for uncorrupted P_i, (1) Input commitment is dummy instead of a commitment of m_i; (2) $\tilde{m}_i$ is not equal to

m_i for uncorrupted P_i. In the remaining part, we show this modification does change the distribution of $\mathcal{A}$'s view.

Let the simulated game be Γ. Consider the mental game Γ_1 of Γ, where the only difference is that for uncorrupted P_i, the input commitment is for $\tilde{m}_i$ instead of first being dummy and reformulating later. The view of $\mathcal{A}$ under this change is identically distributed in Γ, as x_{it}^* in Γ is uniformly at random (thus x_{it} obtained in the reformation is random) in Z_q.

Now we will show that the adversary view in Γ_1 and the variant of Γ_1, where $\tilde{m}_i$ for uncorrupted P_i is replaced by m_i, is indistinguishable. We achieve this via a sequence of game techniques.

First we modify Γ_1 to Γ_2 such that in Phase Two, for each uncorrupted P_i, $D_{ijt} = h^{y_j x_{ijt}} \cdot \beta_{ijt}^{\tilde{m}_j - t} \cdot z_1^{u_{ij}}$, where $\beta_{ijt} \leftarrow G_q$. Note if $\beta_{ijt} = \sigma^{x_{ijt}}$, then Γ_2 becomes Γ_1. We show that the execution in these two games are indistinguishable. If this were not true, we construct an adversary $\mathcal{B}_2$ to break DDH assumption. Given $(h, \sigma, \alpha, \beta)$ (either DH tuple or random tuple), $\mathcal{B}_2$ takes $(\sigma, \alpha_{ljt}, \beta_{ljt}) \leftarrow Rud_0(h, \sigma, \alpha, \beta)$, $1 \leq t \leq N, 1 \leq j \leq n$, for each uncorrupted P_l (algorithm Rud is presented in Appendix A). He then follows Γ_1 (and Γ_2) for input commitment stage. In the second phase, he follows the simulator $\mathcal{S}$ in Γ_2, except that for uncorrupted P_i, $h^{x_{ijt}}$ in defining B_{ijt} is taken as α_{ijt} and that $D_{ijt} = \alpha_{ijt}^{y_j} \cdot \beta_{ijt}^{\tilde{m}_j - t} \cdot z_1^{u_{ij}}$. Finally, $\mathcal{B}_2$ feeds the execution output to the distinguisher and outputs whatever he does. Note if $(h, \sigma, \alpha, \beta)$ is DH tuple, then the simulated game by $\mathcal{B}_2$ is distributed as in Γ_1; otherwise, it is distributed according to Γ_2. Thus, the distinguishability between Γ_1 and Γ_2 implies the non-negligible advantage of $\mathcal{B}_2$, a contradiction to DDH assumption.

We modify Γ_2 to Γ_3 such that in Phase Two, for any uncorrupted *pair* P_i and P_j, $D_{ijt} = \gamma_{ijt} \cdot \beta_{ijt}^{\tilde{m}_j - t} \cdot z_1^{u_{ij}}$ for $\gamma_{ijt} \leftarrow G_q$ (instead of $\gamma_{ijt} = h^{y_j x_{ijt}}$). We show the execution in Γ_2 and Γ_3 is indistinguishable; otherwise, consider a DDH breaker $\mathcal{B}_3$. Upon receiving input (h, μ, ν, γ), $\mathcal{B}_3$ first takes $(\mu_j, \nu_j, \gamma_j) \leftarrow Rud_1(h, \mu, \nu, \gamma)$ (See Appendix A for details) for each uncorrupted P_j and then further takes $(\mu_j, \nu_{ijt}, \gamma_{ijt}) \leftarrow Rud_0(h, \mu_j, \nu_j, \gamma_j)$ for $1 \leq t \leq N$ and each uncorrupted P_i. He follows the simulation in Γ_2 for input commitment except that $h^{x_{jN}}$ is computed as $\mu_j \cdot \prod_{t=1}^{N-1} h^{-x_{jt}}$. Note this simulation is distributed identically as in Γ_2 (and Γ_3) as μ_j is uniform in G_q. In Phase Two, $\mathcal{B}_3$ follows the simulation in Γ_3, except that for any uncorrupted pair P_i, P_j, (1) $h^{x_{ijt}}$ in computing B_{ijt} is defined to be ν_{ijt}; (2) $D_{ijt} = \gamma_{ijt} \cdot \beta_{ijt}^{\tilde{m}_j - t} \cdot z_1^{u_{ij}}$, where $\beta_{ijt} \leftarrow G_q$ as in Γ_2 and γ_{ijt} is the value just derived from Rud. By the property of Rud algorithm, if (h, μ, ν, γ) is a DH tuple, then the simulated game is identically distributed as Γ_2; otherwise, it is according to Γ_3. Thus, the distinguishablity between Γ_2 and Γ_3 implies the distinguishability of DDH, contradiction.

Consider the mental game Γ_4, a variant of Γ_3 while $\tilde{m}_i$ is replaced by m_i (registered by $\tilde{P}_i$ to the ideal functionality). We show that the views of $\mathcal{A}$ between Γ_3 and Γ_4 are identically distributed. Otherwise, commitments for $\{\tilde{m}_i\}_{P_i \notin \Phi}$ and $\{m_i\}_{P_i \notin \Phi}$ using the method in Step 1 can be distinguished. However, this is impossible since the two set of commitments have identical distribution. Here is details. Upon input $\{B_{it}\}_1^N$ for each uncorrupted P_i (either commit of m_i or commitment of $\tilde{m}_i$), a distinguisher $\mathcal{B}_4$ simulates Step 2 normally (as in Γ_3). Phase Two is simulated as follows.

- For each uncorrupted P_i, $B_{ijt} = B_{it} \cdot h^{\Delta_{ijt}}$ for each j,t, where $\Delta_{ijt} \leftarrow Z_q$.
- For each uncorrupted P_i and corrupted P_j, take $D_{ijt} \leftarrow G_q$ for $t \neq \tilde{m}_j$. Fix $P_{i_0} \notin \Phi$. Take $D_{i_0j(\tilde{m}_j)} = B_{i_0j(\tilde{m}_j)}^{y_j} \cdot \sigma^{-R_j y_j} \cdot z_1^{u_{ij}}$, where $R_j = \sharp\{i : P_i \notin \Phi, m_i \geq \tilde{m}_j\}$. For $i \neq i_0$, take $D_{ij(\tilde{m}_j)} = B_{ij(\tilde{m}_j)}^{y_j} \cdot z_1^{u_{ij}}$. Note $y_j = -\sum_{t=1}^N x_{jt}$ and x_{jt} is obtained from the input to functionality $\mathcal{F}_{mor+}$ by $\mathcal{A}$ in Phase One.
- For uncorrupted P_i and P_j, take $D_{ijt} \leftarrow G_q$.

Now we claim that no matter the input to the simulator is commitment of $\{\tilde{m}_i\}$ or $\{m_i\}$, the above simulation is consistent with Γ_4. It suffices to show that $D_{ij(\tilde{m}_j)}$ for uncorrupted P_i and corrupted P_j is according to Γ_4. Notice for corrupted P_j, $\tilde{m}_j$ is ranked r_j for both cases $\{m_i\}_{P_i \notin \Phi}$ and $\{\tilde{m}_i\}_{P_i \notin \Phi}$, it follows that $\sharp\{i : P_i \notin \Phi, m_i \geq \tilde{m}_j\} = \sharp\{i : P_i \notin \Phi, \tilde{m}_i \geq \tilde{m}_j\}$. Thus, we only need to consider the case, where for uncorrupted P_i, $\{B_{it}\}_1^N$ is a commitment of $\tilde{m}_i$. Our key point is that for any j, $\{u_{ij} : P_i \notin \Phi\}$ are uniform in Z_q with only constraint $\sum_{P_i \notin \Phi} u_{ij} = -\sum_{P_i \in \Phi} u_{ij}$ (by Lemma 1). Note that $B_{ijt} = \sigma^{\delta_{it} h^{x_{ijt}}}$, where δ_{it} is the bit ($\tilde{m}_i < t$). Thus, in the simulation, $D_{i_0j\tilde{m}_j} = h^{y_j x_{i_0j\tilde{m}_j}} \times \sigma^{y_j \delta_{i_0 \tilde{m}_j}} \cdot \sigma^{-y_j R_j} \cdot z_1^{u_{i_0j}}$; for other uncorrupted P_i, $D_{ij\tilde{m}_j} = h^{y_j x_{ij\tilde{m}_j}} \times \sigma^{y_j \delta_{i\tilde{m}_j}} \cdot z_1^{u_{ij}}$. Note that $\sigma^{-R_j} \prod_{P_i \notin \Phi} \sigma^{\delta_{i\tilde{m}_j}} \cdot z_1^{u_{ij}} = \prod_{P_i \notin \Phi} z_1^{u_{ij}}$. Let $\tilde{u}_{i_0j} = u_{i_0j} - (R_j - \delta_{i_0\tilde{m}_j})y_j \log_{z_1} \sigma$, and $\tilde{u}_{ij} = u_{ij} + \delta_{i\tilde{m}_j} \log_{z_1} \sigma$ for other uncorrupted P_i. Note that $\sum_{P_i \notin \Phi} \tilde{u}_{ij} = \sum_{P_i \notin \Phi} u_{ij}$, we have that $\{\tilde{u}_{ij}\}_{P_i \notin \Phi}$ can be regarded as another feasible assignment for $\{u_{ij}\}_{P_i \notin \Phi}$. Furthermore, $\{\tilde{u}_{ij}\}_{P_i \notin \Phi}$ is according to the real distribution since $\{u_{ij}\}_{P_i \notin \Phi}$ is uniform at random with the only constraint on their additive sum. Therefore, the simulation is actually distributed as in Γ_4. Our claim follows.

Furthermore, notice adversary view in $\Gamma_1, \cdots, \Gamma_4$ when $\tilde{m}_i = m_i$ for all uncorrupted P_i is still indistinguishable. On the other hand adversary view in Γ_1 with $\tilde{m}_i = m_i$ for all uncorrupted P_i is according to the real execution. Thus, the execution of ideal process by $\mathcal{S}$ is indistinguishable from the execution of the real process by $\mathcal{A}$.   ∎

5   Full Ranking Protocol

Cramer *et al.* [3] demonstrates a transformation which, given any $\sum$-protocol, outputs a secure 3-round multi-party (parallel) $\sum$-protocol. In this section, we realize our building functionalities $\mathcal{F}_{m2dl}, \mathcal{F}_{mor+}, \mathcal{F}_{mrc}$ using their transformation. Then a full ranking protocol can be obtained by (sequentially) composing these realizations with the hybrid protocol in the last section. We start with the notion of $\sum$-protocol.

5.1 $\sum$-Protocol

Let R be a binary relation consisting of pair (x, w), where x is a public string and w is a witness of polynomial length. Consider a 3-round proof of knowledge protocol for $(x, w) \in R$, where x is the common input and w is the private input for the prover. The prover starts with a message a. The verifier responds with a challenge e. Finally, the prover responds with a finishing message z. Then the verifier accepts if and only if $ver(a, e, z, x) = 1$ for a public algorithm ver. Such an interactive proof system is said to be a $\sum$-protocol if it satisfies the following.

- **Completeness.** If the prover is given a private input w such that $(x, w) \in R$, then the verifier always accepts.
- **Special Honest Verifier Zero-knowledge.** For any e, one can efficiently compute (a, e, z) such that (a, e, z) is according to the real distribution with a fixed e.
- **Witness Extraction.** For a fixed x, one can efficiently extract witness w from any two accepting transcripts (a, e, z) and (a, e', z') with $e' \neq e$.

Useful Examples. The protocols in Appendix B are $\sum$-protocols π_{dl}, π_{2dl}, π_{or}, and π_{rc} for relations R_{dl}, R_{2dl}, R_{or} and R_{rc} respectively. These examples will soon be applied to realize our building functionalities.

5.2 Secure Multi-party $\sum$-Protocol

Let $R_1, \cdots, R_v$ be v binary relations. Assume n parties $P_1, \cdots, P_n$. Let $\Lambda = (\Lambda_1, \cdots, \Lambda_n)$, where Λ_i is a collection of numbers from $\{1, \cdots, v\}$ with replacement (i.e., taking two identical numbers is allowed). Let $x_{i,j}$, for $j \in \Lambda_i$ and $1 \leq i \leq n$, be the common input for all parties. Suppose $\mathcal{F}_\Lambda$ be an n-party functionality in which each P_i proves the knowledge of witness w_{ij} s.t. $(x_{ij}, w_{ij}) \in R_j$, for each $j \in \Lambda_i$. The following has been established by Cramer *et al.* [3] (Section 6.3 in their paper).

Lemma 2. *Let $R_1, \cdots, R_v$ be v binary relations. Let π_i be a $\sum$-protocol for relation R_i for $1 \leq i \leq v$. Then there exists a 3-round multi-party protocol π_Λ realizing $\mathcal{F}_\Lambda$. In addition, if π_i $(1 \leq i \leq v)$ has a communication complexity upper bounded by $O(K)$, then the communication complexity of π_Λ is upper bounded by $O(K \sum_{i=1}^n |\Lambda_i|)$.*

Now we are ready to realize $\mathcal{F}_{m2dl}, \mathcal{F}_{mor+}, \mathcal{F}_{mrc}$. From Lemma 2, we we can easily conclude the following result.

Corollary 1. *There exists a 3-round multi-party protocol π_{m2dl} (resp. π_{mor+}, π_{mrc}) realizing the ideal functionality $\mathcal{F}_{m2dl}$ (resp. $\mathcal{F}_{mor+}, \mathcal{F}_{mrc}$). Furthermore, the communication complexity of π_{m2dl} (resp. π_{mor+}, π_{mrc}) is $O(n^2k)$ (resp. $O(nNk)$, $O(nNk)$), where k is the security parameter (i.e., the length of p).*

Proof. In $\mathcal{F}_{m2dl}$, each P_i proves the knowledge of witness for n instances w.r.t R_{2dl}. In $\mathcal{F}_{mor+}$, each P_i proves the knowledge of witness of N instances w.r.t. R_{or} and one instance w.r.t. R_{dl}. In $\mathcal{F}_{mrc}$, each P_i proves the knowledge of witness of one instance w.r.t. R_{rc}. On the other hand, π_{dl}, π_{2dl}, π_{or} and π_{rc} are $\sum$-protocols for R_{dl}, R_{2dl}, R_{or} and R_{rc} respectively. Thus, the first part follows from Lemma 2. The second part follows since π_{dl}, π_{2dl}, π_{or} and π_{rc} have communication complexity $O(k), O(k), O(k), O(Nk)$, respectively. ∎

5.3 Full Ranking Protocol

Now we are ready to state our full ranking protocol. Let $FulRank$ be the ranking protocol in the last section but functionalities $\mathcal{F}_{m2dl}$, $\mathcal{F}_{mor+}$, $\mathcal{F}_{mrc}$ are replaced by π_{m2dl}, π_{mor+}, π_{mrc} respectively.

Theorem 2. *Protocol FulRank realizes the ranking functionality in the non-adaptive but malicious model. In addition, FulRank has a constant round complexity and a communication complexity $O(n^2k(N + n))$.*

Proof. Since $FulRank$ is obtained from the hybrid protocol via sequential compositions, the security follows. It has a constant round complexity since the hybrid protocol, and π_{m2dl}, π_{mor+}, π_{mrc} all are constant round. Since the zero-sharing protocol has the communication complexity $O(n^3k)$ and the main ranking part has $O(n^2Nk)$, it follows that the whole protocol has $O(n^2k(N + n))$. ∎

References

1. M. Bellare, A. Boldyreva, S. Micali, Public-Key Encryption in a Multi-user Setting: Security Proofs and Improvements, *EUROCRYPT'00*, pp. 259-274, 2000.
2. M. Bellare and S. Goldwasser, Verifiable Partial Key Escrow, *ACM CCS'97*, pp. 78-91, 1997.
3. R. Cramer, I. Damgrd, J. Nielsen, Multiparty Computation from Threshold Homomorphic Encryption, *EUROCRYPT'01*, pp. 280-299, 2001.
4. R. Cramer, I. Damgard, and B. Schoenmakers, Proofs of partial knowledge and simplified design of witness hiding protocols, *CRYPTO'94*, LNCS 839, Y. Desmedt (ed.), SpringerVerlag, 1994.
5. R. Cramer, I. Damgard, and U. Maurer, Gemeral secure multi-party computation from any linear secret sharing scheme, *Advances in Cryptology-EUROCRYPT'00*, B. Preneel (Ed.), LNCS 1807, Springer-Verlag, pp 316-334, 2000.
6. M. Franklin, Comlexity and Security of Distributed Protocols, *Ph. D thesis*, Columbia University, 1993.
7. M. Franklin and Haber, Joint encryption and message-efficient computation, *Journal of Cryptology*, 9(4): 217-234, 1996.
8. Z. Galil, S. Haber, and M. Yung. Cryptographic computation: secure fault-tolerant protocol and the public-key model. In *Advances in Cryptology-CRYPTO'87*, C. Pomerance (Ed.), LNCS 293, Spriner-Verlag, New York, 1988.

9. O. Goldreich, S. Micali, and A. Wigderson, How to play any mental game or a completeness theorem for protocols with honest majority, *STOC'87*, pp. 218-229, New York City, 25-27 May, 1987.

10. V. Hadzilacos, J. Halpern, Message-Optimal Protocols for Byzantine Agreement (Extended Abstract), *PODC 1991*, pp. 309-323, 1991.

11. M. Hirt and U. Maurer, Robustness for Free in Uncondidtional Multi-party Computation, *Advances in Cryptology-CRYPTO'01*, J. Killian(Ed.), LNCS 2139, Springer-Verlag, pp. 101-118, 2001.

12. Y. Ishai and E. Kushilevitz, Randomizing polynomials: a new representation with application to random efficient secure computation, *FOCS '00*, pp. 294-304, 2000.

13. M. Jakobsson and A. Juels, Mix and match: secure function evaluation via ciphertexts, *Advances in Cryptology-ASIACRYPT'00*, T. Okamoto (Ed.), LNCS 1976, Springer-Verlag, pp. 162-177, 2000.

14. V. Shoup, On Formal Models for Secure Key Exchange, Available at http://philby.ucsd.edu/cryptolib/1999.html.

15. A. C. Yao, Protocols for secure computations (extended abstract), *FOCS'82*, pp. 160-164, 1982.

Appendix

Appendix A (Diffie-Hellman Self-reduction)

Now we introduce a self-reduction technique $[1, 14]$. Let p, q be two large primes with $p = 2q + 1$; G_q be the subgroup of order q in Z_p^*. And $g \in G_q \setminus \{1\}$. Thus, $\langle g \rangle = G_q$. Given a bit x and a tuple (g, g^a, g^b, g^c), a self-reduction algorithm Rud_x in $[1, 14]$ can efficiently compute a new triple $(g^{a'}, g^{b'}, g^{c'})$ with the properties in Table 3.

Table 3. Properties of Output from Self-reduction Rud_x

	$x = 0$	$x = 1$
$c = ab$	$a' = a$ & b' random in Z_q & $c' = a'b'$	a', b' random in Z_q & $c' = a'b'$
$c \neq ab$	$a' = a$ & b', c' random in Z_q	a', b', c' random in Z_q

For example, if the input is $x = 0$ and a tuple (g, g^a, g^b, g^{ab}), then the output will be $(g^a, g^{b'}, g^{ab'})$, where b' is uniformly random in Z_q. For simplicity, we use $(g^{a'}, g^{b'}, g^{c'}) \leftarrow Rud_x(g, g^a, g^b, g^c)$ to denote a random output of Rud with input x and (g, g^a, g^b, g^c).

Appendix B

In our protocol, $\sum$-protocols for R_{dl}, R_{2dl}, R_{or} and R_{rc} are presented in Tables 4, 5, 6, 7. All these protocols are not new. For example, Tables 6 [2] and Table 7 both are examples of the general OR protocol in [4].

Table 4. The $\sum$-Protocol π_{dl} for relation R_{dl}

Common Input:	$p = 2q + 1, h, X = h^x$.
Auxiliary Input:	x for Prover.

1. Prover takes $x' \leftarrow Z_q$, computes $X' = h^{x'}$. Then he sends X' to Verifier.
2. Verifier takes $e \leftarrow Z_q$ and sends to Prover.
3. Prover computes $r = ex + x'$ and sends r to Verifier.
4. Verifier checks if $h^r = X^e X'$. He accepts if the check is successful.

Table 5. The $\sum$-Protocol π_{2dl} for relation R_{2dl}

Common Input:	$p = 2q + 1, g, h$, and $S = g^x h^y$.
Auxiliary Input:	(x, y) for Prover.

1. Prover takes $x', y' \leftarrow Z_q$, computes $S' = g^{x'} h^{y'}$ and sends it to Verifier.
2. Verifier takes $e \leftarrow Z_q$ and sends back to Prover.
3. Prover computes $r_1 = ex + x', r_2 = ey + y'$ and sends (r_1, r_2) to Verifier.
4. Verifier checks if $S^e S' = g^{r_1} h^{r_2}$. The proof is accepted if the verification is successful.

Table 6. The $\sum$-Protocol π_{or} for relation R_{or}

Common Input:	$p = 2q + 1, h, \sigma, X = h^x$ or σh^x.
Auxiliary Input:	x for Prover.

1. If $X = h^x$, Prover takes $w_1, r_2, c_2 \leftarrow Z_q$, computes $R_1 = h^{w_1}, R_2 = h^{r_2}(X/\sigma)^{-c_2}$.
   If $X = \sigma h^x$, Prover takes $w_2, r_1, c_1 \leftarrow Z_q$, computes $R_1 = h^{r_1} X^{-c_1}, R_2 = h^{w_2}$.
   Then he sends (R_1, R_2) to Verifier.
2. Verifier takes $c \leftarrow Z_q$ and sends back to Prover.
3. If $X = h^x$, Prover computes $c_1 = c - c_2, r_1 = w_1 + c_1 x$.
   If $X = \sigma h^x$, Prover computes $c_2 = c - c_1, r_2 = w_2 + c_2 x$.
   Then he sends r_1, r_2, c_1, c_2 to Verifier.
4. Verifier checks if $c_1 + c_2 = c$, $h^{r_1} = R_1 \cdot X^{c_1}$, $h^{r_2} = R_2 \cdot (X/\sigma)^{c_2}$. He accepts if the verification is successful.

Table 7. The $\sum$-protocol π_{rc} for relation R_{rc}

Common Input: $p = 2q + 1, g, h, \sigma, z_1, \{B_t || \tilde{B}_t || D_t\}_{t=1}^N || U || B$
Auxiliary Input: $\{x_t || \tilde{x}_t\}_{t=1}^N || u || u'$ for Prover.

1. If $B_t = h^{x_t}$, Prover takes $x_{t1}^*, u_{t1}^*, u_{t1}'^*, r_{t2}, \tilde{r}_{t2}, s_{t2}, s_{t2}', e_{t2} \leftarrow Z_q$, computes

$$B_{t1}^* = h^{x_{t1}^*}, \qquad\qquad B_{t2}^* = h^{r_{t2}}(B_t/\sigma)^{-e_{t2}},$$
$$\tilde{B}_{t1}^* = h^{\tilde{x}_{t1}^*}, \qquad\qquad \tilde{B}_{t2}^* = h^{\tilde{r}_{t2}}(\tilde{B}_t/\sigma)^{-e_{t2}},$$
$$D_{t1}^* = (B\sigma^{-t})^{\tilde{x}_{t1}^*} z_1^{u_{t1}^*}, \quad D_{t2}^* = (B\sigma^{-t})^{\tilde{r}_{t2}} z_1^{s_{t2}} D_t^{-e_{t2}},$$
$$U_{t1}^* = g^{u_{t1}^*} h^{u_{t1}'^*}, \qquad\qquad U_{t2}^* = g^{s_{t2}} h^{s_{t2}'} U^{-e_{t2}}.$$

   If $B_t = \sigma h^{x_t}$, Prover takes $x_{t2}^*, u_{t2}^*, u_{t2}'^*, r_{t1}, \tilde{r}_{t1}, s_{t1}, s_{t1}', e_{t1} \leftarrow Z_q$, computes

$$B_{t1}^* = h^{r_{t1}} B_t^{-e_{t1}}, \qquad\qquad B_{t2}^* = h^{x_{t2}^*},$$
$$\tilde{B}_{t1}^* = h^{\tilde{r}_{t1}} \tilde{B}_t^{-e_{t1}}, \qquad\qquad \tilde{B}_{t2}^* = h^{\tilde{x}_{t2}^*},$$
$$D_{t1}^* = (B\sigma^{-t})^{\tilde{r}_{t1}} z_1^{s_{t1}} D_t^{-e_{t1}}, \quad D_{t2}^* = (B\sigma^{-t})^{\tilde{x}_{t2}^*} z_1^{u_{t2}^*},$$
$$U_{t1}^* = g^{s_{t1}} h^{s_{t1}'} U^{-e_{t1}}, \qquad\qquad U_{t2}^* = g^{u_{t2}^*} h^{u_{t2}'^*}.$$

   Then he sends $\{B_{t1}^* || B_{t2}^*,\ \tilde{B}_{t1}^* || \tilde{B}_{t2}^*,\ D_{t1}^* || D_{t2}^*,\ U_{t1}^* || U_{t2}^*\}_1^N$ to Verifier.
2. Verifier takes $e \leftarrow Z_q$ and sends back to Prover.
3. If $B_t = h^{x_t}$, Prover computes $e_{t1} = e - e_{t2}, r_{t1} = x_{t1}^* + e_{t1}x_t, \tilde{r}_{t1} = \tilde{x}_{t1}^* + e_{t1}\tilde{x}_t,$
   $s_{t1} = u_{t1}^* + e_{t1}u, s_{t1}' = u_{t1}'^* + e_{t1}u'.$
   If $B_t = \sigma h^{x_t}$, Prover computes $e_{t2} = e - e_{t1}, r_{t2} = x_{t2}^* + e_{t2}x_t, \tilde{r}_{t2} = \tilde{x}_{t2}^* + e_{t2}\tilde{x}_t,$
   $s_{t2} = u_{t2}^* + e_{t2}u, s_{t2}' = u_{t2}'^* + e_{t2}u'.$
   Then he sends $\{r_{t1} || r_{t2}, \tilde{r}_{t1} || \tilde{r}_{t2}, s_{t1} || s_{t2}, s_{t1}' || s_{t2}', e_{t1}, e_{t2}\}_1^N$ to Verifier.
4. Verifier checks if $e_{t1} + e_{t2} = e$, and if the following for all t:

$$h^{r_{t1}} = B_{t1}^* B_t^{e_{t1}}, \qquad\qquad h^{r_{t2}} = B_{t2}^*(B_t/\sigma)^{e_{t2}},$$
$$h^{\tilde{r}_{t1}} = \tilde{B}_{t1}^* \tilde{B}_t^{e_{t1}}, \qquad\qquad h^{\tilde{r}_{t2}} = \tilde{B}_{t2}^*(\tilde{B}_t/\sigma)^{e_{t2}},$$
$$(B\sigma^{-t})^{\tilde{r}_{t1}} z_1^{s_{t1}} = D_{t1}^* D_t^{e_{t1}}, \quad (B\sigma^{-t})^{\tilde{r}_{t2}} z_1^{s_{t2}} = D_{t2}^* D_t^{e_{t2}},$$
$$g^{s_{t1}} h^{s_{t1}'} = U_{t1}^* U^{e_{t1}}, \qquad\qquad g^{s_{t2}} h^{s_{t2}'} = U_{t2}^* U^{e_{t2}}.$$

   He accepts if the verification is successful.

Notation: $B_t = h^{x_t}$ and $\tilde{B}_t = h^{\tilde{x}_t}$, or, $B_t = \sigma h^{x_t}$ and $\tilde{B}_t = \sigma h^{\tilde{x}_t}$; $U = g^u h^{u'}$;
   $D_t = (B\sigma^{-t})^{\tilde{x}_t} z_1^u.$

Author Index

Lecture Notes in Computer Science

For information about Vols. 1–3744

please contact your bookseller or Springer